EPOXY ADHESIVE FORMULATIONS

EPOXY ADHESIVE FORMULATIONS

Edward M. Petrie

McGRAW-HILL

New York Chicago San Francisco Lisbon London Madrid
Mexico City Milan New Delhi San Juan Seoul
Singapore Sydney Toronto

The McGraw-Hill Companies

Library of Congress Cataloging-in-Publication Data

Petrie, Edward M.
Epoxy adhesive formulations / Edward M. Petrie.
p. cm.
Includes index.
ISBN 0-07-145544-2 (acid-free paper)
1. Epoxy resins—Formability. 2. Epoxy compounds—Formability. 3. Epoxy coatings—Formability. 4. Adhesives—Formability. I. Title.

TP1180.E6P48 2006
668′.374—dc22

2005051088

1 2 3 4 5 6 7 8 9 0 DOC/DOC 0 1 0 9 8 7 6 5

ISBN 0-07-145544-2

The sponsoring editor for this book was Kenneth P. McCombs, the editing supervisor was David E. Fogarty, and the production supervisor was Rick Ruzycka. It was set in Times Roman by International Typesetting and Composition. The art director for the cover was Anthony Landi.

Printed and bound by RR Donnelley.

This book was printed on recycled, acid-free paper containing a minimum of 50% recycled, de-inked fiber.

CONTENTS

PREFACE

The adhesive industry has seen significant changes in recent years. The type and number of these changes have been astounding. They include new substrate materials, regulatory burdens, acquisitions and mergers, new raw materials, new application and curing processes, and a host of volatile technical, commercial, and political issues. However, a staple in the industry throughout this change has been the adhesive formulator.

At first glance, the job of the adhesive formulator appears to be deceptively simple. There are many resources and tools available on the *supply* side of the industry for the formulator to use. The problem is that there seems to be an even more staggering amount of requests for new products on the *demand* side of the industry.

Epoxy technology gives the formulator an almost unlimited number of tools to employ. The type of epoxy polymer backbone, curative, resinous modifiers, and special additives or fillers all serve as degrees of freedom available in developing an adhesive system for a given application. However, often the adhesive must fulfill many different requirements, such as ease of assembly, moderate cure time, resistance to thermal cycling, and resistance to stress at high temperature and humidity. The creation of a functional adhesive system for a new application, therefore, is usually a lengthy undertaking, which depends on the skill and discretion of the formulator.

Unfortunately there is not a lot of information to assist the formulator in the proper or optimal use of the tools and resources available. There are not many forums that provide instruction on the "art" of adhesive formulating. Only a few textbooks have concentrated on the subject. And even though epoxy adhesives are the "workhorse" of the industry and occupy the majority of the structural adhesives market, practical, concentrated information on epoxy adhesive formulation is noticeably absent. This book is an attempt to correct this situation.

I have tried not so much to present new material and the latest developments in epoxy adhesive technology as to provide a useful and organized summary of the many fragmented sources of information that already exist. Most of the information in this text comes from raw materials suppliers, technical papers, conference proceedings, and other scattered details. I have attempted to blend this with my nearly 40 years of experience in formulating, using, and consulting on epoxy adhesive applications. Hopefully, the outcome is (1) a useful starting point and guidance for the novice in the field, (2) a handbook of readily accessible practical information for those working everyday with epoxy adhesives, and (3) a portal to additional technical information and research on the subject.

This book is aimed primarily at the formulator, but it is sometimes difficult to define who the formulator actually is. Often the end user may also need to formulate an epoxy adhesive from raw materials in order to reduce cost or because of the uniqueness of the application. Thus, I hope that the end user as well as those involved in other aspects of the adhesives industry (purchasing agents, designers, manufacturing engineers, etc.) find this book interesting and worthwhile.

When undertaking the job of preparing for this book, I originally intended to provide a collection of epoxy adhesive formulations that could be used for specific applications, and hopefully I have accomplished that task. However, it is also necessary to discuss surface preparation, testing, application processes, and curing methods to properly employ the formulation information. As a result, I also provide "accessory" information that is needed to either manufacture a successful epoxy formulation or use it in practice. For greater detail regarding adhesives in general, bond theory, substrate preparation, test methods, and so forth, the reader is directed to the *Handbook of Adhesives and Sealants* (Edward M. Petrie, McGraw-Hill, 2000).

Epoxy adhesives can be formulated to bond to a variety of substrates and perform over a broad range of conditions. They are used as one- or two-component systems, which are curable under ambient or elevated-temperature conditions. They are available in a variety of forms including liquids, pastes, film, solids, solvent solutions, and water dispersions. They are used in many industries and can be truly considered a workhorse.

The epoxy adhesives are popular with both the formulator and the end user for a number of reasons. When compared to other possible adhesive systems, epoxies offer the following added values:

- Nearly infinite number of ways to engineer an adhesive to provide the required application properties and end-use properties in a joint
- A variety of forms and curing methods to optimize the assembly process
- No evolution of volatiles and low shrinkage during cure
- Good wetting properties resulting in excellent adhesion to most substrates
- Excellent cohesive strength and mechanical properties
- Good moisture, humidity, chemical, and temperature resistance

The main raw materials used in epoxy adhesive formulations (resins and curing agents) can be synthesized in a variety of ways to create many different products. Epoxies react readily via several polymerization mechanisms. The extent of crosslinking is an important determinant of the final properties of the adhesive. Crosslinking can be controlled by the choice of resin and curing agent and by the curing conditions.

Crosslinked epoxy resins are cohesively very strong, but their brittleness reduces their usefulness in many adhesive applications, especially those requiring a high degree of peel and impact strength or thermal cycling. Development in recent years has largely focused on the toughening needed for epoxies to be structural adhesives. Epoxies can be toughened by modifying the chemistry of the raw materials within the formulation or by adding particulates, elastomers, or thermoplastics to the formulation.

Since epoxy adhesive formulation represents a surprisingly broad area of technology, a road map to the use of this book may be valuable. Chapters 1 through 3 discuss the synthesis of raw materials, epoxy chemistry in general, and the physical and chemical properties that are important for an epoxy adhesive. These properties are important during the three primary phases or conditions of an adhesive: (1) uncured, (2) during cure, and (3) fully cured.

Chapters 4 through 10 describe the basic raw materials that are commonly employed in formulating epoxy adhesives. These include the epoxy resins, curing agents and catalysts, solvents and diluents, resinous modifiers, flexibilizers and tougheners, fillers, and adhesion promoters.

Formulation details are then presented in Chapters 11 through 14 for the various possible forms of epoxy adhesive systems: room temperature and elevated-temperature curing liquids, pastes, and solids. The more or less unconventional forms of epoxy adhesives are also identified and discussed, since these are now achieving prominence in industry. These include uv and electron beam radiation curable, waterborne systems, and epoxy adhesives capable of curing via the indirect application of heat or energy.

Chapters 15 and 16 describe formulations that have been specifically developed for use in certain service environments and with certain substrates. The effects of the environmental exposure on epoxy adhesives are described. The substrate properties that enhance or are detrimental to bonding success are identified. Surface preparation procedures for common substrates are identified in the text and detailed processes are defined in Appendix F.

Chapters 17 through 20 describe the various processes and equipment employed in the formulation or end use of epoxy adhesives. Health and safety issues regarding the use of these materials are discussed. The importance of quality control methodologies and specification preparation is also noted. Finally, test methods that are commonly used by both the formulator and the end user are identified.

Throughout this book, specific applications and examples illustrate the concepts being discussed. Due to the broad nature of the subject, the reader will often be directed from one section to another relevant section. References at the end of each chapter direct the interested reader to more in-depth information and understanding regarding specific subjects.

Formulating skills cannot be easily learned in the classroom or from reading books. These sources can provide a starting point and a road map, but that is not the end point. Formulating is somewhat of an "art" and as such can only be effectively learned at the bench by practicing and, unfortunately, often by trial and error. This book can be a means of reducing the time, cost, and potential problems inherent in this process.

ACKNOWLEDGMENT

This book is dedicated to my mother, Katherine. I am sure that I would never have achieved what I have in life without her skills at formulating and applying the *real* basics. She has been both the cornerstone and the shining beacon of our family. Most of all, I am grateful for her dedication and tireless enthusiasm in resolving all the troubles that life throws at us.

EDWARD M. PETRIE

CHAPTER 1
EPOXY ADHESIVES

1.1 INTRODUCTION

Epoxy adhesives are chemical compounds used to join components by providing a bond between two surfaces. Epoxy adhesives were introduced commercially in 1946 and have wide applications in the automotive, industrial, and aerospace markets. Epoxies are probably the most versatile family of adhesives because they bond well to many substrates and can be easily modified to achieve widely varying properties. This modification usually takes the form of

1. Selection of the appropriate epoxy resin or combination of resins of which many are available
2. Selection of curing agent and associated reaction mechanism
3. Simple additions of organic or inorganic fillers and components

Such modification is commonly described as *formulating* or *compounding*. Formulating is necessary to achieve an adhesive that will yield the desired application characteristics and end-use properties at an acceptable cost. As a result, an enormous number of epoxy adhesive formulations are possible.

Therefore, it is a mistake to describe epoxy adhesives in a generic manner as if all these formulations had similar properties. Depending on the type of resin and curing agent used and on the specific formulation, epoxy adhesives can offer the user an almost infinite assortment of end properties as well as a wide diversity of application and curing characteristics.

Because of their good wetting characteristics, epoxy adhesives offer a high degree of adhesion to all substrates except for some low-surface-energy, untreated plastics and elastomers. Cured epoxies have thermosetting molecular structures. They exhibit excellent tensile shear strength but poor peel strength unless modified with a more resilient polymer. Epoxy adhesives offer excellent resistance to oil, moisture, and many solvents. Low shrinkage on curing and high resistance to creep under prolonged stress are characteristics of many high-quality epoxy adhesives. Epoxy resins have no evolution of volatiles during cure and are useful in gap-filling applications.

Commercial epoxy adhesives are composed primarily of an epoxy resin and a curing agent. Various additives and modifiers are added to the formulation to provide specific properties. Example trade names and suppliers of these ingredients are included in App. A. The curing agent may be incorporated into the resin to provide a single-component adhesive, or else it may be provided in a separate container to be mixed into the resin immediately prior to application.

Epoxy adhesives are commercially available as liquids, pastes, films, and solids. Epoxy adhesives are generally supplied as 100 percent solids (no solvents or other volatiles), but some are available as sprayable solvent systems or water emulsions. Epoxy adhesives are also available in a wide range of chemical compositions, which allows for variations in how these materials can be applied and set as well as variations in final performance properties within a joint.

Most commercially available epoxy structural adhesives are either single-component, heat curing adhesives or multiple-component adhesives that cure at either room or elevated temperature. Generally, epoxy systems that cure at elevated temperatures have a higher crosslinking density and glass transition temperature than systems that are formulated to cure at room temperature. This provides the elevated-temperature curing epoxies with better shear strength, especially at elevated temperatures, and better environmental resistance. However, they usually have poor toughness and peel strength due to their rigidity.

Depending on the epoxy resin and curing agent used, room temperature curing adhesive formulations can harden in as little as several minutes at room temperature, but most systems require from 18 to 72 h to reach full strength. The room temperature curing epoxy adhesives can also be cured at elevated temperatures, if faster curing times are required. The curing time is greatly temperature-dependent, as shown in Fig. 1.1.

Once the curing agent is added to the epoxy resin, the adhesive must be used within a time period that is dependent on the resin and curing agent in the formulation and on the ambient conditions. With most room temperature systems, the time period is short, and the maximum time from mixing to application is considered the *pot life* of the system. For most single-component chemistries, the allowable time from mixing the hardener into the resin to application is relatively long and dependent on the storage conditions. This time period is defined as the adhesive's *shelf life*.

An exothermic reaction occurs on curing of epoxy adhesive systems. Exotherm is the heat that is self-generated by the curing reaction. If the exotherm is not controlled, it can lead to greatly reduced working life and even to a dangerous situation where the exotherm can be great enough to cause combustion. The exotherm can be minimized by reducing the reactivity of the mixed components and by limiting the batch size.

Epoxy adhesives are often given a postcure. Room temperature curing and elevated-temperature curing epoxy adhesives are cured to a state in which the resin is hard and the bond is capable of handling (i.e., handling strength). This provides the opportunity of

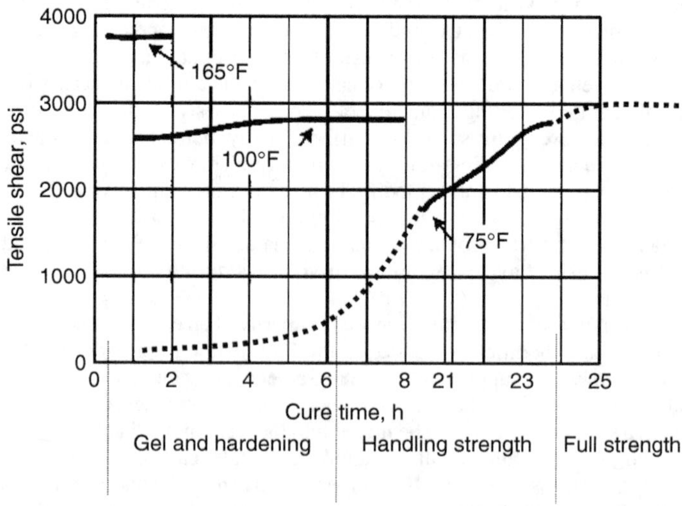

FIGURE 1.1 Characteristics of a particular room temperature curing epoxy adhesive under different cure time and temperature conditions.[1]

releasing the fixturing used to hold the parts in place and moving the bonded component to the next manufacturing stage. Once the handling strength has been reached, the joint is postcured without fixturing by placing it in an oven or other heat source until it achieves full strength. Often postcuring can be combined with some other postbonding process such as paint curing or component drying. The epoxy adhesive manufacturer should be consulted to determine if postcuring is a viable option.

Many epoxy adhesives are capable of being B-staged. A B-staged resin is one in which a limited reaction between the resin and hardener has taken place so that the product is in a semicured but solid state. In the B-staged state, the polymeric adhesive is still fusible and soluble. On additional heat curing, the adhesive will progress from the B stage to a completely cured state. This will usually be accompanied by moderate flow.

The advantage of B-staged resins is that they permit the formulation of one-component solid adhesives such as films, powders, and preforms. The user can then purchase an adhesive product that does not require metering or mixing. There is very little waste associated with B-staged products, and they generally provide better consistency. The B-staged epoxy resins are formulated using curing agents, such as aromatic polyamines, along with solid forms of DGEBA epoxy resins.

Secondary ingredients in epoxy adhesives include reactive diluents to adjust viscosity; mineral fillers to lower cost, adjust viscosity, or modify the coefficient of thermal expansion; and fibrous fillers to improve thixotropy and cohesive strength. Epoxy resins are often modified with other resins to enhance certain properties that are necessary for the application. Often these modifications take the form of additions of elastomeric resins to improve toughness or peel strength.

1.2 IMPORTANCE OF EPOXY RESINS

Epoxy resins are high-performance thermosetting resins, which display a unique combination of properties. Epoxy resins have been commercially available for almost a half-century. Epoxy resins are arguably one of the most versatile polymers with uses across an enormously wide variety of industries. The outstanding physical properties exhibited by epoxy resins include

- Low cure shrinkage
- No volatiles given off during cure
- Compatibility with a great number of materials
- Strength and durability
- Adhesion
- Corrosion and chemical resistance
- Electrical insulation.

Furthermore, epoxy resin systems are capable of curing at either ambient or elevated temperatures, and they require only minimal pressure during the cure. Thus, epoxies can be applied and cured under many adverse conditions including outdoors. These properties provide great added value in many industries engaged in product assembly. Epoxy resins have been commercially available for almost a half-century.

Many new applications are being developed that will ensure the prominence of epoxy resins in the future. For example, epoxies are the material of choice for the reinforcement or consolidation of aging or damaged concrete structure including walls, ceiling beams, bridge columns, and anchoring mechanisms. They are the main binders for graphite-reinforced composites such as those used in manufacturing tennis rackets, fishing rods, skis, snowboards, and golf club shafts. They have also been used for manufacturing high-strength, lightweight carbon fiber pultrusion as replacements for steel rebar in special concrete expansion projects.

Industries using epoxy adhesives include aerospace, civil engineering, automotive, chemical, electrical, marine, leisure, and many others. Table 1.1 lists some of the more common applications for epoxy compounds. The prevailing reason for the broad acceptance of epoxy resins in these important and diverse markets is their capacity to provide a good balance of handling characteristics and ultimate physical properties. They adhere well to a very large variety of substrates, and they generate tough, environmentally resistant films or matrices.

One of the major advantages of epoxy chemistry is the wide latitude it provides the formulator for solving technical problems. Epoxies can be designed to be flexible or rigid; high or low modulus; homogeneous, filled, or foamed; conductive or insulative; fire-retardant; and resistant to heat and chemicals.

The number of raw materials that the formulator has to work with is enormous. These include epoxy resins and modified epoxy resins of all types and forms, fillers and additives,

TABLE 1.1 Common Applications for Epoxy Resins[2]

Aircraft and aerospace: Structural parts of aircraft, spacecraft, and satellites Adhesives Aircraft paints and coatings	Electrical: Switchgear construction and insulation Transformer construction and insulation Turbine alternator insulation Electric motor insulation Cable jointing Coatings for domestic electrical appliances
Automobile: Automotive primers and primer surfaces Sealers Adhesives Structural components Racing car bodies Tooling compounds Ignition coil impregnators Encapsulants for control modules	Electronic: Printed-circuit boards Packaging of active and passive components Encapsulation of electronic modules Adhesives
Construction: Industrial flooring Grouts for roads and bridges Antiskid road surfaces Tooling compounds Repair compounds Adhesives Sealants Pipes Do-it-yourself compounds Maintenance paints Coil coated steel (such as roofing)	Food and beverage: Can and drum coatings Coatings for flexible tubes Marine: Primers and protective coatings for ships and marine structures (such as oil rigs) Leisure: Fishing rods Tennis rackets Gold club shafts Bicycle frames Skis Musical instruments
Chemical: Linings for storage tanks Chemical plant including coatings Pipes and pipe linings Filters	Textile: Equipment parts Glass and carbon fiber sizing agents Light engineering: Adhesives Protective and decorative coatings Composite structures for artificial limbs

modifiers, reinforcements, diluents, and solvents. The possible curing agents that can be used also provide great latitude in formulation. Often the curing agent becomes an integral part of the resulting compound. Its choice is a controlling influence on the curing properties of the mixture and on the performance properties of the cured adhesive. One of the chief advantages of epoxy resins is that they generally allow easy incorporation of additives, and the resulting formulation is one that can be easily adapted to many manufacturing processes, such as

- Pultrusion
- Lamination
- Filament winding
- Molding
- Casting and potting
- Coating
- Adhesive bonding

Epoxy resins are not the lowest-cost resins potentially available for most applications. Thus, epoxy resins must provide added value to justify their additional cost. This added value is usually realized by the incorporation of a special property or combination of properties into the final product.

The epoxy resin production value in the United States, western Europe, and Asia is estimated to be over $2.5 billion. Almost 900,000 metric tons (t) of epoxy resins is consumed in these regions with around 35 percent of the consumption in Asia. The overall global consumption in 2002 was approximately 970,000 t (2.14 billion lb) with a 4 to 5 percent annual growth rate.[3] The 7 to 10 percent growth rate in the 1970s has slowed except in Asia. Future consumption will average about 2 to 4 percent per year from 2004 to 2008 in the United States and western Europe.[4] In North America the demand for epoxy resins is forecast at 690 million lb, valued at $1.4 billion, in 2004.[5]

The three leading producers of epoxy resin account for approximately 75 percent of the world's capacity. These resin manufacturers are Resolution Performance Products (formerly Shell), Dow Chemical, and Huntsman (formerly Ciba). The other 50+ smaller epoxy manufacturers primarily produce epoxy resins only regionally or produce specialty products.

The industry is currently in the midst of a major restructuring period, with producers of epoxy resins announcing mergers, acquisitions, and divestiture of operations. Examples of leading producers of epoxy resins, curing agents, and additives for the adhesives market are included in Apps. A through D. These and other companies also offer a wide range of associated products, including solvents, primers, and chemicals.

Epoxy resins are commercially available as either liquids or solids. The liquids are available as (1) solvent-free resins, ranging in viscosity from waterlike liquids to crystalline solids; (2) waterborne emulsions; and (3) solvent-borne solutions. Generally, the higher the molecular weight of the epoxy resin molecule, the higher the viscosity or melting point.

Epoxy resins are composed of polymeric molecules that are converted to a solid by a chemical reaction. Epoxy systems physically comprise two essential components: a resin and a curative. The curative causes the chemical reaction, which turns the epoxy resin into a solid, crosslinked network of molecules. This polymer is called a *thermoset* polymer because, when cured, it is irreversibly rigid and relatively unaffected by heat. (By contrast, *thermoplastic* polymers are not crosslinked and can be made to flow with the application of heat.)

Cure of the epoxy resins is initiated once the resin is mixed with a curative. The cure of all epoxy resins is an exothermic process where heat is generated as a natural result of the chemical reaction. Success in using most epoxies is dependent on handling the product in the correct way to avoid premature cure and unwanted side reactions.

Epoxy resins are seldom used in their unmodified forms. Formulation is generally a necessity of any epoxy product. There are many reasons for this, including

1. Overcoming the inherent brittleness of cured epoxy resin
2. Reducing material cost
3. Incorporation or enhancement of specific properties in either the uncured (e.g., thixotropy, cure rate) and/or cured epoxy system (e.g., electrical conductivity, chemical resistance)
4. Improving cure requirements (time and temperature)
5. Reducing environmental and safety problems such as volatile organic components and flammability

Formulators in the adhesives industry do not normally manufacture epoxy resins. Generally, formulators buy epoxy resins, modify them with other materials, do similar compounding to the curative, and then package the product as a complete adhesive system ready for the end user. There are many excellent textbooks[6–8] available giving information about the preparation, chemistry, and use of epoxy resins in general applications. It is not the intention here to go into such detail but to focus only on epoxy adhesive systems.

1.3 IMPORTANCE OF EPOXY ADHESIVES

1.3.1 Advantages and Disadvantages of Adhesive Bonding

Almost everything that is made by industry has component pieces, and these have to be fixed together. There are many alternatives to adhesive bonding that a manufacturer can employ such as screws, rivets, and spot welds. Each potential joining process must be considered with regard to its specific requirements. There are times when adhesives are the worst possible option for joining two substrates, and there are times when adhesives may be the best or only alternative. Usually, the choice of joining process is not all black or white. Certain processes will have distinct advantages and disadvantages in specific applications. The choice may involve a tradeoff in performance, production capability, cost, and reliability.

Often, one must consider the time, trouble, and expense that may be necessary to use an adhesive. For example, certain plastics may require expensive surface preparation processes to allow the adhesive to wet the surface. Applications requiring high-temperature service conditions may require an adhesive that necessitates an elevated-temperature cure over a prolonged period.

On the other hand, certain applications could not exist without adhesive bonding. Examples of these are the joining of ceramic or elastomeric materials, the joining of very thin substrates, the joining of surface skin to honeycomb, and numerous other applications. There are also certain applications where adhesives are chosen because of their low cost and easy, fast joining ability (e.g., packaging, consumer products).

Sometimes conventional welding or a mechanical joining process is just not possible. Substrate materials may be incompatible for metallurgical welding due to their thermal expansion coefficients, chemistry, or heat resistance. The end product may not be able to accept the bulk or shape required by mechanical fasteners.

The science of adhesive bonding has advanced to a degree where adhesives must be considered an attractive and practical alternative to mechanical fastening for many applications. Adhesive bonding presents several distinct advantages over other conventional methods of fastening. There are also some disadvantages which may make adhesive bonding impractical. These pros and cons are summarized in Table 1.2.

TABLE 1.2 Advantages and Disadvantages of Adhesive Bonding

Advantages	Disadvantages
1. Provides large stress-bearing area.	1. Surfaces must be carefully cleaned.
2. Provides excellent fatigue strength.	2. Long cure times may be needed.
3. Damps vibration and absorbs shock.	3. Limitation on upper continuous operating temperature (generally 150 to 175°C).
4. Minimizes or prevents galvanic corrosion between dissimilar metals.	4. Heat and pressure may be required.
5. Joins all shapes and thicknesses.	5. Jigs and fixtures may be needed.
6. Provides smooth contours.	6. Rigid process control is usually necessary.
7. Seals joints.	7. Inspection of finished joint is difficult.
8. Joins any combination of similar or dissimilar materials.	8. Useful life depends on environment.
9. Often is less expensive and faster than mechanical fastening.	9. Environmental, health, and safety considerations are necessary.
10. Heat, if required, is too low to affect metal parts.	10. Special training sometimes is required.
11. Provides attractive strength-to-weight ratio.	

The design engineer must consider and weigh these factors before deciding on a method of fastening. However, in many applications adhesive bonding is the only logical choice. In the aircraft industry, for example, adhesives make the use of thin metal and honeycomb structures feasible because stresses are transmitted more effectively by adhesives than by rivets or welds. Plastics, elastomers, and certain metals (e.g., aluminum and titanium) can be more reliably joined with adhesives than with other methods. Welding is usually at too high a temperature, and mechanical fastening destroys the lightness and aesthetics of the final product. Certain examples of less obvious applications where adhesive bonding is a practical method of assembly are shown in Table 1.3.

The design engineer will find that once he or she decides that adhesive bonding is the optimal method of assembly, there are a significant number of adhesive families to choose from and an enormous number of formulations within each family. However, several universal factors should be considered when one is selecting or developing a formulation for an adhesive application. These are generally related to the materials that will be bonded, the bonding processes that are available and practical for use, and consideration of the environments to which the final adhesive joint will be exposed.

- Materials to be bonded:

  Surface energy of the adhesive and the substrate

  Surface preparation processes that are available and practical

  End-use properties of the joint

- Process of bonding:

  Size of gap, viscosity of adhesive formulation

  Off-gassing, flammability, etc., during the curing process

  Pot life: maximum time between metering/mixing and application of the adhesive

- Environmental considerations:

  Service temperature limited by glass transition and thermal degradation of the adhesive

  Influence of water on the adhesive and interface

  Chemical degradation by ultraviolet (uv), radiation, solvents, and other environmental media

TABLE 1.3 Examples of Applications Where Bonding Is Considered to Be a Practical Method of Assembly[9]

Application areas for adhesives	Examples
Dissimilar materials	Combinations of metals, rubbers, plastics, foamed materials, fabrics, wood, ceramics, glass, etc.
Dissimilar materials that constitute a corrosion couple	Iron to copper or brass
Heat-sensitive materials	Thermoplastics, magnetic materials, glass
Laminated structures	Sandwich construction based on honeycomb materials; heat exchangers; sheet laminates, core laminates
Reinforced structures	Stiffeners for wall paneling, boxes and containers, partitions, automobile chassis parts, aircraft body parts
Structural applications	Load-bearing structures in the aircraft fuselage, automotive and civil engineering industries
Bonded inserts	Plug inserts, studs, rivets, concentric shafts; tubes, frame construction; shaft-rotor joints; tools; reinforced plastics with metal inserts; paintbrush bristles
Sealed joints and units	Pipe joining, encapsulation, container seams, lid seals
Fragile components	Instrumentation, thin films and foils, microelectronic components and others where precise location of parts is required
Components of particular dimensions	Where bonding areas are large or there is a need for shape conformity between bonded parts
Temporary fastening	Where the intention is to dismantle the bond later, the use of various labels, surgical and pressure-sensitive tapes, adhesives for positioning and locating parts, in lieu of jigs, prior to assembly by other means

When considered against the criteria listed above, epoxy adhesive systems provide a well-balanced set of properties. This solidifies their competitive position in the adhesives industry.

1.3.2 The Nature of the Epoxy Adhesive Industry

The industries that are most affected by epoxy adhesives consist of four main categories:

1. Base material producers including resins, mineral fillers, extenders, etc.
2. Formulators who take the base materials and combine, process, and package them into adhesive systems providing various levels of performance
3. End users who take the packaged adhesives and sealants and produce assembled products
4. Associated industries such as equipment manufacturers, testing laboratories, and consultants

The base material producers are usually large chemical or material companies that manufacture products for broad markets such as petrochemicals or plastics. When demand warrants, they will produce materials specifically for the adhesive and sealant formulators.

The formulators can range from very small businesses with several employees, addressing small niche markets, to large international companies with several hundred products. Both small and large formulators are generally willing to modify a formulation if they believe that it will improve a customer's performance or production efficiency or will add some other value. However, a minimum volume is usually necessary before formulators will make modifications to a standard formulation or develop a new product for a specific application. Formulators have a significant knowledge base regarding adhesive or sealant systems and how they are to be applied in practice. The properties of several commercial epoxy adhesive formulations appear in App. B.

The trend is for the end user to purchase an adhesive from a formulator, rather than produce it internally. It is increasingly difficult for an end user to keep up with continuing technological changes in adhesives. However, the end user must select a commercial adhesive, substrate, joint design, and processing conditions for specific applications. Once these are selected and verified as to performance and cost, the end user must be vigilant that none of the processes, materials, or other relevant factors in the assembly process change.

Several other industries are also affected by adhesives and sealants. For example, equipment suppliers specialize in producing machinery for application, assembly, curing, surface preparation, etc. Also equipment suppliers specialize in developing and manufacturing testing apparatus that can be used to measure joint strength and processing parameters. Then there are testing laboratories and consultants who provide assistance and services to the formulators and end users on a contractual basis.

1.3.3 Epoxy Adhesive Markets

The inception of epoxy adhesives occurred almost simultaneously with the commercialization of epoxy resins. This is due to the fact that all the unique handling, application, and performance properties of epoxy resins are especially important when it comes to adhesive bonding.

Epoxy adhesives have become the most recognizable structural adhesive type. They have found commercial success in demanding industries such as aerospace, automotive, building and construction, and electrical and electronic. Their ease of use has also encouraged their commercialization in the do-it-yourself markets, which has added to the exemplary reputation of epoxy adhesives.

Adhesive markets represent only a small percentage of the total consumption of epoxy resins. However, epoxy adhesives provide significant value added, so that their prices and profit margins are generally higher than those for other adhesive types. Globally in the year 2002, epoxy adhesives represented about 31 percent (or about 0.7 billion lb) of the total liquid epoxy demand (Fig. 1.2). The U.S. market is estimated to be just less than 200 million lb.[10]

Epoxy adhesives represent a significant part of the overall structural adhesives market (about $1.8 billion). The main competitors to epoxy adhesives are polyurethanes; however, thermosetting acrylics and cyanoacrylate adhesives are also strong challengers in certain market segments.

Single-component epoxy adhesive formulations are the largest type of epoxy adhesives sold, with about 55 percent of the consumption, while two-component formulations account for another 44 percent of the volume. Radiation cure formulations represent the remainder of the market. Epoxy adhesives can also take many forms including solids, solvent-free liquids, solvent-borne systems, and waterborne systems.

Despite epoxy adhesives finding use in many fragmented markets, actual consumption in volume is surprisingly concentrated in a few specific end-use market segments. For example, automotive assembly applications account for nearly 50 percent of the total volume of epoxy adhesives consumed in the United States. The highest-value market areas include structural automotive, aircraft, and many specialty product assembly applications. These market areas are also expected to enjoy the highest growth rates.

Although the overall annual growth rate for epoxy adhesive is in the 3 to 5 percent range, certain regional markets, such as China, are expected to have an annual growth rate

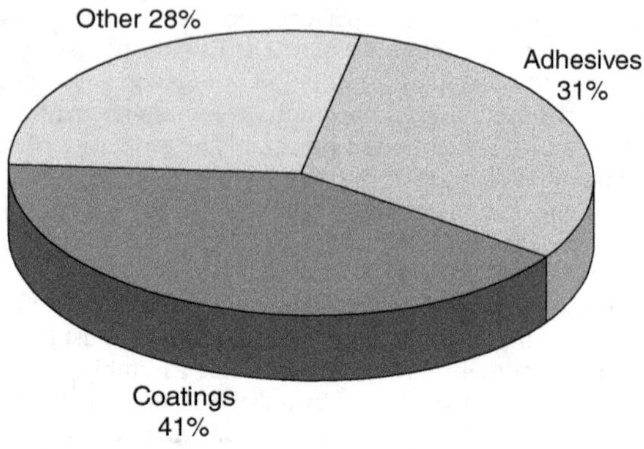

FIGURE 1.2 Global liquid epoxy resin production.[11] 2002 Total Liquid Epoxy Resin Demand: 2.14 billion pounds; 970 thousand metric tons; $2.0 billion; AGR 4–5%. (*Source: DPNA International Inc.*)

of at least 8 percent per year (Fig. 1.3). There are several new developments that are affecting the growth rate for epoxy adhesive systems. These drivers are classified as "technical developments" and "market trends" in Table 1.4.

Analysis of the epoxy adhesive markets is difficult because so many market segments fragment the industry and because it is difficult to confine manufacturers, distributors, and end users into specific categories. However, there are several organizations that periodically publish market reports on the adhesives industry including epoxy adhesives.[14]

Advances in epoxy adhesive technology have meant that an increasing number of assembly and repair problems can be resolved without the use of mechanical fasteners, welding, or other nonadhesive assembly processes. Several of the more important developing market segments for epoxy adhesives are described in the following sections.

Automotive. The automotive industry has a wide variety of applications for adhesives and sealants. Nonstructural applications dominate, and they include all interior trim and much of the exterior trim such as side molding, wheel covering emblems, front and rear

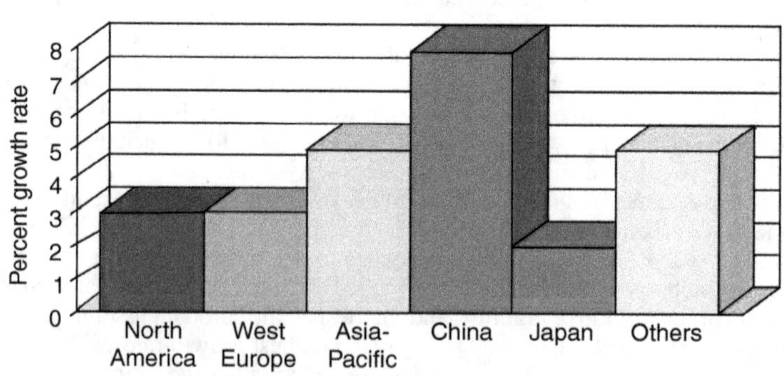

FIGURE 1.3 Epoxy adhesive regional growth rate.[12] (*Source: DPNA International Inc.*)

TABLE 1.4 Developments Affecting Epoxy Adhesive Demand[13]

Technical developments	Business macro-trends and issues	Market trends and opportunities	Global market driving forces
Toughened epoxies High-performance, underwater curing Fast (<5 min) room temperature cure One-pack systems Hybrid polymers Dispensing equipment Curing systems Expanding assembly applications (e.g., modularity)	e-Commerce Rising energy costs Shrinking profits Backward-facing support Partnering for success Consolidation/ globalization Niche marketing	Fastening options Changing substrates New polymers Market dislocations	Economic recession in United States and Europe Stagnant growth in Japan Recent price increases Fluctuating oil prices War in Middle East Continued growth in China Environmental and heath concerns (Europe/Japan) Customer and market relocation

taillight assemblies, side-view mirrors, and ornamentation. Adhesives and sealants are also being used in many under-the-hood applications. However, structural adhesives are making the largest gains in this industry. After years of field experience, structural adhesive bonding has proved that it can provide high-volume assemblies with greater stiffness and better crash performance. Being viscoelastic, most structural organic adhesives also provide improvements in vehicle *noise, vibrations, and harshness* (NVH). This became possible only after the successful development of new generations of adhesives and specifically with the development of new toughening additives. As a result, adhesive bonding is now accepted in the transportation industry as a "classic" joining technique.

To understand the popularity of adhesive bonding in the automotive industry, one needs to take a look at the characteristics that define the current market situation for autos.

- Although steel, including high-strength types and coated types, will continue to be the dominant body material, nonferrous metals (e.g., aluminum) and composites are increasingly being used.
- Combinations of different materials (e.g., metals and composites) are being utilized for design variation and for reduced weight.
- Increasing emphasis is being placed on multifunctional application (e.g., vibration damping plus joining) to reduce costs and improve production times.
- The need to improve crash capability and to reduce weight is more important today than ever before to meet safety and environmental standards.
- Fatigue and durability of assembled parts are more important because of increasing warranty coverage.
- Modular design concepts are being introduced to maximize design versatility and to minimize costs. These modular designs are conveniently assembled with adhesives.

Because adhesive formulators resonate well with these requirements, adhesives are bound to play an increasingly important role in the future of the automotive industry.

Also, several new bonding processes and products have gained the attention of the automotive industry. Weldbonding (see Chap. 14) has improved the production situation by eliminating the need for fixturing and being easily adapted to robotic processing methods. New structural adhesives such as toughened epoxy (Chap. 8), thermosetting acrylics, and polyurethanes are providing a greater degree of crash protection in automobiles, and these materials bond well to nonmetals and poorly prepared surfaces.

Just as their counterparts rejuvenated the aircraft industry decades ago, modern epoxy structural adhesives are providing automobile designers and manufacturers with innovative possibilities that could not have been considered only a short time ago. As a result, structural automotive adhesives will have an average annual growth rate of greater than 7 percent over the next 5 years.[15] This growth rate is additive to any growth that will occur in the automotive industry itself.

Electrical and Electronic. There are many industries where adhesives must perform several special functions in addition to their primary function of bonding or joining. These industries often have test methods, specifications, and nomenclature that are unique and somewhat foreign to adhesive formulators. One such industry is the electrical and electronic industry.

The electrical and electronic industries are generally lumped into one since it is difficult to place boundaries between the two, and they often have similar, if not identical, requirements. The electrical industry is comprised of manufacturers of motors, transformers, meters, etc. It is a large, stable industry with demands for many types of adhesive products.

The electronic industry, on the other hand, is growing significantly, and new products are forthcoming at a rapid rate. Demand for electronic polymer products is forecast to expand 6.4 percent per year to $3.5 billion in 2005. Adhesives used in this industry will grow at a rate of 8.8 percent, faster than the industry itself. This will require 485 million lb of resin valued at $1.6 billion. Of this, about 123 million lb of adhesives will be used in the electronics market.[16,17]

Adhesives are used in the electrical and electronic industries in a variety of different ways, from holding microcomponents in place on a circuit board to bonding coils in large power transformers. Reliability is always a concern, since bond failure could lead to component failure, which in turn leads to equipment failure and then possibly to a massive system failure. A "system" in this industry could be a commercial aircraft's electrical system or the power distribution system in an urban city.

In the electrical and electronic equipment market, adhesives compete with other joining methods such as mechanical fasteners, brazing, welding, soldering, and thermocompression bonding. Major uses of adhesives in the electrical and electronic equipment market are shown in Table 1.5. Adhesives are generally the preferred method of joining whenever

- Costs must be minimized
- The adhesive can provide a secondary function(s)
- Assembly speed is essential
- Adhesives are the only method by which a product can be made, such as the assembly of microminiature components

Thermosetting adhesives, such as epoxies, are the workhorses of the industry. However, there are persistent challenges from various higher-priced specialty resins.

In general, the use of adhesives in this industry assumes that the adhesive must provide a dielectric function in addition to providing the mechanical joining function. The dielectric function is often related to some degree of electrical insulation, but could also be related to electrical conductivity, depending on the application. An excellent example of an insulating function is the bonding of conductive foil to plastic films and laminates for the fabrication of printed-circuit boards and flexible wiring harnesses. Here the adhesive must

TABLE 1.5 Representative Adhesive Applications for the Electrical and Electronic Industries

Electrical	Electronic
Assembly of motors and relays	Bonding embossed or die-stamped printed-circuit boards
Bonding of nonstructural insulators	Bonding copper foil to dielectric material (laminates or flexible film)
Assembly of magnetic recording heads	Fabrication of electronic coils
Sealing stainless steel shells on submerged pump motors	Surface-mounted conductive adhesives
Bonding stranded wire ends	Electrically conductive parts
Bonding thermal and electronic insulators	Adhesives for static dissipative, interconnecting, and shielding applications
Sealing capacitor cases	Thermally conductive adhesives for bonding heat sinks
Felt and rubber gasketing in appliances	
Adhering sound-deadening materials	
Bonding embossed or die-stamped printed-circuit boards	
Bonding voice coil leads to speaker cones	
Adhesive coating of copper wire, coil windings, etc.	
Adhesives systems for electrical tapes	

adhere the conductive copper to the laminate. It must also provide a degree of surface insulation once conductive paths are etched from the bonded specimen.

In addition to joining, adhesives in electrical applications may be required to conduct heat, conduct or isolate electricity, provide shock mounting, seal, protect substrates, etc. Thermal and chemical resistance, weathering, and structural compatibility must also be considered in diverse electrical and electronic applications. Of course, the choice of adhesive will also be governed by application methods, cure temperature, processing speed, and overall economic cost.

Epoxy resin can have varying amounts of inorganic material remaining in the product after its synthesis. One of these is sodium chloride, which is formed by the reaction of sodium hydroxide with epichlorohydrin molecules, as described in Chap. 2. This is important in electrical applications because the sodium salt can be hydrolyzed and can degrade electrical properties. Some epoxy resin manufacturers will offer special ultrapure electrical grades or low-chlorine grades of epoxy resins for application where this is an important factor.

Construction. Polymeric materials such as adhesives, sealants, and composites have been used considerably in the last several decades for the construction, repair, and rehabilitation of our transportation infrastructures. Even though most processes were experimental until recently, they have evolved to the point where many are now standardized and well accepted. Table 1.6 lists several common applications for advanced polymeric materials (as well as the polymeric resins that are most commonly employed). In the construction or repair of roads and bridges, epoxy adhesives have primarily been used for bonding concrete and for bonding stiffening members or repair structures to degrading concrete facilities.

Commonly used adhesives include epoxy, rubber, acrylic or vinyl emulsions, and urethanes. Epoxies are high-priced, but they have better chemical resistance and durability than the others, and they have dominated the market in outdoor applications. Significant advantages of the epoxy-based adhesives are that they have no solvents and, therefore, exhibit little shrinkage. They cure relatively fast and, therefore, are not as exposed to inclement

TABLE 1.6 Common Applications and Polymeric Resins Used in the Construction and Repair of the Transportation Infrastructure

Common applications:	Common resins used:
Airfield runways and aprons	Epoxy
Highway and bridge joints	Polysulfide
Bridge abutments	Polyurethane
Concrete lined canals, traverse and longitudinal joints	Polyester
Multilevel parking lot joints	Silicone
Underground tunnel construction joints and sections	Neoprene
Attaching metal studs in concrete	Nitrile
Bonding pancake lighting systems on airport runways	Acrylic
Bonding traffic markers on roadways and road dividers	Polyvinyl acetate
	Butyl
	Asphalt, bituminous resins

weather as are slower-curing systems. Their physical properties do not change significantly on aging in the field.

One problem with early epoxy formulations is that they cured to a relatively brittle material. By using reactive flexibilizers, such as polysulfides, epoxy adhesive formulators have obtained the flexibility required for many applications in this industry. Polyamides and even coal tars have also been used to provide flexibility to epoxy base resins.

Another problem is that many two-component, room temperature curing epoxies cure with a high exotherm. This can cause the mixed uncured epoxy to get very hot. Excessive heat of exotherm provides a safety hazard and significantly shortens the working life of the mixed adhesive. This problem is generally solved by mixing small batches or using continuous mixing equipment.

Cured concrete can be bonded to cured concrete, as in the installation of precast buttons to a highway surface. Steel bridge railings can also be bonded to the concrete surface of a bridge sidewalk. In the case of deteriorated concrete, the adhesive can be used to rebuild the structure to its former line and grade. Epoxy adhesives are also commonly used on other roadway materials, such as asphalt and brick; however, the predominant application is concrete substrates. The most frequent combinations of substrates that are bonded with adhesives in this market segment are

- Cured concrete to cured concrete
- Cured concrete to other materials
- New concrete to cured concrete
- New concrete to other materials

In the bonding of cured concrete to cured concrete, the adhesive is applied directly to the substrates. In the case of new concrete, the bonding agent may be incorporated into the concrete formulation, or the adhesive may be applied to the nonconcrete surface.

There are many applications for bonding cured concrete to another material. These include bonding of plastic reflectors to road surfaces or dividers, attachment of railings or metal structures, and bonding of floor covering materials. An interesting application that has received much attention because of the aging road infrastructure in many regions is the repair and strengthening of existing concrete structures. Concrete beams can be strengthened by attaching steel plates and composites with adhesives.

For applying new concrete over old concrete, such as in the case of a patch, epoxy is often the best solution as an alternative for the new concrete. In these formulations an epoxy mortar is made using aggregate similar to the aggregate used in concrete mixtures. Adhesion is excellent and the epoxy mortar has generally better properties than the concrete; however, it is a relatively high-cost solution. The old concrete surface should be cleaned and sandblasted down to sound concrete before the epoxy patch is applied.

There are fewer standards for highway adhesives than there are for sealants. However, a very useful reference, published by the American Concrete Institute (ACI), provides the engineer, contractor, or architect with a description of the various types of polymer adhesives most frequently used for bonding of concrete.[18] The guide emphasizes the factors that should be considered when one is selecting a structural adhesive, including characteristics during installation and when in service. The benefits and limitations of adhesive bonding are also discussed for each application.

In the future, the use of adhesives in this market is expected to grow significantly due to the aging infrastructures. Also, the use of epoxy adhesive systems with fiber reinforcement is expected to increase significantly. Such adhesives will be applied to bridge piers, beams, and other elements requiring repair. These adhesives can be used not only to repair deteriorating structures but also to provide added strength to new structures.[19,20]

Medical. Many types of medical devices rely on adhesives for assembly. Medical device manufacturing requires that the final product exhibit maximum reliability and performance under many conditions that are not common in other industries. As a result, manufacturers of medical devices require significant testing and verification prior to choosing an adhesive. The standards and regulations in this industry are much different from those of other industries. The nature of the medical device market also dictates that the adhesive be economical and amenable to high-volume manufacturing methods.

This section focuses on adhesives that are used for the assembly of medical devices. In medical device assembly, the primary substrates are plastics, elastomers, and metals. The total "medical adhesive" market is much larger since it encompasses a broader definition of products. For example, medical adhesives can be used for bonding human tissue, transdermal drug delivery systems, dental restoration, and wound care in addition to medical device assembly.

Adhesives are currently used in a wide range of medical devices, and the market for these specialized adhesives is expected to grow significantly over the next decade. Factors that will contribute to this growth include longer human life expectancy and the development of new medical technology.

Just about any medical apparatus or diagnostic device may have applications for medical-grade adhesives. Traditionally, there are three classes of medical devices that have been assembled with adhesives:

- Disposables (e.g., syringe, catheters, and oxygenators)
- Reusables (e.g., surgical instruments, diagnostic equipment)
- Implantables (e.g., pacemakers)

These products generally come into contact with blood or bodily fluids, and they must be subjected to a sterilization process prior to use. Nonsterile medical devices are also used (e.g., diagnostic devices, medical electronics). These generally do not come into direct contact with the patient.

Recently two additional types of medical devices have emerged:

- Sterile reusables (e.g., endoscopes, laparoscopes)
- Resposables (devices originally intended as a disposable but now considered for reuse)

Sterile reusable devices have undergone significant market growth over the past several years due to advances in less invasive surgery. To the adhesive manufacturers, resposables may be considered to have the same requirements as sterile reusables.

Implantables are less likely to use adhesives because of their longevity, performance, and toxicity requirements. With these parts, self-assembly (snap-fit or interference fit) or solvent and heat welding processes are commonly used for assembly.

End users of adhesives and medical devices expect the device to withstand the rigors of sterilization, exposure to fluids, and occasional abuse. Because of the criticality of the product, the medical device manufacturer is generally more motivated than most to pay attention to adhesive selection criteria and the requirements of good bonding practice.

Choosing an adhesive for a medical application follows the same process as choosing an adhesive for any other purpose. Criteria include the particular substrates to be joined, strength requirements, type of loading, impact resistance, temperature resistance, fluid resistance, and processing requirements. However, the function of many medical devices requires at least two other important criteria: resistance to sterilization and low toxicity. As a result, several important standards and regulations have been developed in the industry.

Adhesives used in medical devices are tested for their effect on cells (cytotoxicity), blood constituents (hemolysis), and adjacent tissues, and for overall systemic effect. Several classes of biocompatibility testing exist. Adhesive suppliers, however, generally test to the following guidelines that have been established for toxicological properties and biocompatibility:

- United States Pharmacopoeia (USP), Class VI Standard
- International Standards Organization (ISO), ISO-10993

These guidelines were originally developed for testing the suitability of plastics used in medical devices that may come into contact with bodily fluids, but they have been extended to adhesives as well. Generally, products are tested by an independent laboratory. The results are typically provided to device or adhesive manufacturers in the form of certifications of compliance on an as-requested basis. Meeting these standards verifies that the successfully tested products are nontoxic and biologically inert in the cured state.

The two standards specify slightly different tests and follow different methodologies. The USP Class VI test method consists of acute systemic (over the tissue), intracutaneous (under the skin), and muscle implantation (in the muscle) tests. Establishing a USP Class VI rating has little bearing on whether the product will win approval from the Food and Drug Administration (FDA). The Class VI rating merely states that the products exhibit a low level of toxicity under the test conditions.

The ISO-10993 standard is generally recognized as the standard of choice for companies operating globally. It is widely recognized by North American, European, and Asian countries. It is also somewhat more extensive than the USP Class VI standard. The ISO-10993 biocompatibility testing includes

- Intracutaneous injection tests to evaluate the irritation potential of the material
- Acute systemic injection tests to evaluate the material for potential toxic effects as a result of single-dose systemic injections
- Cytotoxicity tests to determine the biological reactivity of monolayer cell cultures to the material
- Hemocompatibility tests to evaluate the hemolytic potential of the material with rabbit blood (In vitro hemocompatibility tests ensure that the test material extract does not adversely affect the cellular components of the blood.)

Such biocompatibility tests are only a guideline, and more extensive and specific testing may be required for certain device manufacturers. The extent and nature of the testing can also vary from adhesive supplier to adhesive supplier.

Another important criterion to recognize for medical device adhesives is the resistance to sterilization. Most disposable and reusable medical devices go through some type of sterilization process prior to use. Some products (endoscopes, surgical instruments, etc.) may be subjected to multiple sterilization cycles, and the adhesive must resist these processes and continue to perform its primary function.

The most common sterilization processes are gaseous ethylene oxide (EtO), gamma radiation, and autoclave sterilization. The resistance of adhesives to these processes will be dependent on the conditions of the sterilization cycle as well as on the chemistry of the adhesive.

Autoclave sterilization is one of the most difficult common sterilization environments for a medical adhesive, and it is commonly used in hospitals and health care facilities for reusable devices. Autoclaves sterilize with high-pressure steam. Temperatures inside the sterilization chamber typically can reach 130°C with pressures above ambient. Certain adhesive systems, such as polyurethanes, may show hydrolytic degradation in such environments especially after multiple cycles. Epoxies perform the best under multiple autoclave exposures. However, on certain substrates, light-cured acrylics and cyanoacrylates will also perform fairly well.

Epoxies cure to a thermosetting molecular structure. As a result, they provide optimum thermal, chemical, and environmental resistance compared to the other medical adhesives. Epoxy adhesives can be used at temperatures over 150°C, and they are resistant to most fluids and chemicals. Depending on their formulation, epoxy adhesives can vary significantly in modulus and toughness. However, most epoxy systems are relatively rigid when cured so that their peel strength is less than those of adhesives with a greater degree of elongation (e.g., polyurethanes).

Both room temperature and heat-curable epoxy adhesives are used in medical applications. The room temperature curing systems require metering and mixing, and the cure time is generally slow (several hours). Elevated-temperature cures could affect temperature-sensitive substrates.

Epoxies can cure in deep sections and are useful in potting and deep-section sealing applications. They adhere well to different substrates and therefore are used in the general assembly of many medical devices. A clear, medical-grade, low-viscosity epoxy adhesive has proved useful in the fabrication of access ports that are implanted beneath the skin of patients who require multiple infusions.[21]

Aircraft and Aerospace. Adhesives have always played a significant role in the aircraft and aerospace industries primarily because they offer a low-weight, fatigue-resistant, and aerodynamically sound method of assembly. Adhesive bonding is also less labor- and cost-intensive when applied to large structures such as those commonly utilized in the aerospace industry. Structural adhesives account for the greatest market share of all the adhesives used in aerospace applications.

In the aerospace market a distinction is made between primary and secondary structural applications. Joint failure in a primary structure will result in the loss of the aircraft, whereas failure in a secondary structure will result in only localized damage. Structural adhesives are used in both applications. Figure 1.4 illustrates the degree to which adhesive bonding is used in modern aircraft.

The primary types of epoxy adhesives that are used in the aerospace industry are structural film adhesives and liquid and paste adhesives. The film applications include metal-to-metal and honeycomb bonding. Advanced composite-to-composite bonding using film adhesive is gaining in significance because of the rapid growth rate and use of these materials. Liquid and paste epoxy adhesives are used for a variety of assembly operations including liquid shimming and aircraft maintenance and repair.

The early aerospace adhesives were primarily based on epoxy resin chemistry. However, unique applications requiring high temperatures and fatigue resistance have forced the development of epoxy-phenolic, epoxy-nitrile, epoxy-nylon, and epoxy-vinyl adhesives specifically for this industry. The aerospace industry has led in the development and utilization of these epoxy-hybrid adhesives.

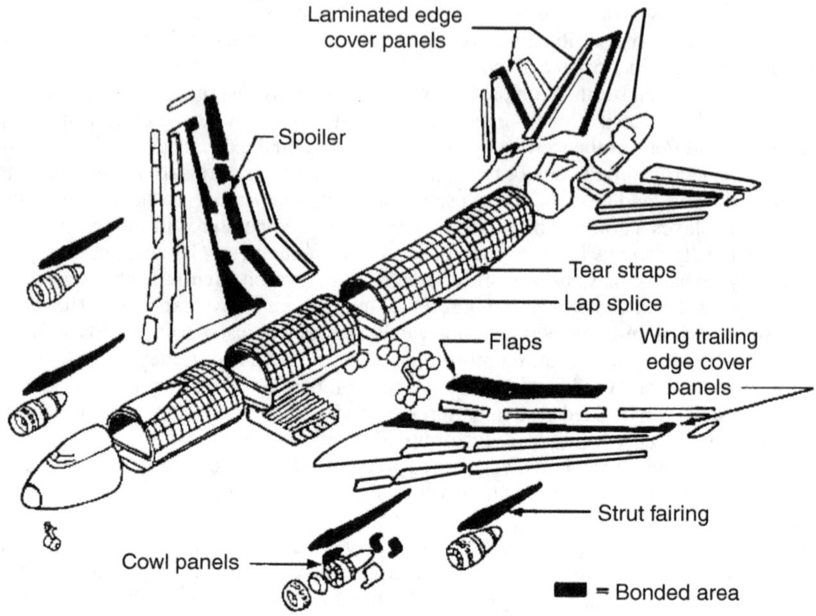

FIGURE 1.4 Schematic showing bonded area on a modern aircraft.[22]

Over the past several decades, significant advances have been made in developing epoxy-based adhesives having improved performance over these early adhesive systems. These improvements were made possible by (1) the incorporation of toughening additives into epoxy resin formulations and (2) the use of multifunctional epoxy resins primarily for high-temperature applications. These innovations are discussed in later chapters.

Structural adhesive bonding in the aerospace industry has primarily been through the initiative and development of applications for military. For example, in 1975 the U.S. government sponsored a program known as the Primary Adhesively Bonded Structures Technology (PABST) to extend the use of adhesive bonding for aerospace and to optimize manufacturing. Furthermore, many of the standards and specifications that are used in the industry come from military and/or government sources. These include Mil-A-25463 (bonding honeycomb structures) and MMM-A-132 (bonding metal-to-metal airframe structures). These standards define the physical (temperature, fatigue, etc.) and chemical environments (e.g., salt spray, humidity, jet fuel, hydraulic oil) to which aerospace adhesives will be exposed and their minimum requirements.

1.4 FORMULATING EPOXY ADHESIVES

1.4.1 The Job of the Adhesive Formulator

The adhesive formulator has many challenges in meeting the demands of the marketplace, and many of the tools that the formulator had utilized in the past are no longer acceptable. These internal and external pressures place the adhesive formulator in a special place among other professions.

At first glance, the job of the adhesive formulator appears deceptively simple. There are many resources on the *supply side* of the industry for the formulator to use. The problem is that there seems to be an even more staggering amount of requests for new products on the *demand side* of the industry.

The formulator has the unenviable task of responding to the latest issues impacting the end users of adhesives and devising the technology and resources to come up with practical solutions. It may seem that as soon as a product is developed, tested, and commercialized, the demands change. This is, of course, not necessarily the fault of the end user. There are many outside influences that affect the end users' processes, just as there are many influences affecting the type of raw materials and equipment that can be utilized for formulating adhesives and sealants.

This job has not gotten any easier in the last few years. Substrates continue to undergo significant transformations. Today's formulator is faced with the challenge of adhering to surfaces with a broad range of surface energies and to products that have been developed and characterized only recently. A competitive marketplace also necessitates the use of the most economical substrate available. These substrates may contain extenders and cost-reducing additives or processes that are not compatible with adhesives. Often substrate changes are made without notice to the end user or adhesive formulator who must work together for an assembly solution.

In an age of increasing environmental concerns, many of the tools that formulators had utilized in the past are no longer acceptable. Use of solvents has been drastically curtailed. Plasticizers must often be used in limited amounts. Regional regulations place demands that total VOC emissions be kept to a minimum. Certain catalysts and additives are discouraged because of safety or health issues.

Another megatrend in this competitive business environment is that all production processes need to be "world-class" and "lean." This means that processes must be automated; lead times shortened; rework, scrap, and waste minimized or eliminated; and less product must appear in inventory. This places a burden on the adhesive formulator in that the products not only must perform in service (e.g., creep resistance, peel strength) but also must "perform" in the assembly process (e.g., fast cure time, minimal scrap, low energy cure).

The job of the adhesive formulator has been made particularly difficult by the lack of practical information on the topic. There are only several forums that provide an introduction to adhesive formulating. One such forum is the Adhesive and Sealant Council, which offers short courses on adhesive formulating. There are also a few books that offer information and guides on adhesive formulations.[23–27] Information specific to epoxy adhesives is usually found in a chapter or section within the work. There has been no book devoted solely to epoxy adhesive formulations, although several have focused on the more general topics of epoxy resins and their applications.[28,29]

However, valuable information can be gained from literature provided by raw materials suppliers, who often develop a formulary using their materials. (Many of the formulations that appear in this book are from such sources.) Several Internet web sites have recently been established that cater specifically to the adhesive formulator. They provide information on raw materials, formulations, and end markets. One of the best of these sites is SpecialChem4Adhesives.com.

1.4.2 The Basics of Adhesive Formulation

As shown in Table 1.7, the epoxy resin is rarely used unmodified as an adhesive system. Rather, it is used in the form of a compound containing various modifiers and additives to improve properties, such as strength, flow, and heat resistance, and to add or advance other properties that are demanded by the specific application.

TABLE 1.7 Compounding Ingredients of Epoxy Resins and Their Roles[30]

Constituents	Ingredients	Roles
Resin content	Epoxy resin	The bisphenol A type is most common. However, there are many other types of epoxy resins having differing properties.
	Curing agents and catalysts	Curing agents and catalysts react with epoxy groups to form a three-dimensional network structure by crosslinking.
Modifying ingredients	Flexibilizers	Elasticity agents flexibilize compounds to improve their peel strength, impact resistance, and elongation.
	Tougheners	Tougheners eliminate brittleness from epoxy resin to prevent cracks and decrease distortion. Unlike flexibilizers, they do not reduce crosslink density.
	Heat-resistant additives	Generally these resins provide increased heat resistance by nature of their multifunctionality and/or aromaticity.
	Fillers	Fillers increase the weight in order to decrease cost and improve various types of properties (application and performance).
	Diluents	Diluents reduce viscosity. Reactive diluents have epoxy groups, and nonreactive diluents have no epoxy groups available for reaction.
	Thixotropic agents	Thixotropic agents impart thixotropy to the adhesive to control flow and increase viscosity.
	Other agents	Pigments, coupling agents, defoaming agents, film formers, etc. are used for specific properties.

The following section offers a brief description of how certain additives and formulation parameters are used to control the characteristics of adhesive systems. The roles of specific additives are defined in Chaps. 6 through 10.

The customary techniques used by the formulator to overcome several everyday problems are also discussed below. These techniques include formulation processes and materials to control flow, extend temperature range, improve toughness, match thermal expansion coefficients, reduce shrinkage, increase tack, and modify electrical and thermal conductivity.

This description of additives is admittedly generic and brief. There are literally thousands of additives that can be used in adhesive systems. The choice depends on the composition of the adhesive system, how it is to be used, system cost, and the properties that need to be obtained.

Adhesive Composition. Modern-day adhesives are often fairly complex formulations of components that perform specialty functions. Very few polymers are used without the addition of some modifying substance such as plasticizer, diluent, or inert filler. The selection of the actual ingredients will depend on the end properties required, the application and processing requirements, and the overall cost target of the adhesive. The various components that constitute an adhesive formulation can include the following:

- Epoxy resin base or binder
- Catalyst or curing agent

- Accelerators, inhibitors, and retarders
- Solvents
- Diluents
- Extenders
- Fillers
- Carriers and reinforcements
- Plasticizers
- Tackifiers
- Thickeners and thixotropic agents
- Film formers
- Antioxidants, antifungal agents, and other stabilizers
- Soaps, surfactants, and wetting agents

The adhesive *base* or *binder* is the principal component of an adhesive. The binder provides many of the main characteristics of the adhesive such as wettability, curing properties, strength, and environmental resistance. The binder is often by weight the largest component in the adhesive formulation, but this is not always the case, especially with highly filled adhesives or sealant systems. The binder is generally the component from which the name of the adhesive is derived. For example, an epoxy adhesive may have many components, but the primary material or base is an epoxy resin. Once the binder is chosen, the other necessary ingredients can be determined. Chapter 4 describes in detail the various polymeric resins that are commonly used as bases or binders in epoxy adhesive formulations.

A *curing agent* or *hardener* is a substance added to an adhesive to promote the curing reaction by taking part in it. Curing agents affect curing by chemically combining with the base resin. They are specifically chosen to react with a certain resin. They will have a significant effect on the curing characteristics and on the ultimate properties of the adhesive system. Reactive polyamide resins are an example of common curing agents used in two-part epoxy systems.

The criticality of the amount of curing agent used in the adhesive formulation is dependent on the chemistry of the specific reaction involved. By over- or underusing a polyamide curing agent, for example, the resulting formulation will have more or less of the characteristics of the curing agent, but a usable adhesive system generally results. Ten percent by weight additional polyamide in a two-part epoxy formulation will result in a system with greater flexibility and peel strength but lower temperature and environmental resistance due to the flexible nature of the polyamide molecule. Ten percent less polyamide will provide higher shear strength and temperature resistance but poorer peel strength. Note, however, that certain curing agents (e.g., amines, acids, and anhydrides) do have critical mixing requirements, and deviation from the manufacturers' mixing instructions could drastically affect the adhesive system.

Catalysts remain unchanged in the curing reaction, causing the primary resin to crosslink and solidify. Only small quantities are usually required to influence curing. Unlike with hardeners, the amount of catalyst used is critical, and poor bond strengths can result when resins are over- or undercatalyzed.

An *accelerator, inhibitor,* or *retarder* is sometimes incorporated into an adhesive formulation to accelerate or decelerate the curing rate. These are critical components that control the curing rate, storage life, and working life of the adhesive formulation.

Solvents are sometimes needed to disperse the adhesive to a consistency that is more easily applied such as by brush or spray. Solvents are also used to aid in formulating the adhesive by reducing the viscosity of the formulation so that additions of other components and uniform mixing may be more easily achieved. Solvents used with synthetic resins and elastomers are generally organic, and often a mixture of solvents is required to achieve the desired properties.

Polar solvents are required with polar resins; nonpolar solvents with nonpolar resins. When solvents are used in the adhesive formulation, they must be completely evaporated from the bond line prior to cure, or otherwise bubbles could form in the bond line, causing a weak joint. The substrate must also be tested so that the solvents that are used in the adhesive formulation do not attack or degrade it.

Water is sometimes used as a solvent for water-soluble resins. Certain epoxy adhesives are available as water-based emulsion or latex formulations. In the early 1970s, during the time of the petroleum crisis, water-based adhesives were thought of as a possible replacement for solvent-based adhesives systems. However, water-based adhesives never met the lofty expectations primarily because of the time and energy required to remove water from the bond line, the corrosion that the water causes in drying ovens, and the poor moisture resistance of cured water-based adhesives.

An ingredient added to an adhesive to reduce the concentration of base resin or binder is called a *diluent*. Diluents are principally used to lower the viscosity and modify the processing conditions of some adhesives. The degree of viscosity reduction caused by various diluent additions to a conventional epoxy adhesive is shown in Fig. 1.5. Diluents do not evaporate as does a solvent, but they become part of the final adhesive. Reactive diluents react with the resin base during cure, so that the final adhesive characteristics are determined by the reaction product of the binder and diluent. Nonreactive diluents do not react with the resin or curing agent and, therefore, more seriously weaken the final properties. Coal and pine tar are common nonreactive diluents.

Extenders are substances that usually have some adhesive properties and are added to reduce the concentration of other adhesive components and, thereby, the cost of the overall formulation. Extenders also have positive value in modifying the physical properties of

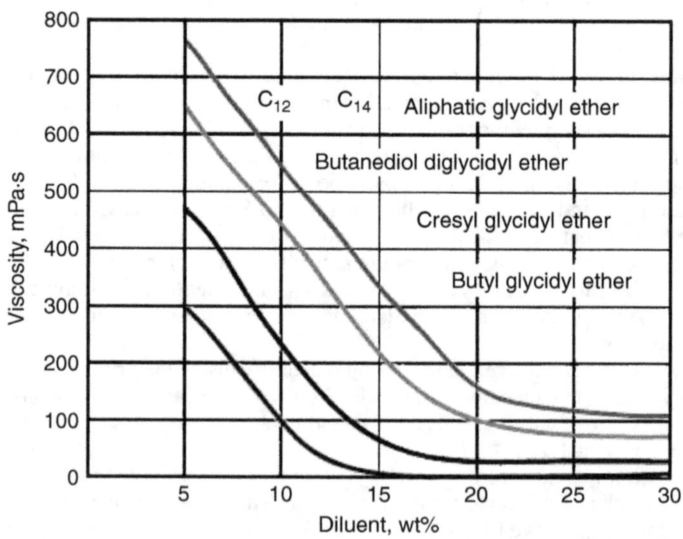

FIGURE 1.5 Viscosity reduction of a diglycidyl ether of bisphenol A (DGEBA) epoxy resin by reactive diluents.[31]

the adhesive. Common extenders are coal tar, soluble lignin, and pulverized partly cured synthetic resins.

Fillers are relatively nonadhesive substances added to the adhesive formulation to improve its working properties, strength, permanence, or other qualities. The improvements resulting from the use of fillers are listed in Table 1.8. Fillers are also used to reduce material cost. By selective use of fillers, the properties of an adhesive can be changed significantly. Thermal expansion, electrical and thermal conduction, shrinkage, viscosity, and thermal resistance are only a few properties that can be modified by the use of fillers. Common fillers are wood flour, silica, alumina, titanium oxide, metal powders, china clay and earth, slate dust, and glass fibers. Some fillers may act as extenders.

A *carrier* or *reinforcement* is usually a thin fabric, cloth, or paper used to support the semicured adhesive composition to provide a tape or film. In tapes, the carrier is the backing on which the adhesive is applied. The backing may be used for functional or decorative purposes. In epoxy films or structural tape, the carrier is usually porous and the adhesive saturates the carrier. Glass, polyester, and nylon fabric are common carriers for supported B-staged epoxy adhesive films. In these cases, the carrier provides for a method of applying the adhesive and also may act as reinforcement and a internal "shim" to control the thickness of the adhesive.

Plasticizers and *flexibilizers* are incorporated into an adhesive formulation to provide it with flexibility and/or elongation. Plasticizers may also reduce the melt viscosity of hot melt adhesives or lower the elastic modulus of a solidified adhesive. Similar to diluents, plasticizers are nonvolatile solvents for the base resin, and by being incorporated into the formulation, they separate the polymer chains and enable their deformation to be more easily accomplished. Plasticizers generally affect the viscoelastic properties of the base resin whereas diluents simply reduce the viscosity of the system. Whereas diluents result in brittle, hard adhesive systems, plasticizers result in increased flexibility and lower modulus. The temperature at which polymers exhibit rubbery properties (i.e., the glass transition temperature) can also be modified by incorporating plasticizers.

Certain resinous materials that act as plasticizers are well noted for increasing the tack of the formulation. Traditional *tackifiers* were based on naturally occurring resins such as pine tar. Today, tackifiers used in modern adhesive formulations include aliphatic and aromatic

TABLE 1.8 Adhesive Properties That Can Be Modified by the Use of Fillers

Properties capable of modification
Material cost
Flow properties (viscosity, thixotropy)
Bond line thickness
Coefficient of thermal expansion
Shrinkage
Conductivity (electrical and thermal)
Electrical properties
Specific gravity
Mechanical strength properties
Heat and chemical resistance properties
Adhesion properties
Working life and exotherm
Fire resistance
Color

hydrocarbons, terpenes, and rosin esters. Tackifiers are useful in pressure-sensitive adhesives or adhesives that require aggressive tack or "green strength" to assist in assembly of the product. In addition to increasing tack, increased peel strength and decreased shear strength also result from the addition of tackifiers in the adhesive formulation.

Thickeners and *thixotropic agents* are used to maintain reasonable thickness of the glue line through viscosity adjustment. Thixotropic fillers are materials which when added to the adhesive increase the viscosity when it is under rest (i.e., not undergoing stress). Therefore, thixotropic agents provide sag resistance and the ability for an adhesive to remain in place on a vertical substrate. However, when a slight force is applied, such as in the act of stirring or extruding the adhesive, the system then acts as if it had a lower viscosity and flows with relative ease.

Thickeners, fillers, and thixotropic agents are also commonly used to control flow and the bond line's thickness within a joint. Scrims, carriers, and woven reinforcements are other methods commonly used to control bond line thickness.

Many polymers have a limited life and are subject to aging processes even before they are used in production. To delay these aging processes to a usable time period, *antioxidants, antihydrolysis agents*, and *stabilizers* are added to adhesive formulations. *Antifungal agents* or *biocides* are used in many water-based adhesive systems.

Aqueous suspensions comprise a wide range of adhesives. These will contain as additives the various *soaps, surfactants,* and *wetting agents* necessary to stabilize the emulsion or latex. Additives are also incorporated into aqueous formulations to provide system stability under repeated freeze-thaw cycles during storage.

The removal of solvent from a true solution of a polymer leaves a film of the polymer. If the same polymer is available as an emulsion, removal of the liquid by evaporation does not necessarily leave a coherent film. The individual globulars of polymer will merge to form a film only if the polymer is well above its glass transition temperature. Drying of some emulsions will produce a powder unless its glass transition temperature has been first lowered by the addition of a plasticizer or *film former* to the emulsion. Film formers are carefully chosen to lower the glass transition temperature without appreciably lowering the strength of the film.

REFERENCES

1. Austin, J. E., and Jackson, L. C., "Management—Teach Your Engineers to Design Better with Adhesives," *SAE Journal*, October 1961.
2. *The Long and Short of Epoxy Resins*, Shell Chemical, Houston, TX, 1992.
3. Nick, D. P., *The World of Epoxy Adhesives*, TRFA Annual Meeting, Philadelphia, PA, November 11, 2003.
4. Michaud, P., "Epoxy Resins in Composites," *JEC—Composites*, no. 12, October-November 2004, pp. 24–25.
5. Freedonia Group, Cleveland, OH, "Epoxy Resins Demand to Reach 690 Million Pounds in 2004," *Paint and Coating Industry*, June 2000.
6. Lee, H., and Neville, K., *Handbook of Epoxy Resins*, McGraw-Hill, New York, 1967.
7. May, C. A., and Tanaka, Y., eds., *Epoxy Resins: Chemistry and Technology*, Marcel Dekker, New York, 1973.
8. Oldring, P. K. T., ed., *Waterborne and Solvent Based Epoxies and Their End User Applications*, 2d ed., vol. 2, SITA Technology Ltd., Edinburgh, UK, 1999.
9. Schields, J., *Adhesives Handbook*, 3d ed., Chapter 1, Butterworths, London, 1984.
10. Smith, R., "The U.S. Epoxy Adhesives Business," *Composites News*, June 30, 2004.
11. Nick, *The World of Epoxy Adhesives*.
12. Nick, *The World of Epoxy Adhesives*.

13. Nick, *The World of Epoxy Adhesives.*
14. Frost and Sullivan, ChemQuest Group, The Freedonia Group, Impact Marketing Consultants, AMA Research (United Kingdom).
15. ChemQuest Group, Inc., Cincinnati, OH, 2001.
16. "U.S. Demand for Electronic Polymer Products to Expand 6.4% Annually Through 2005," *Market Trends*, May 2002, www.adhesivesmag.com.
17. Schwartz, J., *Electrical Outlet*, Adhesives Age/Adhesive and Sealant Council Custom Publication, New York, 2002.
18. "Guide for the Selection of Polymer Adhesives with Concrete," *ACI Materials Journal*, January-February, 1992, pp. 90–104.
19. Meier, U., "Composite Material in Bridge Repair," *Applied Composite Materials*, vol. 7, 2000, pp. 75–94.
20. "Field Test Increases Viability of Composites in Bridge Repair," *Engineering News Record*, vol. 237, no. 12, September 16, 1996, p. 22.
21. Estes, R. H., "The Suitability of Epoxy Based Adhesive for Use in Medical Devices," Technical Paper GB-63, Epoxy Technology, Pembroke, MS, 1999.
22. Boeing Company, Seattle, WA; also in Politi, R. E., "Structural Adhesives in the Aerospace Industry," *Handbook of Adhesives*, 3d ed., I. Skiest (ed.), van Nostrand Reinhold Publishing, New York, 1990.
23. Petrie, E. M., *Handbook of Adhesives and Sealants*, McGraw-Hill, New York, 2000.
24. Pizzi, A., and Mittal, K. L, eds., *Handbook of Adhesive Technology*, 2d ed., Marcel Dekker, New York, 2003.
25. Skiest, I., ed., *Handbook of Adhesives*, 3d ed., van Nostrand Reinhold Publishing, New York, 1990.
26. Flick, E. W., *Adhesive and Sealant Compound Formulations*, 2d ed., Noyes Publications, New York, 1984.
27. Ash, M., and Ash, I., *Handbook of Adhesive Chemical and Compounding Ingredients*, Synapse Information Resources, Inc., New York, 1999.
28. May, C., and Tanaka, Y., *Epoxy Resins: Chemistry and Technology.*
29. Lee, and Neville, *Handbook of Epoxy Resins.*
30. Osumi, Y., "One Part Epoxy Resin," *Technical News*, Three Bond, October 1, 1987.
31. Behm, D. T., and Gannon, J., "Epoxies" in *Adhesives and Sealants*, vol. 3, *Engineered Materials Handbook*, ASM International, Materials Park, OH, 1990.

CHAPTER 2
EPOXY RESIN CHEMISTRY

2.1 INTRODUCTION

Proper formulation of epoxy adhesives requires knowledge of the chemical reactions that lead to polymerization as well as the chemical and physical properties of both the uncured mixture and the cured material. This chapter reviews the general principles of epoxy resin chemistry including synthesis of the epoxy monomer itself and its possible polymerization reactions.

The general properties of cured and uncured epoxy resins are reviewed in Chap. 3. The chemical structures of the resin and curing agent will determine these physical properties. They will also determine, to a great extent, the surface chemistry and adhesion properties of the final product.

2.2 EPOXY RESIN CHARACTERISTICS

The term *epoxy, epoxy resin*, or *epoxide* (Europe) refers to a broad group of reactive compounds that are characterized by the presence of an oxirane or epoxy ring, shown in Fig. 2.1. This is represented by a three-member ring containing an oxygen atom that is bonded with two carbon atoms already united in some other way.

An epoxy resin can be any molecule containing more than one of these epoxy groups. The number of epoxy groups per molecule is the *functionality* of the resin. The groups can be situated internally, terminally, or on cyclic structures. Epoxy groups are capable of reacting (1) with suitable curing agents or (2) catalytically (homopolymerized) to form higher-molecular-weight polymers. Once cured, the epoxy polymers have a densely crosslinked, thermosetting structure with high cohesive strength and adhesion properties. However, the term *epoxy* can also be used to indicate an epoxy resin in the thermoplastic or uncured state.

A general formula for an epoxy resin can be represented by a linear polyether with terminal epoxy groups and secondary hydroxyl groups occurring at regular intervals along the length of the chain. The epoxy resin structure and properties influenced by the various chemical groups are illustrated in Fig. 2.2.

Several important statements can be made relative to the structure of the epoxy molecule:

- The epoxy groups at both terminals of the molecule and the hydroxyl groups at the midpoint of the molecule are highly reactive.

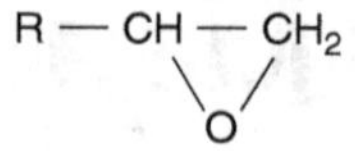

FIGURE 2.1 The epoxy or oxirane ring structure.

- The outstanding adhesion of epoxy resins is largely due to the secondary hydroxyl groups located along the molecular chain (see Fig. 2.3); the epoxy groups are generally consumed during cure.
- The large part of the epoxy resin backbone contains aromatic rings, which provide a high degree of heat and chemical resistance.
- The aliphatic sequences between ether linkages confer chemical resistance and flexibility.
- The epoxy molecule can be of different molecular weight and chemistry. Resins can be low-viscosity liquids or hard solids. Low viscosity can be obtained at 100 percent solids, which results in good penetration and wetting.
- A large variety of polymeric structures can be obtained depending on the polymerization reaction and the curing agents involved. This can lead to versatile resins that can cure slowly or very quickly at room or at elevated temperatures.
- No small molecules such as water are liberated during the curing process. Thus, epoxies exhibit low shrinkage, and they can be cured under very low pressure. This provides an adhesive joint with a very low degree of internal stress when cured.

Epoxy resins that are commercially produced are not necessarily completely linear or terminated with epoxy groups. Some degree of branching occurs, with the end groups being either epoxy or hydroxyl. The amount and degree of branching vary from resin to resin and from supplier to supplier. Epoxy resins are not completely difunctional. Tri-, tetra-, and polyfunctionality are possible. Various end groups can be introduced as a consequence of the manufacturing process.

The ratio of the main ingredients used in the synthesis of epoxy resins (epichlorohydrin: bisphenol A) determines the extent of the reaction and the molecular weight (or value of *n* repeating units in the molecular chain). The addition of bisphenol A to the reaction mix will advance the molecular weight of the resin and the value of *n*. As *n* increases, the viscosity or melting point of the resin also increases. Also as the value of *n* increases, the number of hydroxyl groups increases while the number of epoxy groups remains constant.

An important term that is used in formulating epoxy adhesive compositions is *epoxy equivalent weight* (EEW). This is defined as the weight of resin in grams that contains one equivalent of epoxy. As the resin's molecular weight increases, the EEW will also increase.

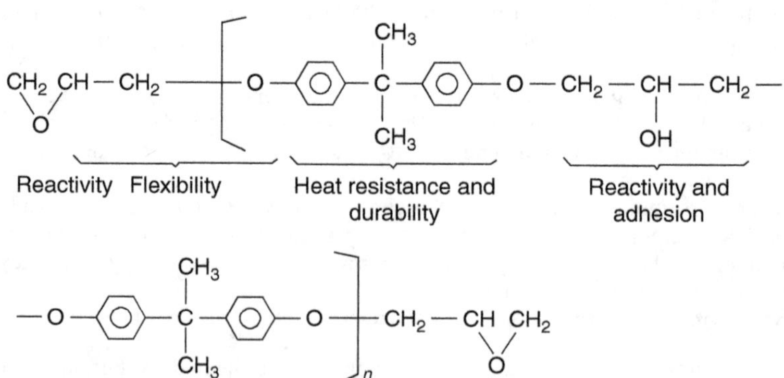

FIGURE 2.2. The structure and properties of an epoxy resin.[1]

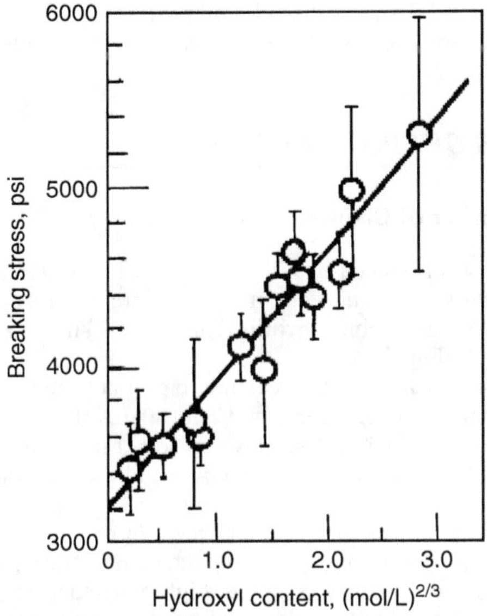

FIGURE 2.3 Relation between the adhesive strength of epoxy resins and their hydroxyl content.[2]

If the resin chains are assumed to be linear with no side branching, and it is further assumed that an epoxy group terminates each end of the molecule, then the epoxide equivalent weight is one-half the average molecular weight of the epoxy resin.

The use of EEW in determining curing agent concentrations is described in Sec. 2.5. Generally, the higher the molecular weight and EEW of the epoxy resin, the less curing agent is required because there are less functional epoxy groups per unit weight of resin.

There are many terms used to refer to the number of epoxy groups per molecule in addition to EEW, and as a result there is a lot of confusion around this issue. Other terms that can be used interchangeably with EEW and are numerically equivalent are *weight per epoxy* (WPE) and *epoxy molar mass* (EMM). Sometimes the term *epoxy value* or *epoxy group content* is also used. This is the fractional number of epoxy groups per 100 g of resin. Dividing the epoxy value into 100 gives the epoxide equivalent weight.

The use of the expression *epoxy group content* (EGC), expressed in millimoles per kilogram (mmol/kg), is a somewhat more recent trend. Note that as the EGC increases, the molecular weight decreases. This is in contrast to the other expressions used. In summary,

$$\text{EEW} = \text{EMM} = \text{WPE} = \frac{10^6}{\text{EGC}}$$

Most of the epoxy resins that are used in the formulation of adhesives have EEWs in the range of 180 to 3200, corresponding to a molecular weight range from 250 to 3750. This applies equally to epoxy resins that have undergone chemical reaction or modification to be used as a base resin in an adhesive formulation.

Similarly, the hydroxyl equivalent weight is the weight of the resin containing one equivalent of hydroxyl group, and it may also be expressed as equivalents per 100 g. Several studies have demonstrated that the hydroxyl content greatly influences the adhesion of the epoxy resin.[3,4] These results are best explained by the polar character of the hydroxyl groups. Figure 2.3 shows the relation between adhesive strength and hydroxyl content for a series of epoxy resins bonding stainless steel substrates.

2.3 SYNTHESIS OF EPOXY RESINS

2.3.1 Diglycidyl Ether of Bisphenol A (DGEBA)

Epoxy resins were first introduced by Ciba-Geigy, Ltd., in 1946 although their development began decades earlier. The most common commercial epoxy resin is formed from the reaction of bisphenol A and epichlorohydrin. This resin is known as the *diglycidyl ether of bisphenol A* or DGEBA (Fig. 2.4).

The addition of NaOH catalyzes the reaction to produce the chlorohydrin intermediate, acts as the dehydrohalogenating agent, and neutralizes the formed HCl. The reaction generally occurs over 16 h at 110°C. The caustic is added slowly as a 30% aqueous solution. The resulting organic is separated, dried with sodium sulfate, and fractionally distilled under vacuum.

The reaction is always carried out with an excess of epichlorohydrin so that the resulting resin has terminal epoxy groups. By varying the manufacturing conditions and the concentration of epichlorohydrin, resins of low to high molecular weight (MW) (low to high value of *n*) can be produced. If a large excess of epichlorohydrin (10 to 20 mol per mole of bisphenol A) is used, the likelihood of producing a low-molecular-weight resin, where *n* is near to 0, is considerably improved. Table 2.1 shows the effect of varying reactant ratios on the molecular weight of epoxy resins. Very low-molecular-weight epoxy resins can be obtained by fractionation (distillation) of the liquid epoxies.[6] As *n* increases, the molecular weight increases, and the resin moves from a low-viscosity liquid to a semisolid and then to a brittle solid. When *n* is between 0 and 1, the product is a liquid; when *n* is greater than 1, it is a solid. Solid resins generally need to be melted or dissolved in solvent prior to compounding or processing.

Commercially useful grades of DGEBA are relatively low-MW weight resins in which *n* ranges from 0 to 4, although *n* could be as high as 20. Most commercial resins are mixtures of polymers so that the value of *n* is an average for a given product. Typical properties of the DGEBA resins are shown in Table 2.2. Note that as *n* increases, the epoxy content is decreased and the hydroxyl content is increased.

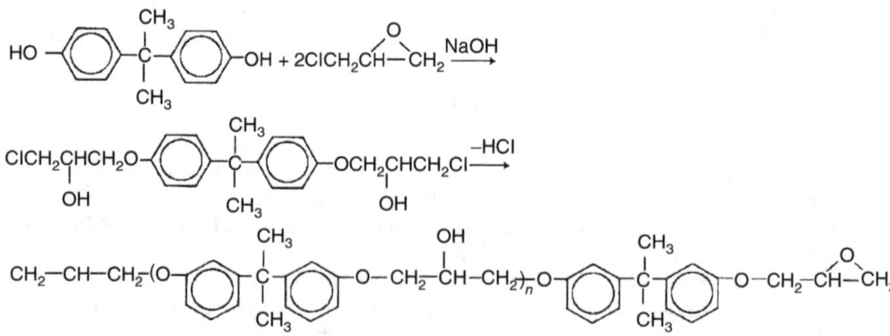

FIGURE 2.4 Synthesis of DGEBA epoxy resin from bisphenol A and epichlorohydrin.[5]

TABLE 2.1 Effect of Varying Reactant Ratios on Molecular Weight of Epoxy Resins[7]

Mole ratio: epichlorohydrin/ bisphenol A	Mole ratio: NaOH/ epichlorohydrin	Softening point, °C	Molecular weight	Epoxide equivalent	Epoxy groups/ molecule
2.0	1.1	43	451	314	1.39
1.4	1.3	84	791	592	1.34
1.33	1.3	90	802	730	1.10
1.25	1.3	100	1133	862	1.32
1.2	1.3	112	1420	1176	1.21

The low-MW, liquid epoxy resins are generally the most desirable to use in adhesive formulations because of easier compounding capability and application via liquid dispensers. They also have relatively high reactivity and crosslink density. The high concentration of epoxy rings per unit of mass provides the high reactivity. This allows the resin to be cured at low temperatures and with conventional curing agents. The crosslink density results in a material with a high glass transition temperature having good heat, chemical, and solvent resistance.

Solid, higher-MW epoxy resins are often used for adhesive formulations that are applied as solids (e.g., film and powder) or a solvent solution. The higher-MW DGEBA resins are also used where improved toughness, flexibility, and adhesion are required. These resins have a greater number of hydroxyl groups along the chain and, thus, can provide better adhesion and additional reaction mechanisms.

Extremely high-MW (MW equal to about 30,000 and n equal to about 100) DGEBA resins have also been produced. These resins are known as *phenoxy* compounds. They are not necessarily epoxy-terminated, and when they are epoxy-terminated, they have a very low epoxy equivalent weight. Phenoxy resins are essentially linear molecules that are thermoplastics. However, they can be crosslinked through their epoxy groups and, more conveniently, through the hydroxyl groups (through reaction with isocyanates or melamine formaldehyde resins) along the molecular chain. These hydroxyl groups also provide excellent adhesion to many polar surfaces. In adhesive formulations, phenoxy resins can be processed into powders, films, and solvent solutions. Phenoxy resins have considerable mechanical strength by themselves and are often used as unformulated thermoplastic hot melt adhesives.

TABLE 2.2 Typical Properties of DGEBA Epoxy Resins[8]

Resin designation*	Average value of n	Melting point, °C	Epoxide content, mmol/kg	Epoxy equivalent weight	Hydroxyl content, mmol/kg
828	0.1	Liquid (125 cps @ 25°C)	5350	188	300
1001	2	50–60	2000	475	2200
1004	4	80–100	1200	830	3300
1007	9	100–120	600	1775	3600
1009	12	120–150	300	3200	3800

* Resolution Performance Products LLC: EPON Resins.

2.3.2 Other Epoxy Resins

In addition to the DGEBA resins, there are several other types of epoxy resins of commercial significance. The most common of these are epoxy novolacs, glycidyl ether of tetraphenolethane, bisphenol F–based resins, and aliphatic and cycloaliphatic resins.

Epoxy Novolac and Other Phenols. Resins of greater functionality than DGEBA can be produced in several ways. Polyols having more than two hydroxyl groups per molecule (e.g., phenol novolac resins) can be reacted with epichlorohydrin to produce epoxy novolac resins with a structure shown in Fig. 2.5.

The epoxy novolac resins are synthesized by reaction of phenolic or cresol novolacs with epichlorohydrin in the same fashion as the bisphenol A resins. The number of epoxy groups per molecule is dependent on the number of hydroxyls in the phenol novolac molecule and to the extent to which they are reacted. Complete epoxidation can be accomplished, but this will lead to steric factors, which could limit the useful size of the cured polymer. Thus, selective epoxidation is often practiced.[9]

The epoxy novolacs are high-viscosity liquids or semisolids, and they are often mixed with other epoxy resins to improve handling and formulation properties. They cure more rapidly than DGEBA epoxy resins and have higher exotherms. The epoxy novolac resins have much greater functionality (generally 2.3 to 6.0) than conventional DGEBA resins, and the increased crosslink density results in greater temperature and chemical resistance.

The inherent thermal stability of the phenol formaldehyde chemistry is preserved but with the crosslinking characteristics of the epoxy groups. However, epoxy novolacs also form very rigid and brittle polymers when fully cured because of their high crosslink density. For this reason, they are often used as modifiers in epoxy adhesive systems rather than as the base polymer.

Glycidyl Ether of Tetraphenolethane. A large number of polyhydric phenols have been used to prepare diglycidyl ethers. The polyphenol, 1:1,2:2-(*p*-hydroxyphenol)ethane, is used to prepare a tetrafunctional epoxy resin, tetraglycidyl ether of tetraphenolethane. The functionality of commercial resins (e.g., EPON Resin 1031, Resolution Performance Products, LLC) is about 3.5. This forms a solid resin (melting point of 80°C) with a structure as shown in Fig. 2.6. Commercial products are solid resins and solutions.

Similar to epoxy novolac resins, tetraglycidyl ether of tetraphenolethane resin is usually employed in a blend with lower-MW liquid resins. The high aromatic ring content combined with polyfunctionality provides increased thermal stability for high-temperature applications. The chemical resistance and humidity resistance of cured tetraglycidyl ether of tetraphenolethane epoxy resins are also outstanding.

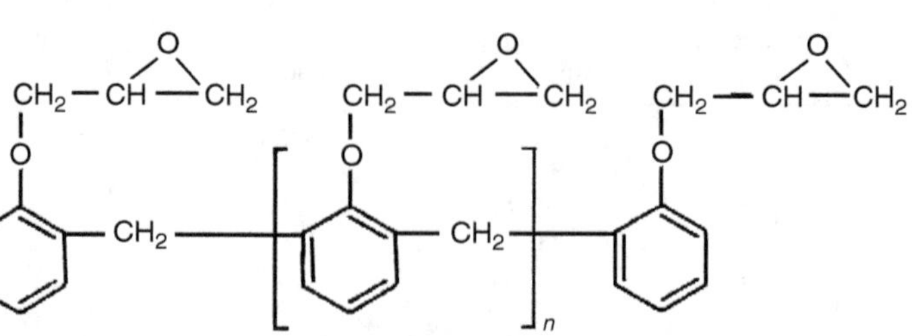

FIGURE 2.5 Chemical structure of epoxy novolac resin.

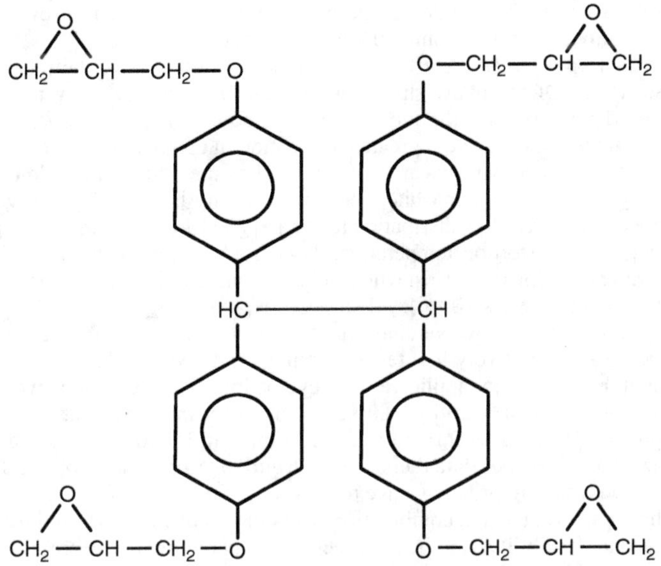

FIGURE 2.6 Chemical structure of glycidyl ether of tetraphenolethane epoxy resin.

Bisphenol F Resins. Diglycidyl ether resins based on bisphenol F (DGEBF) have been developed to provide cured epoxy resins with greater flexibility and lower softening temperatures than conventional DGEBA epoxy resins. The preparation of bisphenol F resins is from formaldehyde and phenol. Three isomers are possible because substitution can occur at the ortho-, meta-, or para- positions. The proportion of isomers depends upon the pH of the reaction medium.

The DGEBF resins have much lower viscosity (generally an order of magnitude) than their DGEBA counterparts for the same *n* value. The lower viscosity of DGEBF resins provides for greater latitude in formulation. The major disadvantage of bisphenol F epoxy resins is their significantly higher cost than DGEBA resins.

Bisphenol F–based epoxy resins (Fig. 2.7) are analogous to DGEBA-based epoxy resins in most respects. They use the same curing agents and reaction mechanisms. Bisphenol F epoxies are often used in blends with DGEBA resins to lower the viscosity or to modify certain properties.

Aliphatic and Cycloaliphatic Resins. Aliphatic and cycloaliphatic epoxy resins have been produced from the epoxidation of olefinic compounds. The epoxidation process involves the use of an olefinic or polyolefinic compound and a peracid (e.g., peracetic acid) or other

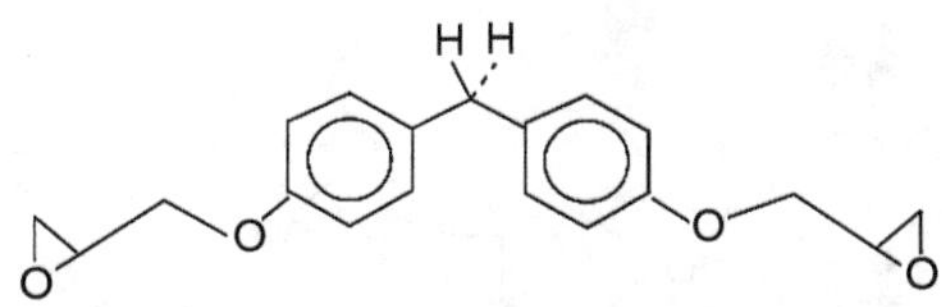

FIGURE 2.7 Chemical structure of bisphenol F resin (4,4 isomer).

oxidizing substances such as hydrogen peroxide, molecular oxygen, or even air. Several common cycloaliphatic epoxy resin structures are illustrated in Fig. 2.8.

Two types of epoxy resins are formed by this process: (1) cycloaliphatic resins and (2) aliphatic resins. Of the many structures that can be synthesized by this process, the cycloaliphatic diepoxies offer the most interesting combination of properties. However, the aliphatic epoxy resins have the greatest utilization in epoxy adhesive formulation.

The cycloaliphatic epoxy resins are characterized by the saturated ring in their chemical structure. They are almost water-white, very low-viscosity liquids. They provide excellent electrical properties such as low dissipation factor and good arc-track resistance, good weathering, and high heat distortion temperature. They are also free of hydrolyzable chlorine, sometimes present in DGEBA resins, which adversely affects certain electronic applications.

Because of their low viscosity, cycloaliphatic epoxies are often used to dilute other epoxy resins. These resins, however, have not achieved general importance in adhesive formulations because of relatively low tensile strength and because they do not cure well at room temperature. One major application for cycloaliphatic epoxies, however, is for adhesives and coatings that can be cationically cured by exposure to uv light.

The aliphatic epoxy resins formed from reaction with hydrogen peroxide or peracetic acid include epoxidized polybutadiene, epoxidized soya or linseed oil, and epoxidized polyglycols. The resulting products have too low a functionality for use as base polymers. They are almost always used in combination with other epoxy resins to improve properties such as cure rate, flexibility, and heat deflection temperature. Therefore, these resins are often considered to be reactive diluents and flexibilizers.

Glycidyl Ethers of Aliphatic Polyols. Glycidyl ether epoxy resins based on polyols provide greater flexibility and lower softening temperatures for the final cured epoxy system. The polyol is reacted with epichlorohydrin to produce these resins. These resins are generally not used alone because of water sensitivity and overall lack of toughness. However, they serve as modifiers for DGEBA-based epoxy resins. An idealized structure for a flexible resin based on this chemistry is shown in Fig. 2.9.

Special grades of glycidyl ether epoxies have been made with components such as glycerol, polyglycols, pentaerythritol, and cashew nut oil. Other epoxy-polyglycidol resins have been produced from the reaction of epichlorohydrin and polyester polyol based on

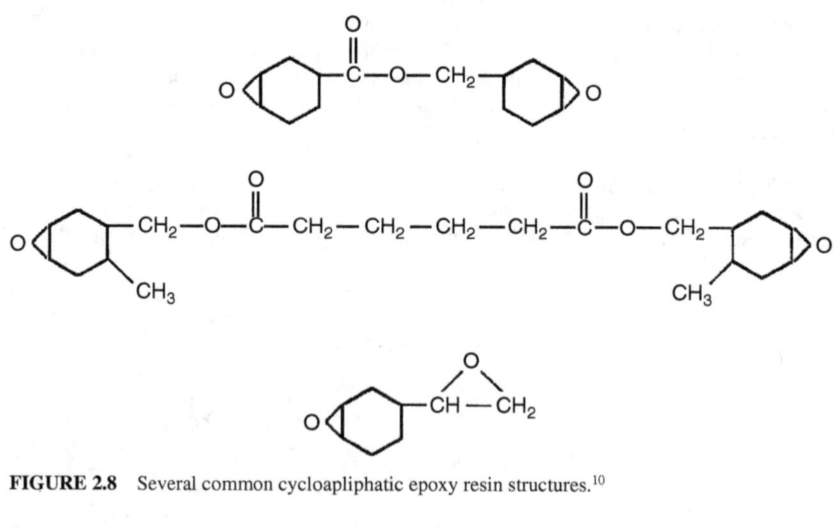

FIGURE 2.8 Several common cycloapliphatic epoxy resin structures.[10]

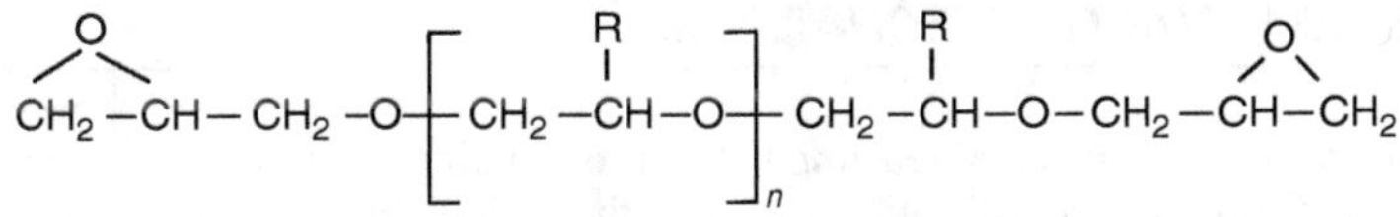

R-an organic group or hydrogen

FIGURE 2.9 An idealized structure of a flexible epoxy resin made from a polyol.[11]

ethylene or propylene oxide. These resins are produced by the reaction of epichlorohydrin with the polyol in the presence of a catalyst, similar to the phenolic-based glycidyl ethers.

The resins based on glycerol and pentaerythritol are water-soluble and have low viscosity. They can have greater functionality and reactivity than conventional DGEBA resins. Resins based on polytaerythritol are claimed to have excellent adhesive properties including the ability to adhere to wet surfaces. They cure between 2 and 8 times faster than DGEBA epoxy resins and reduce the viscosity of DGEBA by 50 percent when used in concentration of 20 pph.

The polyglycol diepoxies are used primarily as flexibilizers and reactive diluents. Commercial products are available where *n* varies from 2 to 7. Generally, these are used in the range of 10 to 30 percent by weight with DGEBA or epoxy novolac resins to improve flexibility without a significant loss in physical properties.

Other products that may be epoxidized in this way include dihydric and trihydric phenols, aliphatic polyols such as glycerol, and simple alcohols such as butanol or alyl alcohol. These products, especially the monofunctional glycidyl ethers, are used at relatively low percentages to reduce the viscosity of formulations containing DGEBA resins. In this way, they act as reactive diluents.

Brominated Epoxy Resins. The conventional DGEBA epoxy resins are flammable when cured. In an adhesive, flammability is generally not considered critically important because the mass of adhesive in any one area is relatively small. However, in certain applications (printed-circuit manufacture, aircraft interiors, furniture, etc.) nonflammability is an important criterion. Flame-retardant additives and chlorinated curing agents have been used to impart nonflammability to epoxy resins.

However, the optimum degree of flame resistance can only be achieved by making the resin itself nonflammable. A halogenated form of bisphenol A, such as tetrabromobisphenol A, has accomplished this. This resin is formed from the reaction product of epichlorohydrin and brominated bisphenol A. These resins acquire their fire-retardant characteristics through bromine substitution on the phenyl rings of the bisphenol A.

Tetrabromobisphenol A epoxy resins are available as viscous liquids with several molecular weight ranges. As the bromine content increases, the flame resistance increases but the viscosity of the resin generally increases as well. They are primarily used as an additive in formulations producing epoxy laminates and adhesives that require improved resistance to ignition.

Epoxy Adducts. The epoxy resin itself may be reacted with di- or polyfunctional reactants containing more than one epoxy group. The number of potential epoxy adducts is very large. Typical are phosphorous-containing epoxy resins prepared by reaction of diepoxies with acid dichlorides of phosphoric acids. This is accomplished at an acid : diepoxy ratio of 1 : 2 in the presence of $FeCl_3$ catalyst. Adducts may also be prepared by the reaction of chain hydroxyls.

Epoxy-amine adducts are produced to provide curing agents that have reduced skin and eye irritation and better flexibility than unmodified amines. These epoxy amines are formed by the reaction of either a primary or secondary amine with an epoxy resin. They are more fully described in Chap. 5.

2.4 EPOXY CURING MECHANISMS

The epoxy resins are capable of reacting with various curing agents or with themselves (via a catalyst) to form solid, crosslinked materials with considerable strength and adhesion. This transformation is generally referred to as *curing* or *hardening*. This ability to be transformed from a low-viscosity liquid (or thermoplastic state) into a tough, hard thermoset is the most valuable single property of epoxy resins.

This transformation or conversion is accomplished by the addition of a chemically active compound known as a curing agent or catalyst. Depending on the particular details of the epoxy formulation, curing may be accomplished at room temperature, with the application of external heat, or with the application of an external source of energy other than heat such as ultraviolet (uv) or electron beam (EB) energy.

Two primary types of epoxy curing reactions are discussed in this section:

1. Polyaddition reactions
2. Homopolymerization reactions

Both polyaddition and homopolymerization reactions can result in increased molecular weight and crosslinking. Both types of reaction occur without the formation of by-products. The curing reactions are exothermic, and the rates of reaction increase with temperature.

Since the epoxy resin cures primarily by a ring-opening mechanism, it exhibits a smaller degree of cure shrinkage than other thermosetting resins. In these reaction processes, the epoxy group may react in one of two different ways: anionically and cationically. Both are of importance in epoxy resin chemistry. In the anionic mechanism, the epoxy group may be opened in various fashions to produce an anion, as shown in Fig. 2.10.

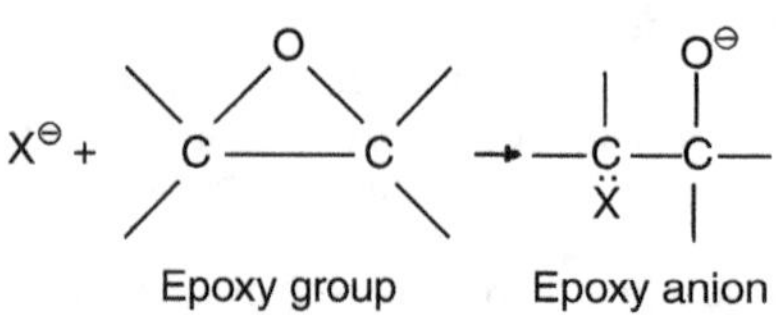

FIGURE 2.10 Anionic mechanism of epoxy reaction.[12]

The anion is an activated species capable of further reaction. In the cationic mechanism, the epoxy group may be opened by active hydrogen to produce a new chemical bond and a hydroxyl group. This reaction may proceed in a number of different ways.

Depending on the curing agent and the epoxy resin, curing can take place at ambient or elevated temperatures. Room temperature curing generally cannot achieve the same performance as is obtained by curing the epoxy adhesive at elevated temperatures. However, room temperature curing does allow sufficient strength and durability for many applications.

The homopolymerization reactions generally require elevated-temperature cure. However, most curing agents or catalysts will react, at least partially, with the resin at room temperature if given a long enough time. Thus, once the curing agent or catalyst is mixed with the resin, a finite pot life or working life is realized. The working life of a particular adhesive can greatly affect the time required to mix, apply, and cure the adhesive.

The curing reactions are exothermic, and the rates of reaction are increased by increases in temperature. The heat formed by the exothermic reaction can lead to a considerable rise in temperature of the system. The actual temperature levels reached depend on the reactivity of the system and the rate at which heat can be transferred to the surroundings.

Since epoxy formulations are generally good thermal insulators, the exotherm will depend on the mass of the system. A high rate of exotherm is needed with some epoxy adhesive systems to achieve practical curing rates. However, excessively high exothermic temperatures can result in bubble formation, thermal degradation, and even a potentially hazardous situation. Control of the exotherm is, therefore, a very important factor in formulating epoxy adhesives.

2.4.1 Polyaddition Reactions

When the curing is based on the reaction of the epoxy molecule and other kinds of reactive molecules with or without the help of a catalyst, the reaction is an addition reaction. The cured structure is a heteropolymer that is composed essentially of the epoxy resin molecules linked together through the reactive sites of the curing agent molecules. The curing agent can be thought of as a comonomer in the polymerization reaction.

The compound that reacts with the epoxy resin and actually forms part of the final epoxy network is generally known as a *curing agent* or *hardener*. It is usually a polyfunctional coreactant, so that stoichiometric amounts need to be considered when one is developing a formulation. The curing agent makes up a significant portion (several percent to over 50 percent by weight) of the epoxy formulation.

Significant latitude in mix ratios can be provided with most curing agents because they are locked within the polymeric structure after cure. As a result, the type and concentration of curing agent have a significant effect on both the curing chemistry and the final properties of the cured resin. Curing agents that have flexible molecules between reactive groups will provide tough, flexible cured adhesives.

The polyaddition reaction is the most commonly used type of reaction for the cure of epoxy resins. The curing agents used in this type of reaction have an active hydrogen compound, and they include amines, amides, and mercaptans. With this reaction mechanism, the most important curing agents for adhesives are primary and secondary amines containing at least three active hydrogen atoms and various di- or polyfunctional carboxylic acids and their anhydrides.

A generalized polyaddition reaction of an active hydrogen compound and an epoxy molecule is represented in Fig. 2.11. Hydroxyl groups, especially phenolic hydroxyls, and tertiary amines catalyze this reaction. The bulk and nature of substituent groups on the epoxy and curing agent molecules also play a major role in the reaction rate. Thus, low-molecular-weight curing agents and resins are more reactive and produce more densely crosslinked structures than high-molecular-weight curing agents and resins. This is why low-molecular-weight DGEBA resins will react much more readily at room temperature than will epoxy novolac resins.

The exotherm caused by reactive, low-MW resins and curing agents is generally much more severe than with high-MW systems. Although the exotherm may result in a faster reaction, it could also result in an objectionably short working life and even a dangerous situation if the exothermic temperature rises above the flame point of the resin. Thus, the formulator will often use steric hindrance as a tool for controlling exotherm to provide systems that will cure readily at elevated temperature but will react sluggishly at room temperature.

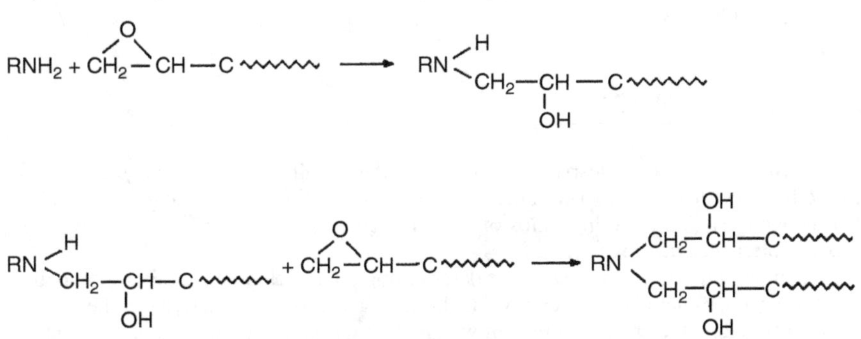

FIGURE 2.11 Polyaddition reaction of an epoxy resin with an active hydrogen compound.

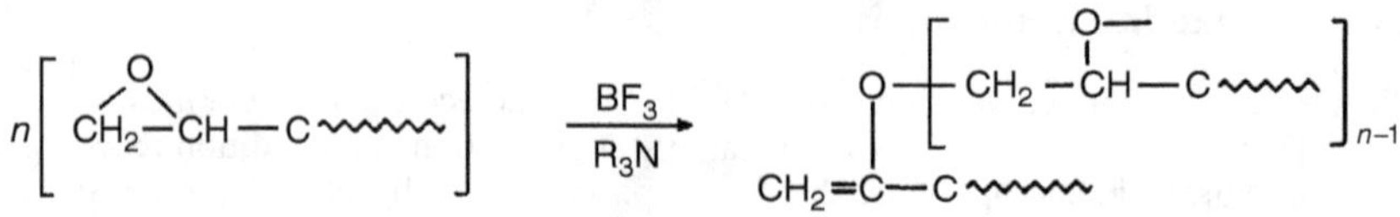

FIGURE 2.12 Homopolymerization reaction of an epoxy resin.

2.4.2 Homopolymerization Reactions

When the curing process is based on the reaction between the reactive epoxy molecules only, the reaction is homopolymerization (see Fig. 2.12). The cured structure is essentially made up only of the original epoxy molecules linked together through their own reactive sites. The reactive compound that initiates the homopolymerization reactions is generally known as a *catalyst*. (*Note*: Sometimes the term *hardener* is used interchangeably for *catalyst*, although this is not the convention followed in this book.)

The catalyst does not make up part of the final epoxy network structure or have a significant effect on the final properties of the cured resin. Thus, the final cured properties of the epoxy system are primarily due to the nature of the epoxy resin alone. Homopolymerization normally provides better heat and environmental resistance than polyaddition reactions. However, it also provides a more rigidly cured system, so that toughening agents or flexibilizers must often be used. In adhesive systems, homopolymerization reactions are generally utilized for heat cured, one-component formulations.

The catalytic curing agents commonly used include tertiary amines, Lewis acids and bases, and dicyandiamide. Since their function is truly catalytic, the catalyst is added at relatively low concentrations (0 to 5% by weight) to the epoxy formulation. Homopolymerization generally requires both the presence of catalysts and elevated temperatures for the reaction to proceed. Like the polyaddition reaction, the homopolymerization reaction is accelerated by hydroxyl groups or tertiary amines.

Reactions between resin and curing agent or between resin and catalyst are generally much more complex than the reactions shown above. A number of side reactions can occur, and these reactions can occur at different rates and reaction sequences depending on the types of materials employed and the curing conditions. Therefore, a significantly different chemistry could be achieved by curing an epoxy system at two progressive temperature stages than by curing at only one stage. These differences in chemistry are, of course, noticeable in the final properties provided by the cured epoxy structure.

As epoxy resins can vary enormously as suggested above, so can the means for curing these resins. Curing reactions specific to specific types of curing agents or catalysts are described in Chap. 5.

2.5 STOICHIOMETRY

With formulated epoxy adhesives, the supplier will generally establish the correct mix ratios to use for specific end properties and/or production situations. Depending on the resin and curing agent used, the mixing ratios of these ingredients could vary substantially, or tight mixing tolerances may be required.

The most important parameters for determining the stoichiometric mixing ratio are the reactivity (functionality) of the epoxy and the curing agents. The reactivity of the epoxy can be determined by the epoxy equivalent weight. The amount of curing agent needed to cure epoxy resins is the same for epoxy resins of similar equivalent weight.

The calculation of stoichiometric quantities of curing agents differs for amine and anhydride curing agents, and these are discussed below. The stoichiometric calculation has no meaning for catalytic curing agents since their action is truly catalytic. Catalytic curing agents are generally added at low levels (0 to 5%).

Mixing ratios of resin to curing agent do not necessarily have to be stoichiometric ratios (i.e., the exact ratio depending on chemical functionality of resins and curing agent). With certain curing agents, especially the higher-molecular-weight flexible curing agents (e.g., polyamides, amidoamines) and acid anhydrides, the actual stoichiometric equivalency does not always produce a cured adhesive with optimum properties. The balance between curing agent and epoxy resin will affect the toughness and high-temperature strength of the adhesive, the former being favored by low degrees of crosslinking and the latter by high degrees of crosslinking.

When one is using polyamide curing agents, for example mixing ratios that employ curing agents at a greater than stoichiometric concentration will provide flexibility and elongation to the resulting adhesive. This is a result of the higher concentration of the flexible polyamide molecule between epoxy linkages. Employing less than the stoichiometric concentration of the curing agent will result in a more brittle system, but it will have greater tensile strength and heat and chemical resistance.

When one is using primary amines, for example diethylene triamine (DETA), the mix ratio should be closer to stoichiometric amounts. However, there is always a percentage of homopolymerization in practice, especially at the temperature of reaction. Smaller amounts of DETA than stoichiometric, therefore, will cause a complete cure. But this will generally occur at the expense of increasing brittleness.

Petterson[13] has shown that the use of only 50 percent of the stoichiometric amount of hexamethylenediamine imparts to a cured DGEBA adhesive the maximum butt-joint strength, bulk tensile strength, and flexural moduli, whereas higher proportions of the diamine give lower properties.

In the case of the adhesive supplier or the manufacturing engineer who finds it necessary to internally formulate an epoxy adhesive, stoichiometric ratios will need to be determined so that one can find a safe mixing ratio to start formulating and to understand the implications of various reactive proportions.

The amount of curing agent selected should be fine-tuned for the specific application. This can be done by varying the concentration over a narrow range and then testing performance to find which concentration is optimal. The optimized concentration should then be tested in the actual product and under end-use conditions since there may be other variables that affect performance (i.e., the mass and geometry of the joint, cure conditions, combinations of environmental conditions).

Of course, in the case of both curing agents and catalysts, suitable adjustments will have to be made for the presence of nonreactive fillers and modifiers. Such ingredients can be liquids such as a solvent, a hydrocarbon resin, or a plasticizer. Since they do not contribute any epoxide functionality, they should not be considered when one is determining stoichiometry. However, if the additives have epoxy functionality, such as in the case of reactive diluents, the stoichiometric calculations will have to take these materials into consideration, by calculating ratios similarly as with an epoxy resin.

2.5.1 Amine Concentration

Amines are one of the most important curing agents for epoxy resins. They provide fast cures with a relatively high crosslink density. Unmodified amine cured epoxy resins are generally too brittle for adhesive applications, and so there are many derivatives that have been developed. More information on amine curing agents can be found in Chap. 5.

The stoichiometric quantity of amine to be used to cure an epoxy resin is easily calculated from the molecular weight of the amine, the number of active hydrogens in the amine, and the epoxy equivalent weight of the resin. This calculation is shown in Table 2.3.

When the chemical structure of the amine is not known (as with proprietary curing agents), the amine equivalent or active hydrogen equivalent weight is generally provided

TABLE 2.3 Calculations of Stoichiometric Concentrations of Amine and Anhydride to Be Used with Epoxy Resins

Procedure for calculating the stoichiometric quantity of amine curing agent

For example, using an amine of structure NH_2-CH_2-CH_2-NH-CH_2-CH_2-NH_2 (D.E.H. 20 from Dow Epoxy Resins) and DGEBA epoxy resin having an EEW of 189 (D.E.R. 331 from Dow Epoxy Resins).

1. Calculate the amine equivalent weight.

$$\text{Amine equivalent weight} = \frac{\text{MW of amine}}{\text{no. of active hydrogens}}$$

$$\text{Amine equivalent weight} = \frac{103.2}{5} = 20.6$$

2. Calculate the stoichiometric ratio of amine to use with the epoxy resin.

$$\text{pph of amine} = \frac{\text{amine equivalent weight} \times 100}{\text{epoxy equivalent weight of resin}}$$

$$= \frac{20.6 \times 100}{189} = 10.9$$

Procedure for calculating the stoichiometric quantity of amine to use with a blend of materials

For example, using the following materials in the formulation of an epoxy adhesive base resin:

100 parts by weight DER 331	Average EEW 189
100 parts by weight DER 337	Average EEW 240
30 parts by weight BGE (diluent)	Average EEW 130
230 parts filler	No epoxy functionality

1. Calculate the EEW of the total mix.

$$\text{EEW of mix} = \frac{\text{total weight}}{\text{Wt}_a/\text{EEW}_a + \text{Wt}_b/\text{EEW}_b + \text{Wt}_c/\text{EEW}_c + \ldots}$$

$$= \frac{460}{100/189 + 100/240 + 30/130} = 391$$

2. $\text{pph of amine} = \frac{20.6 \times 100}{391} = 5.27$

Procedure for calculating the stoichiometric quantity of anhydride curing agent

1. Calculate the anhydride equivalent weight.

$$\text{Anhydride equivalent weight} = \frac{\text{MW of anhydride}}{\text{no. of anhydride groups}}$$

2. Calculate the stoichiometric ratio of anhydride to use with the epoxy resin.

$$\text{pph of anhydride} = \frac{\text{anhydride equivalent weight} \times 100}{\text{epoxy equivalent weight of resin}}$$

by the supplier. Nomographs are also often provided by the resin or curing agent supplier to simplify the calculation.

2.5.2 Anhydride Concentration

After the primary amines, acid anhydrides are the next most important class of curing agents, although these are not used as often in adhesive systems as they are in casting compounds, encapsulants, molding compounds, etc. More information on anhydride curing agents can be found in Chap. 5.

The stoichiometric quantity of anhydride to be used to cure an epoxy resin may be calculated from the molecular weight of the anhydride, the number of anhydride groups per mole, and the epoxy equivalent weight. An example of this calculation is shown in Table 2.3. However, it must be recognized that when one is working with anhydrides, one anhydride group provides two carboxylic acid groups, and it is these carboxylic acid groups that react with the epoxy resin. Thus, anhydrides have a functionality of 2 in calculating the stoichiometry of the epoxy-anhydride reaction.

Accelerators are often used to increase the reactivity of epoxy-anhydride systems. The proper amount of cure accelerator is empirically determined and varies according to the material being used as the accelerator, the anhydride and epoxy resin used, and the desired cure speed and end properties.

As with amine cured epoxies, frequently the optimum properties of anhydride-epoxy systems are obtained at other than the calculated stoichiometric amount. This is because competing reactions occur, such as the reaction of the anhydride group with a hydroxyl group. Typically, optimum anhydride concentrations occur at 85 to 95% of stoichiometric concentrations.

REFERENCES

1. Osumi, Y., "One-Part Epoxy Resin," *Three Bond Technical News*, Three Bond, October 1, 1987.
2. Dannenberg, H., and May, C., "Epoxide Adhesives," in *Treatise on Adhesion and Adhesives*, vol. 2, R. L. Patrick, ed., Marcel Dekker, New York, 1969.
3. DeBruyne, N. A., *Journal of Applied Chemistry*, London, vol. 6, 1956, p. 303.
4. May, C. A., "The Physical Significance of Acid Anhydride Cures on Epoxy Adhesive Properties," *SPE Transactions*, vol. 3, 1963, p. 251.
5. Schwartz, S. S., and Goodman, S. H., *Plastics Materials and Processes*, van Nostrand Reinhold Company, New York, 1982, p. 351.
6. Dowd, R. T., "General Chemistry of Bisphenol A Based Epoxy Resins," Chapter 1 in *Epoxy Resin Technology*, P. F. Bruins, ed., Interscience Publishers, New York, 1968.
7. Newey, H. A., and Shokal, E. C., Glycidyl Ether Compositions and Method of Using Same, U.S. Patent 2,575,558.
8. Coyard, H., et al., *Resins for Surface Coatings*, vol. 1: *Acrylics and Epoxies*, John Wiley & Sons, New York, 2001, p. 142.
9. Grover, "New Developments in 100% Solids Epoxy Adhesives," Society of Plastic Engineers, 15th ANTEC, 1957.
10. Stevens, J. J., "Cycloaliphatic Epoxy Resins," Chapter 2 in *Epoxy Resin Technology*, P. F. Bruins, ed., Interscience Publishers, New York, 1968.
11. Meath, A. R., "Epoxy Resin Adhesives," Chapter 19 in *Handbook of Adhesives*, 3d ed., I. Skeist, ed., van Nostrand Reinhold, New York, 1990.
12. Lee, H., and Neville, K., *Handbook of Epoxy Resins*, McGraw-Hill, New York, 1967, p. 5.2.
13. Petterson, C. M., *Kolloid Z. – Z. Polymers*, vol. 222, 1968, p. 148.

CHAPTER 3
IMPORTANT PROPERTIES OF EPOXY ADHESIVES

3.1 INTRODUCTION

Physical chemistry is an important factor, which leads to the exceptionally good performance properties of epoxy adhesives. Physical chemistry deals with the physiochemical and surface chemistry aspects of the system. Here *system* is defined as the adhesive, its constituents, the substrates, and the interface region of the joint. The physical chemistry can determine the success or failure of an epoxy adhesive application.

These properties have a profound effect on the processing properties of the uncured adhesive and on the end properties of the fully cured product. The properties determined by physical chemistry affect both the cohesive strength of the adhesive film as well as the degree of adhesion to the substrate. They also affect the permanence and durability of the adhesive bond once it is placed into service.

To understand the overall effect of physical chemistry, one must consider the adhesive system in three stages:

1. Uncured
2. As it cures
3. When finally cured

In all these stages the adhesive properties will be determined primarily by the molecular structures of the epoxy resin and curing agent or catalyst employed. Additives and modifiers, of course, will also affect these properties, and they are discussed in later chapters.

All three stages markedly influence overall joint performance. The degree of interfacial adhesion is greatly determined by the first two stages. The final two stages determine the degree of cohesive strength within the cured adhesive film. From a practical standpoint, the last stage is where the ultimate performance of the adhesive or sealant is measured. It includes aging of the joint "system" in the service environment as well as exposure to any stresses required of the application.

There are processes going on in all three stages that ultimately will affect the adhesion and the performance of the adhesive or sealant. It is important to understand these processes and the effects that they will have on the quality of the joint. They are summarized in Table 3.1 and described in the sections that follow.

The discussion centers on how the epoxy adhesive properties depend on the processes that occur during each stage. Special attention is given to the three requisites for good adhesion: wetting, solidification, and avoidance of internal stresses. These are fundamental and well-known factors that affect the quality of the adhesive bond. The objective here is not to

TABLE 3.1 Important Properties of Epoxy Adhesives during Various Stages of Cure

Stage	Property	Process
Uncured adhesives	Vapor pressure	Toxicity, skin irritation Gassing during vacuum Boiling of low-MW components
	Viscosity	Ease of compounding Dispensing and flow Bond line thickness control
	Wetting	Adsorption at the interface Intimate contact with surface Displacement of air at interface
Curing adhesive	Shrinkage	Internal stress within joint
	Thermal expansion coefficient	Internal stress within joint
	Localized discontinuities	Internal stress Reduction in total bond area
Cured adhesive	Modulus	Tensile strength increases and peel and impact decrease with increasing crosslink density
	Crosslink density	Increases rigidity Increases heat and chemical resistance Tensile strength increases and peel and impact decrease with increasing crosslink density

go into theory, definitions, or detailed dissertations on these subjects, since they are well covered in many textbooks and essays on adhesive bonding. However, these properties are the reason that epoxy adhesives have become so well accepted in many markets.

3.2 PROPERTIES OF THE UNCURED EPOXY SYSTEM

The properties of epoxy adhesives in their uncured condition will determine primarily how easily the adhesive can be processed, applied, and cured. They will also determine, to some extent, the performance characteristics of the cured joint. The properties of the individual components as well as that for the mixed formulation are important.

Properties often used by epoxy resin manufacturers to specify particular grades of resin include the epoxy content, viscosity or softening point, and color. In addition, properties such as density, vapor pressure, flash point, refractive index, solubility characteristics, and hydroxyl content are often reported. The important properties of uncured epoxy formulations with regard to most adhesive applications are

- Vapor pressure
- Viscosity
- Wetting of the substrate (surface tension)
- Reactivity

These are essential considerations for a practical adhesive system. The adhesive composition must have sufficient reactivity to cure under conditions that are convenient and practical to the end user. It must have viscosity that allows easy mixing and application. Once on the substrate, the adhesive must be able to flow over the substrate and come into intimate contact with it—a process called *wetting*.

3.2.1 Vapor Pressure

The vapor pressure of an epoxy resin system will have an indirect effect on the final properties of the cured adhesive. However, vapor pressure of the epoxy adhesive and its components may have a direct effect on the health and safety of those who manufacture or apply these products.

Vapor pressure is defined as the pressure of the gaseous phase of a substance in equilibrium with its liquid phase. Vapor pressure increases with temperature. When the vapor pressure equals the surrounding or atmospheric pressure, the component will boil. Thus, high-vapor-pressure materials will boil at relatively low temperatures, and their vapors are likely to be present in the surroundings at lower temperatures.

Vapor pressure is largely determined by the molecular weight of the epoxy resin or curing agent. Low-molecular-weight polymers have a high vapor pressure, and high-molecular-weight polymers have a low vapor pressure. Vapor pressures of low-molecular-weight curing agents are important considerations because they are generally higher than those of epoxy resins. Figure 3.1 shows the vapor pressure for several common amine curing agents as a function of temperature.

High-vapor-pressure materials are more likely to be present in vaporous form in the surroundings. Thus, health and safety issues may be a concern if these materials are toxic or cause skin irritation. The occurrence of such events is always exacerbated by high temperatures because of the increase in vapor pressure.

In hot mixing or elevated-temperature curing of an epoxy system, vapor pressure could also be of concern relative to the quality of the adhesive bond. If the components in an epoxy system become too hot, boiling can occur, resulting in gas bubbles. If gas bubbles become trapped in the cured adhesive film, they can lead to reduction of cohesive strength and stress risers. For many adhesive applications, particularly those in the electrical and electronic industries (due to possible ionization of air voids), complete removal of any gas bubbles from the epoxy is essential.

Vapor pressure must also be taken into account if vacuum is used to remove entrapped air from the formulation after mixing. High-vapor-pressure components, such as certain curing agents, can be evaporated and lost during the vacuum processing operation. The result can be a mixed formulation with lower than expected curing agent concentration.

3.2.2 Viscosity

Epoxy resins and curing agents must have a relatively low viscosity so that formulation compounding can be accomplished easily and without a great deal of energy or degradation of the components. *Viscosity* is defined as the resistance of a liquid material to flow. It is usually measured in fundamental units of poise (P) or centipoise (cP). Table 3.2 shows a relationship between various common fluids and their viscosity as measured in centipoise.

Viscosities of liquid epoxy systems are usually measured with a rotating spindle instrument, such as a Brookfield viscometer. Solid resins are usually dissolved in solvent for viscosity measurement by these instruments. Temperature and spindle speed are important

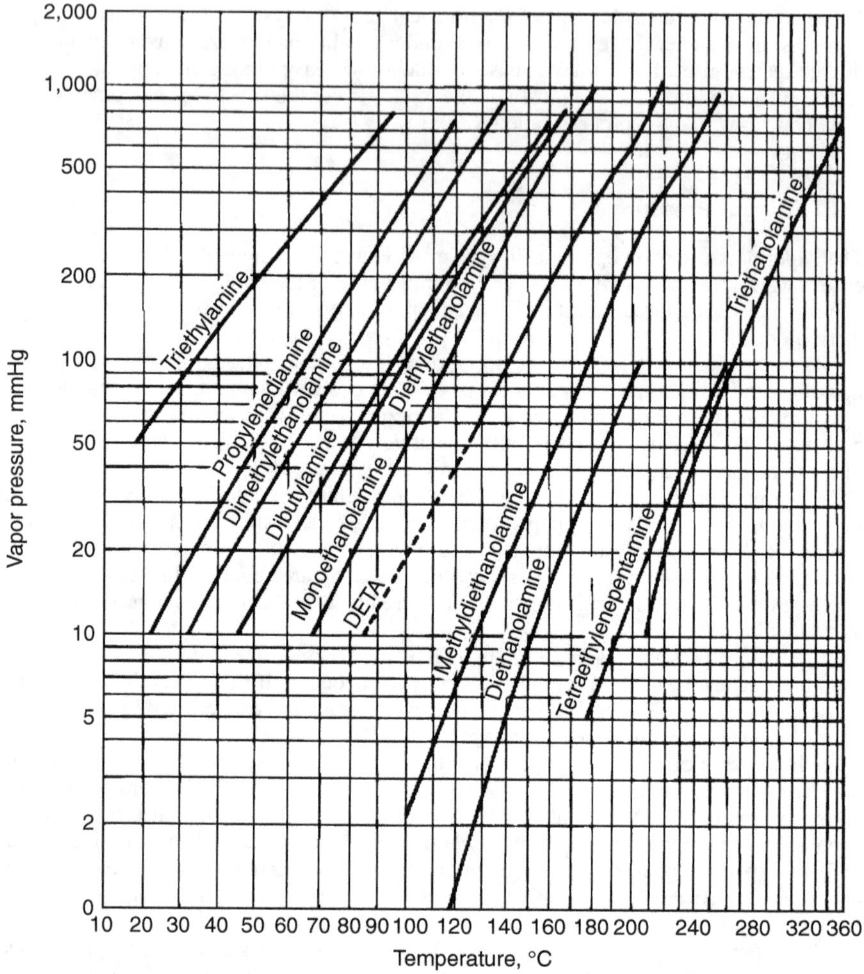

FIGURE 3.1 Vapor pressure of various amines.[1]

TABLE 3.2 Viscosity of Common Fluids

Viscosity, cP	Fluid
1.0	Water
400	#10 Motor oil
1,000	Castor oil
3,500	Karo syrup
4,500	#40 Motor oil
25,000	Hershey chocolate syrup
100,000	Peanut butter

parameters in this test. Brookfield viscosity measurements as well as other methods of measuring viscosity are described in Chap. 20. Several viscosity measurement methods are described in ASTM D 1084 for free-flowing adhesives, and ASTM D 2556 describes a method for thixotropic systems.

The viscosity of the epoxy resin is only one factor in determining the final viscosity of the formulated system. Some curing agents and resinous modifiers produce little effect on mixed viscosity; however, others can have a significant effect by either increasing or decreasing viscosity. Fillers, in general, increase viscosity in direct relationship to their concentration in the system. Diluents can be used to decrease viscosity.

Controlling flow or viscosity is an important part of the adhesive formulation process. If the adhesive has a propensity to flow easily before and during cure, then one risks the possibility of a final joint that is starved of adhesive material. If the adhesive flows only with the application of a great amount of external pressure, then one risks the possibility of entrapping air at the interface and getting too thick of a bond line. These factors could result in localized high-stress areas within the joint and reduction of the ultimate joint strength.

The viscosity of epoxy resins and curing agents can be used to control the bond line thickness within the adhesive joint. But the bond line can also be regulated by the incorporation of fillers, by the use of scrim cloth or woven tapes as "internal shims" within the adhesive itself, or by the careful regulation of the cure cycle.

In certain cases it is necessary for the adhesive formulator to reduce the viscosity of the adhesive system to achieve better application and wetting characteristics. Wetting (see below), as measured by the contact angle that the adhesive makes on the surface, is not governed by the viscosity of the adhesive. However, the rate and ease with which the adhesive wets the surface of the substrate and fills in the peaks and valleys on the surface are a function of viscosity. Viscous adhesives could require an impractical amount of time to adequately wet the surface of a substrate.

It is often difficult to make generalizations regarding the viscosity of epoxy resin components based on chemical structure. The relationships described below between structure and viscosity are based on unmodified epoxy resins and curing agents. However, one must be careful since there are many exceptions to these generalities.

1. For a given resin or curing agent, viscosity increases in proportion to the high-molecular-weight species present. Epoxy molecules having n greater than 2 are semisolid or solid.
2. Linear resins or curing agents of the same molecular weight often give higher viscosity than do branched or cycloaliphatic resins.
3. Aromatic epoxies with three or more groups per molecule are semisolid or solid at room temperature.
4. Epoxies based on ring structures (cycloaliphatic epoxies) give lower viscosities than comparable resins based on aromatic rings (DGEBA).
5. In general, the effect of substitution is to decrease the viscosity compared to unsubstituted product.

The viscosity of an epoxy resin is dependent primarily on its molecular weight (MW). Low-MW resins typically have a viscosity in excess of 6000 cP, and conventional DGEBA epoxy resins (EEW = 190) have a viscosity around 12,000 cP. Therefore, for applications requiring relatively low viscosity, it is necessary to include other types of epoxy resin or to use diluents to achieve the desired properties.

The viscosity of epoxy resin systems decreases rapidly with temperature, as shown in Fig. 3.2. Several problems, however, can occur by using heat to reduce the viscosity of the resin system. The heating can increase the reactivity of the system and, thereby, reduce the working life. Heating can also increase the exotherm generated by the system on curing.

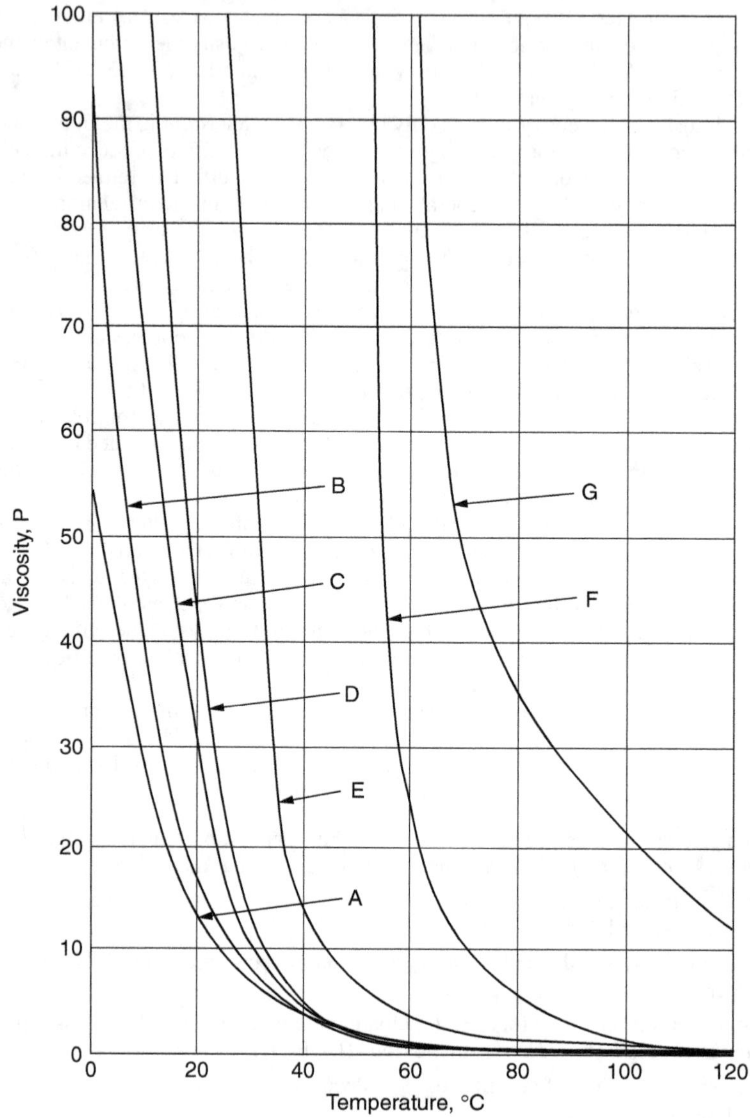

FIGURE 3.2 Variation of liquid epoxy resin viscosity with temperature (A, flexibilizing resin based on dimerized fatty acid; B, liquid diglycidyl ether plus *n*-BGE reactive diluent; C, liquid diglycidyl ether plus glycidyl ester of a tertiary carboxylic acid as a reactive diluent; D, liquid diglycidyl ether plus dibutyl phthalate as a plasticizer; E, unmodified diglycidyl ether; F, semisolid diglycidyl ether; G, epoxy novolac).[2]

3.2.3 Wetting

Initially, the adhesive must be either a liquid or a readily deformed solid so that it can be applied and formed to the required geometry within the assembled joint. It is necessary for the adhesive to flow and conform to the surface of the adherends on both micro- and macroscales. Small air pockets caused by the roughness of the substrate must be easily displaced with adhesive.

While it is in the liquid state, the material must "wet" the substrate surface. The term *wetting* refers to a liquid spreading over and intimately contacting a solid surface, as shown in Fig. 3.3.

As explained under the adsorption theory of adhesion,[3] an adhesive must first wet the substrate and come into intimate contact with it. (A brief description of the adsorption theory of adhesion is presented in the section below.) The result of good wetting is simply that there is greater contact area between adherend and adhesive over which the forces of adhesion (e.g., van der Waals type of forces) may act. For good wetting, the surface free energy (surface tension γ_{LV}) of the liquid adhesive must be less than that (critical surface tension γ_C) of the solid adherend, or

$$\gamma_{LV} < \gamma_C \qquad \text{for good wetting}$$

Table 3.3 provides surface tensions γ_{LV} for common adhesive liquids and critical surface tension γ_C for various solids. A brief discussion of how these properties are measured follows.

Wetting is favored when the substrate surface tension γ_{SV} or its critical surface energy γ_C is high and the surface tension of the wetting liquid γ_{LV} is low (that is, $\gamma_{C\ \text{substrate}} > \gamma_{\text{adhesive}}$). Low-energy polymers, therefore, easily wet high-energy substrates such as metals. Conversely, polymeric substrates having low surface energies will not be readily wet by most other materials and are useful for applications requiring nonstick, passive surfaces.

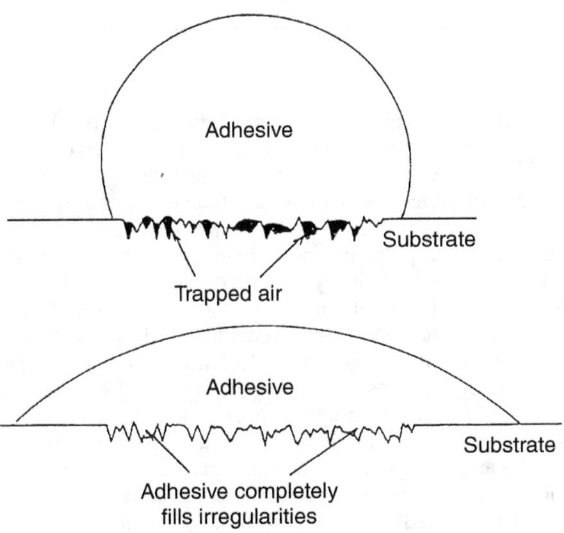

FIGURE 3.3 Illustration of poor (top) and good (bottom) wetting by an adhesive spreading over a surface.

TABLE 3.3 Surface Tension of Several Liquids Including Epoxy Adhesive Formulations (Top) and Critical Surface Tension of Various Substrate Materials (Bottom)

Adhesive	Surface tension, dyn/cm
Epoxy resin	47
Fluorinated epoxy resin	33
Epoxy resin + DETA	44
Epoxy resin + DEAPA	33
Silicone	24
Polyolefins	31
Water	72
Substrate	**Critical surface tension, dyn/cm**
Aluminum	~500
Copper	~1000
Glass	~1000
Acetal	47
ABS	35
Epoxy	47
Nylon	46
Polycarbonate	46
PPS	38
PET	43
Polyimide	40
Polystyrene	33
Polysulfone	41
PTFE	18
PVC	39
Polyethylene	31
Silicone	24

Most common adhesive liquids readily wet clean metal surfaces, ceramic surfaces, and many high-energy polymeric surfaces. However, epoxy adhesives do not wet low-energy surfaces such as polyethylene and fluorocarbons. The fact that good wetting requires the adhesive to have a lower surface tension than the substrate explains why organic adhesives, such as epoxies, have excellent adhesion to metals, but offer weak adhesion on many untreated polymeric substrates, such as polyethylene, polypropylene, and the fluorocarbons.

Figure 3.4 provides a simple view of the relationship of wetting and surface energies. Here the contact angle of a drop of an epoxy adhesive on a variety of surfaces is shown. The surface energy of a typical epoxy resin is about 42 mJ/m^2 (dyn/cm). The expected bond strengths would increase as the contact angle decreased. Therefore, the bond strength of the epoxy adhesive on an epoxy substrate would be expected to be the greatest, followed by polyvinyl chloride, polyethylene, and polytetrafluoroethylene in that order.

The wetting of surfaces by adhesives can be described by two activities: (1) a lateral spreading of the film and (2) a penetration of the fluid adhesive into the surface cavities that are characteristic of the inherent surface roughness. The first activity is controlled by the relative surface energies of the adhesive and substrate as explained above. The second activity is controlled mainly by the viscosity of the adhesive and the time it is in the liquid state.

The surface energetics that control wetting are largely related to the general chemical composition of the epoxy polymer molecule. However, the surface tension of an epoxy

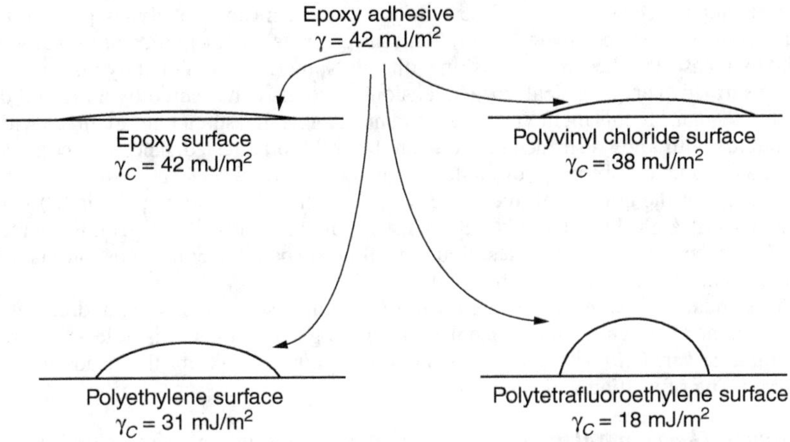

FIGURE 3.4 Contact angle of an uncured epoxy adhesive on four substrates of varying, critical surface tension. Note that as the critical surface tension of the substrates decreases, the contact angle increases, indicating less wetting of the surface by the epoxy adhesive.[4]

system can be modified somewhat by the choice of curing agent, as shown in Table 3.4. Certain amines are surprisingly effective in reducing the surface tension of epoxy resins.[5] The surface tension of a specific product is also generally inversely dependent on temperature.

Epoxy resins have relatively low surface tension as well as good processing properties, strength, and durability, and thus they make good adhesives. The good adhesion of epoxy resins to high-energy surfaces is attributed to interfacial hydrogen bonding, which can be considered a special form of crosslinking. The relatively good wetting and adhesion properties of epoxy resins can be explained by the pendant secondary hydroxyl groups, which occur along the molecular chain, and which are strongly adsorbed onto oxide and hydroxyl surfaces.

Note that good wetting is necessary for bond formation. However, it is not the sole criterion for a strong adhesive joint. Several other important parameters, as noted in the sections that follow, strongly affect the adhesive strength of epoxy systems.

Measurement of Surface Energy Properties. The surface tension and surface energy of liquids are numerically equivalent. Surface energy is generally given in millijoules per

TABLE 3.4 Surface Tension of Liquid Epoxies and an Epoxy-Amine Mixture[6]

Material	Temperature, °C	Surface tension, dyn/cm
Epoxy resin (DER 332LC, Dow)	25	47.2
Epoxy resin (EPON 828, Resolution Performance Products)	25 40 60 80 100	46.2 44.4 42.5 40.7 39.0
Epoxy resin (DER 332LC, Dow) and 7 pph diethylaminopropylamine	25	38.8
Diethylaminopropylamine	25	23.6

square meter (mJ/m^2), while surface tension is given in units of dynes per centimeter (dyn/cm) or newtons per meter (N/m). The surface tensions of liquids are readily determined by measuring the surface tension with a duNouy ring[7] or Wilhelmy plate.[8]

The surface energy (critical surface tension) of solids is measured by a method developed by Zisman.[9] In this method a series of contact angle measurements are made with various liquids with known surface tensions on the solid to be tested. The contact angle θ is plotted as a function of the γ_{LV} of the test liquid. The *critical surface tension* is defined as the intercept of the horizontal line $\cos\theta = 1$ (i.e., when the contact angle is 0°) with the extrapolated straight-line plot of $\cos\theta$ against γ_{LV} of the liquids. The γ_{LV} at this intersection point (i.e., where a hypothetical test liquid would just spread over the substrate) is defined as the critical surface tension of the solid.

The critical surface tension value for most inorganic solids is in the hundreds or thousands of dynes per centimeter. For polymers and organic liquids, it is at least an order of magnitude lower. Critical surface tension is an important concept that leads to a better understanding of wetting and adhesion.

Adsorption Theory of Adhesion. The adsorption theory states that adhesion results from molecular contact between two materials and the surface forces that develop. Adhesion results from the adsorption of adhesive molecules on the substrate and the resulting attractive forces, usually designated as secondary or van der Walls forces. For these forces to develop, the respective surfaces must not be separated more than 5 angstroms (Å) in distance. Therefore, the adhesive must make intimate, molecular contact with the substrate surface.

The process of establishing continuous contact between an adhesive and the adherend is known as *wetting*. Figure 3.3 illustrates good and poor wetting of an adhesive spreading over a surface. Good wetting results when the adhesive flows into the valleys and crevices on the substrate surface; poor wetting results when the adhesive bridges over the valley. Obtaining intimate contact of the adhesive with the surface is essentially ensuring that interfacial flaws are minimized or eliminated. At a minimum, poor wetting causes (1) less actual area of contact between the adhesive and adherend and (2) stress risers at the small air pockets along the interface. This results in lower overall joint strength.

Wetting can be determined by contact angle measurements. It is governed by the Young equation, which relates the equilibrium contact angle θ made by the wetting component on the substrate to the appropriate interfacial tensions:

$$\gamma_{LV}\cos\theta = \gamma_{SV} - \gamma_{SL}$$

The term γ_{SV} is the interfacial tension of the solid material in equilibrium with a fluid vapor; γ_{LV} is the surface tension of the fluid material in equilibrium with its vapor; and γ_{SL} is the interfacial tension between the solid and liquid materials. Complete, spontaneous wetting occurs when $\theta = 0°$ or when the material spreads uniformly over a substrate to form a thin sheet. A contact angle of 0° occurs with pure water droplet on a clean, glass slide. Therefore, for complete spontaneous wetting, $\cos\theta > 1.0$ or when

$$\gamma_{SV} > \gamma_{SL} + \gamma_{LV}$$

After intimate contact is achieved between adhesive and adherend through wetting, it is believed that permanent adhesion results primarily through forces of molecular attraction. Four general types of chemical bonds are recognized as being involved in adhesion and cohesion: electrostatic, covalent, and metallic, which are referred to as *primary bonds*, and van der Walls forces, which are referred to as *secondary bonds*.

3.2.4 Reactivity

Reactivity of the formulated adhesive is determined primarily by the epoxy resin and curing agent or catalyst that is used. The structure of the molecule and the number and type of functional groups greatly influence reactivity. The reactivity of the epoxy resin molecule is dependent primarily on the number of reactive epoxy groups on the molecule. The number of reactive epoxy groups can be determined by the weight per epoxy or epoxy equivalent weight (EEW), as explained in Chap. 2. Epoxy equivalent weights can be determined by chemical or physical methods or by the use of infrared spectroscopy. These methods are described in Chap. 20.

The type of epoxy group and its location within the molecule influence reactivity. Some epoxy resin structures prefer to react with acid curing agents, and others with basic curing agents. Certain epoxy structures are extremely reactive with specific catalysts, and others are virtually inactive. Several types of epoxy resins are reactive with almost all classes of curing agents.

Reactive groups other than epoxy rings may be present in the molecule. Hydroxyl and olefinic groups are the most common. These can affect reactivity as well as the course and sequence of the polymerization reactions. Their number and location could accelerate or retard the overall reaction rate and lead to significantly different three-dimensional cured polymeric structures. Catalytic sites (e.g., tertiary nitrogen) on the epoxy molecule can also influence the reactivity and favor certain reactions.

The hydroxyl content is generally expressed as hydroxyl equivalent weight, the weight of resin in grams containing one gram equivalent of hydroxyl, or as hydroxyl equivalent per gram. The hydroxyl equivalent weight is used similarly to the epoxy equivalent weight, and it is also measured by physical and chemical processes as well as by infrared spectroscopy.

Steric factors influence reactivity by blocking possible reaction sites. Bulky side chains that are developed during the course of polymerization may also inhibit reaction. Reactivity is also dependent on the temperature that occurs during the curing process. The temperature that the epoxy resin will experience during cure is reliant on (1) the external curing process temperature (e.g., oven temperature for an elevated-temperature curing epoxy) and (2) the internal exotherm generated during the process of cure.

Exotherm is defined as the increase in temperature of the mixed compound above the external cure temperature due to energies released as the epoxy groups react. Unlike external heating, exothermic heating originates at the center of the mass, and the epoxy is heated from inside to outside. The degree of exotherm is very much dependent on the mass of the epoxy due to its relatively low thermal conductivity. As mass increases, so does the exotherm.

With epoxy adhesive systems, exothermic heating can be both good and bad. The exothermic temperatures accelerate the reactivity of the system. Many epoxy adhesives rely on the exotherm for complete curing and dense crosslinking under moderate external temperature conditions. However, exothermic temperature rise can be significant. For example, a 1-lb mass of epoxy could reach a temperature in excess of 350°C depending on the curing agent. Amine curing agents because of their high reactivity have high exotherms, whereas anhydride curing agents generally have low exotherms. Exotherm temperatures can degrade heat-sensitive substrates, reduce the working life of the adhesive, and even cause smoldering or burning of the epoxy resin itself.

Reactivity can also be increased by externally heating the epoxy formulation to a preselected curing temperature. Epoxy resin reactions roughly obey Arrhenius' law that for every 10°C rise in temperature, the reaction rate doubles. Certain epoxy resin systems must be heated for any reaction to take place at all. This is beneficial in that these "latent" adhesive formulations are one-component products that do not require metering or mixing yet have long, practical shelf lives.

In a heat-cured epoxy resin system, the hydroxyls generally react with either epoxy or acid groups. Hence, heat-cured diepoxies can be regarded as being potentially tetrafunctional, rather than difunctional. Conventional diepoxy resins, therefore, yield much more highly crosslinked structures with higher heat and chemical resistance after heat cure than after room temperature cure.

Reactivity is often measured by the working life (ASTM D 1338) of a predetermined amount of mixed resin at a defined temperature. ASTM D 1338 uses two methods of determining the working life. One method uses the viscosity change, and the other method uses the shear strength development as the criterion for determining when the effective working life has expired. However, a measure of the reactivity can be made by simply picking at the reacting mass with the end of a toothpick and determining when penetration is no longer possible.

Cure rate of an actual adhesive film can also be determined by several useful analytical methods. With these methods, fundamental properties of the adhesive, such as dielectric loss, mechanical damping, or exotherm, are measured as a function of time and temperature as the adhesive cures. Several of these test methods are described in Chap. 20.

3.3 PROPERTIES OF THE CURING EPOXY SYSTEM

Once the adhesive is applied to the substrate, it must harden into a thermosetting film having high cohesive and adhesive strength. Several processes occur during the curing or solidification of epoxy resins that can lead to fundamental problems with adhesive systems. These problems are generally due to internal stresses that develop within the joint as the result of the curing processes, but they can also be due to side reactions that occur prior to joint assembly. Such problems are primarily caused by

- Weak boundary layers due to preassembly reactions
- Localized stresses due to trapped gases or voids
- Shrinkage due to polymerization
- Thermal expansion differences between the adhesive and the substrate (mainly associated with adhesives that cure at temperatures different from their normal service temperatures)

These internal stresses often can have a degrading effect on the adhesive properties but little or no effect on the cohesive properties of the adhesive film. They mainly affect the interface area of the joint.

3.3.1 Preassembly Reactions

The viscosity increase in a epoxy resin–curing agent system could result in poor wetting of the substrate surface, resulting in suboptimal adhesion. Several reaction mechanisms can also occur to an epoxy adhesive once it is mixed and applied to a substrate but before the substrates are mated. These mechanisms can result in a weak boundary layer, which will prevent optimal wetting and reduce the strength of the adhesive.

Many epoxy resin–curing agent mixtures are hygroscopic and will adsorb moisture from the ambient humidity. This is primarily due to the properties of the curing agent. One example is a polyamide curing agent. If the mixed adhesive is applied to the substrate and allowed to wait until the substrates are mated, water can be adsorbed onto the surface of the adhesive mainly through the polyamide molecules. Assemblies that are most prone to this effect are those that have a long time period between adhesive application and joint closure and where relative humidity conditions are high. The very thin layer of water that forms on the adhesive surface will present a weak boundary layer between the adhesive and the mating substrate when the joint is closed. When in service, the boundary layer fails under stress. The failure mode that is noticed in these cases is generally an adhesion failure at the interface.

Another possible preassembly reaction mechanism has been noted with regard to amine cured epoxy resins.[10] A variability and reduction in the rate of conversion of epoxy groups in DGEBA epoxy resin cured at room temperature with diethylene triamine (DETA) was noticed. This is due to a side reaction of the amine with air, resulting in bicarbonate formation. As a result, the adhesive strength decreased drastically when the uncured epoxy amine was exposed to ambient air for a significant period of time.

One concludes that when room temperature curing systems are to be used as adhesives, the assemblies should be joined quickly to preclude various reaction mechanisms from taking place that could degrade the strength of the final bond.

3.3.2 Localized Stresses due to Gas or Air Pockets

Loss of theoretical adhesive strength can also arise from the action of internal stress concentrations caused by trapped gas and voids. Griffith[11] showed that adhesive joints may fail at relatively low stress if cracks, air bubbles, voids, inclusions, or other surface defects occur as a result of the curing process.

These trapped gas pockets could be formed due to poor wetting or from evaporation of high-vapor-pressure components. If the gas pockets or voids in the surface depressions of the adherend are all nearly in the same plane and not far apart (as is shown in Fig. 3.5, upper adherend), cracks can rapidly propagate from one void to the next. However, a variable degree of roughness, such as shown in Fig. 3.5 lower adherend, provides barriers to spontaneous crack propagation. Therefore, not only is surface roughening important, but the degree and type of roughness may be important as well.

Since real surfaces are not smooth or perfectly flat and most epoxy adhesives are viscoelastic fluids, it is necessary to understand the effects of surface roughness on joint strength. A viscous liquid can appear to spread over a solid surface and yet leave many gas pockets or voids in small surface pores and crevices. Even if the liquid does spread spontaneously over the solid, there is no certainty that it will have sufficient time to fill in all the voids and displace the air. The gap-filling mechanism is generally competing with the setting mechanism of the liquid.

This problem occurs when the liquid solidifies rapidly after being applied. An example is a fast-curing epoxy. Very fast-reacting epoxy adhesive systems that set in several minutes generally do not have the high adhesion strength that slower-curing epoxy systems have. One reason for this (there are others primarily related to the chemistry of these

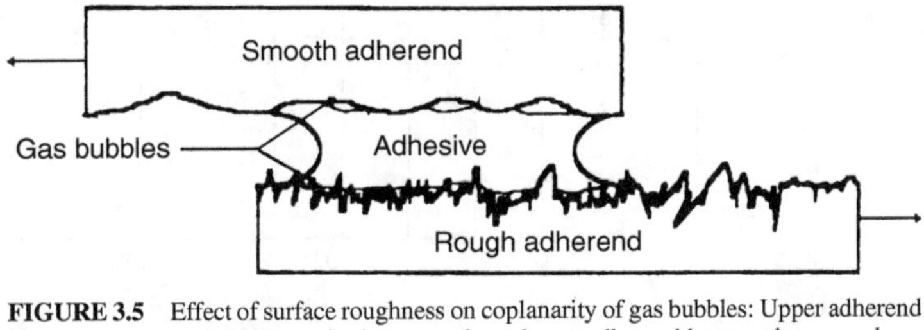

FIGURE 3.5 Effect of surface roughness on coplanarity of gas bubbles: Upper adherend is smooth, and gas bubbles are in the same plane; lower adherend has roughness, and gas bubbles are in several planes.

fast-acting systems) is that the curing reaction does not provide sufficient time for the adhesive to fill the crevices on the substrate surface. This effect can also be shown by the differences in adhesive strength that can be achieved with an epoxy system that is applied to the substrate after varying time periods between curing agent addition and joint closure. A lower degree of adhesion accompanies the system that is applied after a longer "induction" period, presumably due to increase in viscosity during the early stages of cure.[12]

Since slower-curing epoxy adhesives systems flow over and wet high-energy surfaces very well, there is little chance for air to become trapped at the interface. As a result, mechanical abrasion is often recommended as a substrate surface treatment prior to application of the epoxy adhesive. The added surface area and the mechanical bonding provided by the additional peaks and valleys on the surface will enhance adhesive strength. If the adhesive does not wet the substrate surface well, such as in the case of epoxy resin on polyethylene, mechanical abrasion is not recommended since it will only encourage the probability of gas voids being trapped at the interface.

It has also been shown[13] that a concentration of stress can occur at the point on the free meniscus surface of the adhesive (edge of the bond line). This stress concentration increases in value as the wetting becomes poorer and the contact angle θ increases. At the same time, the region in which the maximum stress concentration occurs will move toward the adhesive-adherend interface. Thus, poor wetting will be associated with a weak spot at the adhesive-adherend interface and at the edge of the adhesive joint with a consequent likelihood of premature failure at this region.

The stress concentration factors for a lap joint in shear bonded with adhesives having a contact angle from 0 to 90° are illustrated in Fig. 3.6. For contact angles less than 30° (i.e., good wetting) the maximum stress occurs in the free surface of the adhesive away from the edges, and the stress concentration is not much greater than unity. These localized stress effects can be reduced by improving the wetting of the adhesive, optimizing surface preparation processes, or modifying the adhesive with a toughener to stop the propagation of cracks due to these phenomena.

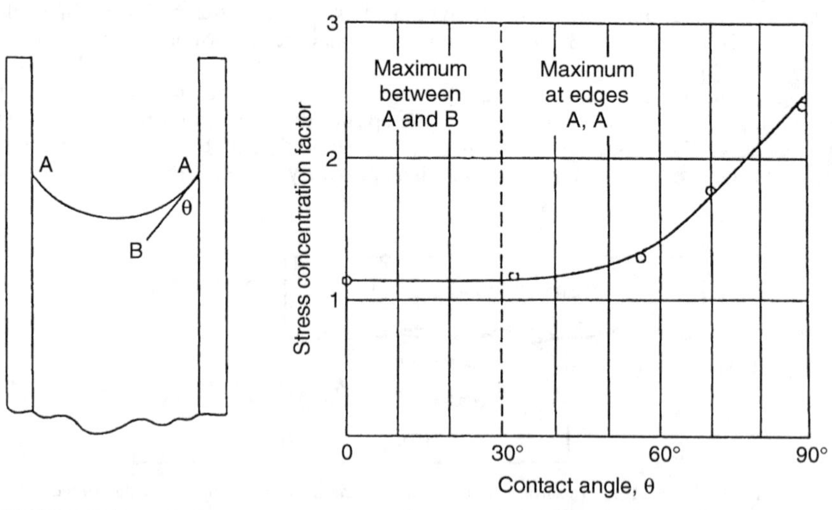

FIGURE 3.6 Maximum stress concentration in a lap joint. Poor wetting of the adherend produces maximum stress concentration at point of contact of adhesive, adherend, and atmosphere.[14]

TABLE 3.5 Surface Tension of Common Organic Solvents

Solvent	Surface tension, dyn/cm	Temp. of test, °C
Hexane	20	25
Isopropanol	20.8	25
Acetone	22.6	25
n-Propanol	23.8	20
Methyl ethyl ketone	24.6	20
Trichloroethane	25.1	20
n-Butyl acetate	27.6	27
Toluene	28	NA
Xylene	28	NA
Ethylene glycol	48.4	20
Polypropylene glycol	72	25
Water (distilled)	72.8	20

Solvents can also be used to reduce the surface tension of the adhesive formulation. The surface tensions of common solvents are shown in Table 3.5. Of course, when using solvents, one needs to make sure that they evaporate from the bond line before cure. Solvent solutions do not change the equilibrium surface energetics of the system. They only provide lower viscosity so that wetting is established at a faster rate.

3.3.3 Shrinkage

Nearly all polymeric materials (including adhesives and sealants) shrink during solidification. Sometimes they shrink because of escaping solvent or volatile by-products, leaving less mass in the bond line. Even 100 percent reactive adhesives, such as epoxies, with no formation of by-products during cure experience some shrinkage because their solid polymerized mass occupies less volume than the liquid reactants.

Table 3.6 shows typical percentage volumetric shrinkage for various reactive adhesive systems during cure. One of the reasons for the great acceptance of epoxy resins as adhesive materials is their low degree of shrinkage on cure relative to other reactive adhesives.

Typical linear shrinkage values are shown in Table 3.7 for various highly filled epoxy adhesive formulations and cure conditions. Notice that the incorporation of filler significantly reduces the shrinkage of the epoxy systems. Higher shrinkage values are generally noticed when the cure is carried out at higher temperatures. It is proposed that this is due to greater crosslinking density that occurs during an elevated-temperature cure.

TABLE 3.6 Volume Shrinkage of Common Adhesives[15]

Adhesive type	Percent shrinkage
Acrylics	5–10
Anaerobic	6–9
Epoxies	4–5
Urethanes	3–5
Polyamide hot melts	1–2
Silicones	<1

TABLE 3.7 Effect of Curing Agents on the Shrinkage of Highly Filled Epoxy-Amine Systems during Cure[16]

Curing agent		Cure schedule				Linear shrinkage, percent
		Cure		Postcure		
Type	g	h	°C	h	°C	
Aromatic amine eutectic	20	20	25	24	65	0.21
	20	20	25	22	150	0.31
Triethylenetetramine	10	20	25	24	565	0.35
Diethylenetriamine	8	20	25	24	65	0.35
	8	145 days	25	—	—	0.13

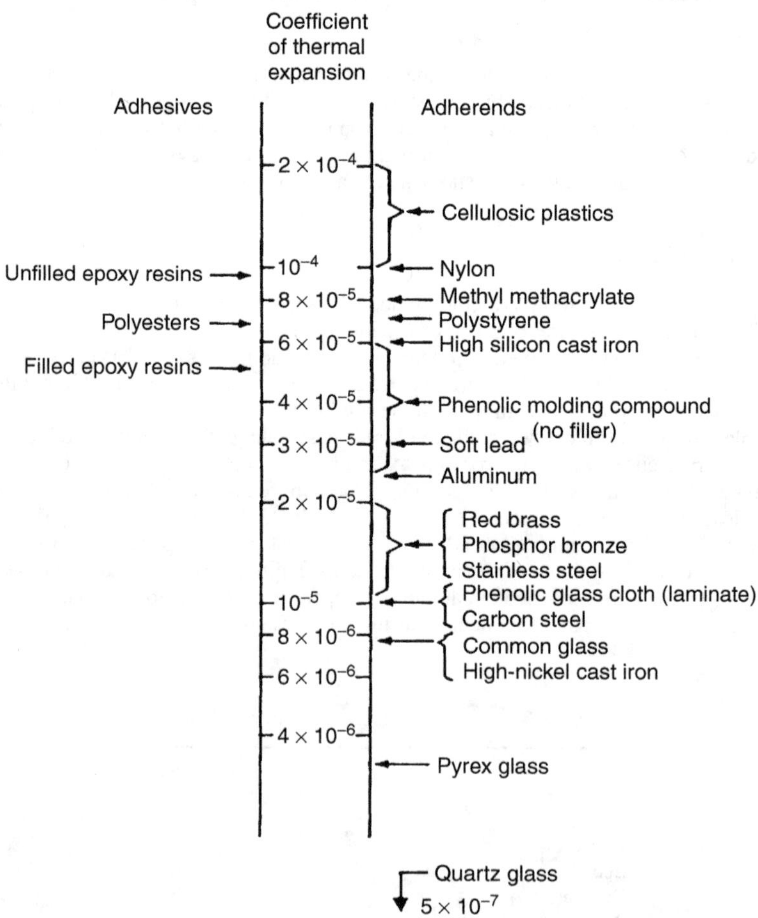

FIGURE 3.7 Thermal expansion coefficients of common materials.[18]

The result of such shrinkage is internal stresses at the adhesive-substrate surface and the possible formation of cracks and voids within the bond line itself. Formulators are often able to adjust the adhesive in several ways to minimize stress from shrinkage:

1. Elastic adhesives deform when exposed to such internal stress and are less affected by shrinkage. Thus, flexibilizers or tougheners may be added to the formulation to distribute the stresses over a larger area (Chap. 8).
2. Inorganic fillers may be added to the adhesive composition to reduce the concentration of epoxy resin in the formulation as the degree of polymeric shrinkage is dependent on the volume of epoxy in the joint (Chap. 9).
3. Certain epoxy monomers[17] have been developed that expand slightly during cure. These formulations are expected to provide very high adhesive properties when blended with conventional epoxy resins to provide zero net shrinkage; however, to date, their commercialization has been limited due to cost.

3.3.4 Thermal Expansion Differences

One of the most common causes of internal stress is due to the difference in the thermal expansion coefficients of the adhesive and the adherend. These stresses must especially be considered when the adhesive solidifies at a temperature that is different from the normal temperature that it will be exposed to in service.

Figure 3.7 shows that thermal expansion coefficients for some common adhesives and substrates are more than an order of magnitude apart. This means that the bulk adhesive will be required to move more than 10 times as far as the substrate when the temperature changes, thereby causing stress at the interface. In general, the more highly flexibilized the epoxy resin system, the higher the expansion rate, and the more highly filled the epoxy, the lower the expansion rate.

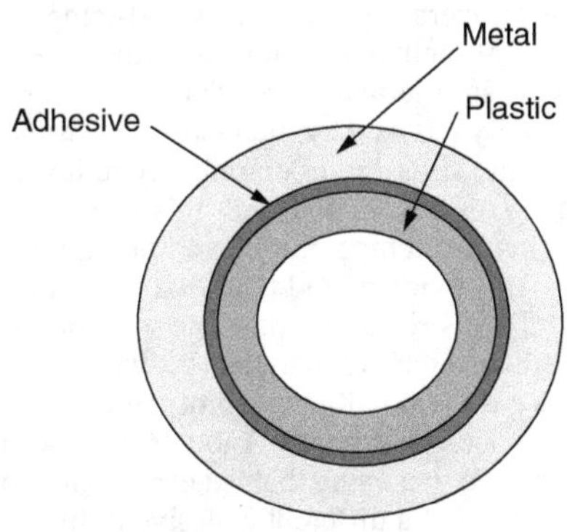

FIGURE 3.8 Journal bearing application. Outer cylinder (stainless steel) is bonded to inner cylinder (polyamide-imide) with an epoxy adhesive. Exposure to low temperatures causes significant stress on bond due to differences in coefficient of thermal expansion.

The stresses produced by thermal expansion differences can be significant. Take, e.g., an annular journal bearing where a polyamide-imide sleeve is bonded to the internal diameter of a stainless steel housing (Fig. 3.8). Further, assume that the epoxy adhesive used is one that cures at 125°C. At the cure temperature, all substrates and the gelled adhesive are in equilibrium. However, when the temperature begins to reduce as the system approaches ambient conditions after cure, stresses in the adhesive develop because the polyamide-imide substrate wants to shrink faster and to a greater extent than the steel. At room temperature, these stresses may be significant but not high enough to cause adhesive failure. Now further assume that the bonded bearing is to be placed in service with operating temperatures that will vary between 125 and –40°C. At 125°C the internal stresses due to the mismatch in thermal expansion are reduced to zero, since we are back at the equilibrium condition (assuming that there was no shrinkage in the adhesive as it cured or other stresses in the joint). However, when the service temperature reaches –40°C, the thermal expansion

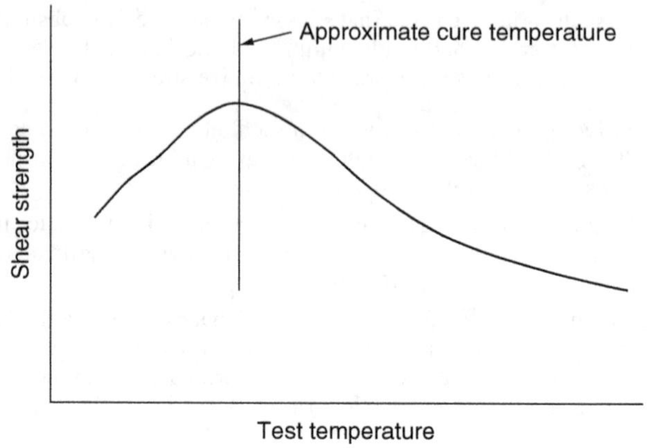

FIGURE 3.9 Plot of tensile shear strength of an aluminum joint bonded with an elevated-temperature curing epoxy adhesive as a function of test temperature.

differences create internal stresses in addition to those already there due to curing, and a failure could result.

A similar example is evident by a typical graph (Fig. 3.9) of tensile shear strength versus temperature for an elevated-temperature cured epoxy adhesive. Notice that the bond strength actually increases with temperature to a maximum, and then it falls off with increasing temperature. This is similar to the case above where the internal stresses are actually reduced by the increasing service temperature. At some elevated temperature, the internal stresses are completely relieved, and the bond strength reaches a maximum. The test temperature at which this occurs is usually very close to the curing temperature. At higher test temperatures, additional stresses develop or the effects of thermal degradation become evident, and the bond strength then decreases with increasing test temperature.

There are several possible solutions to the expansion mismatch problem. One is to use a resilient adhesive that deforms with the substrate during temperature change. The penalty in this case is possible creep of the adhesives, and highly deformable adhesives usually have low cohesive strength. Another approach is to adjust the thermal expansion coefficient of the adhesive to a value that is nearer to that of the substrate. This is generally accomplished by selection of a different adhesive or by formulating the adhesive with specific fillers to "tailor" the thermal expansion. A third possible solution is to coat one or both substrates with a primer. This substance can provide either resiliency at the interface or an intermediate thermal expansion coefficient that will help reduce the overall stress in the joint.

3.4 PROPERTIES OF THE CURED EPOXY SYSTEM

In addition to good adhesion, an adhesive must have satisfactory cohesive strength and durability once it is cured. The physical and chemical properties of the cured product are determined by

- The structure of the resin, particularly the number, location, and reactivity of the epoxy and other groups present
- The curing agent, stoichiometry, and cure conditions

These parameters establish the rigidity, thermal stability, chemical resistance, and overall usefulness of the adhesive joint.

Factors that can affect cohesive strength include (1) the molecular weight and the nature of the molecules between crosslinks and (2) the degree of crosslinking, Epoxy polymers are network polymers with three-dimensional molecular linkages formed throughout the bulk polymer. In these types of polymers, a great deal of importance falls on the functionality of monomer units and the degree to which crosslinking takes place. These factors determine the most important properties such as rigidity, heat resistance, strength, and chemical resistance.

The structure of a crosslinked epoxy resin can be represented by a ladder-type model (Fig. 3.10) in which the dimensional network is formed by tying together a relatively high-MW backbone polymer with random and relatively short crosslinking segments. The properties of the network are then determined by the nature of the backbone polymer chain segment, crosslink segments, and branch points. As any one, or any combination, of the structural changes shown in the left-hand column of Fig. 3.10 occurs, all the properties shown in the right-hand column are changed in the direction shown. Notice that certain properties can be obtained at the expense of other properties. It becomes apparent, therefore, that the formulator must perform a balancing act to provide the correct set of properties for the application. Fortunately, the nonresinous components in the formulation (filler, modifiers, etc.) help provide the tools to do this.

3.4.1 The Problem with Rigidity in Epoxy Adhesives

Much of the development in epoxy adhesive technology has been toward attempts to combine the advantages of rigid and flexible epoxy systems and to minimize their disadvantages. The property tradeoffs are illustrated in Fig. 3.11. Generally, the rigid structural adhesive that displays excellent tensile or shear strength does not perform well when tested in peel, impact, or fatigue. This reason is that the elongation of the adhesive distributes stresses

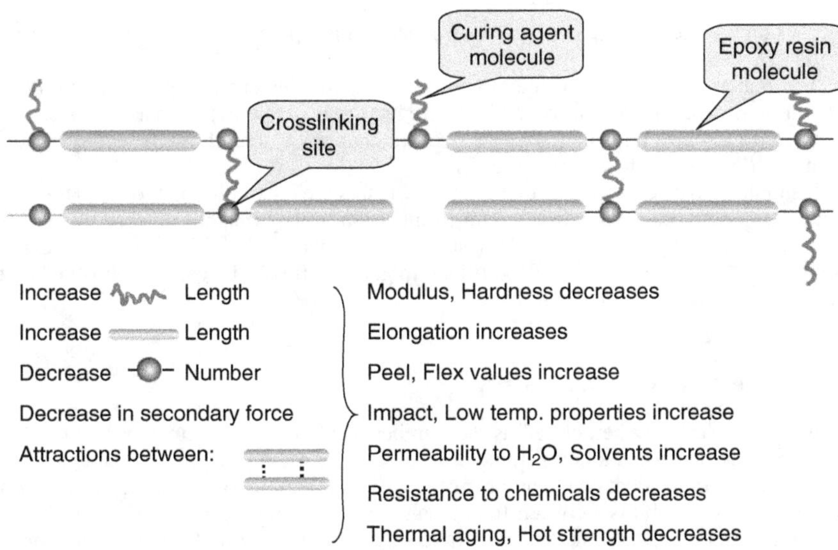

FIGURE 3.10 Relationship between cured epoxy molecule in physical properties.

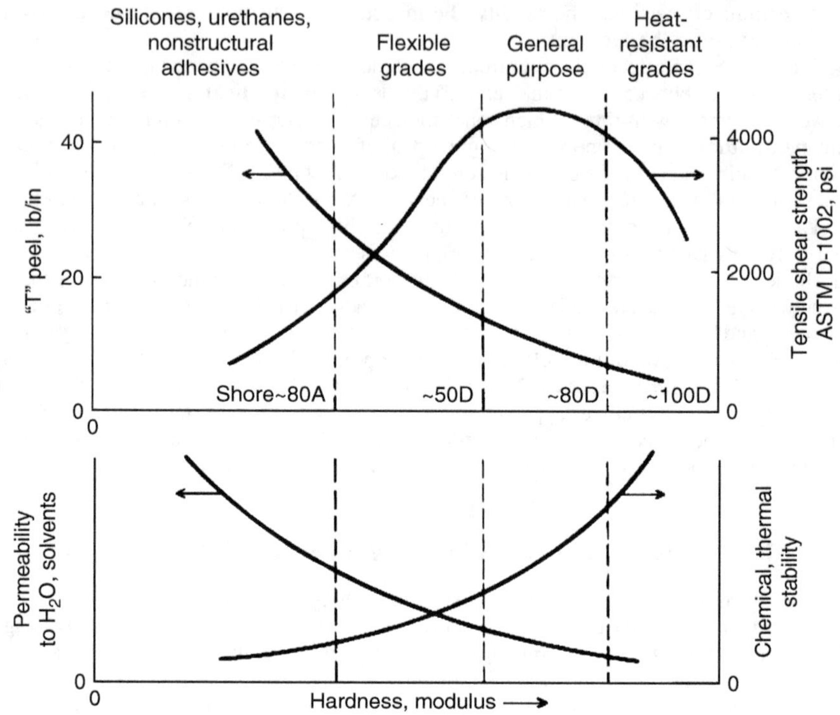

FIGURE 3.11 Property tradeoffs for single-phase, thermosetting adhesives.[19]

over a much larger area (reducing stress concentration points) than if the adhesive were brittle.

The major problem is that the attainment of properties such as peel, flexibility, and toughness is generally accompanied by the reduction in properties such as heat resistance, chemical resistance, and shear strength. Future chapters discuss how the epoxy adhesive formulator can merge these properties.

It should also be emphasized that the epoxy resins are only building blocks in the development of epoxy-based adhesives. The formulation will ultimately involve one or more epoxy resins; curing agents; modifiers, such as plasticizers, flexibilizers, accelerators, stabilizers, and flow control agents; and fillers. In general, flexibility can be controlled by the parameters shown in Table 3.8.

3.4.2 Effect of Crosslink Density

Crosslink density may be defined as the number of effective crosslinks per unit volume. The crosslink density is a key parameter in determining the properties of an epoxy resin after cure. It is dependent on the number of reactive sites (functionality), the molecular distance and chain mobility between functional sites, and the percentage of these sites that enter into reaction. Crosslink density is inversely related to the molecular weight between crosslinks M_c.

TABLE 3.8 Effect of Various Formulation Parameters on the Flexibility of an Epoxy Adhesive

Increasing parameter	Effect on flexibility
Molecular weight	Generally increases
Crosslink density or reactivity	Decreases
Chain branching	Generally increases
Crystallinity	Decreases
Glass transition temperature	Decreases
Filler concentration	Decreases
Plasticizer concentration	Increases
Flexibilizers	Increase

Epoxy adhesives have cohesive properties that significantly depend on the crosslink density. For example, torsional butt shear strength of epoxy aluminum joints has been noticed to decrease as M_c increased.[20] However, flexibility and toughness are increased with increasing M_c. Figure 3.12 shows the general physical relationship between M_c and the physical state of epoxy resins.

The crosslink density ultimately defines the rheological and mechanical properties of the polymer. Polymers that have a high crosslink density are thermosets and are infusible, insoluble, and dimensionally stable under load. These properties make epoxy resin systems useful as structural adhesives as well as important materials in other applications. Polymers that have a low crosslink density are more flexible and show greater resistance to stress concentration, impact, and cold.

In the DGEBA product ($n = 1$) shown in Fig. 3.13, the epoxy groups are separated by seven units (the aromatic ring is counted as one unit). In other resins the groups may be

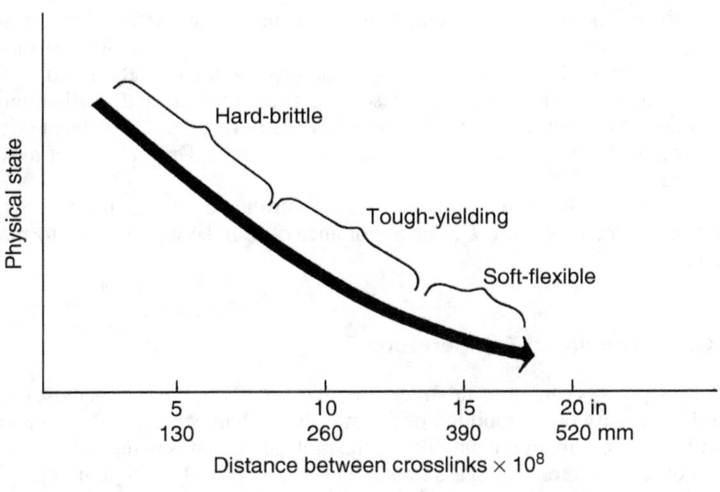

FIGURE 3.12 Effect of molecular weight between crosslinks on the physical state of epoxy resins.

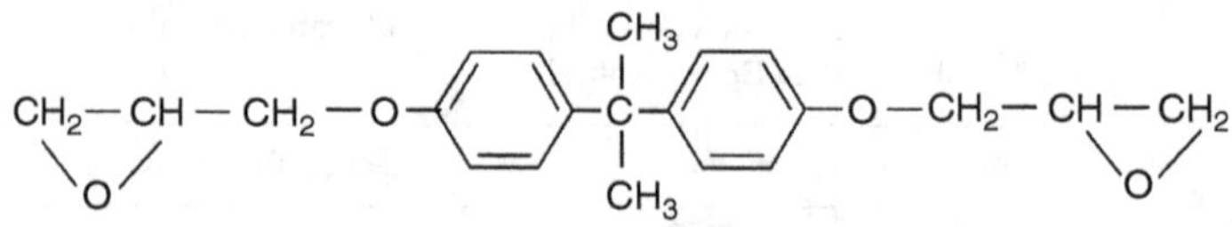

FIGURE 3.13 Idealized molecular structure of a diglycidyl ether of biphenol A (DGEBA) epoxy resin.

separated by 30 to 40 units. In general, short spacing (2 to 10 units) gives rigidity, medium spacing (10 to 30 units) gives semirigid properties, and long spacing (>30 units) gives flexibility or softness.

However, crosslink density is not the only factor affecting the rigidity or flexibility of the cured resin. The nature of the resin or curing agent molecules between the reactive groups, whether it is rigid or flexible, also has a direct influence on physical characteristics. The increased molecular mobility brought about by heating gives the molecules a greater opportunity to undergo collision and bond formation. A small proportion of nonreactive diluent may also serve to improve molecular mobility and lead to a lesser degree of crosslinking and greater flexibility. Decreases in crosslink density can also be achieved by (1) using monofunctional reactive diluents as chain stoppers or (2) using resins and curing agents with widely spaced functional groups.

Curing of epoxy thermosets requires a knowledge of the chemical kinetics and the crosslinking reactions. This information is necessary to optimize the cure cycle. The parameters that define the cure cycle ultimately determine the crosslink density and the final physical properties of the polymer. In addition to temperature, these parameters include the rate of temperature increase, the number of stages in the cure, the hold temperature at each stage, the pressure at which cure takes place, and the time allotted for the cure cycle. These parameters are usually determined empirically. Once the kinetics are understood and the actual chemistry behind the cure is established, these cure cycle parameters can be chosen based on the desired end properties. Usually the cure cycle seeks to establish a certain degree of cure that is in line with the expected final properties.

Many different methods can be used to measure the degree of crosslinking within an epoxy specimen. These methods include chemical analysis and infrared and near infrared spectroscopy. They measure the extent to which the epoxy groups are consumed. Other methods are based on the measurements of properties that are directly or indirectly related to the extent and nature of crosslinks. These properties are the heat distortion temperature, glass transition temperature, hardness, electrical resistivity, degree of solvent swelling and dynamic mechanical properties, and thermal expansion rate. The methods of measurement are described in Chap. 20.

Perhaps the most significant property that is controlled by the degree of crosslinking is the glass transition temperature T_g. The importance of T_g in epoxy adhesive formulations is discussed next.

3.4.3 Glass Transition Temperature T_g

When chain segments can move relatively freely in cured polymers, it is most likely due to low crosslink density or the mobility of the molecular chain structure. The glass transition temperature is a measure of the mobility of the molecular chains in the polymer network as a function of temperature. The glass transition is the reversible change in a polymer from (or to) a rubbery condition to (or from) a hard and relatively glassy state condition (Fig. 3.14). This transition occurs at a temperature called the *glass transition temperature* or T_g. It is

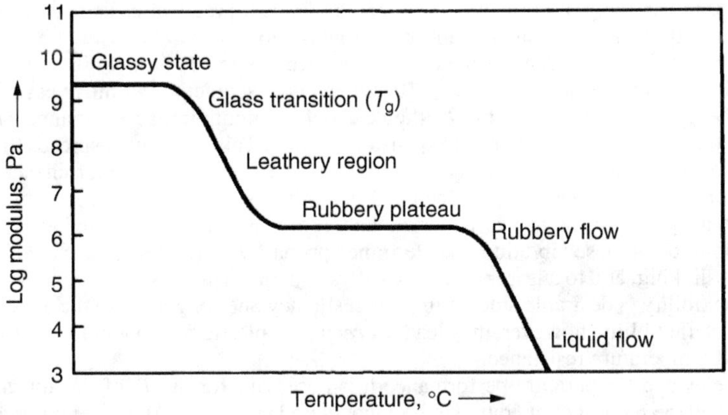

FIGURE 3.14 Relationship between elastic modulus and temperature showing the glass transition region.[22]

usually associated with the onset of long-range motion in the polymer backbone due to temperature effects.

As shown in Fig. 3.14, the glass transition temperature is usually a narrow temperature range rather than a sharp point, as is the freezing or boiling point. Molecular motion at this point does not involve entire molecules, but in this region deformation begins to become nonrecoverable as permanent set takes place.

Generally T_g is measured as the temperature at which the slope of a temperature-volume plot undergoes a sudden upward change, as shown in Fig. 3.15. It is the temperature at which there is significantly more molecular mobility than at lower temperatures (i.e., the molecules have sufficient thermal energy to be considered mobile).

T_g is a property of a polymer that depends on its chemical composition and the degree of crosslinking or molecular interaction. Often the glass transition temperature is used as a measure of the degree of crosslinking in a specifically defined epoxy system. Adhesive

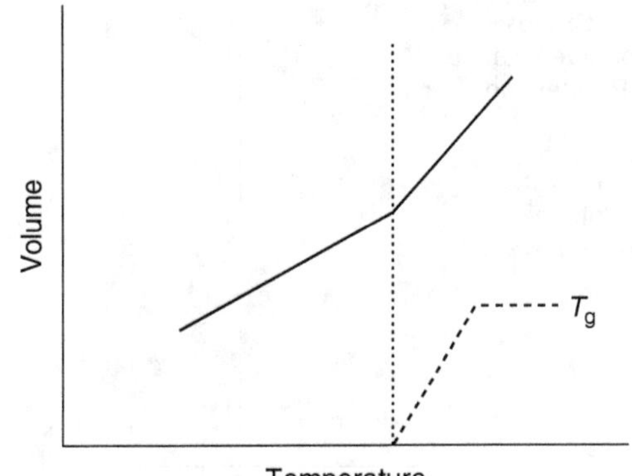

FIGURE 3.15 The effect of temperature on the total volume of a polymer.

formulators use the glass transition temperature T_g as a practical basis for compounding products with the appropriate amount of internal motion.

Many properties of adhesive bonds are influenced by the T_g or mobility of the molecular chain structure, as shown in Fig. 3.16. When chain segments can move easily, such as when the temperature exceeds the T_g, they can deform under impact or assume new alignments under mechanical or thermal expansion stresses. This movement spreads the applied energy over a greater number of atoms and thus gives the bond a better chance to resist stress. Brittleness is, therefore, reduced and flexibility is increased.

Like all polymers, adhesive materials undergo constant thermally induced vibration. The amplitude of these vibrations is determined primarily by temperature, chain flexibility, and crosslinking, and to a lesser extent by fillers and physical stresses. A certain amount of chain flexibility is desirable since it imparts resiliency and toughness to the adhesive film. Too much flexibility, however, may lead to *creep* (i.e., plastic flow under load) or very poor elevated-temperature resistance.

For elevated-temperature performance it is necessary for the T_g of the adhesive to be higher than the highest temperature to be encountered in service. At temperatures above the T_g of any polymer strength and stiffness diminish rapidly. Bond strength at elevated temperatures can be increased by raising the degree of crosslinking or using more thermally resistant (aromatic) epoxy base resins.

However, flexibility and peel strength at room temperature will be low if the T_g is high. It is very difficult to provide an adhesive that has high peel strength with good cohesive strength and environmental resistance. This problem and potential solutions are discussed in detail in the following chapters. A high T_g may also limit the low-temperature properties of the adhesive.

Typical glass transition temperatures for adhesive resins are shown in Table 3.9. Note, however, that the T_g for epoxy adhesives can vary significantly with their formulation. Glass transition temperatures for several epoxy formulations are shown in Table 3.10.

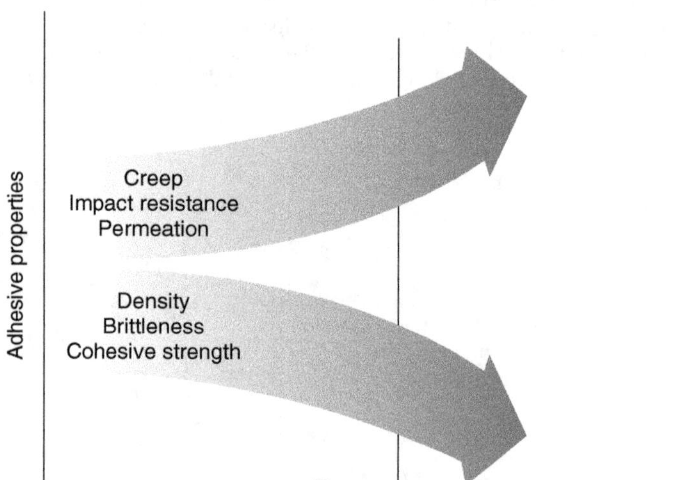

FIGURE 3.16 General trends of some adhesive properties related to temperature or molecular mobility. The change at the T_g reflects the abrupt increase in molecular motion.[23]

TABLE 3.9 Glass Transition Temperature of Common Polymers[24]

Adhesive type	Glass transition temperature, °C
Silicone elastomer	−90
Natural rubber	−70
Neoprene	−50
Polyamide thermoplastic	60
Epoxy	100

3.4.4 Properties Resulting from Elevated versus Room Temperature Cure

Room temperature curing cannot, in principle, achieve the same degree of crosslinking as is obtained by curing at elevated temperatures, although it does achieve sufficient properties for many adhesive applications. At elevated temperatures, the epoxy resin and curing agent molecules are mobile, and there is a greater potential for reaction than at room temperature. Figure 3.17 shows the glass transition temperatures of common epoxy formulations cured at both room and elevated temperatures.

Another reason for higher T_g with elevated-temperature curing systems is that the hydroxyls along the epoxy chain can react. At ambient temperature, reaction between an epoxy group and a hydroxyl group proceeds very slowly. Hence, the diepoxy group, which is formed when the epoxy group reacts with an amine, generally cannot enter into further reaction with a second epoxy ring at room temperature. As a result, heat-cured diepoxies can be regarded as being potentially tetrafunctional, rather than difunctional.

Postcuring at elevated temperatures after a room temperature cure is a common process in epoxy technology, and this can moderately increase the T_g in some systems.[26] Such effects could be due to secondary reactions (irreversible) or to free volume effects (reversible). These effects could also be realized during the normal aging of the epoxy system in service. One should be careful, however, in assuming that a low-temperature cure followed by an elevated-temperature postcure (cure condition 1) will provide properties equivalent to only a high-temperature cure (cure condition 2). As explained in the previous section, the types

TABLE 3.10 Thermal Properties of Various Epoxy Formulations[25]

		Curing agent			
Property	Method	BF_3: amine complex	DTA	MPD	Aliphatic diamine
Heat distortion temperature, °C at 264 psi	ASTM D 648	120	91	131	47
Modified Vicat penetration, °C		125	93	136	51
Coefficient of thermal expansion, in/in · °C	ASTM D 696				
• Below T_g		6.38×10^{-5}	6.00×10^{-5}	5.83×10^{-5}	7.79×10^{-5}
• Above T_g		16.4×10^{-5}	17.9×10^{-5}	20.8×10^{-5}	20.0×10^{-5}
Glass transition temperature, °C	ASTM D 696	141	122	190	47

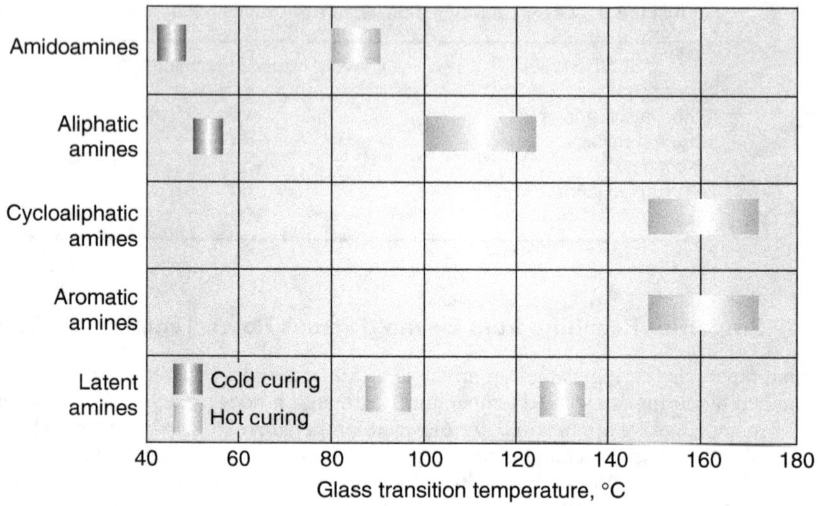

FIGURE 3.17 Glass transition temperatures of several epoxy formulations cured at room and elevated temperatures.[27]

and sequences of reactions that occur are temperature-dependent, so the properties of an epoxy subjected to these two cure conditions could be significantly different.

It is not generally desirable to rely on the service environment for completion of the cure reaction and establishing optimum adhesive strength. Incomplete reaction is undesirable since the presence of reactive, polar groups increases the susceptibility to uptake of moisture and other small molecules. This could be detrimental to the long-term durability of the adhesive. Undercure should not be used as a potential source of flexibility for these reasons. Rather, another adhesive formulation should be chosen that produces the desired level of flexibility in its fully reacted state.

REFERENCES

1. Lee, H., and Neville, K., *Handbook of Epoxy Resins*, McGraw-Hill, New York, 1967, p. 17.11.
2. Potter, W. G., *Epoxide Resins*, Springer-Verlag, New York, 1970, p. 30.
3. Petrie, E. M., *Handbook of Adhesives and Sealants*, McGraw-Hill, New York, 2000, pp. 59–62.
4. Pocius, A. V., *Adhesion and Adhesives Technology*, Chapter 6, Hanser Publishers, New York, 1997.
5. Schornhorn, H., and Sharpe, L., *Journal of Polymer Science Part B: Polymer Letters*, vol. 2, no. 7, 1964, p. 719.
6. Dannenberg, H., and May, C. A., "Epoxide Adhesives," in *Treatise on Adhesion and Adhesives*, R. L. Patrick, ed., Marcel Dekker, New York, 1969, p. 18.
7. DuNouy, P., and Lecounte, J., *General Physiology*, vol. 1, 1919, p. 521.
8. Wilhelmy, L., *Annals of Physics*, vol. 119, 1863, p. 177.
9. Zisman, W. A., "Relation of Equilibrium Contact Angle to Liquid and Solid Constitution," Chapter 1 in *Contact Angle, Wettability, and Adhesion*, R. F. Gould, ed., American Chemical Society, Washington, 1964.

10. Bell, J. P., et al., "Amine Cured Epoxy Resins: Adhesion Loss due to Reaction with Air," *Journal of Applied Polymer Science*, vol. 21, 1977, pp. 1095–1102.
11. Griffith, A. A., *Philosophical Transactions of the Royal Society of London*, ser. A., vol. 221, no. 163, 1920.
12. Chessin, N., and Taylor, G., "Working Life of Room Temperature Curing Epoxy Adhesives," *Adhesives Age*, vol. 10, no. 9, 1967, p. 29.
13. Mylonas, C., *Proc. Seventh International Congress of Applied Mechanics*, London, 1948.
14. Baier, R. E., et al., "Adhesion: Mechanisms That Assist and Impede It," *Science*, vol. 162, December 1968, pp. 1360–1368.
15. Schneberger, G. L., "Basic Bonding Concepts," *Adhesives Age*, May 1985.
16. Dannenberg, and May, "Epoxide Adhesives," in *Treatise on Adhesion and Adhesives*.
17. Sadhir, R. K., and Luck, R. M., *Expanding Monomers: Synthesis, Characterization, and Applications*, CRC Press, Boca Raton, FL, 1992.
18. Perry, H. A., "Room Temperature Setting Adhesive for Metals and Plastics," *Adhesion and Adhesive Fundamentals and Practice*, J. E. Rutzler and R. L. Savage, eds., Society of Chemical Industry, London, 1954.
19. Bolger, J., "Structural Adhesives State of the Art," in *Adhesives in Manufacturing*, G. L. Schneberger, ed., Marcel Dekker, New York, 1983, p. 143.
20. Lin, C. J., and Bell, J. P., "The Effect of Polymer Network Structure upon the Bond Strength of Epoxy-Aluminum Joints," *Journal of Applied Polymer Science*, vol. 16, 1972, p. 1721.
21. Schneberger, G. L., "Polymer Structure and Adhesive Behavior," in *Adhesives in Manufacturing*.
22. Baker, A. M. M., and Mead, J., "Thermoplastics," *Modern Plastics Handbook*, C. A. Harper, ed., McGraw-Hill, New York, 2000.
23. Schneberger, "Polymer Structure and Adhesive Behavior," *Adhesives in Manufacturing*.
24. Schneberger, "Basic Bonding Concepts."
25. Potter, *Epoxide Resins*, p. 93.
26. Brewis, D. M., Comyn, J., and Fowler, J. R., *Polymer*, vol. 20, 1979, p. 1548.
27. Air Products, "Curing Agents Center for Epoxy," SpecialChem4Polymer.com, December 2003.

CHAPTER 4
EPOXY RESINS

4.1 INTRODUCTION

The syntheses of commercial epoxy resins that are commonly used in many applications were discussed in Chap. 2. Additional information is provided in this chapter with regard to the physical and chemical properties of certain epoxy resins relative to their use in adhesive systems.

Many types of epoxy resins can be used in adhesive formulations. These are characterized in Table 4.1. The most commonly used type is the resin-based diglycidyl ether of bisphenol A (DGEBA). Epoxy novolac, flexible epoxy, high-functionality, and film-forming epoxy resins are also used in specialty applications.

The epoxy resin is a primary component in any epoxy adhesive formulation, and it is often referred to as the *base polymer*. However, it is certainly not the only or even not always the predominant component in influencing desirable end properties. Epoxy resins by themselves are often too rigid to provide the required properties such as flexibility, peel and impact strength, and thermal cycling resistance. As a result, they are often modified with other components or hybridized with other types of polymeric resins to provide these functions.

Similarly, lower-molecular-weight diluents may be added to reduce viscosity. Viscosity reduction is often necessary to allow for easier compounding of the ingredients into the formulation or to provide specific application properties. Accelerators are also frequently added to speed the cure rate of epoxy resins at or below room temperature.

Thus, the epoxy resin is only the base foundation on which a complete formulation must be constructed. The formulation freedom and opportunities that it provides are some of the most valuable features of epoxy technology.

This chapter reviews the various types of epoxy resins and their characteristics in adhesive systems. These resins are classified by generic type as one of the following:

- DGEBA epoxy resins
- Epoxy novolac and other epoxy resins
- Flexibilized epoxies
- Waterborne epoxy
- Epoxy acrylate

Future chapters describe the other raw materials that contribute to the epoxy adhesive formulation (curing agents and catalysts, Chap. 5; solvents and diluents, Chap. 6; hybrid resins, Chap. 7; flexibilizers and tougheners, Chap. 8; fillers, Chap. 9; and adhesion promoters, Chap. 10). Complete adhesive formulations are then discussed in subsequent chapters.

TABLE 4.1 Epoxy Resin Types

Epoxy resin type	Characteristics	Primary curing requirements*
Diglycidyl ether of bisphenol A (DGEBA)	General-purpose use	RT, ET
Bisphenol F diluted DGEBA	Very low viscosity	RT, ET
Epoxy novolac	High heat and chemical resistance	ET
Flexibilizing	High elongation and impact resistance	RT, ET
Brominated DGEBA	Flame resistance	RT, ET
Aliphatic	High reactivity and low viscosity	RT, ET
Cycloaliphatic	Good electrical characteristics and chemical resistance, low viscosity	ET
High functionality	High heat and chemical resistance	ET
Film formers	Adhesion promotion, high resiliency	RT, ET

*RT = room temperature, ET = elevated temperature.

4.2 DIGLYCIDYL ETHER OF BISPHENOL A EPOXY RESINS

Epoxy resins based on diglycidyl ether of bisphenol A are not only the oldest type of epoxy resins but also the most valuable in adhesive formulations. The primary reasons for their popularity have been

1. The relatively low cost of raw materials used to synthesize these resins
2. Their ability to be cured by a large number of different crosslinking agents at both room and elevated temperatures
3. Their relatively good mechanical and chemical properties that are exhibited by the cured resins

The DGEBA family of epoxy resins provides superior overall performance in a variety of end-use applications. With proper additives or modifiers, these base polymers enable formulators to achieve practically any desired performance level, while optimizing processability and durability in many environments.

Properties of several commercially available DGEBA epoxy resins from various suppliers are listed in App. C. Generally they can be divided into the following classifications:

- Low-molecular-weight liquids
- High-molecular-weight semisolids and solids
- Brominated resins

Figure 4.1 shows a comparison of the physical and curing properties for liquid and solid DGEBA epoxy resins.

These resins are most often characterized by their epoxy equivalent weight (EEW), molecular weight (number of repeating units n), and viscosity. Table 4.2 shows the relationship between EEW and viscosity. These DGEBA epoxy resins can be used alone or in blends with other DGEBA resins, other epoxy resins, or even other types of polymeric resins. Very often commercial epoxy resin products are actually blends of resins having a broad molecular weight distribution.

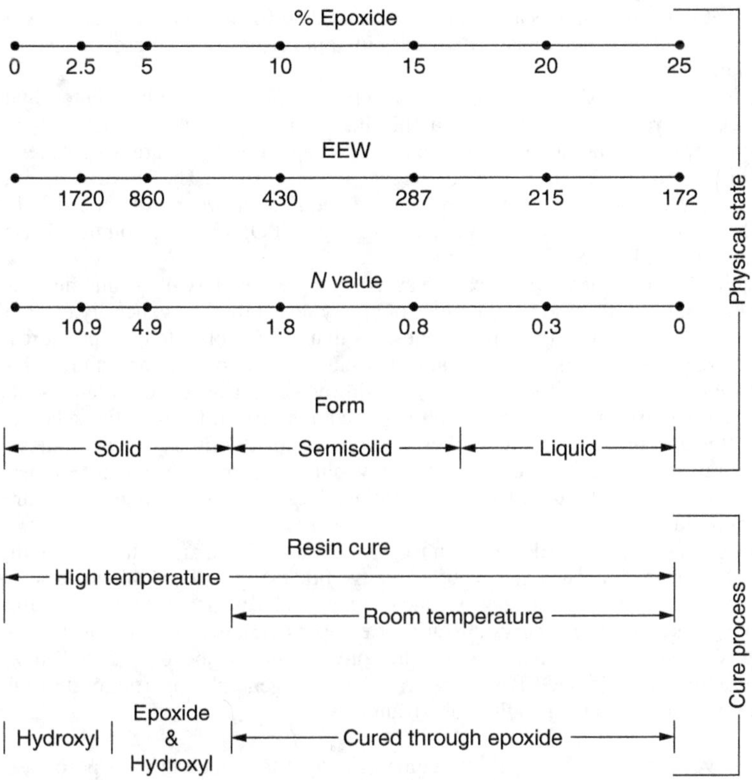

FIGURE 4.1 Comparative physical and curing properties for DGEBA epoxy resins.[1]

4.2.1 Liquid DGEBA Resins

A majority of the world's epoxy resin market consists of the DGEBA type. The liquid DGEBA resins are typically used where low viscosity, high reactivity, and high crosslink density are required. The low viscosity allows them to be conveniently and easily compounded with fillers and other additives. The relatively high functionality allows them to

TABLE 4.2 DGEBA Epoxy Resin Properties

Number of repeating units *n*	Epoxy equivalent weight	Viscosity, poise at 25°C	Melting point, °C	Molecular weight	Approximate number of groups	
					Epoxy	Hydroxyl
0	175–210	50–225		380	2	0
2	450–525		65–75	900	2	2
3.7	870–1025		95–105	1400	2	4
8.8	1650–2050		125–135	2900	2	9
12.0	2400–4000		145–155	3750	2	12

be cured at low temperatures and with conventional curing agents. This reactivity produces reasonably high crosslink density necessary in applications requiring high heat, chemical, or solvent resistance.

DGEBA epoxy resins are formed by reacting bisphenol A with epichlorohydrin (see Chap. 2). The reaction proceeds via a chlorination intermediate to produce bisphenol A diglycidyl ether. Variation of the bisphenol A and epichlorohydrin ratios enables resins to be produced with different molecular weights and viscosities. The DGEBA having a molecular weight of approximately 380 or an epoxy equivalent weight of about 190 is the primary liquid epoxy resin in commercial use (e.g., EPON 828, Resolution Performance Products LLC; DER 331, Dow Plastics).

Liquid DGEBA epoxy resins can be cured by a great variety of curing mechanisms, as is shown in Chap. 5. These are the types of epoxy resins that are generally used in formulating conventional one-component adhesives that cure at elevated temperatures or two-component adhesives that can cure at either elevated or room temperatures. The liquid form enables convenient formulation with additives and fillers. Formulated systems can be easily mixed prior to application and applied as uniform films to the substrates to be joined. Because the resulting adhesives are high-viscosity liquids or pastes, they will wet the substrate well. Yet they do not readily flow out of the joint area to cause adhesive starvation. Pressure is not a requisite to cure these adhesives as only contact pressure is generally applied.

The lower-viscosity grades have an EEW of about 175 and are virtually pure diglycidyl ethers of bisphenol A. They are so pure, however, that they will crystallize on storage. The crystals melt on warming above 40°C, and heating can be used to restore a crystallized resin to its previous form. Special crystallization-free resins systems have been formed by blending low-viscosity DGEBA resins with more conventional bisphenol A–based epoxy resins.

These low-viscosity DGEBA resins provide all the general properties of the higher-MW epoxies with the following additional advantages:

- Better wetting of fillers and reinforcements and better penetration of porous substrates due to low viscosity
- Increased crosslink density (increased T_g and heat distortion temperature)
- Greater chemical uniformity (often preferred in electronic applications because of purity)
- Pale, almost water white color

Higher-molecular-weight DGEBA resins are used where improved toughness, flexibility, and adhesion are required. As the molecular weight and chain length increase, the number of hydroxyl groups along the chain also increases. These hydroxyl groups provide better adhesion characteristics and allow the resin to be cured with mechanisms other than those related to the epoxy ring (e.g., with polyisocyanate curing agents). Also, these hydroxyl groups can cure by reacting with epoxy groups in the presence of tertiary amine catalysts.

As the EEW increases for these resins, the following characteristics are noted for the cured epoxy product:

- Reactivity is reduced. Thus, pot life is increased, and the exothermic temperature is reduced. Cure may be required at elevated temperatures.
- The degree of crosslinking is generally less, leading to a lower glass transition temperature, heat distortion temperature, and chemical resistance.
- Adhesion properties are improved.
- Property increases occur in flexibility, impact, and elongation.

4.2.2 Solid and Semisolid DGEBA Resins

The higher-MW semisolid (EEW of 225 to 280) and solid epoxy resins (EEW > 450) may be blended into lower-MW resins to improve flexibility and decrease reactivity. They also improve adhesion due to the higher concentration of hydroxyl groups along the molecular chain. Ten percent of a higher-MW epoxy resin blended into a conventional liquid epoxy resin (EEW of 190) can significantly improve flexibility, but a reactive diluent frequently needs to be added to the formulation to counteract the increased viscosity caused by the addition.

The highest-MW DGEBA epoxy resins are termed *phenoxy* resins. They are highly linear molecules that are used primarily as thermoplastic coating resins. However, they can be blended with lower-MW epoxy resins for the improvement of specific properties such as flexibility, impact and fatigue resistance, and thermal cycling. Phenoxy resins are sometimes used alone as a thermoplastic hot melt adhesive generally in film form.

The solid and semisolid DGEBA resins are often employed as coatings or adhesive systems where solvents are added to the formulation to reduce viscosity and allow for easy application onto the substrate. Blends of ketone solvents (methyl ethyl ketone and methyl isobutyl ketone) with aromatics (xylene and toluene) are generally suitable for thinning these systems. Esters are also often used as solvents for epoxy-based systems. Higher-boiling-point solvents such as glycol ethers can be used in amounts of 5 to 15 percent by weight to improve flow and film surface properties. Solvents and diluents for epoxy resins are discussed in Chap. 6.

However, one of the greatest advantages of epoxy resin chemistry is that epoxy resins can provide low-viscosity formulations without solvents. As a result, these 100 percent solids systems elude environmental regulations and production precautions that are required for a flammable material.

Solvents are used, however, for special applications. For example, solvents may be added to reduce viscosity and assist penetration on porous substrates. On certain polymeric substrates, solvents may be added to improve adhesion by assisting the diffusion of the adhesive molecules into the substrate. On nonporous substrates, volatile solvents must be evaporated before cure because the solvent could interfere with the degree of crosslinking, and under certain curing conditions, gaseous bubbles could form in the bond line and degrade joint strength.

Solid epoxy resins are usually formulated as solvent solutions and blends with lower-MW resins for the production of liquid adhesive systems. However, solid epoxy resins are also often employed in the manufacture of adhesive systems having solid form. There are several forms of solid epoxy adhesives that find application. The most common are supported or unsupported film, powder, and solder stick. Formulations for these adhesives are detailed in Chap. 13.

Epoxy adhesives can be manufactured into a film form. This is most conveniently done with solid epoxy resins in solution. Epoxy film adhesives can be thermoplastic (e.g., linear ultrahigh molecular weight phenoxy resin) hot melts, but more commonly they are formulated, thermosetting materials.

Latent catalysts, fillers, and additives are added to the epoxy resin solution, and the formulation is cast into a free film or onto a supporting medium such as a glass or polymeric fabric. Once the solvent evaporates in a drying oven, the film can be wound into rolls with a separator sheet and applied directly to the joint without mixing or metering. The joint is then assembled and placed in an elevated-temperature oven or autoclave to cure under light pressure. At the curing temperature, the solid epoxy melts, wets the surfaces of the substrate, and eventually cures to a thermosetting structure. Epoxy films are commonly used to bond large-area parts such as aerospace, transportation, or paneling components.

Similarly prepared solid epoxy films can be ground into powder. The powder can be pressed into preformed pellets and applied directly to the adhesive joint as annular rings or via another shape. This type of adhesive is conveniently employed in mass-produced articles that have concentric joint designs (e.g., flared tubing joints). The joint configuration will hold the solid preform until it melts under the application of heat and forms a reservoir for the cured adhesive. A powder epoxy adhesive can also be applied to a substrate by electrostatic methods.

The primary advantage of solid epoxy adhesives is that they avoid the disadvantages of working with liquids. Waste and cleanup are minimized, and health problems are reduced because the end user handles only a solid substance. Since they are essentially one-component adhesives, they also eliminate the need to meter and mix individual components.

4.2.3 Brominated DGEBA Epoxy Resins

Brominated epoxy resins are the reaction product of epichlorohydrin and brominated bisphenol A. They are primarily used in applications where ignition resistance is a requirement, such as printed-circuit boards and other products that need to be flame-retardant. Tetrabromobisphenol A is the largest flame retardant in terms of commercial use at present. It is used in an estimated 95 percent of all flame-retardant printed wiring boards and is used in many flame-retardant surface-mounted adhesives.[2] It is manufactured by several producers and is priced as a commodity product.

The synthesis of brominated epoxy resins was discussed in Chap. 2. The resulting resins are available primarily as semisolids or solids in solvent solutions. They have properties similar to those of other DGEBA epoxies except that the high bromine content (18 to 21 percent) in the finished resins provides outstanding flame ignition resistance. Tetrabromo diphenylolpropane (Fig. 4.2) is an example of a commercially brominated epoxy resin.

Standard epoxy resins are usually blended with brominated epoxy resins to provide the concentration of bromine required to provide ignition temperature resistance. Brominated resins have also been used as flame-retardant additives in thermoplastic compounds.

Brominated epoxy resins function by liberating acid halide gases as the product thermally breaks down at the temperatures incurred in a fire. These halide gases act as extinguishers to significantly increase the ignition temperature of the cured epoxy resin. The volume of gases liberated and the degree of flame resistance are dependent on the bromine content of the cured epoxy.

Environmental activists, especially in Europe and Japan, tend to oppose the commercial use of halogen compounds, such as bromine-containing flame retardants and tetrabromobisphenol A. The German Environmental Department (UBA) has issued a position paper in which it recommends phasing out tetrabromobisphenol A because it has been detected in the "food chain."[3] As a result, newer epoxy formulations are employing nonbrominated flame-retardant additives. These are discussed in Chap. 9. However, one of the greatest advantages of using a halogenated epoxy resin rather than an additive is the ability to maintain physical properties.

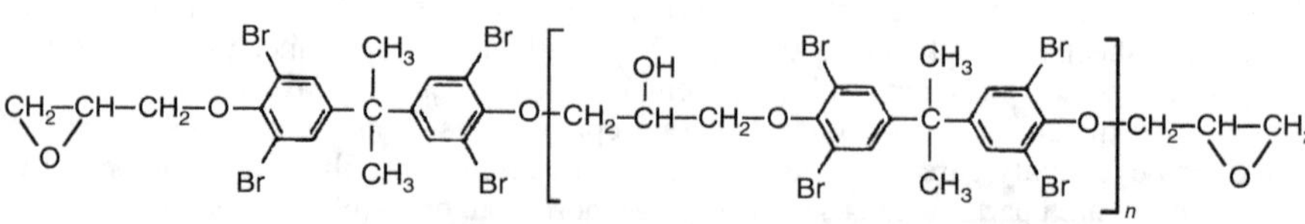

FIGURE 4.2 Chemical structure of epoxy resin based on tetrabromo diphenylolpropane.

4.3 EPOXY NOVOLAC AND OTHER EPOXY RESINS

Epoxy novolac resins are polyglycidyl ethers of a novolac resin. They are prepared by reacting epichlorohydrin with a novolac resin (see Chap. 2). The most common epoxy novolacs are based on medium-MW molecules with phenol and *o*-cresol novolacs. They generally have significantly different properties from DGEBA epoxies because of the presence of the phenolic structure.

Epoxy novolac resins also differ from standard DGEBA-based epoxy resins in their multifunctionality, which is about 2.5 to 6.0. The multiplicity of epoxy groups allows these resins to achieve increased crosslink density. The commercial epoxy novolac resins (e.g., DER 438, Dow Plastics, and EPON 164, Resolution Performance Products LLC) are semisolid to solid resins with EEW in the range of 170 to 230. Recently low-viscosity epoxy novolac resins have been produced (18,000 to 28,000 cP) to provide easy processing; however, these generally have lower epoxy content.

When cured with any of the conventional epoxy curing agents, epoxy novolacs generally produce a product with better elevated-temperature performance, chemical resistance, and adhesion than those of the bisphenol A–based resins. Thus, epoxy novolacs are used in structural adhesive systems that require high-temperature performance. Table 4.3 shows the thermal stability, as measured by weight loss, of epoxy novolac systems having a functionality of 2.5 and 4.0. Notice that significant improvements over standard DGEBA epoxy resins occur only when the functionality of the epoxy novolac is above 2.5.

To develop the properties of an epoxy novolac to its fullest extent, a high-temperature cure is necessary. With room temperatures cures, the properties of the final product are similar to those of conventional DGEBA systems. The thermal stability of most epoxy novolac systems is affected markedly by the length of the cure cycle.

Bisphenol F epoxy resins are produced by condensing phenol with formaldehyde, resulting in a mixture of isomers and higher-MW condensation products. Bisphenol F epoxy resins have lower viscosity then DGEBA epoxy resins for the same molecular weight (or number of repeating units *n*). Cured bisphenol F epoxy resins also have increased solvent resistance.

Bisphenol F resins are often mixed with conventional DGEBA epoxy resins because of the relatively high cost of the bisphenol F product. When mixed with bisphenol A resins, the two form crystallization-free resins of moderate viscosity.

TABLE 4.3 Thermal Weight Loss, Weight Percent, of Nadic Methyl Anhydride Cured Epoxy Novolac Resin (Cured 2 h at 90°C, 4 h at 165°C, and 16 h at 200°C)[4]

Thermal aging conditions				
Temperature, °C	Aging time, h	DGEBA epoxy	2.5 Functional epoxy novolac	4.0 Functional epoxy novolac
160	100	0.66	0.21	0.13
	300	1.50	0	+0.05
	500	1.80	0	+0.12
210	100	0.56	0.67	0.33
	300	1.60	1.3	0.56
	500	1.80	1.55	1.08
260	100	5.6	—	5.2
	200	10.2	—	9.2

Tetraglycidyl ether of tetraphenolethane is an epoxy resin that is noted for high-temperature and high-humidity resistance. It has a functionality of 3.5 and thus exhibits a very dense crosslink structure. It is useful in the preparation of high-temperature adhesives. The resin is commercially available as a solid (e.g., EPON Resin 1031, Resolution Performance Polymers). It can be crosslinked with an aromatic amine or a catalytic curing agent to induce epoxy-to-epoxy homopolymerization. High temperatures are required for these reactions to occur.

Diglycidyl ether of resorcinol–based epoxy resins provide the highest functionality in an aromatic diepoxide. It is one of the most fluid of epoxy resins, with a viscosity of 300 to 500 cP at 25°C. Because of its high functionality, it is a very reactive resin and cures more rapidly than DGEBA epoxies with most conventional curing agents.

Cycloaliphatic and *heterocyclic epoxy* have better weather resistance and less tendency to yellow and chalk than do aromatic epoxy resins. These resins possess excellent electrical properties and are often used in electrical and electronic applications. They are generally formulated into casting and filament winding compounds.

Their use in adhesive systems is minimal because they are relatively brittle and higher in cost than aromatic resins. However, cycloaliphatic epoxy resins are used in cationically cured epoxy adhesive formulations. These are cured via uv or electron beam (EB) radiation.

Glycidyl amine epoxy resins are reaction products of aromatic amines and epichlorohydrin. They have high modulus and high glass transition temperature. These resins find use in aerospace composites and high-temperature adhesive formulations.

4.4 FLEXIBLE EPOXY RESINS

Epoxy adhesive formulators have generally addressed the problem of improving flexibility by adding chemical groups to the epoxy structure—via either the base resin or the curing agent—or by adding separate flexibilizing resins to the formulation to create an epoxy-hybrid adhesive system. This section mainly addresses the flexibility achieved through the epoxy resin structure. The flexibility achieved via additives, hybrid resins, and curing agents is addressed in later chapters.

Flexibility can be provided through the epoxy resin constituent by incorporating large groups in the molecular chain, which increases the distance between crosslinks. Gross changes in the flexibility of a resin system can be obtained only through a major change in the structure of the cured resin. Such changes include shifting from an aromatic structure to a more aliphatic hydrocarbon or by reducing functionality. Insertion of long hydrocarbon side chains also can provide additional flexibility to the epoxy molecule.

Examples of epoxy resins modified for improved flexibility include the following:

- Epoxy resins derived from acid functional oils (dimer acid, cashew nut oil, and Castor oil)
- Epoxy resins derived from polyalkylene glycol (polyethylene or polypropylene glycol)
- Epoxy resins having increased molecular weight while maintaining the same number of reactive sites

The reduction in crosslink density increases the flexibility (elongation) of the resulting molecule, but at the expense of a lowering of the glass transition temperature. This, in turn, generally results in a significant decrease in tensile and shear strength as well as a decrease in other performance properties, such as chemical and heat resistance.

TABLE 4.4 Properties of an Epoxy Resin Formulation Based on a Blend of Polyglycol Diepoxy Flexibilized Resin and Standard DGEBA Resin, Cured with Methylene Dianiline[5]

Property	70 Parts DGEBA : 30 parts flexible resin (EEW: 320)	70 Parts DGEBA : 30 parts flexible resin (EEW: 190)	100 Percent DGEBA
Viscosity, cP at 70°C	58	50	100
Heat distortion temperature, °C	84	103	157
Flexural strength, psi	14,060	14,870	16,970
Flexural modulus	2.76×10^5	2.6×10^5	2.27×10^5
Compressive strength, psi	24,320	27,640	32,000
Tensile strength, psi	9160	10,310	8150
Ultimate elongation, %	8.1	7.0	3.8
Izod impact strength, ft · lb/in notch	1.16	0.94	0.44
Hardness, Rockwell M	97	100	106

Long-chain aliphatic epoxy resins provide flexible molecules with high elongation but little toughness. They are generally based on polyglycol or vegetable oils reacted with epichlorohydrin. Because of their lack of hydrolytic stability and lack of strength, they are generally not used alone but are blended as modifiers with other epoxy resins. Generally when used in a concentration range of 10 to 30%, they improve flexibility without greatly impairing other properties.

Glycidyl ethers of aliphatic polyols based on polyglycol, glycerin, and other polyols are flexible epoxy resins. They are used as reactive diluents and flexibilizers for solvent-free epoxy resin formulations. Epoxy-polyglycol resins that are produced from the reaction of epichlorohydrin and polyester polyols based on ethylene or propylene oxide are the most common of these types of flexible epoxy resins. Examples of typical commercial aliphatic epoxy resins are shown in App. C.

Commercial products are available where n varies from 2 to 7. The flexible epoxy resins based on polyglycol also make excellent reactive diluents because they have a viscosity of 100 cP at 25°C. Table 4.4 shows the effect of two polyglycol diepoxides on the physical properties of a cured epoxy system. Another type of flexible epoxy resin is derived from dimerized unsaturated fatty acid, cashew nut oils, and other vegetable oils. Other flexible epoxy resins can be made with thiols, aliphatic acids, and hydroxyl-terminated compounds.

Applications where flexible epoxy resins are valued include

1. Adhesives to laminate safety glass
2. Adhesives and sealants to dampen vibration and sound in addition to providing joining
3. Encapsulants for electrical components and other delicate components where thermal cycling is expected

4.5 WATERBORNE EPOXY RESINS

Waterborne epoxy dispersions have been employed effectively for many years in the coatings market, primarily for the surface protection of concrete and metals. These products were developed in response to environmental regulations to reduce solvent levels in coatings. Since these dispersions adhere well to a wide variety of substrates and provide a high degree of strength and resistance to service environments, they have naturally found applications as adhesives.

Epoxy resins are hydrophobic and consequently are not, by themselves, dispersible in water. However, water dispersibility can be conveyed to epoxy resins by two general methods:

1. "Chemical modification" of the epoxy resin
2. The process of emulsification

Both of these processes are applicable to waterborne epoxy adhesives and coatings, although the emulsification process is generally used with adhesives.

Chemical modification of the epoxy resin includes either attaching hydrophilic groups to the epoxy resin or attaching the epoxy resin to hydrophilic polymers. This is most often done by grafting. For example, one of the largest volume uses for waterborne epoxy is the coating of metal cans. In this application the epoxy resin is rendered water-dispersible by the grafting of the epoxy resins to acrylic polymer.

The emulsification method is primarily used for waterborne epoxy adhesive systems and is the focus of this section. The epoxy resin is made water-dispersible by partitioning the epoxy resin within a micelle, effectively separating the resin from the water. This emulsification can be achieved by a suitable surfactant.

The choice of surfactant and processing parameters determines the long-term mechanical and chemical stability of the dispersion. The use of improper surfactants results in epoxy dispersions of relatively large and unstable particle size, which exhibits noticeable settling within as little as 1 day. Also, the possibility of hydrolysis of the epoxy group is present unless the surfactant provides a "protective" function. Through the use of surfactant technology, novel waterborne epoxy dispersions can be manufactured that, with few exceptions, contain no organic cosolvents, are mechanically and chemically stable, and are formaldehyde-free, and several are FDA-acceptable.[6]

The surfactant selection determines the emulsion properties, such as stability, particle size, viscosity, and internal phase content. A correct balance between the hydrophobic and hydrophilic character of the emulsifier is necessary for minimizing the surfactant concentration at the resin-water interface. The surfactants used in resin emulsification can be ionic (in most cases anionic), nonionic, polymeric, or a combination of these.

To emulsify liquid epoxy resins, high-MW, nonionic emulsifiers can be used to develop high solids (typically 55 to 60 percent) epoxy resin emulsions with a particle size of less than 1 micrometer (μm).[7] A typical surfactant concentration for this application is about 8 to 10 percent by weight based on the resin. The physical and chemical characteristics of several nonionic surfactants are given in Table 4.5. The emulsification can be done without the use of any solvents, although small concentrations of solvent are often employed in waterborne epoxy adhesives and coatings for good film-forming properties.

In addition to the surfactant and epoxy resin, the parameters of the emulsification process will significantly influence the properties of the final emulsion. To obtain the smallest achievable droplet size with a narrow droplet size distribution, it is essential to optimize process parameters such as temperature of emulsification and mix ratio of surfactants when more than one surfactant is used.

The emulsification process is simple but must be carefully controlled. Epoxy resin is loaded into a high-speed disperser, and the surfactant is added. A defoamer is generally added to prevent excessive aeration, and high-shear mixing is employed. Water is then slowly added to the mixture. The system at this stage has the epoxy resin as the continuous phase and the water as the dispersed phase. As the water addition continues, the ratio of the dispersed phase to the continuous phase increases until a phase inversion occurs. The inversion occurs at about 65 percent volume ratio of dispersed to continuous phase and is accompanied by a rapid reduction in viscosity. Water addition is then continued until the desired solids concentration is achieved. Additional additives and modifiers can be incorporated into the formulation at this stage.

TABLE 4.5 Physical and Chemical Characteristics of Surfactants Recommended for Epoxy Resins[8]

Chemistry	EO/PO block copolymer	EDA EO/PO block copolymer
Class	Nonionic	Nonionic
Active content, % by weight	>99	>99
Physical form	Flake	Flake
Color	White	Bright yellow
Water content, % by weight	<0.75	<0.75
Hydroxyl value (mg KOH/g)	8.5–11.5	17.3–19.5
pH 2.5% aqueous	5–7.5	9–11
Approximate molecular weight	12,000	12,500
Viscosity, cP at 77°C	3120	675
Melting point, °C	56.2	51
Solubility in water	Soluble	Soluble
Solubility in other solvents	Ethanol, toluene	No data
Flash point, °C	255	>150
Specific gravity at 77°C	1.05	1.06

There are several epoxy resin chemistries that are appropriate for waterborne adhesives. These can be broadly classified into three types:

- Difunctional bisphenol A type, based on reaction products of bisphenol A and epichlorohydrin
- Polyfunctional epoxies
- Modified epoxies consisting of hybrids

The base epoxy resin can be either liquid or solid. As molecular weight increases, the epoxy equivalent weight and the number of hydroxyl groups available for reaction increase. Waterborne epoxy adhesives provide excellent adhesion to metals and other high-energy substrates. Modified waterborne epoxy adhesives can also provide good adhesion to substrates such as vinyl and flexible plastic film. Characteristics of these epoxy dispersions are summarized in Table 4.6.

TABLE 4.6 Characteristics of Major Classes of Epoxy Dispersions Used in Adhesive Applications

	Bisphenol A	Polyfunctional	Modified
WPE (weight per epoxy, based on resin solids)		195–2200	
Viscosity, cP		500–20,000	
Functionality, epoxy groups per molecule	2	3–8	Depends on base resins
Properties	Higher MW provides greater flexibility. Lower MW provides greater crosslinking density.	Higher T_g and crosslinking density than bis-A systems.	Modifications are generally for greater toughness, flexibility, and adhesion to substrates such as vinyl.

TABLE 4.7 Typical Effect of Dilution and Shear on Viscosity of a Waterborne Epoxy Resin[9]

Epoxy resin, % by weight	Brookfield viscometer spindle speed, rpm	Viscosity, cP
44	5	3,000
	10	2,500
	20	2,000
	50	1,500
50	5	25,000
	10	20,000
	20	12,000
	50	7,500

Epoxy resin emulsions are commercially available from several sources. As a group, the typical particle size of the dispersion is in the 0.5- to 3.0-μm range. Solids typically range from 50 to 65 percent, and viscosity from 10,000 to 12,000 cP. The dispersions, in general, are thixotropic as supplied. There is also a dramatic decrease in viscosity of the system with the addition of water. Table 4.7 shows the effect of dilution and Brookfield viscosity spindle speed (thixotropy) on a typical epoxy emulsion.

4.6 EPOXY ACRYLATE RESINS

There are basically two types of epoxy acrylate resins used in formulating adhesive systems. One is a vinyl ester resin that is used in two-component adhesive formulations much as a DGEBA epoxy or a polyester resin is. The other is a special type of resin that is used in radiation cure processes. This latter type of epoxy acrylate does not have any free epoxy groups, but reacts through its unsaturation.

Epoxy acrylate resins or "vinyl esters" are made from the esterification of epoxy resin via their terminal group with an unsaturated acid, such as methacrylic acid derived from epoxy resin. A typical reaction sequence is illustrated in Fig. 4.3. The resultant polymer is usually dissolved in a reactive monomer such as styrene.

Three types of vinyl ester resins are commercially available:

1. Conventional or resilient grades based on bisphenol A type of epoxies
2. Fire-retardant grades based on tetrabromobisphenol A epoxies
3. High-heat grades based on novolac epoxies

These resins are primarily used as a resin matrix in reinforced plastic composites and in structural adhesives.

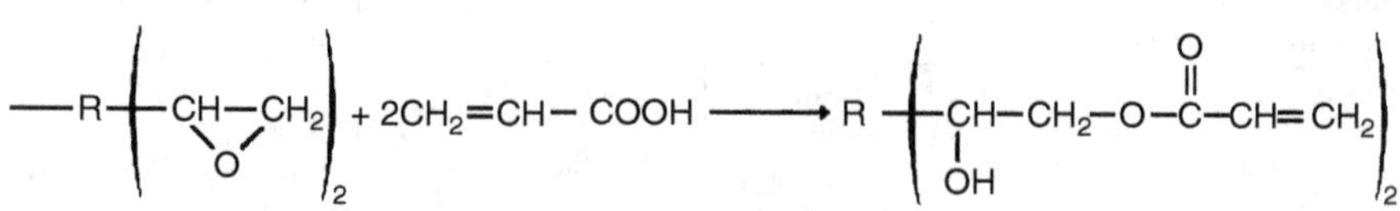

FIGURE 4.3 Reaction of epoxy resins with acrylic to produce epoxy acrylate resin.[10]

These resins act more as polyester resins than they do as epoxy resins. They are easily processed, have fast cure rates at room temperature, and can be cured with peroxides. They effectively wet out glass fiber reinforcement and cure quickly. Therefore, vinyl ester resins are often used in the manufacture of composites such as filament- wound and pultruded structures.

However, once crosslinked, vinyl esters have much greater strength, stiffness, and toughness than conventional polyester resins. They have excellent chemical resistance and mechanical properties at both room and elevated temperatures. Bisphenol A vinyl esters are relatively tough (4 percent elongation), and heat distortion temperatures can range from 93 to 260°C depending on the cure agents and processing conditions.

In adhesive systems, vinyl esters impart low-viscosity, flexibility, and superior wetting characteristics to DGEBA-type epoxy resins. Because of their good chemical resistance they are often found in construction applications such as flooring and grouting. However, their shrinkage is greater than that of any conventional epoxy resins, and often the formulator will have to counteract this.

TABLE 4.8 Various Types of Epoxy Acrylates That Are Commonly Used for UV/EB Curing Adhesives and Coatings[11]

Epoxy acrylate type	Properties
Aromatic difunctional epoxy acrylates	Members of this group have very low molecular weight, which gives them attractive properties such as high reactivity, high gloss, and low irritation. Common applications for these resins include overprint varnishes for paper and board, wood coatings for furniture and flooring, and coatings for compact discs and optical fibers. Aromatic difunctional epoxy acrylates have limited flexibility, and they yellow to a certain extent when exposed to sunlight. The aromatic epoxies are viscous and need to be thinned with functional monomers. These monomers are potentially hazardous materials.
Acrylated oil epoxy acrylates	These epoxy acrylates are essentially epoxidized soybean oil acrylate. These resins have low viscosity, low cost, and good pigment wetting properties. They produce relatively flexible coatings. Acrylated oil epoxy acrylates are used mainly in pigmented coatings or to reduce cost.
Epoxy novolac acrylates	These are specialty products. They are mainly used in the electrical and electronics industries because of their excellent heat and chemical resistance. However, they provide rigid coatings with relatively high viscosity and high costs.
Aliphatic epoxy acrylates	Several varieties are available as difunctional and trifunctional or higher. The difunctional types have good flexibility, reactivity, adhesion, and very low viscosity. Some difunctional types can be diluted with water. The trifunctional or higher types have moderate viscosity and poor flexibility but excellent reactivity. Aliphatic epoxy acrylates have higher cost than the aromatic epoxy acrylates and are generally used in niche applications.
Miscellaneous epoxy acrylates	This group consists mainly of oligomers with fatty acid modification. They provide good pigment wetting properties and higher molecular weight but lower functionality than other aromatic epoxy acrylates. They are used in printing inks and pigmented coatings.

Epoxy acrylates are also commonly used as oligomers in radiation-curing coatings and adhesives. However, their name often leads to confusion. In most cases, these epoxy acrylates have no free epoxy groups left but react through their unsaturation. These resins are formulated with photoinitiators to cure via uv or electron beam (EB) radiation. The reaction mechanism is generally initiated by free radicals or by cations in a cationic photoinitiated system. The uv/EB cured epoxy formulations are discussed in Chap. 14.

Epoxy acrylate oligomers that are used in uv/EB curing are very low-viscosity systems with high vapor pressures. Within this group of oligomers, there are several major subclassifications: aromatic difunctional epoxy acrylates, acrylated oil epoxy acrylate, novolac epoxy acrylate, aliphatic epoxy acrylate, and miscellaneous epoxy acrylates. Characteristics of these various classes are summarized in Table 4.8.

REFERENCES

1. Meath, A. R., "Epoxy Resin Adhesives," Chapter 19 in *Handbook of Adhesives*, 3d ed., I., Skeist, ed., van Nostrand Reinhold, New York, 1990.
2. Weil, E. D., and Levchik, S., "A Review of Current Flame Retardant Systems for Epoxy Resins," *Journal of Fire Sciences*, vol. 22, January 2004, pp. 25–40.
3. Leisewitz, A., et al., Chapter V, Summarized Substance Evaluation: 2. Tetrabromobisphenol A (TBBPA), UBA Report 204 08 542, *Substituting Environmentally Relevant Flame Retardants: Assessment Fundamentals*, Oko-Recherche, Frankfurt am Main, Germany, 2001.
4. Meath, A. R. "Chemistry, Properties, and Applications of Epoxy Novolac, Flexible Epoxy, and Flame Retardant Epoxy Resins," Chapter 3 in *Epoxy Resin Technology*, P. F. Bruins, ed., Interscience Publishers, New York, 1968, p. 34.
5. Meath, "Chemistry, Properties, and Applications of Epoxy Novolac, Flexible Epoxy, and Flame Retardant Epoxy Resins," p. 38.
6. Buehner, R. W., and Atzinger, G. D., "Waterborne Epoxy Dispersions in Adhesive Applications," Epoxy Resin Formulators Conference, San Francisco, February 20–22, 1991.
7. Uniqema Synperonic Surfactants, "Resin Emulsification for Waterborne Coating and Adhesives," Uniqema, Wilmington, DE, 1999.
8. Uniqema Synperonic Surfactants, "Resin Emulsification."
9. Buehner and Atzinger, G.D., "Waterborne Epoxy Dispersions in Adhesive Applications."
10. Savia, M., "Epoxy Resin Adhesives," Chapter 26 in *Handbook of Adhesives*, 2d ed., I. Skeist, ed., van Nostrand Reinhold, New York, 1977, p. 438.
11. Radiation Cure Center, www.SpecialChem4Coatings.com, 2004.

CHAPTER 5
EPOXY CURING AGENTS AND CATALYSTS

5.1 INTRODUCTION

A variety of curing agents and catalysts will react with epoxy resins to provide crosslinked adhesives. The curing agents generally react with the available epoxy or hydroxyl groups. The catalysts initiate homopolymerization of the epoxy groups. Six main classifications of curing agents are commonly utilized with epoxy adhesive formulations, and these can be further divided into several subclassifications.

- Aliphatic amines and modified aliphatic amines
- Polyamides
- Aromatic amines and modified aromatic amines
- Anhydrides
- Catalytic and latent hardeners
- Polysulfides and mercaptans

Some curing agents can also be used to form adducts with epoxy resins, and these adducts, in turn, offer unique curing properties with other epoxy resins. Adducts are commonly used in adhesive formulations to reduce the vapor pressure of the system, to modify the reactivity of the system, to improve mix ratios so that they are closer to equal parts of the resin component and of the curing agent component, and to provide a certain degree of flexibility into the end product.

The choice of a particular curing agent or catalyst depends on the processing requirements (e.g., viscosity, pot life, application method, curing temperature, reactivity, mix ratio) and the end-use requirements (e.g., thermal and chemical resistance, shear strength, toughness) of the cured adhesive. The curing agents along with the epoxy resin determine the type of chemical bonds and the degree of crosslinking that will occur.

The advantages, disadvantages, and applications for the major types of epoxy curing agents are summarized in Table 5.1. The required mix ratios, curing temperatures, and the resulting heat distortion temperatures of the cured product are provided in Table 5.2.

Curing agents can be used to improve the flexibility of inherently rigid epoxy resins. Certain common epoxy formulations can be flexibilized by altering the stoichiometric mix ratio of resin to curing agent or by changing the type of curing agent to one that has a more flexible molecule. For example, by changing from hexahydrophthalic anhydride to hexamine-ethylenediamine, one can double the impact resistance of a resin system and increase its tensile elongation at break.[4]

TABLE 5.1 Advantages, Disadvantages, and Applications of Common Epoxy Curing Agents[1,2]

Curing agent	Advantages	Disadvantages	Applications
Polyamides	• Convenience • Room temperature cure • Low toxicity • Good bond strength and flexibility • Moderately high peel and impact strength	• High formulation cost • Long cure times at room temperature • High viscosity • Low heat and chemical resistance	• General-purpose adhesives • Casting and encapsulation
Polysulfides and mercaptan	• Moisture resistance • Quick set time • Flexible	• Odor • Poor elevated-temperature performance • Poor tensile strength	• Adhesives and sealants • Civil engineering • Casting and encapsulation • Coatings
Aliphatic amines	• Convenience • Room temperature cure, fast elevated-temperature cure • Low viscosity • Low formulation cost • Moderate chemical resistance	• Critical mix ratios • Strong skin irritant • High vapor pressure • Short working life, exothermic • Poor bond strength above 80°C • Rigid, poor peel and impact properties	• Adhesives and sealants • Casting and encapsulation • Coatings
Amidoamines	• Reduced volatility • Convenient mix ratios • Good toughness	• Poor elevated-temperature performance • Some incompatiblity with certain epoxy resins	• General-purpose adhesives • Construction adhesives • Concrete bonding • Toweling compounds
Aromatic amines	• Moderate heat and chemical resistance	• Solid at room temperature • Rigid • Long elevated-temperature cures	• Composites • Electrical encapsulation • Adhesives
Dicyandiamide	• Latent cure • Good elevated-temperature properties • Good chemical resistance • Good combination of tensile and peel strength	• Long elevated-temperature cure • Insoluble in resin	• One-component adhesives • Powder coatings • Film and solid adhesives • Laminates and other composites
Anhydride	• Good heat and chemical resistance	• Long elevated-temperature cure • Critical mix ratio • Rigid	• Composites • Electrical encapsulation • Adhesives
Catalytic curing agents (e.g., tertiary amines)	• Long pot life • High heat resistance • Can be used as an accelerator or as the sole curative	• Long elevated-temperature cure • Poor moisture resistance • Rigid	• Adhesives • Electrical encapsulation • Laminates • Powder coatings

TABLE 5.2 Characteristics of Curing Agents Used with Epoxy Resins in Adhesives Formulations[3]

Curing agent	Physical form	Amount required*	Cure temperature, °C	Pot life at 23°C†	Complete cure conditions	Max. use temperature, °C
Triethylenetetramine	Liquid	11–13	20–135	30 min	7 days at 25°C	70
Diethylenetriamine	Liquid	10–12	20–95	30 min	7 days at 25°C	70
Diethylaminopropylamine	Liquid	6–8	27–150	5 h		85
Metaphenylene diamine	Solid	12–14	66–200	8 h	1 h at 85°C 2 h at 163°C	150
BF_3-MEA complex	Solid	1–4	135–200	6 months	3 h at 163°C	163
Nadic methyl anhydride	Liquid	80–100	120–200	5 days	3 h at 160°C	163
Triethlyamine	Liquid	11–13	20–95	30 min	7 days at 25°C	83
Polyamides:						
Amine value 80–90	Semisolid	30–70	20–150	5 h	5 days at 25°C	‡
Amine value 210–230	Liquid	30–70	20–150	5 h	5 days at 25°C	‡
Amine value 290–320	Liquid	30–70	20–150	5 h	5 days at 25°C	‡

*Per 100 parts by weight for an epoxy resin with an EEW of 180 to 200.
†Five hundred grams per batch; with a DGEBA epoxy with an EEW of 180 to 200.
‡Highly dependent on curing agent concentration.

5.2 ALIPHATIC AMINES

Depending on the number of amine groups in the molecule, the amine can be a mono-, di-, tri-, or polyamine. Aliphatic amines can also be classified by their molecular structure as linear, branched, aliphatic, or containing aromatic groups. However, the most valuable method of classification is by functionality.

The functionality of an amine is determined by the number of reactive hydrogens present on the molecule. The amines typically have greater than three reactive sites per molecule that facilitate the formation of the crosslinked epoxy structure. The difference between primary, secondary, and tertiary amines is reflected by the number of hydrogens that are bound to each nitrogen atom:

- *Primary amine*: Two hydrogens are bound to each nitrogen atom.
- *Secondary amine*: One hydrogen is bound to each nitrogen atom.
- *Tertiary amine*: No hydrogens are bound to each nitrogen atom (will not react readily with an epoxy group, but will act as a catalyst to accelerate epoxy reactions).

The primary and secondary amines are discussed in this section. The secondary amines are derived from the reaction product of primary amines and epoxies. They have rates of reactivity and crosslinking characteristics that are different from those of primary amines. The secondary amines are generally more reactive toward the epoxy group than are the primary amines, because they are stronger bases. They do not always react first, however, due to steric hindrance. If they do react, they form tertiary amines. Tertiary amines are primarily used as catalysts for homopolymerization of epoxy resins and as accelerators with other curing agents.

The chemical structures of important amines for curing epoxy resins in adhesive systems are identified in Fig. 5.1. Diethylenetriamine (DETA), triethylenetetramine (TETA), *n*-aminoethylpiperazine (AEP), diethylaminopropylamine (DEAPA), *m*-phenylenediamine (MPDA), and diaminodiphenyl sulfone (DDS) are the most commonly used members of this class. They are all primary amines. They give room or elevated temperature cure at near stoichiometric ratios. Ethylenediamine is too reactive to be used in most practical adhesive formulations. Polyoxypropyleneamines (amine-terminated polypropylene glycols) impart superior flexibility and adhesion.

5.2.1 Primary and Secondary Aliphatic Amines

Amine curing agents are one of the most common types of curing agents for epoxy resins and can be primary or secondary. Aliphatic amines, such as diethylenetriamine or triethylenetetramine, were the first epoxy curing agents. Although they provided systems with acceptable properties, wider usage of epoxy adhesives has led to the synthesis of amine derivatives, mixtures, and epoxy adducts with a wide range of improved properties to meet specific user requirements.

Primary and secondary aliphatic amines react relatively rapidly with epoxy groups at room or lower temperature to form three-dimensional crosslinked structures. The resulting cured epoxies have relatively high moisture resistance and good chemical resistance, particularly to solvents. They also have moderate heat resistance with a heat distortion temperature in the range of 70 to 110°C. Thus, short-term exposures of cured adhesive joints at temperatures up to 100°C can generally be tolerated.

These aliphatic amines can also be cured at elevated temperatures to provide a more densely crosslinked structure with better mechanical properties, elevated-temperature performance, and chemical resistance. Table 5.3 illustrates the effect of curing temperature on the bond strength of DGEBA epoxy with two different aliphatic amines.

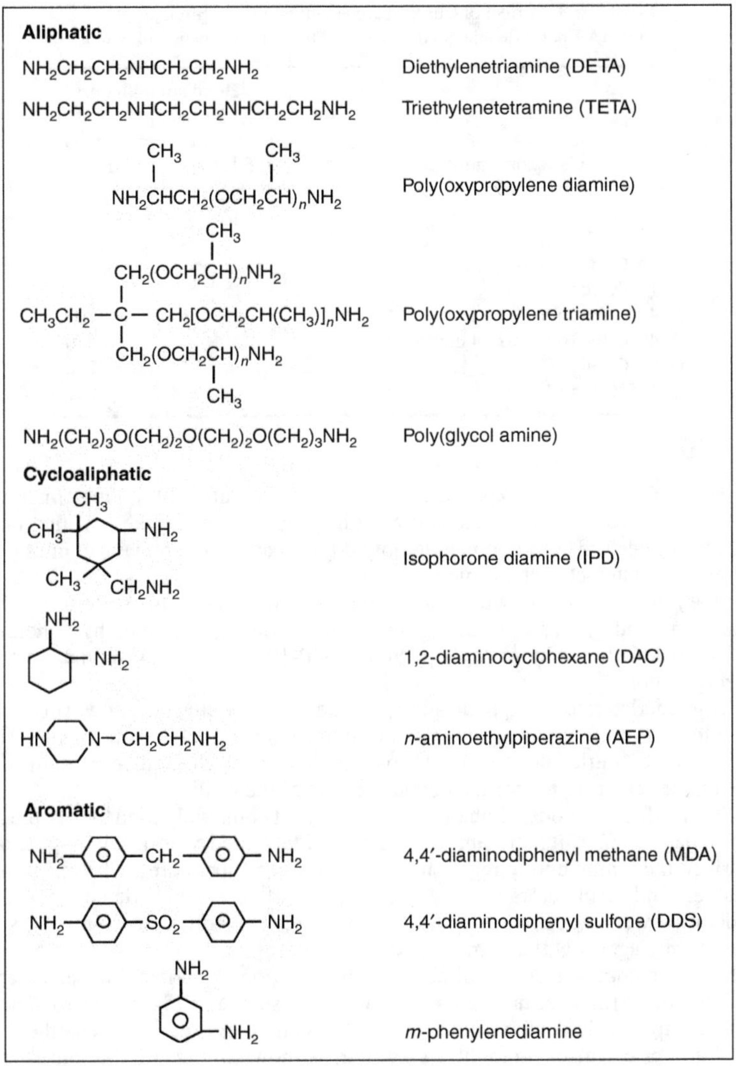

FIGURE 5.1 Common polyamines for curing epoxy resins.[5]

Other amines, such as aromatic or cycloaliphatic, are less reactive and generally require elevated-temperature cures that result in higher heat distortion temperatures (140 to 150°C). However, aromatic amine adducts of liquid epoxies can be accelerated to cure at room temperature. Aliphatic amines can also be accelerated.

Epoxy resin formulations containing aliphatic amines will blush and provide an oily surface under very humid conditions. This is due to a reaction of the amine primary hydrogen atoms with carbon dioxide. Resistance to blushing is more important for coatings than for

TABLE 5.3 Effect of Curing Temperature on Bond Strength of DGEBA Epoxy Resin Cured with Two Different Aliphatic Polyamines[6]

Cure conditions	Bond strength on aluminum, psi TETA	DEAPA
RT for 3 days	1162	
RT for 15 days	1690	
95°C for 30 min	3172	
145°C for 30 min	3426	
40°C for 16 h		702
40°C for 16 h plus 14 days at RT		840
95°C for 5 h		3236
145°C for 30 min		4056

adhesives. Amines that are exposed to the ambient conditions for long times will also absorb moisture from the air. This can result in a weak boundary layer if the moisture is trapped at the interface before joint assembly. Modifications of aliphatic amines have been developed to counteract such problems.

A tertiary amine is often used as an accelerator in primary amine systems to correct for these problems and to adjust the rate of cure. Ortho-(dimethylaminomethyl) phenol, DMP-30, and tris-(dimethylaminomethyl) phenol, DMP-10, are common commercial tertiary amine accelerators.

The fast reaction rate of aliphatic amines at room temperature can, however, lead to various problems. Pot life is short, which can lead to an unacceptable working life. The cured resin is relatively brittle due to its high crosslink density. High exotherms in thick sections or large masses can lead to thermal decomposition of the resin.

In adhesive formulations, aliphatic amines are most commonly used to cure the DGEBA type of epoxy resin. Aliphatic amines are not widely used with the non–glycidyl ether resins, since the amine-epoxy reaction is slow at low temperatures. The reaction usually requires heat and accelerators for an acceptable rate of cure. Aliphatic amines are primarily used with lower-viscosity DGEBA resins because of the difficulty in mixing such low-viscosity curing agents with the more viscous epoxy resins.

In the well-recognized epoxy–aliphatic amine reaction, the primary or secondary amine adds to the epoxy ring, forming a tertiary amine, as shown in Fig. 5.2 (top). The formed hydroxyl groups accelerate the amine curing, and with excess epoxy present, the secondary hydroxyl groups can also add to the epoxy ring, as shown in Fig. 5.2 (bottom).

The resin and polyfunctional amine must interact in approximately stoichiometric amounts to obtain the best balance of properties. Thus, mix ratios and their precision are considered critical for optimum performance. It is assumed that the functionality of the amino group toward the epoxy is one active amino hydrogen for each epoxy ring. (See Sec. 2.5 for calculating the stoichiometric mix ratios.) Since the tertiary amines weakly catalyze the reaction of epoxy groups with each other, slightly less than stoichiometric ratios of amine curative to epoxy groups are generally used when a tertiary amine catalyst is present.

The complicated reaction mechanism described above leads to the conclusion that curing epoxy adhesives with amines can be highly complex. Considerable study and skill must be exercised in (1) selecting the amine curing agent, (2) determining the curing conditions, and (3) maintaining the curing conditions from adhesive joint to joint to ensure performance consistency.

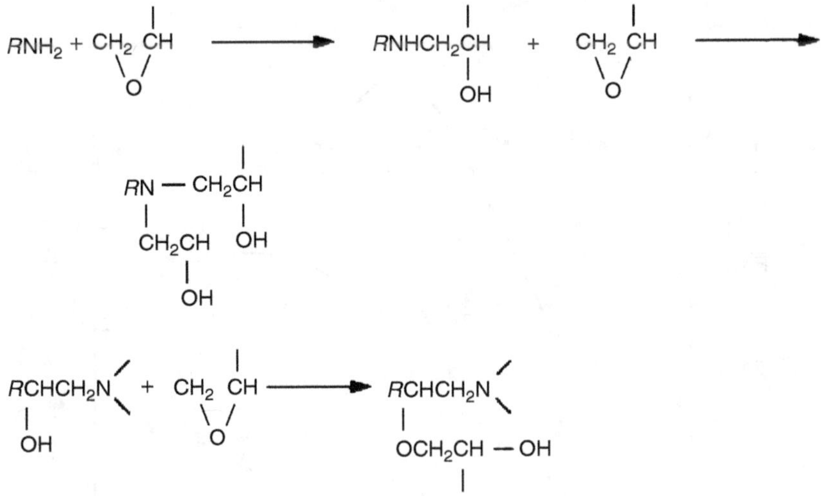

FIGURE 5.2 Top: Addition of primary or secondary amine to epoxy ring. Bottom: Addition of secondary hydroxyl groups to epoxy ring.[7]

Aliphatic polyamines are corrosive and may result in skin sensitization upon prolonged exposure. The high vapor pressure is a significant disadvantage with this type of curing agent. Evaporation of the curing agent due to heating or vacuum degassing can lead to lower than anticipated amine concentrations. The high vapor pressure can lead to gassing in adhesive joints that are cured at elevated temperatures. High vapor pressure also magnifies the toxicity and skin irritation properties of the amines.

Diethylenetriamine and Triethylenetetramine. Diethylenetriamine (DETA) and triethylenetetramine (TETA) are very reactive, low-viscosity liquids that are widely used with DGEBA epoxy resins. The application characteristics and cured properties of adhesive formulations prepared with these two curing agents are very similar. The lower vapor pressure of TETA generally favors its use.

With liquid DGEBA epoxy resins, DETA is normally used at the stoichiometric concentration of 10 to 11 parts per hundred (pph), and TETA is used at a concentration of 14 pph. However, both curing agents can be used at mix ratios as low as 70 to 75 percent of stoichiometry for greater toughness and increased pot life at the sacrifice of heat and chemical resistance. The effect of the mix ratio of DETA and TETA on the heat deflection temperature of castings is shown in Fig. 5.3.

Pot lives of DETA and TETA adhesives are on the order of 20 to 30 min at room temperature. When mixed with DGEBA epoxy resins in large batches, the exotherm can be significant due to the reactivity. This generally limits the amount of mixed adhesive that can be prepared at one time, and it also limits the amount (mass) of adhesive that can be applied to a joint [although often thin bond line and the thermal conductivity of the substrate (e.g., metals) will diminish exotherm effects].

Typical cure schedules for DETA and TETA systems are 4 to 7 days at room temperature or 1 h at 100°C. With these adhesive systems, a certain degree of mechanical strength sufficient to handle the joint will occur within several hours at room temperature.

Good all-around mechanical and chemical resistance properties are obtainable from DETA or TETA cures at room temperature. However, these properties will degrade rapidly

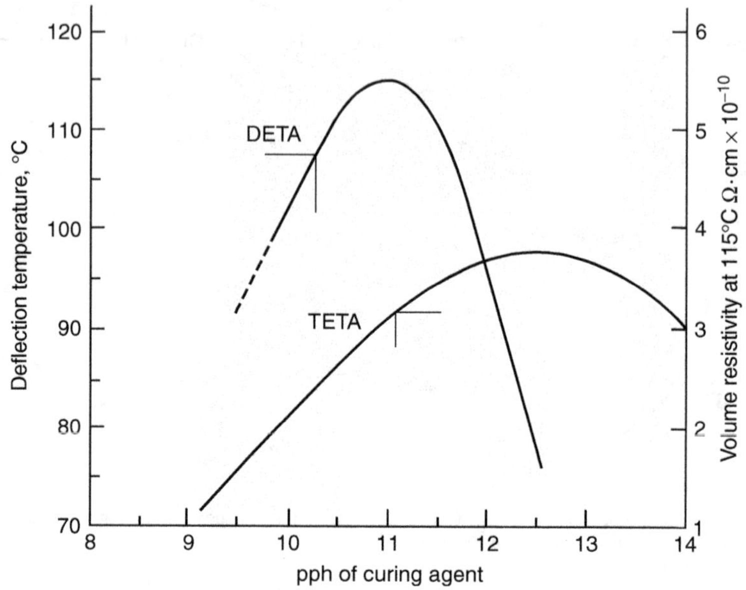

FIGURE 5.3 Effect of concentration of DETA and TETA curing agents on heat deflection temperature of DGEBA.[8]

at operating temperatures greater than 50°C. For maintenance of properties at higher temperatures, the aromatic amines are preferred as curing agents.

Diethylaminopropylamine. Diethylaminopropylamine (DEAPA) is an aliphatic polyamine that is used for curing epoxy adhesives where extended pot lives and low heat exotherms are required. This is a reactive primary amine with two active hydrogens as well as a tertiary amine with catalytic activity.

The stoichiometry for DEAPA cured systems is less critical than with DETA or TETA since DEAPA acts as both a crosslinking agent and a catalyst. Due to the reactivity of the tertiary amine group, the concentration required is also less (4 to 8 pph in DGEBA epoxy resins with an EEW of 190).

Working lives of DEAPA mixtures with DGEBA epoxy resins are 3 to 4 h at room temperature. DEAPA reacts slowly with liquid DGEBA resins; hence, moderate heat (115°C) is generally required to speed cures in an adhesive system. In castings, the exotherm developed is generally sufficient to provide reasonable degrees of cure at room temperature. In adhesive systems, an elevated-temperature cure is generally required.

DEAPA cured epoxies have a less densely crosslinked structure than do DETA or TETA cured epoxies. This results in lower heat and chemical resistance and less hardness; however it also improves the toughness and peel strength. The other physical properties are very similar to those of DETA or TETA cured epoxies.

DEAPA was used in several early commercial epoxy adhesive formulations. Schonhorn and Sharpe[9] have shown that this amine is surprisingly effective in reducing the surface tension of epoxy resins (Table 5.4). It is speculated that the utility of DEAPA adhesives is in part due to better wetting than other epoxy formulations.

Other Common Unmodified Aliphatic Amines. *Cycloaliphatic amines* provide good adhesion and very good chemical resistance. They are also noted for excellent low color

TABLE 5.4 Surface Tension of Several Epoxy Resin Formulations

Formulation	Surface tension, dyn/cm
DGEBA liquid epoxy resin alone (e.g., EPON 828, Resolution Performance Products, LLC)	44–49
DGEBA epoxy and DETA curing agent	44
DGEBA epoxy and DEAPA curing agent	33

and color stability. Cycloaliphatic amines cure well at low temperatures, even under damp conditions. A range of working lives and cure schedules are possible depending on the type of cycloaliphatic amine that is used in the epoxy adhesive formulation. When cured at elevated temperatures, cycloaliphatic amines provide high glass transition temperatures, excellent chemical resistance, and strong mechanical properties. However, they are relatively brittle and exhibit low peel strength and poor impact characteristics in unmodified systems.

Aminoethylpiperazine (AEP) is a cycloaliphatic amine possessing primary, secondary, and tertiary amine groups. Only the primary and secondary groups are involved in the DGEBA curing mechanism. AEP is a clear, high-boiling-point liquid with a low vapor pressure. It is often used instead of DETA or TETA when improved toughness is required. The Izod impact strength of AEP cured DGEBA resins is about 2 to 3 times greater than that of epoxy systems cured with DETA or TETA.

Optimum cures with AEP catalyzed DGEBA are obtained using 20 to 22 pph. However, the crosslink density is not great, and the cured product does not attain a high degree of thermal or chemical resistance. The pot life and exotherm are similar to those of DETA and TETA, but a postcure (2 h or more at 100°C) is required to develop properties fully.

Other cycloaliphatic diamines such as isophorone diamine, bis-*p*-aminocyclohexylmethane and 1,2-diaminocyclohexane are used as epoxy resin curing agents for both ambient and heat cured epoxy resin systems. While they have advantages, such as light color and good chemical resistance, they provide rather sluggish cure rates at low temperatures.

Dimethylaminopropylamine (DMAPA) provides similar cure characteristics and properties to DEAPA, but it has a slightly shorter pot life.

Secondary amines can be represented by piperidine and diethanolamine. Secondary amines when used alone may be considered a special class of tertiary amine in that after the secondary hydrogens have reacted, the resulting crosslinking is believed to occur via the tertiary amine mechanism. These curing agents generally have limited temperature resistance, poorer chemical resistance, and equivalent mechanical properties to the primary amines. The curing temperature significantly influences the reactivity, heat generation, and properties of the cured resin. In adhesive formulations, they are used as blends with primary amines for providing specialized properties.

5.2.2 Modified Aliphatic Amines

There are several reasons why unmodified aliphatic amines, such as those described above, are less widely used than other curing agents in epoxy adhesive systems. These include

- Objectionable skin and respiratory irritant
- Objectionable odor
- Requirement of an inconvenient and critical mixing ratio
- Short pot life
- Cure to yield a rigid, glassy, and easily fractured adhesive

Most of these objections can be overcome through modification of the aliphatic amine by reducing the density of active hydrogens on the molecule. As a result, amines have been converted to adducts of higher molecular weight and viscosity by reaction with mono- or polyfunctional glycidyl compounds. These adducts have generally decreased volatility and irritancy, more convenient mix ratios, and lower reactivity. They also produce cured epoxy systems with somewhat greater toughness.

The aliphatic amines can be adducted with mono- and diepoxies that are prepared from ethylene and propylene oxide, from acrylonitrile, from aldehydes, and from a variety of other compounds. When cured with DGEBA epoxy resins, these curing agents do not offer significantly different end properties other than the properties mentioned above for unmodified amines. However, because there are longer molecular chains between active hydrogen groups, the flexibility and toughness of the cured resin are improved at the disadvantage of a lower crosslinking density.

Glycidyl Adducts of Aliphatic Amines. An aliphatic amine such as diethylenetriamine can be partially reacted with an epoxy, such as a DGEBA resin, to produce a low-volatility adduct. In a typical reaction, the epoxy is added slowly to a large excess of DETA. The reaction is maintained at 75°C by cooling. The reaction products are continuously agitated effectively to provide for good contact and uniform concentration effects. At the end of the reaction, excess DETA is vacuum-distilled away from the adduct.

The adduct gives a viscosity of about 8000 cP at 25°C. It is used at about 25 pph to cure DGEBA epoxy resins. The pot life and exotherm that is generated are similar to those of DETA or TETA cured epoxies. Curing agents having viscosities ranging from viscous to solid can be produced by adducting a wide variety of primary amines with epoxy resins.

The advantages of these adducts are low volatility, higher mixing ratios, and faster cure rates. The reason for the faster cure rate is that the adduct is already partially reacted, thus less reaction is required to reach a gel. In addition, the presence of hydroxyl groups causes an acceleration of cure. Faster cure is a distinct advantage in adhesive formulations where thin coatings of adhesive must set rapidly.

Ethylene and Propylene Oxide Amine Adducts. Polyamines, such as DETA, react readily with ethylene oxide in the presence of water to yield mono- or dihydroxyalkyl derivatives, depending on the ratio of reactants. As the extent of the reaction progresses, the resulting compound contains fewer and fewer active hydrogens. The most common commercial product in this class is hydroxyethyldiethylenetriamine. Several others are noted in Fig. 5.4.

The advantages of these products are low skin irritation potential and low viscosity. Physical properties and reactivity suffer somewhat due to the reduction of active amine hydrogens of DETA to four in the case of the monoadduct of ethylene oxide and three when the bisadduct is formed.

These compounds tend to be hydroscopic and must be stored in tightly closed containers. High humidity will interfere with cure, particularly in thin films. The slow cure can be overcome by the addition of bisphenol A, which acts as an acid accelerator.

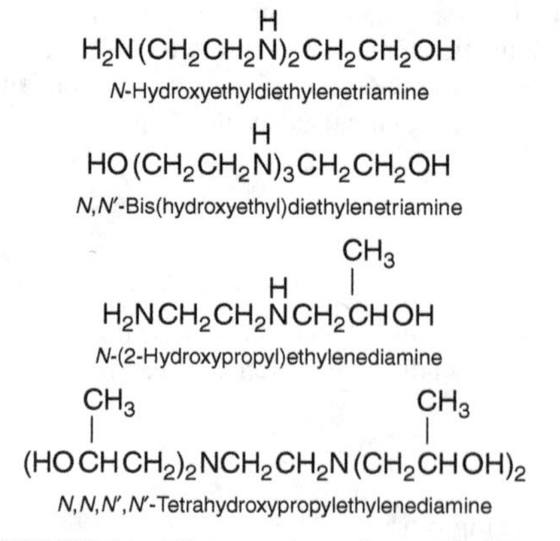

FIGURE 5.4 Ethylene and propylene oxide adducts of amines.

5.3 POLYAMIDES AND AMIDOAMINES

Two curing agents that have found their way into many epoxy adhesive formulations are the polyamides and amidoamines. These are commonly used in the "hardware store variety" two-part epoxy resins that cure at room temperature. Both are reaction products of aliphatic amines, such as diethylenetriamine, and should be included under the subclassification of modified amines. However, these products have such widespread and popular use, they are addressed here as a separate classification.

When compared to other curing agents, polyamides and amidoamines offer the following three unique features.

1. They can be used over a broad range of noncritical concentrations including one-to-one.
2. They are less volatile, and as a result they have fewer odors and lower skin-irritating potential than do other room temperature curing hardeners.
3. They provide a modest amount of flexibility that is directly related to the concentration of curing agent used in the epoxy formulation.

In addition to these processing properties, polyamides and amidoamines offer moderately good shear strength, temperature resistance, and environmental resistance. This balance of properties is why these curing agents are used in many general-purpose epoxy adhesives. DGEBA epoxy resins cured with these materials are widely used in adhesive formulations in the general assembly and construction industries.

5.3.1 Polyamides

Polyamide curing agents are the reaction products of dimerized fatty acids and aliphatic amines such as diethylenetriamine. This introduces a bulky, oil-compatible, C_{36} carbon group between the amine sites. Similar to the diglycidyl ether adducts of aliphatic amine, they are manufactured by adding the fatty acid to an excess of amine. They are available in a range of viscosities that can be achieved by varying the amine/acid molar ratio in the reaction.

The reactive sites are the terminal primary and internal secondary amine groups along the backbone of the polyamide. The amide groups for all practical purposes are not reactive. As a result of their relatively large molecular weight, the amide groups add flexibility to the final crosslinked structure. Generally, polyamides are used as room temperature curing agents.

These products have higher molecular weight than do primary amines and, therefore, exhibit lower vapor pressure, resulting in lower irritation potential and odor. The various molecular weight (MW) derivatives show different degrees of compatibility with epoxy resin. Some high-MW polyamides might show incompatibility with epoxies unless a partial reaction is accomplished. This is known as an *induction period*, and it ensures compatibility. Fortunately, the polyamide curing agents offer a long pot life and low exotherm, so that the induction period is not usually a detriment.

Polyamide mix ratios can be very forgiving and are less critical than with aliphatic polyamines. They are generally used at a 50- to 100-pph level with DGEBA epoxy resins. A mixture consisting of 50 parts polyamide and 50 parts DGEBA epoxy can provide moderately good physical properties in most environments. Increasing the curing agent levels yields increased flexibility and adhesion but reduces heat distortion temperature and chemical resistance.

Polyamide cured DGEBA epoxies provide improved flexibility, moisture resistance, and adhesion over aliphatic amines alone. However, polyamide cured epoxies are generally inferior in thermal resistance and shear strength due to the reduction in crosslink density. Polyamide cured epoxies lose structural strength rapidly with increasing temperatures and

TABLE 5.5 Properties of DGEBA Epoxy Cured with an Amidopolyamine from Tall-Oil Fatty Acid and TETA[10]

	Amidopolyamine, pph			
Property	50	75	100	150
Pot life, min	75	63	57	60
Gel time, min	78	66	61	135
Exotherm, °C	209	190	169	133
Tensile strength, psi, cured 7 days at 25°C and 70 h at 100°C	8400	6000	1780	200
Elongation, %	9	12	70	93
Hardness, Shore D after 24-h cure	80	74	56	10
Ultimate hardness/elapsed time, days	80/1	76/2	60/2	15/2

are generally limited to applications under 65°C. A further disadvantage of polyamide curing agents in certain applications is that they have a much darker color than polyamines.

The *Versamid* series of curing agents were the original and best known polyamide curing agents. Current commercial polyamide curing agents and their properties are shown in App. D. More recently developed polyamides provide lower viscosity, better compatibility with epoxy resins, and better cure profiles under adverse conditions.

5.3.2 Amidoamines

Amidoamine or polyamidoamine curing agents have reactivity with DGEBA epoxy resins that is similar to the polyamides. However, they are lower-viscosity products and are also lower in color. Amidoamines are derivatives of monobasic fatty carboxylic acids and aliphatic polyamines. Since the amidoamines have only one amide group per molecule, they are lower in molecular weight, viscosity, and amine functionality than the polyamides.

Amidoamines have noncritical mixing ratios with large curing agent/resin ratios similar to those of polyamide curing agents. Similarly, the physical properties of the cured product can be varied significantly by altering the mix ratio. Table 5.5 shows properties of DGEBA epoxy that is cured with varying concentrations of an amidopolyamine derived from tall-oil fatty acid and TETA.

Both the polyamide and amidoamine curing agents can be accelerated by the addition of a tertiary amine such as DMP-10, tris(dimethylamino methyl) phenol.

Some products in the amidoamine group are manufactured to contain a significant amount of imidazoline structures. This is accomplished by a high reaction temperature that converts the open amide structure to the cyclic imidazolin with loss of water. This conversion leads to lower viscosities since the concentration of the polar amide group is reduced.

Amidoamines exhibit very good adhesion characteristics, particularly to porous substrates such as concrete and wood. They also cure extremely well under humid conditions. They are much less corrosive than aliphatic amines and provide less skin irritation.

5.4 AROMATIC AMINES

Aromatic amines are widely used as curing agents for epoxy resins. However, they are not used as widely in adhesive formulations as they are in composites, molding compounds, and castings. They offer cured epoxy structures with good heat and acid resistance.

The primary advantages of aromatic amines over aliphatic amines for curing epoxy resins are the longer pot life as well as the development of higher heat resistance and greater chemical resistance. The major disadvantages are that they are solids at room temperature and generally require heat for processing as well as for cure. The added heat required for mixing and cure increases the dermatitis and toxicity potential by releasing irritating vapors. Modifications of aromatic amines are available that cure at room temperatures.

The lower reactivity of the aromatic amines in adhesive formulations is an advantage in that epoxy resin mixtures can be *B-staged* at room temperature (react to a glassy but fusible and thermoplastic intermediate structure) and will not fully cure for months. In this way, dry films and solid powders can be formulated as elevated-temperature curing, one-component adhesives with long shelf life.

The color of aromatic amines is poor (dark), and they stain easily. They are generally solid materials that require some formulating at elevated temperatures to produce a product that can be easily handled. The vapors resulting from elevated temperatures can cause staining, and their irritancy can be a problem. Certain aromatic amines such as diaminodiphenylmethane are carcinogenic.

When compared to aliphatic amines, aromatic amines generally have reduced exotherm and reactivity. Elevated temperatures are required to achieve optimum properties. In certain cases aromatic amines can be cured at room temperature with catalysts such as phenols, BF_3 complexes, and anhydrides.

Examples of aromatic amines are shown in Fig. 5.5. Of these compounds the most common are *meta*-phenylene diamine (MPDA), methylene dianiline (MDA), and eutectics of the two.

5.4.1 Metaphenylene Diamine

Metaphenylene diamine (MPDA) is one of the most common of the aromatic amines used to cure epoxies. This product is amber to very dark in color. It is a solid that melts at 65°C and is generally mixed with the epoxy resin at that temperature. The molten liquid or vapors from MPDA can stain the skin and nearby structures rather badly. The para-isomer is reported to be carcinogenic, but the meta-isomer is free from this disadvantage.

MPDA has four active hydrogens and is used stoichiometrically with DGEBA at 14.5 pph. Once it is in solution within an epoxy resin, the resulting mixture has excellent

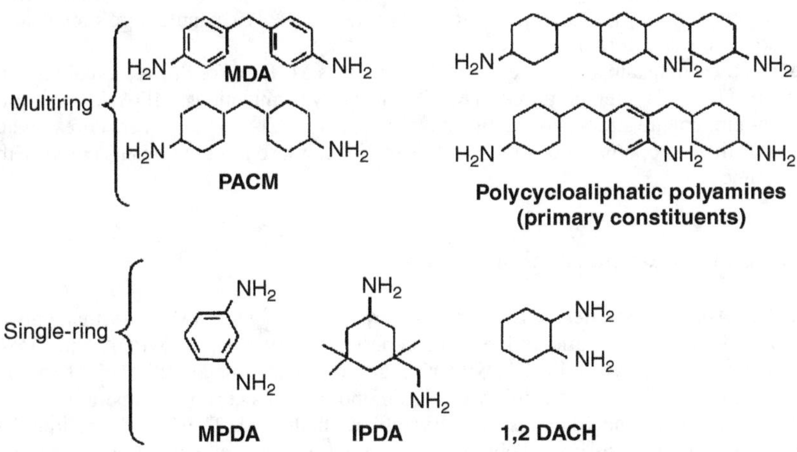

FIGURE 5.5 Chemical structures of common aromatic amine curing agents.

handling characteristics and low viscosity. There are three methods of mixing this curing agent with epoxy resins.

1. Heat both the MPDA and the epoxy resin to 65°C; the two components can then be blended.
2. Heat the resin to about 80°C and then, while stirring continuously, dissolve the curing agent into the resin.
3. Use technical-grade MPDA that can be heated to about 80°C and then supercooled slowly to room temperature to obtain a liquid, which is stable for at least 9 months at room temperature.

In all of the above cases, a considerable amount of objectionable staining and irritating fumes will be given off during the mixing operation. After mixing, the pot life will be about 6 h at 25°C.

Once the pot life is exceeded at room temperature, the MPDA–epoxy mixture will *B-stage.* This technique is used to produce dry filament winding materials (prepreg) and solid molding compounds. In adhesive compounding this technique can be used to produce one-component, dry adhesives in the form of solid stick, powder, or film. To cure the B stage, the product is exposed to temperatures in the range of 150 to 175°C, which causes the B-staged material to flow and then cure. The adhesive is then postcured at 175°C for optimal property formation. The B stage can also be dissolved in solvent and used to impregnate reinforcement of a carrier.

5.4.2 Methylene Dianiline

Methylene dianiline (MDA) is also a solid diaromatic amine. Similarly to MPDA, MDA is not often used in adhesive formulations because of the difficulty in compounding and curing practical epoxy formulations and the resulting brittleness of cured structures. MDA also has relatively low polarity so its adhesion properties would be suspect.

MDA has a melting temperature of about 90°C. It has four active hydrogens, and its stoichiometric mixing ratio with DGEBA epoxy resin is in the range of 27 to 30 pph. The resulting viscosity is somewhat greater for MDA–epoxy resin mixtures than for MPDA. The mixing and curing procedure is similar to that described above for MPDA except that higher temperatures are required. The pot life of MDA mixtures with DGEBA epoxy resins (20 h) are somewhat longer, and the reaction rates are slower than they are for MPDA. MDA B-stages epoxy resins similarly to MPDA, and is used to manufacture stick solder and other solid adhesive forms.

MDA does not stain skin or equipment as badly as MPDA. For this reason along with its lower cost, MDA is often preferred to MPDA in epoxy formulations. MDA does not provide the high-temperature strength or chemical resistance of MPDA for equivalent cure conditions. However, these properties are significantly superior to those of epoxy resins cured with primary amines.

5.4.3 Other Aromatic Amines

Aromatic Amine Eutectics. There are several curing agents available that consist of eutectics of various aromatic amines. These perform very much as MPDA and MDA do. However, the eutectics are liquids with viscosity of approximately 2000 cP at room temperature. They are readily miscible with liquid epoxy resins at room temperature.

The aromatic amine eutectics may crystallize on storage. They can be reliquefied by heating to 40°C with stirring. This liquefaction can be accomplished without sacrificing either the curing properties or the final physical and chemical properties of the cured resins.

Solvent Solutions. Certain solvent solutions of aromatic amines have been noticed to polymerize epoxy resins at room temperature.[11] The effect of the solvent is probably to allow sufficient mobility of the polymer chains for an adequate degree of crosslinking to occur before the viscosity becomes so high that the molecules are immobilized. The aromatic amine solutions are usually used with a cure accelerator to achieve practical cure rates at room temperature.

However, the incorporation of the solvent in the cured resin will significantly lower the glass transition temperature and thermal resistance. When cured at room temperatures, these solutions give properties more similar to those of the polyamide curing agents, but they do have the advantage of low viscosity and adjustable cure rate.

Diaminodiphenylsulfone. Diaminodiphenylsulfone (DADS) is another solid aromatic amine that is primarily used in elevated-temperature applications. DADS provides the best strength retention after prolonged exposure to elevated temperatures of any amine curing agent. The curing agent can be used with DGEBA epoxy resins of various molecular weights. It is used with higher-functionality solid epoxy resins (e.g., tetrafunctional resins of the epoxy novolac type) for maximum crosslink density, thermal resistance, and heat distortion temperatures. It has been reported that a mixture of 100 parts EPON 1031 (tetrafunctional bisphenol A) and 30 parts DADS, cured for 30 min at 180°C, will provide bond strength in excess of 1000 psi at 260°C.[12]

DADS melts at 135°C and is employed stoichiometrically with DGEBA at 33.5 pph. Fortunately, it is relatively unreactive so it can be mixed with epoxy resin at elevated temperatures. It can also be used in epoxy solutions to provide an adhesive formulation for manufacturing supported or unsupported film with long shelf life. Because of the low reactivity of the system, DADS is generally employed at a concentration that is about 10 percent greater than stoichiometry, or an accelerator, such as BF_3-MEA, is employed at about 0.5 to 2 pph. When DADS is mixed with liquid DGEBA resin, it provides a pot life of 3 h at 100°C and requires a rather extended high-temperature cure to achieve optimal physical properties.

5.5 ANHYDRIDES

After the primary amines, acid anhydrides are the next most important class of epoxy curing agents, although these are not used as often in adhesive systems as they are in casting compounds, encapsulants, molding compounds, etc. The most common types of anhydrides are hexahydrophthalic anhydride (HHPA), phthalic anhydride (PA), nadic methyl anhydride (NMA), and pyromellitic dianhydride (PMDA), although there are several others. Chemical structures of several anhydrides are illustrated in Fig. 5.6.

These compounds do not readily react with epoxy resins except in the presence of water, alcohol, or some other base, called an *accelerator*. Tertiary amines, metallic salts, and imidazoles often act as accelerators for anhydride cured epoxy systems. The reaction between acid anhydride and epoxy resins is illustrated in Fig. 5.7.

The reaction of anhydrides with epoxy groups is complex, with several competing reactions capable of taking place. The most significant reaction mechanisms are as follows:

1. The opening of the anhydride ring with an alcoholic hydroxyl forms the monoester.
2. Subsequent to the opening of the ring, the nascent carboxylic groups react with the epoxy to provide an ester linkage.
3. The epoxy groups react with nascent or existing hydroxyl groups, catalyzed by the acid, producing an ether linkage.

At low elevated-temperature cures, the ether and ester reactions take place at the same frequency. At higher temperatures the ester linkage is predominant. The presence of a catalyst

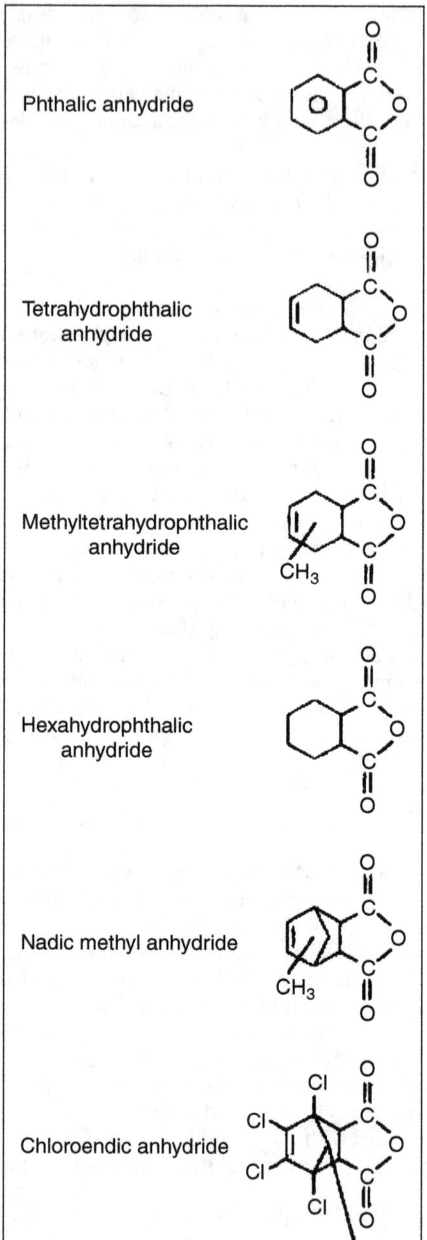

FIGURE 5.6 Chemical structures of acid anhydride curing agents.

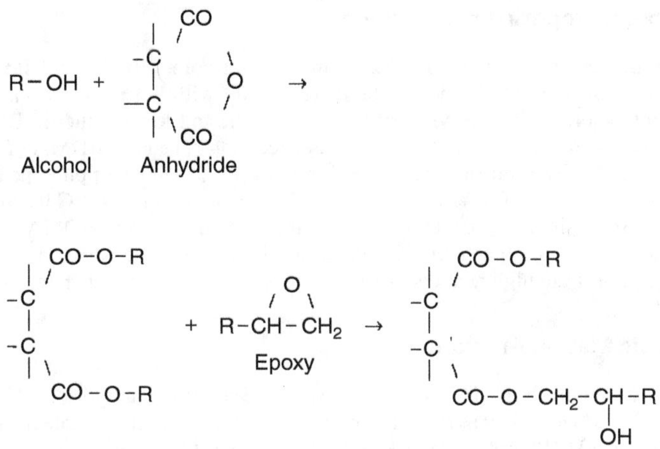

FIGURE 5.7 Anhydride epoxy reaction.

can change the balance of ester-ether linkages. The presence of ester linkages is believed to result in reduced elevated-temperature performance.

Liquid and solid anhydrides are used extensively to cure epoxy resins in such applications as casting, potting, and reinforced plastics. They are valued in these applications because of their relatively high heat distortion temperature, good physical properties, low exotherm, and long pot life when mixed in large masses. However, their use in adhesive formulations is generally limited except for high-temperature applications because of the rather slow reactivity, long and elevated-temperature cure requirement, and inferior adhesion compared to amine cured epoxies.

Anhydrides are sometimes used in epoxy adhesives to provide specific properties or to provide improvements in handling strengths. The most important anhydride in epoxy adhesive formulations is pyromellitic dianhydride (PMDA), which provides very high temperature properties.

The mix ratio of anhydride to epoxy resin is less critical than with amines and can vary from 0.5 to 0.9 equivalent of epoxy. The specific ratio is generally determined experimentally to achieve desired properties. Compared to aliphatic amine cures, the exotherm generated by anhydride cured epoxies is low. Elevated-temperature cures up to 200°C and postcures are required to develop optimal properties.

The high elevated-temperature cures are damaging to adhesive systems due to a mismatch in thermal expansion coefficient that can occur between the epoxy and the substrate. The difference in rate of expansion when returning to room temperature from the cure temperature can lead to significant internal stress within the adhesive joint, which results in poor adhesion.

The reactivity of the epoxy-anhydride reaction is slow; therefore, an accelerator is often used at 0.5 to 3 percent to speed the gel time and cure. Most often the accelerator is a tertiary amine, and the optimum concentration is dependent on the anhydride, the resin used, and the cure conditions. The accelerator concentration, like the anhydride concentration, is usually determined experimentally based on a specific set of end properties.

Anhydrides are hygroscopic materials and should not be allowed to remain exposed to the atmosphere for extended periods. Absorption of moisture from the air or from fillers causes hydrolysis of the anhydride to the acid. When used to cure epoxy resins, this moisture absorption results in variable pot life, reduced thermal resistance, and other problems. As a result, drying of fillers is particularly recommended for anhydride systems.

5.5.1 Hexahydrophthalic Anhydride

Hexahydrophthalic anhydride (HHPA) is a low-melting-point (36°C) solid. It is liquefiable at temperatures of 50 to 60°C and can be mixed easily with hot epoxy resins. The mixed resins are characterized by low viscosity, long pot life, and low exotherm. Because of its low reactivity HHPA is generally used with an accelerator, usually BDMA or DMP-30.

HHPA is generally used in a concentration between 55 and 80 pph depending on the nature of the epoxy resin. The viscosity is generally about 200 cP at 40°C when mixed with a DGEBA epoxy resin. A typical cure schedule for a 0.5 to 2 percent BDMA catalyzed system is 2 h at 80°C plus 1 h at 200°C. Typical of all the anhydride curing agents, the cured epoxy will demonstrate high heat distortion temperatures and excellent chemical resistance.

5.5.2 Nadic Methyl Anhydride

Nadic methyl anhydride (NMA) is the most versatile of all the anhydrides. NMA is a liquid of viscosity about 200 cP at room temperature, and it is readily soluble in epoxy resins. The mix ratio is 60 to 90 pph when used with a liquid DGEBA epoxy resin. At 60 pph and with no catalyst, the working life is about 2 months; and with the incorporation of 0.5 pph DMP-30 as an accelerator, the working life reduces to 4 to 5 days.

The balance of properties for the final crosslinked resin can be varied over a wide range by altering the resin/curing agent ratio, changing the type and concentration of the accelerator, and modifying the cure conditions. In general, the highest degree of crosslinking and hardness is obtained by using stoichiometric mixtures of NMA and long cure schedules at high temperatures. Decreasing the amounts of anhydride and cure temperature leads to an improvement in toughness but at a reduction in heat resistance.

Wide variations in cure schedule are possible. To attain the highest heat resistance, a cure of about 2 h at 220 to 260°C is required. A 90-pph NMA concentration plus an imidazole accelerator results in a more practical cure of 2 h at 80 to 100°C plus 4 h at 140 to 150°C.

5.5.3 Pyromellitic Dianhydride

Pyromellitic dianhydride (PMDA) is a solid having a melting point of 286°C. It contains two anhydride groups symmetrically attached to a benzene ring. Because of the compactness of the molecule, PMDA achieves very high crosslink densities and, therefore, high heat and chemical resistance. PMDA cured epoxy adhesives have a heat distortion temperature on the order of 280 to 290°C.

PMDA is insoluble in DGEBA at room temperature but quite soluble at elevated temperatures. However, it is very reactive in DGEBA, and mixing techniques must be carefully considered so as not to induce gellation during mixing. Several mixing techniques have been used for incorporating PMDA into epoxy resins.

- The reactivity of PMDA may be reduced by replacing a proportion of it with a monofunctional anhydride (usually maleic, but sometimes phthalic). This blend can then be mixed with epoxy resin at 70°C. With this technique, up to 65% PMDA may be used in the mixture; higher concentrations will produce impractically short pot lives.
- PMDA may be dissolved in acetone at reflux temperatures; the solution is stable for 7 days and can be used for prepreg manufacture.
- On interaction of PMDA with a glycol, a resin-soluble adduct is obtained.
- The PMDA can be suspended in liquid resin at room temperature; the elevated-temperature cure then promotes solution followed by reaction. With this method, reduced amounts of PMDA are generally used (0.4:1.0 to 0.5:1.0).

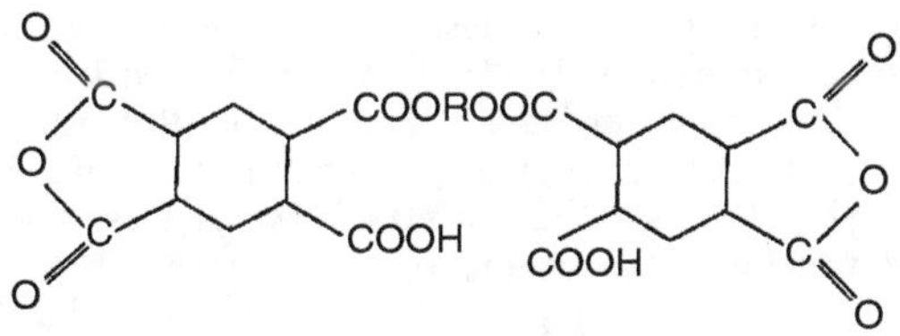

FIGURE 5.8 Idealized PMDA-glycol adduct.[13]

The final two techniques described above are generally used in the preparation of PMDA cured adhesives.

PMDA can be reacted with glycols to produce an adduct having the general structure shown in Fig. 5.8. These adduct resins form in the presence of solvent, under dry nitrogen. The reaction is continued until a clear solution is obtained. Such PMDA adducts are used in the formulation of high-temperature adhesive films.

A PMDA dispersion is prepared by mixing finely powdered PMDA into liquid epoxy resins at room or slightly elevated temperature by stirring. No noticeable settling will occur in resins having an initial viscosity greater than 5000 cP.

Because of its high functionality, PMDA can also be used in monoepoxy resins. These systems produce heat distortion temperatures on the order of 150°C. Cure times are relatively long, but may be accelerated by the addition of glycols or acidic accelerators.

5.5.4 Other Anhydride Curing Agents

Other acid anhydride curing agents are used for optimization of specific properties such as electrical strength. *Dodecyl succinic anhydrides* (DDSA) and adducts of DDSA with polyglycols give long pot life formulations with epoxy resins. When used to crosslink epoxy resins, they provide good heat resistance and excellent electrical properties. DDSA cured epoxies are also useful for bonding many plastics. They have been found to provide especially high adhesion to plastics such as polyethylene terephthalate film (e.g., Dupont's Mylar) as well as polycarbonate.[14]

5.6 CATALYSTS AND LATENT CURING AGENTS

Catalytic curing agents achieve crosslinking by initially opening the epoxy ring and causing homopolymerization of the resin. The resin molecules react directly with one another, and the cured polymer has essentially a polyether structure. The catalysts do not themselves participate in the epoxy polymerization reactions, as do the curing agents described above which provide a polyaddition reaction mechanism.

Therefore, catalysts merely act as an initiator and promoter of epoxy resins curing reactions. The amounts of catalyst used with epoxy resins are usually determined empirically and are chosen to give the optimum balance of properties under the required processing conditions. Generally, only several parts per hundred of catalyst is used with an epoxy resin. Excess amounts of catalyst can result in poor physical properties and degraded resin.

Catalysts can be used with epoxy resins in any of two primary ways:

1. As a sole crosslinker (no other curing agent)
2. As an accelerator in conjunction with another catalyst or with a curing agent such as polyamine, polyamide, or anhydride

This section reviews the function of a catalyst as a sole crosslinker in epoxy resin systems. Their function as accelerators is covered under the sections related to the specific curing agent.

Some catalysts, such as certain Lewis acids, are so reactive that they can provide extremely short gel times with epoxy resins at room temperature.[15] For example, BCl_3, can polymerize epoxy resins, resulting in gel times of less than 60 s. However, these reactive systems result in a very rigid adhesive with low peel strength properties and poor impact strength. As a result, less reactive catalysts are commonly employed in adhesive formulations.

The most popular catalysts for epoxy resins are tertiary amines, tertiary amine salts, boron trifluoride complexes, imidazoles, and dicyandiamide. Many of these catalysts provide very long pot lives (months) at room temperatures and require elevated temperatures for reaction with the epoxy groups. These catalysts are often referred to as *latent hardeners*.

5.6.1 Tertiary Amines

Tertiary amines are a type of Lewis base catalyst and are, perhaps, the most widely used catalyst. Two of the most widely used tertiary amines are

- DMP-10: tris-(dimethylaminomethyl) phenol
- DMP-30: *o*-(dimethylaminomethyl) phenol

The DMP designation comes from the original manufacturer of these chemicals, Rohm and Haas, and continues today with their manufacturer, Resolution Performance Polymers LLC. There are other tertiary amine catalysts such as benzyldimethylamine (BDMA), primarily salts of the above, and substituted imidazoles.

Generally, when used as a sole catalyst, tertiary amines are used only in specialty applications where short pot life can be tolerated and where maximum physical or chemical properties are not required. DMP-10 and DMP-30 are used at concentrations of 4 to 10 pph with liquid DGEBA epoxy resins. They achieve fairly fast cures overnight, even at room temperatures since the hydroxyl groups present in the epoxy molecule enhance the catalytic activity of the tertiary amine groups.

Tertiary amine salts of DMP-30 provide extended room temperature pot life (6 to 10 h at 20°C) when used at concentrations of 10 to 14 pph in liquid DGEBA epoxy resins. They cure at moderately elevated temperatures (4 to 8 h at 60°C), or even at room temperature with a heat bump. The acid moiety blocks the tertiary amine centers and deactivates them. The salt then dissociates on heating, freeing the amine groups, which are then able to react with the epoxy group.

The tertiary amine salts are claimed to provide epoxy formulations with very good adhesion to metal. The cured resins also show a hydrophobic effect when in contact with water or at high humidities. The strength, toughness, and elongation (4.7 percent) of the cured epoxy resin are very good. However, heat distortion temperature is only in the range of 70 to 80°C, and chemical resistance is relatively poor for an epoxy. The physical properties fall off rapidly with any rise in temperature.

Benzyldimethylamine (BDMA) is another tertiary amine that can be used as either a sole catalyst or an accelerator with other curing agents. It is used with DGEBA epoxy resins at 6 to 10 pph. The pot life is generally 1 to 4 h, and the cure will be complete in about 6 days at room temperature. When used by itself, BDMA can provide epoxy adhesive formulations with high-temperature resistance (Chap. 15). However, BDMA is mostly used as an accelerator for anhydride and dicyandiamide cured epoxy resins.

5.6.2 BF_3-Monoethylamine

Boron trifluoride monoethylamine (BF_3-MEA) is a Lewis acid catalyst. Lewis acids are electron pair acceptors that function as curing agents by coordinating with the epoxy oxygen, facilitating transfer of the proton (Fig. 5.9). BF_3-MEA is the only Lewis acid that has achieved broad commercial use in epoxy resin systems. BF_3-MEA is an effective catalyst for the polymerization of linear and cycloaliphatic epoxies as well as for the glycidyl ethers.

BF_3-MEA is a complex that is formed between boron trifluoride gas and monoethylamine. It is a solid melting at close to its dissociation temperature (80 to 85°C). BF_3-MEA shows hydroscopic tendencies. On exposure to moist air, it hydrolyzes into a viscous liquid that is unsuitable as a curing agent.

The BF_3-MEA must be melted and dissolved in liquid epoxy resins. For small batches, the procedure is to heat the resin to 85°C and stir in the curing agent. For larger batches, about 3 parts by weight of the catalyst is stirred into about 5 parts by weight of the resin preheated to 50°C. This produces a smooth paste, which then may be added to the remainder of the resin heated to 85°C. Alternatively, the BF_3-MEA can be dissolved in a solvent such as furfural alcohol, which will also dissolve the epoxy resin.

Although it is formed from a very reactive catalyst (BF_3 gas), the monoethylamine blocks the reactions sufficiently that BF_3-MEA can be considered to be a latent catalyst. It provides a pot life of 6 to 12 months at room temperature. It does not show significant curing activity until temperatures of 100 to 125°C have been reached.

In most formulations, the concentration of BF_3-MEA in DGEBA epoxy resins is on the order of 2 to 4 pph. Curing temperatures are usually 2 h at 105°C followed by a postcure at 150 to 200°C for 4 h for optimum properties. The rate of cure is very sensitive to temperature; below 100°C the rate is negligible, and at 120°C it is very rapid and accompanied by a significant exotherm.

The epoxy product cured with BF_3-MEA is densely crosslinked and has excellent physical properties at high temperatures (150 to 175°C). When reacted with an unmodified epoxy resin, the resulting product is very hard and brittle. The chemical resistance, however, is only fair and somewhat less than that of epoxies that are cured with aliphatic amines.

5.6.3 Imidazoles

The 2-ethyl-4-methyl-imidazole (EMI) is not a tertiary amine; however, it is used in the same manner as a single catalyst or as an accelerator. EMI is a substituted imidazole that is a liquid at room temperature (4000 to 8000 cP at 25°C) with a high boiling point.

EMI cured epoxy adhesive formulations are claimed to have outstanding adhesion to metals, and for this reason it is added as a co-curing agent in many compositions. It is an excellent anhydride accelerator providing higher thermal resistance than typical tertiary amine accelerators.

When used as a single catalyst in concentrations of about 10 pph, the mixed epoxy formulation shows a very low viscosity, which is ideal for the incorporation of high filler contents.

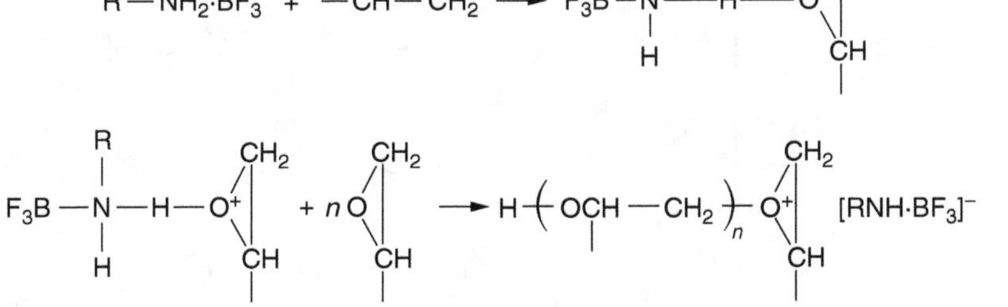

FIGURE 5.9 Reactions of BF_3-MEA with an epoxy resin.

The catalyzed resin has a pot life of 8 to 10 h at 25°C and a normal cure of 6 to 8 h at 60°C. EMI cures to a densely crosslinked structure with liquid DGEBA epoxy resins. These mixtures can cure at relatively low temperatures (60°C) or at higher temperatures (170°C) in very short times.

It has been recognized that the imidazole becomes permanently attached to the polymer chain in the epoxy curing reaction. Figure 5.10 suggests a possible reaction sequence.[16] The imidazole is thus an effective crosslinking agent, operating both through the secondary and tertiary amines.

Compared with other catalysts that homopolymerize epoxies, the imidazole offers improved thermal properties and retention of mechanical properties at more elevated temperatures. The cured resin has a heat distortion temperature between 85 and 130°C, which can be further increased by a postcure to about 160°C.

Latent imidazole catalysts have also been developed to provide cure rates considerably faster than those of dicyandiamide cured epoxy resins.[18] They also exhibit excellent adhesive characteristics and heat and chemical resistance.

A unique feature of these imidazole catalysts is that they do not have the high exotherm that dicyandiamide produces when cured in epoxy resins. Thus, they do not char or burn when exposed to high cure temperatures for fast cure. This is an important factor for adhesives that are cured via induction or dielectric heating. These adhesive systems are also much safer to ship via air freight than conventional dicyandiamide catalyzed epoxy formulations due to their low exotherm.

5.6.4 Dicyandiamide

Dicyandiamide (DICY) is a solid latent catalyst that reacts with both the epoxy terminal groups and the secondary hydroxyl groups. DICY has the advantage that it only reacts with the epoxy resins on heating beyond an activation temperature, and once the heat is removed, the reaction stops. It is widely used with epoxy resins where long shelf life (up to

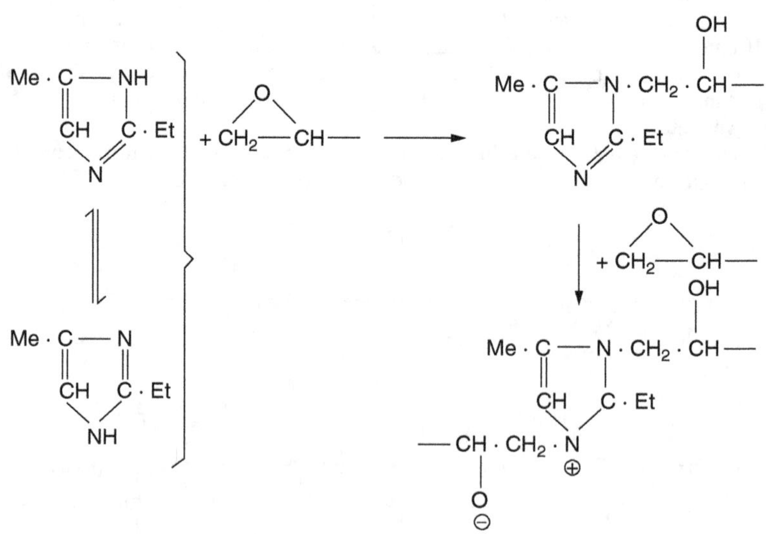

FIGURE 5.10 Imidazole reaction with epoxy resins.[17]

12 months) is required prior to curing. Significantly longer shelf lives can be obtained by storage under refrigeration until use.

As a result of the latency and excellent properties produced by DICY cured epoxies, DICY is used in many B-staged supported film adhesives. DICY is also probably the leading catalyst used in one-component, elevated-temperature curing epoxy adhesives.

DICY is considered a catalyst and polymerizes epoxy resin through the homopolymerization mechanism. But DICY has also shown behavior with epoxies that indicates some breakdown at cure temperatures to produce a curing agent that contributes to the polyaddition reaction mechanism.

The early reaction mechanism of DICY with epoxy resin consists of the epoxy reaction with all four hydrogen atoms on DICY and the epoxy-to-epoxy reaction that is catalyzed by the tertiary amines. The final curing mechanism is between hydroxyl groups in the partly cured resins and DICY cyano groups. This results in the disappearance of the cyano groups to form amino groups. This step is also catalyzed by tertiary amines.

DICY is used at about 5 to 7 pph of liquid epoxy resins and 3 to 4 pph for solid epoxy resins in adhesive formulations. It is generally ball-milled into the epoxy resin. DICY forms very stable mixtures with epoxy resins at room temperature because the catalyst is not soluble at low temperatures. However, on being exposed to temperatures greater than 140°C, the DICY becomes soluble in the epoxy resin, and cure progresses rapidly.

The particle size and distribution are important for maximizing the shelf life of epoxy–DICY systems. Generally optimum properties are produced when the particle size of the DICY is less than 10 μm. Usually fumed silica is used to keep the DICY particles in suspension and evenly distributed in the epoxy resin.

When formulated into one-component adhesive systems, the product is stable when stored for 6 months to 1 year at room temperature. It will then cure when exposed to 145 to 160°C for about 30 to 60 min. Since the reaction rate is relatively slow at lower temperatures, the addition of 0.2 to 1 percent benzyldimethylamine (BDMA) or other tertiary amine accelerators is common to reduce cure times or cure temperatures. Other common accelerators are imidazoles, substituted urea, and modified aromatic amine.

Substituted DICY derivatives have been developed to increase solubility and lower the required activation temperatures. These techniques can reduce the activation temperatures for DICY–epoxy resin mixtures to as low as 125°C.

Epoxy resins cured with DICY exhibit a good balance of physical properties with heat and chemical resistance. The glass transition temperature of a DGEBA liquid epoxy resin cured with 6 pph of DICY is on the order of 120°C, whereas an elevated-temperature curing aliphatic amine would provide a glass transition temperature of no greater than 85°C. Tougheners can be added to the adhesive formulation to achieve relatively high levels of peel strength and impact strength.

5.6.5 Other Latent Catalysts

A significant amount of development is currently occurring relative to latent catalysts because of interest in their long shelf life, high reactivity, and single-component adhesive formulations. Present technologies involve absorption of acidic or basic catalysts in molecular sieves, formation of Lewis acid salts or other amine salts, microencapsulation of amines, and other novel segregation methods.

5.7 Mercaptan and Polysulfide Curing Agents

Polymercaptans and polysulfides are aliphatic oligomers containing sulfohydro (–SH) groups that will react with epoxy groups at room temperature to form cured epoxy structures and epoxy adducts. A generalized structure is shown in Fig. 5.11.

$$R\!-\!\!\left[O-\!\left(C_3H_6O\right)_m CH_2CH(OH)-CH_2-SH\right]_n$$

$$HS-\!\left(C_2H_4OCH_2OC_2H_4-S-S\right)_n C_2H_4OCH_2OC_2H_4-SH$$

FIGURE 5.11 Generalized structure of polysulfide used in epoxy technology.

Polymercaptans, which cure at 0° to –20°C, are attracting attention in low-temperature curing adhesive formulation. At normal room temperature, polymercaptan has a pot life of 2 to 10 min and reaches handling strength in 10 to 30 min. An example of a typical mercaptan and its reaction sequence with an epoxy group is shown in Fig. 5.12.

An epoxy-polymercaptan reaction that is catalyzed with a tertiary amine is used in the standard two-component "5-min curing" epoxy which can be found in the hardware stores. These fast-curing products, however, have a tendency to be somewhat brittle and may perform quite poorly under peel stress. The standard 5-min cure is obtained with the accelerated mercaptan, such as Capcure 3830-81 (Cognis Corporation). The fastest polymercaptan has a gel time of 40 s in a 25-g mass.

The chemistry of epoxy/mercaptan systems involves the tertiary amine catalyst forming a salt with the mercaptan to generate a mercaptide anion, which is a strong nucleophile. The mercaptide will readily open the epoxy ring. Reaction with another mercaptan group can regenerate the mercaptide anion, as shown in Fig. 5.13.

The polymercaptans can also be used to accelerate the curing of epoxy resins systems blended with polyamines, amidoamines, or amines. The other curatives serve as the base to accelerate mercaptans, and the mercaptans react rapidly, generating the heat to accelerate the cure with the other hardener.

The major disadvantages of polymercaptan curing agents are their odor, skinning, and low heat deflection temperature. Progress has been made in the areas of odor and skinning through additives to the adhesive formulation. However, the low heat resistance is an artifact of the epoxy-mercaptan chemistry.

Polysulfide resin, on the other hand, does not have fast and low-temperature curing properties. It is used as more of a flexibilizer than a curing agent. However, polysulfide resins do have mercaptan groups, and these enter into the epoxy cure reaction.

Commercial liquid polysulfide resins are available with varying molecular weight from Toray (Japan). Several manufacturers have recently left the market, which has caused some concern regarding the availability and resulting price of these materials. The liquid polysulfides are mercaptan-terminated. The mercaptan end groups are acidic enough to react with epoxy resins, but usually an additional curing agent is employed in epoxy/polysulfide adhesive formulations.

The aliphatic chain contributes lower viscosity to the adhesive formulation and greater flexibility in the cured state. The reaction of these materials when used alone is very sluggish at room temperature. However, the reaction proceeds at a practical rate by the addition of a base, which acts as an accelerator. Tertiary amines, such as DMP-10 or 30, triethylenetetramine (TETA), and diethylenetriamine (DETA) are commonly used for this application.

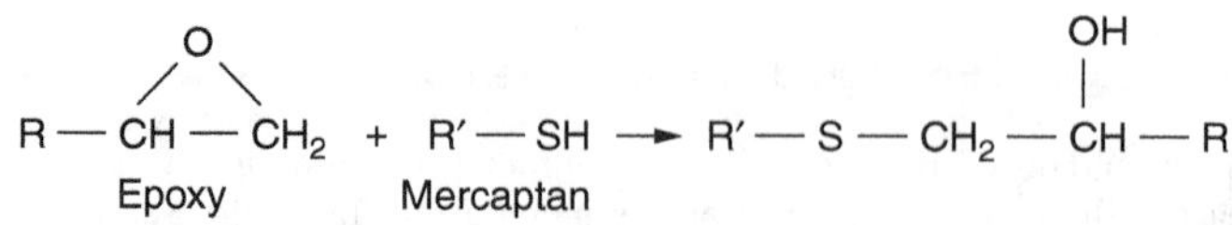

FIGURE 5.12 Reaction of mercaptan with an epoxy resin.[19]

• **Activation**

$$RSH + R'_3N \rightleftharpoons RS^- + R'_3N^+H$$

• **Propagation**

$$RS^- + \text{epoxy} \longrightarrow RSCH_2CH(O^-)CH_2OR''$$

$$RSCH_2CH(O^-)CH_2OR'' + RSH \longrightarrow RSCH_2CH(OH)CH_2OR'' + RS^-$$

FIGURE 5.13 Polymercaptan-epoxy chemistry accelerated with a tertiary amine.[20]

The tertiary amines and TETA are used at about 10 pph, and the liquid polysulfide is used at about 50 to 100 pph.

Polysulfides are typically used at ratios of 1:1 or less with epoxy resins, and can be used as co-curing agents with aliphatic amines. As the proportion of liquid polysulfide polymer to epoxy increases, the cured compound becomes softer and has a higher degree of elongation. Similarly, the tensile strength, heat resistance, and chemical resistance are reduced. Often it is better to think of these materials as co-resins or co-curing agents in hybrid systems.

Epoxy-polysulfide systems do not generate significant amounts of exotherm, and their cure rate does not rely on the exothermic reaction. Thus, epoxy-polysulfide compositions have a cure rate that is relatively insensitive to temperature, and they can cure at very low temperatures (below room temperature).

Stoichiometric quantities of aliphatic amine and 25 to 50 parts by weight of polysulfide will react with 100 parts by weight of epoxy resin to yield a relatively flexible product with good tensile strength at ambient temperatures. As a result of the excellent flexibility, epoxy-polysulfide resin systems have been used more as sealants and coatings than as adhesives systems. Because of their excellent adhesion to metals and glass and their cold-weather curing properties, they are often used in the construction industry as barriers to moisture penetration. Although the properties are quite good at room temperatures, some degree of flexibility is lost on thermal aging.

One of the advantages of having an adhesive or sealant with such a low degree of hardness is that the bond can easily be broken either by high shear stress or by simply cutting with a sharp instrument. Thus, it is easy to salvage expensive components from an assembly.

REFERENCES

1. Meath, A. R., "Epoxy Resin Adhesives," Chapter 19 in *Handbook of Adhesives*, 3d ed., I. Skeist, ed., van Nostrand Reinhold, New York, 1990, p. 350.
2. Cranley, P. E., "Epoxy Adhesives," *Paint and Coatings Industry*, April 1994, pp. 42–47.
3. Burgman, H. A., "Selecting Structural Adhesive Materials," *Electrotechnology*, June 1965.
4. *Epon Resin Structural Resin Manual—Additive Selection*, Resolution Performance Polymers LLC, Houston, Tx, 2001, p. 8.
5. Belm, D. T., and Gannon, J., "Epoxies," in *Adhesives and Sealants*, Engineered Materials Handbook, ASM International, Materials Park, OH, 1990.
6. Houwink, R., and Salomon, G., eds., *Adhesion and Adhesives*, Elsevier, New York, 1965, p. 247.

7. Belm and Gannon, "Epoxies," p. 97.
8. Allen, F.-J., and Hunter, W. M., "Some Characteristics of Epoxide Resin Systems," *Journal of Applied Chemistry*, October 1956.
9. Schonhorn, H., and Sharpe, L., "Surface Energetics, Adhesion, and Adhesive Joints II," *Journal of Polymer Science Part B: Polymer Letters*, vol. 2, no. 7, 1964, p. 719.
10. Schwartz, S. S., and Goodman, S. H., *Plastics: Materials and Processes*, van Nostrand Reinhold, New York, 1982, p. 357.
11. Potter, W. G., *Epoxide Resins*, Springer-Verlag, New York, 1970, p. 75.
12. Hopper, F. C., and Naps, M., Patent to Shell Chemical Co., U.S. Patent No. 2,915,490, 1959.
13. Lee, H., and Neville, K., *Handbook of Epoxy Resins*, McGraw-Hill, New York, 1968, p. 12.25.
14. Bolger, J. C., "Structural Adhesives: State of the Art," Chapter 3 in *Adhesives in Manufacturing*, G. L. Schneberger, ed., Marcel Decker, New York, 1983, p. 182.
15. Wright, C. D., and Muggee, J. M., *Structural Adhesives—Chemistry and Technology*, S. R. Harshorn, ed., Plenum Press, New York, 1986, p. 128.
16. Farkas, A., and Strohm, P. F., "Imidazole Catalysis in the Curing of Epoxy Resins," *Journal of Applied Polymer Science*, vol. 12, 1968, p. 159.
17. Potter, *Epoxide Resins*, p. 83.
18. Bolger, J. C., U.S. Patent to Amicon Corp., No. 4,066,625, 1978.
19. Meath, "Epoxy Resin Adhesives," p. 349.
20. Frihart, C., et al., "Less Odor and Skinning with Stabilized Mercaptans for Curing Epoxies," *Adhesives and Sealants Industry*, January 2001.

CHAPTER 6
SOLVENTS AND DILUENTS

6.1 INTRODUCTION

Solvents and diluents are used to lower the viscosity of epoxy resins systems either to permit easy compounding with other ingredients or to aid in application of the adhesive onto a substrate. Both solvents and diluents are low-molecular-weight liquid compounds that are chemically and physically compatible with epoxy resins and their curing agents. They differ primarily by their vapor pressure.

Solvents have a relatively high vapor pressure and will evaporate given a specific set of environmental conditions. Certain solvents will evaporate quickly at room temperature and atmospheric pressure, and others may require heating to elevated temperatures and pressures that are even lower than atmospheric.

Diluents have much lower vapor pressures and generally do not evaporate at ambient conditions. However, they do have a finite vapor pressure, and given the right set of conditions (time, temperature, and pressure) they will vaporize. Two distinct classes of diluents are used with epoxy resins: nonreactive diluents and reactive diluents. Reactive diluents will enter into the crosslinking reaction with the primary resin, and nonreactive diluents will not. Nonreactive diluents primarily act as low-molecular-weight plasticizers for the epoxy composition.

Epoxy adhesive formulations demand a great variety of solvents and diluents with a wide range of evaporation rates, solvent strengths, and dispersion powers. These variations are required due to (1) the many types of epoxy resins, curing agents, and possible organic additives that can be used within a formulation and (2) the many different possible methods that can be used to apply the epoxy to the substrate (brush, spray, trowel, etc.).

The choice of solvent or diluent is made with regard to the solubility of individual components and to the viscosity, drying times, and wetting characteristics required of the final product. All these properties affect the bond performance of the resulting epoxy adhesive formulation.

6.2 SOLVENTS

Solvents are employed to temporarily lower the viscosity of the epoxy system by imparting a degree of added mobility to the polymeric resins used in the formulation. They are employed for one or several of the following reasons:

1. To aid in dispersing, mixing, and wetting of components in the resin system at the formulation stage

2. To lower the viscosity to provide easier mixing of multicomponent adhesive systems and dispensing at the application stage
3. To liquefy solid epoxy resins and hardeners for application to a supporting carrier or for deposition as a film onto the substrate

The solvent used in any adhesive formulation may need to solubilize more than one component (e.g., resin, curing agent, or polymeric additive). Solvents are generally organic, and often a blend of solvents is necessary to achieve the required degree of solubility or to provide for certain processing conditions (e.g., drying time). Polar solvents are required with polar resins; nonpolar solvents, with nonpolar resins.

Water is also sometimes used as a solvent for water-soluble resins. In the case of epoxy resins, water is generally used to disperse epoxy particles in an emulsion. These waterborne epoxy adhesives are discussed in Chaps. 4 and 14.

When used in epoxy adhesive systems, solvents are generally employed for reducing the viscosity for formulation or application purposes. Once the adhesive is applied to the substrate, the solvent must evaporate prior to cure. Otherwise, bubbles or vapor pockets could form in the bond line, causing a physically weak joint with poor adhesion. The solvent in the adhesive formulation must not adversely affect the substrate to which it is applied. Plastics, elastomers, and polymeric foams are especially sensitive to certain solvents used in epoxy adhesives.

Examples of solvents typically used in epoxy adhesive systems are acetone, methyl ethyl ketone (MEK), toluene, xylene, glycol ethers, and alcohols. Generally, the efficiency of the solvent declines with increasing epoxy molecular weight (MW). Table 6.1 presents viscosity data for DGEBA (molecular weight 1000 to 3000) that is dissolved in various solvent systems at a concentration of 40% by weight. The viscosity versus resin concentration for several DGEBA solvent solutions is illustrated in Fig. 6.1.

Solvent blends are often used to provide for a specific evaporation rate and degree of solvency. These blends normally consist of combinations of fast-evaporating and slow-evaporating solvents. Fast-evaporating solvents are based on low-boiling-point ketones such as acetone or MEK. Slow-evaporating solvents are based on higher-boiling-point (lower-vapor-pressure) compounds.

"True" solvents are considered to be those solvents that provide resin solutions which can be diluted to infinity without resin precipitation. True solvents for epoxies include MEK, diacetone alcohol, methylcyclohexanone, and most glycol ethers and their acetates. Acetone, not a true solvent, can be used to prepare 40% solutions of high-MW DGEBA epoxy resin, but not a 20% solution. Aromatic solvents such as toluene and xylene, as well as simple alcohols, such as isopropyl and *n*-butyl alcohol, are not active solvents for DGEBA resins; however, they can be used in combination with other solvents to improve solubility.

The choice of solvent or solvent blend for a particular epoxy adhesive formulation will primarily depend on

1. The method of application (brush, spray, trowel, etc.)
2. The nature of the epoxy resins and any copolymerizing resins in the formulation
3. Any possible reaction between the solvent and constituents in the formulation
4. The required viscosity and solids content of the final product
5. The sensitivity of the substrate surface to the solvent or solvent blend

Because of these multiple criteria, the determination of a proper solvent balance is somewhat of an art. Improper selection of a solvent can cause problems such as vapor entrapment in the cured adhesive and poor film-forming properties. Many solvent systems

TABLE 6.1 Solubility Characteristics of DGEBA Epoxy Resins*

Solvent	DGEBA epoxy resin, molecular weight 1000	1500	3000
Ketones:			
Acetone			200
Methyl ethyl ketone	10	10	250
Methyl isobutyl ketone	10	125	1100
Diacetone alcohol			>1500
Isophorone			>1500
Esters:			
Ethyl acetate			>1500
n-Butyl acetate			1300
Cellosolve acetate	125	450	>1500
Ether alcohols:			
Methyl Cellosolve			>1500
Ethyl Cellosolve	125	450	>1500
Butyl Cellosolve			>1500
Ethyl carbitol			>1500
Butyl carbitol			>1500
Chlorinated solvents:			
Trichloropropane			>1500
Chloroform	220	1050	>1500
Mixed solvents:			
Toluene/acetone (1/1)			300
Toluene/methyl ethyl ketone (1/1)			350
Toluene/methyl isobutyl ketone (1/1)			600–900
Toluene/diacetone alcohol (1/1)	25	200	>1500
Toluene/isophorone (1/1)			>1500
Toluene/isopropyl alcohol (1/1)			>1500
Toluene/Cellosolve acetate (1/1)	10	300	>1500

*Values shown are viscosity, cP, for a 40% solvent solution.

can be formulated to a specific resin system. One typical solvent blend for a DGEBA with molecular weight of 950 consists of the following:

Xylene	32 pph
Methyl isobutyl ketone	32 pph
Cellosolve	32 pph
Cyclohexanol	4 pph

Solutions made from this solvent blend were found to provide good leveling and flow properties and resulted in a practical evaporation rate.[2]

By proper combination of solvents, it is possible to reduce the viscosity of a given resin system while preserving the solids content at a given percent. For instance, a typical formulation consisting of 70% DGEBA with 30% of solvent system A (below) provides a viscosity

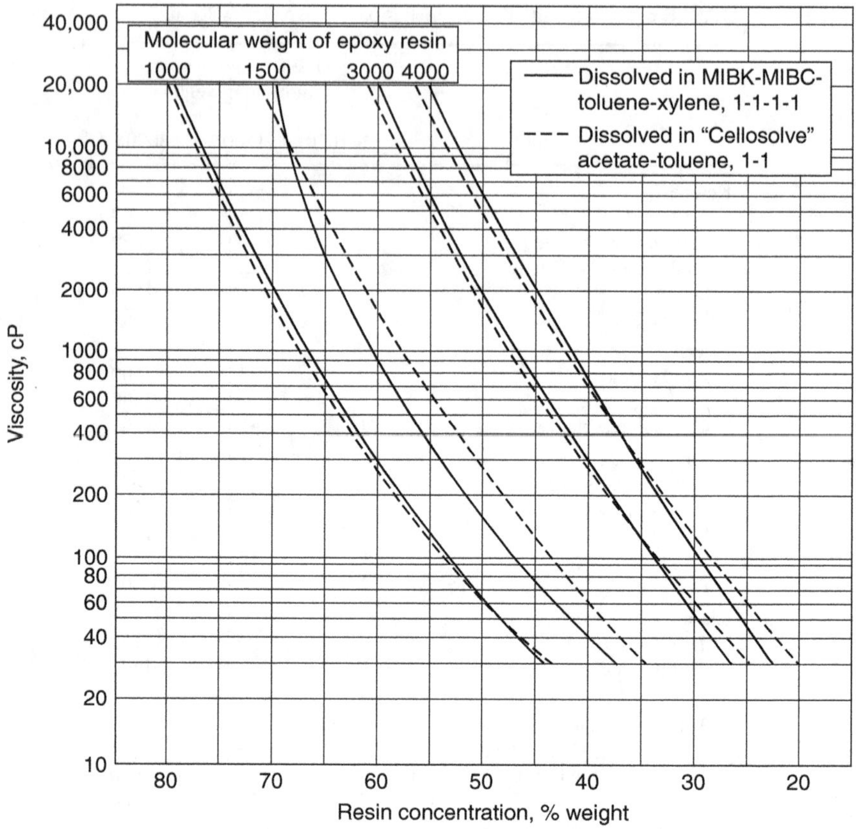

FIGURE 6.1 Viscosity of DGEBA resins in solution at 25°C.[1]

of 5000 cP that is suitable for brush application, whereas the same resin concentration with solvent system B provides a viscosity of 2400 cP that is suitable for spraying.[3]

Solvent system A		Solvent system B	
Methyl isobutyl ketone	32 pph	Methyl isobutyl ketone	45 pph
Cellosolve	33 pph	Butyl Cellosolve	5 pph
Xylene	34 pph	Toluene	50 pph

Even with proper solvent balance and using fast-drying solvents, it is probable that some solvent will remain entrapped in the cured resin unless the cure is preceded by an elevated-temperature exposure to eliminate the solvent. For example, MEK retention in amine cured DGEBA films was noticed after 9 days at room temperature.[4]

Some of the solvents that are commonly used in epoxy resins can present a flammability hazard and special health hazards. Contact with solvents will cause drying of the skin, which may result in an increased probability of skin irritation, especially when one comes in contact with curing agents. Solvents also have the ability to dissolve epoxy resin system components and carry them through the skin in liquid form or into the respiratory system in vapor form. The inhalation of solvent vapors or mist may cause respiratory irritation and problems with the central nervous system. Chapter 18 discusses the safety and health-related issues of epoxy formulation components including solvents.

The solvent industry has made significant strides in developing newer grades and blends of solvents for a variety of applications. "Safety" solvents are being developed that are low in volatility (vapor pressure), low in toxicity, and biodegradable. However, these newer solvents are finding commercial acceptance mainly as cleaning solvents rather than as a dilution medium for epoxy resins. For adhesives, rather than replace the solvent, the trend has been to develop waterborne emulsions.

With regard to cleaning solvents, the industry has developed solvents that are either biodegradable and/or environmentally compatible. New low-volatility solvents are taking the place of the older, less environmentally safe solvents in the adhesive and sealant industries. Substitute solvents, such as those shown below, manufactured by Inland Technology, Inc., have been found acceptable for some industrial applications.

Solvent	Low-volatility, nonchlorinated substitutes
Methyl ethyl ketone	Citra-Safe, EP 921
1,1,1-Trichloroethane	Teksol EP, X-Caliber
Perchloroethylene and petroleum solvents	Iso Prep
Toluene/xylene	Safety Prep

There is also an excellent web site, SAGE (http://clean.rti.org), that provides a comprehensive guide to pollution prevention information on solvent and cleaning process alternatives. The U.S. EPA Air Pollution Prevention and Control Division developed SAGE (Solvent Alternatives Guide).

The effect of solvent type on the curing rate of epoxy reactions has been well defined. Hydroxyl compounds, such as alcohols, act as catalysts and accelerate curing. However, these solvents are not serious competitors with amines for reacting with the epoxy ring. Water, functioning as a hydroxyl compound, also accelerates the reaction, even more than alcohols. Aprotic solvents, such as aromatic hydrocarbons or mineral spirits, have no effect on the amine-epoxy resin and behave as inert diluents. Carbonyl solvents, such as acetone and methyl ethyl ketone, retard the reaction.

Acceleration by the hydroxyl groups will affect the pot life, penetration, film formation, adhesion, and other critical properties. Figure 6.2 shows that the viscosity buildup for

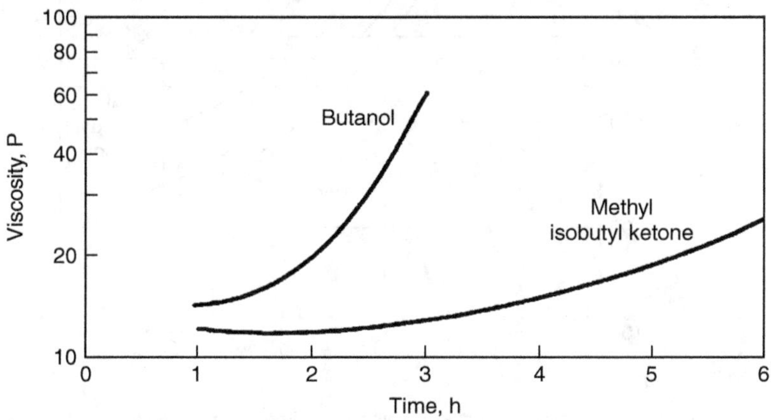

FIGURE 6.2 Effect of solvent type on the pot life of a hydrogenated bisphenol A diglycidyl ether cured with a polyamide.[5]

a solubilized epoxy formulation is much more rapid when the solvent is butanol than when it is methyl isobutyl ketone.

6.3 DILUENTS

Diluents are higher-MW components than solvents that are also added to the epoxy adhesive formulation to lower the viscosity and modify processing conditions. The primary function of a diluent in an epoxy resin formulation is to reduce its viscosity to make it easier to compound with fillers, to improve filler loading capacity, or to improve application properties. Solvents, certain curing agents, and flexibilized epoxy resins can also lower the viscosity of epoxy adhesive formulations, but this is not their primary function. The effect of various diluents on the initial viscosity of a diglycidyl ether of bisphenol A (DGEBA) epoxy resin is illustrated in Fig. 6.3.

Lower viscosity is important in applying the adhesive because it determines what type of mixers and dispensers are required and if the epoxy can be trowled, brushed, or sprayed.

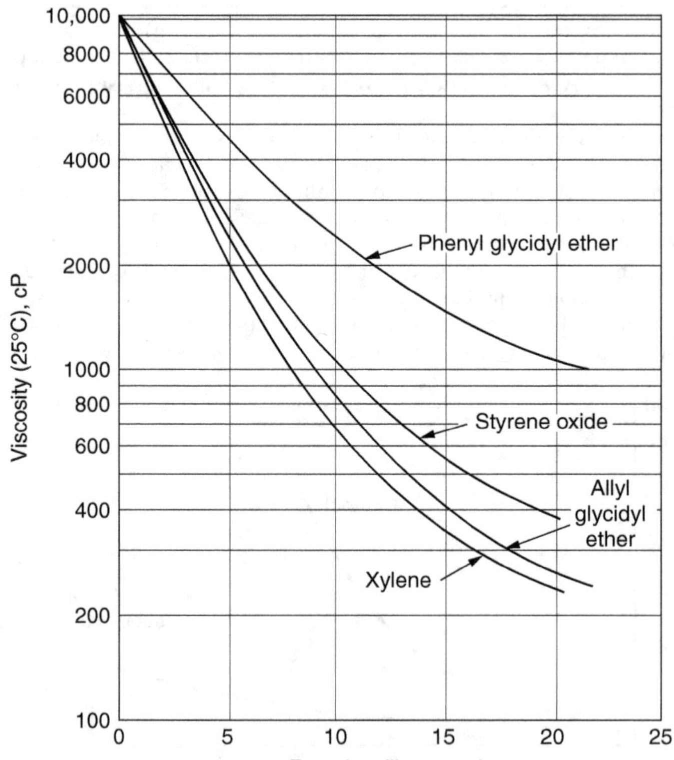

FIGURE 6.3 Effect of various diluents on the viscosity of a standard DGEBA liquid epoxy resin.[6]

Lower viscosity is also important in achieving good adhesion in that it allows greater penetration of porous substrates and faster wetting of the microroughness on nonporous surfaces. Diluents also increase the working life of the catalyzed epoxy system by (1) increasing the time that the mixture's viscosity is below a workable limit and (2) decreasing the reactivity of the curing agent primarily because of dilution of the resin.

Although most diluents decrease the reactivity of the epoxy system, reactive diluents may increase the exotherm because of the heat release of the high number of epoxy groups per gram in the diluent. Diluents containing alcoholic hydroxyl groups accelerate the curing rate in the presence of amine curing agents. However, when low percentages of diluent are used, these effects on reactivity and exotherm are generally minor.

Diluents will also affect the performance properties of the adhesive. Diluents generally lower the degree of crosslinking and degrade the physical properties of the cured epoxy. This reduction in crosslink density increases the resiliency of the adhesive, but it also reduces tensile strength as well as heat and chemical resistance. These effects are more pronounced at elevated temperatures than at room temperature. The degree of these effects will depend on whether the diluent has epoxy functionality (reactive diluents) or whether the diluent is incapable of reacting with the epoxy system (nonreactive diluents).

Reactive diluents are generally preferred over nonreactive diluents because they are chemically linked into the epoxy network. However, they still degrade the physical properties because their functionality is lower than that of the resin. Nonreactive diluents can be thought of as plasticizers. Because nonreactive diluents are relatively mobile, they can be more easily driven off on heating or vacuum degassing. If this occurs during cure, the result is greater shrinkage and appearance of vapor bubbles. This results in reduced adhesive strength because of the internal stresses created in the joint. Entrapped nonreactive diluents can also migrate out of the cured adhesive during service conditions, thereby causing a change in properties.

Both reactive and nonreactive diluents should be used sparingly if the properties of the cured system are to be preserved. An amount of 5 to 10 pph is best and generally provides a sharp reduction in viscosity. Concentrations greater than 20 pph are seldom used in adhesive formulations.

Diluents are generally not as much a problem as solvents in causing skin irritation, but they are low-viscosity, relatively high-vapor-pressure compounds and could lead to irritation in certain applications. Many of the diluents, especially those containing epoxy groups, are more severe skin irritants than the epoxy resins themselves. This is due to their lower molecular weight and high vapor pressures. The viscosity reduction capability (and skin irritation tendency) for the diluents is directly related to their molecular size. With the larger molecules the skin hazard potential is less, but so is the viscosity-reducing-efficiency.

6.3.1 Nonreactive Diluents

Nonreactive diluents do not react with the resin or curing agent and, therefore, generally dilute the final physical properties of the epoxy structure. In essence, they act as plasticizers to the epoxy network. In addition to lowering viscosity, nonreactive diluents are often used to balance the mix ratio proportions in certain epoxy systems. However, nonreactive diluents have not found wide acceptance in epoxy adhesive technology, primarily because most are incompatible with the cured resin and because similar effects can be achieved by proper selection of a long-chain curing agent or a reactive diluent.

High-boiling-point solvents such as xylene can be used as a nonreactive diluent; however, this is not normally done because of the high vapor pressure of the solvent and the probability that solvent remaining after cure would degrade the physical properties and adhesion. Thus, higher-MW organic compounds are more appropriate as nonreactive diluents in epoxy formulations.

Coal and pine tar are examples of common nonreactive diluents from natural substances. These are interesting nonreactive diluents because of their relatively low cost. They are often used as extenders in epoxy systems to reduce the cost. Coal tar is widely used because of its excellent compatibility with epoxy resins and relatively small sacrifice in cured properties. Nonyl phenol, furfural alcohol, and dibutyl phthalate are also common nonreactive diluents for epoxy systems. Dibutyl phthalate is also used as a plasticizer in many thermoplastics, such as polyvinyl chloride.

Since nonreactive diluents do not enter into the crosslinking reaction, they can be lost due to volatilization, especially when exposed to the elevated temperatures of the exotherm or curing cycle. If vaporization does occur, shrinkage of the adhesive film can result in internal stresses being generated within the joint. These internal stresses reduce the degree of adhesion that is realized on final cure.

Most nonreactive diluents are used in concentrations of 5 to 20 percent by weight of the epoxy resin. The general effect of incorporating nonreactive diluents is to increase the working life and decrease the peak exotherm. The effect on cured properties is generally negative, although at low additions the effect is relatively small. In the cured resin, nonreactive diluents will lead to an increase in adhesion and impact strength but a decrease in thermal and chemical resistance and tensile strength. Nonreactive diluents do not generally increase the flexibility or elongation of the cured resin systems, but they do tend to reduce tensile strength and hardness.

Dibutyl phthalate (Fig. 6.4) is a commonly used nonreactive diluent because it does not exhibit migratory tendencies on aging. It is generally incorporated into the DGEBA epoxy with heating. When it is used at about 17 pph, the viscosity of the resin can be reduced from 15,000 to 4000 cP. Dibutyl phthalate also provides added flexibility by virtue of its side chains and the resulting reduction in crosslinking density of the resin.

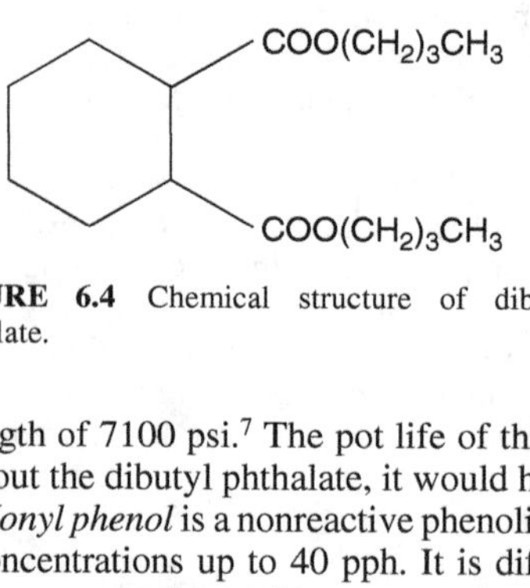

FIGURE 6.4 Chemical structure of dibutyl phthalate.

The improved flexibility results in improved adhesion and thermal shock resistance, but at the sacrifice of elevated-temperature performance. A TETA cured epoxy (EEW = 190) plasticized with 17 pph of dibutyl phthalate exhibited a heat distortion temperature of only 52°C and a tensile strength of 7100 psi.[7] The pot life of this system is 55 min at room temperature, whereas without the dibutyl phthalate, it would have been on the order of 20 to 30 min.

Nonyl phenol is a nonreactive phenolic diluent that can be added to DGEBA epoxy resins in concentrations up to 40 pph. It is different from dibutyl phthalate in that the phenolic groups can accelerate the epoxy amine curing reaction. Nonyl phenol is most commonly used with aliphatic primary amine and polyamide curing agents to reduce the viscosity of the system and accelerate the curing reaction. The gel time is reduced, and the exotherm increases with the addition of nonyl phenol to the DGEBA resin. Figure 6.5 shows the effect of nonyl phenol [(3-pentadecyl)-phenol] concentration on the viscosity and gel time of DGEBA epoxy resin. Similar to dibutyl phthalate, nonyl phenol reduces the tensile strength and hardness of the cured resin.

Some nonreactive diluents have been used to impart special properties on the cured epoxy in addition to lowering the viscosity of the uncured system. For example, *chlorinated diluents* have been used with antimony oxide to impart flame resistance to cured epoxy systems. A typical formulation of this type based on DGEBA employs about 15 pph chlorinated

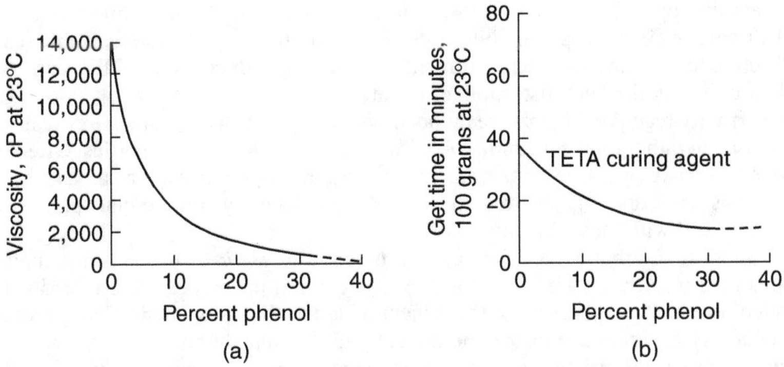

FIGURE 6.5 (*a*) Viscosity versus nonyl phenol concentration in DGEBA epoxy. (*b*) Gel time versus nonyl phenol concentration in catalyzed DGEBA epoxy.[8]

phenol and 5 pph antimony oxide to produce a self-extinguishing system. Another nonreactive diluent, *polymethyl acetal*, can be used to improve tensile shear strength in room temperature curing epoxy adhesive formulations.[9]

6.3.2 Reactive Diluents

Reactive diluents enter into the polymerization reaction of the epoxy resin and the curing agent. In this way the final adhesive characteristics are determined by the reaction product of the binder and the diluent. The most common reactive diluents used for epoxy adhesive formulations are shown in Table 6.2.

Most reactive diluents are mono- or difunctional. They are made by reacting epichlorohydrin with an alcohol, a phenol, or a polyol to produce a mono- or polyglycidyl ether resin. However, there are some nonepoxy diluents that are used as well. These nonepoxy diluents generally react with the curing agent or other functional groups in the epoxy chain.

In addition to viscosity reduction, the presence of a reactive diluent generally leads to a faster rate of cure and a higher crosslink density than with an undiluted resin. This is due

TABLE 6.2 Reactive Diluents for Epoxy Adhesives[10]

Diluent	Viscosity, cP at 25°C
Butyl glycidyl ether	<2
2-Ethylhexyl glycidyl ether	1–4
t-Butyl glycidyl ether	2–5
Phenyl glycidyl ether	4–7
o-Cresyl glycidyl ether	2–10
C12-C14 alkyl glycidyl ether	6–10
Diglycidyl ether of 1, 4 butanediol	14–18

to the fact that by lowering the viscosity the molecules become more mobile in the early stages of cure. This increased mobility results in a more reactive system. Table 6.3 shows the effectiveness of various reactive diluents as viscosity reducers for DGEBA epoxy resins and their effect on the heat distortion temperature.

The primary reactive diluents are monoepoxy low-molecular-weight epoxy resins. These may be used at rather high concentration with little effect on cured properties. Often they are employed to make selective improvements on certain properties such as adhesion, thermal cycling resistance, and impact strength. Some lower-viscosity commercial epoxy resins are already reduced with these diluents.

The monofunctional epoxy diluents are essentially *chain stoppers* since they inhibit crosslinks from forming. The extent to which the cured properties are affected is directly dependent on the concentration of the diluent added to the epoxy resin. The general effect is to reduce viscosity and improve the impact and thermal shock resistance while slightly reducing the thermal resistance. The thermal expansion of the cured resin is increased, as it is also with nonreactive diluents. This can lead to internal stress on the bond line depending on the thermal expansion of the substrate material.

Monofunctional epoxy diluents are used primarily with DGEBA epoxy blends. The most common monofunctional diluents are butyl glycidyl ether and phenyl glycidyl ether. The effect of butyl glycidyl ether and other reactive diluents on the viscosity of epoxy resin is shown in Fig. 6.3. Because the monofunctional diluents reduce crosslink density, they are used at relatively low levels to avoid degrading heat and chemical resistance or other properties of the adhesive.

The reaction of polyols with epichlorohydrin produces polyglycidyl ethers, such as butanediol diglycidyl ether. These are only slightly less effective viscosity reducers than butyl glycidyl ether, but they are less volatile and contribute less to worker irritation problems. They also produce less loss in physical properties.

Difunctional epoxy diluents are low-viscosity, low-MW epoxy resins. These diluents may be used at very high concentrations, and they do not greatly affect the properties of the cured resin. In some instances the difunctional epoxy diluents will actually improve properties. These materials include butadiene dioxide, vinyl cyclohexane dioxide, and diglycidyl ether of resorcinol.

Long-chain polyether polyols, such as polypropylene glycol, when reacted with epichlorohydrin produce a diepoxy resin with an internal polyether chain that serves both to

TABLE 6.3 Viscosity Reducing Power of Reactive Diluents and Their Effect on Heat Distortion Properties[11]

Diluent	Amount, wt%, required to achieve 800 cP*	Heat distortion temperature, °C, for 800-cP mixture using the following curing agents†	
		m-Phenylenediamine	Diethylenetriamine
None	—	143	118
Butyl glycidyl ether	11	101	74
Diglycidyl ether	38	128	104
Phenyl glycidyl ether	18.5	105	82
Butadiene dioxide	10	162	127
Vinyl cyclohexene dioxide	20	148	108

* Amount required to dilute original DGEBA epoxy resin of 12,400 cP.
† Stoichiometric amount, cured 4 h at 150°C.

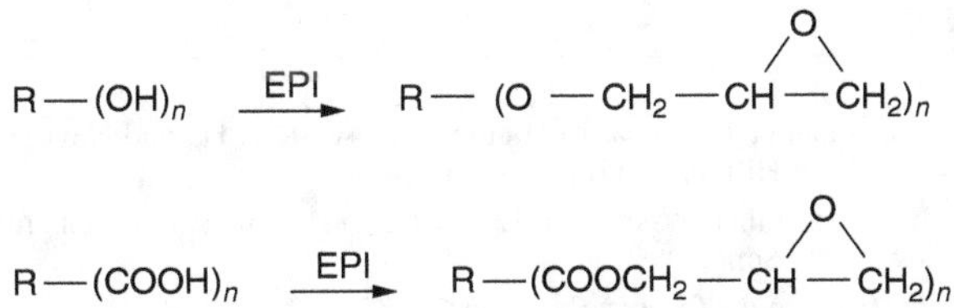

FIGURE 6.6 Top: Reaction of a polyol with epichlorohydrin to produce a mono- or polyglycidyl ether resin. Bottom: Reaction of an acid with epichlorohydrin to form glycidyl ethers.[12]

decrease viscosity and to increase flexibility. Other reactive diluents that provide flexibility to the cured epoxy structure can be synthesized from reacting epichlorohydrin with an organic acid or polybasic acid to form glycidyl esters. The reactions of a polyol and acid with epichlorohydrin to form a reactive diluent are illustrated in Fig. 6.6.

Since many of the difunctional epoxy diluents are added to epoxy adhesive formulations as flexibilizers as well as viscosity reducers, they are discussed in greater detail in Chap. 8 with other flexibilizing additives.

There are also several reactive diluents that do not contain epoxy groups. These are represented by triphenyl phosphite and γ-butyrolactone (Fig. 6.7). The phosphite is a low-viscosity colorless liquid, which is sensitive to moisture. It reacts with hydroxyl groups in the resin. The γ-butyrolactone is a very effective viscosity reducer. It can reduce the viscosity of a liquid DGEBA from about 15,000 to 2000 cP with only 10 pph. In the curing reaction with amines (Fig. 6.8), the lactone forms an amide, which can then crosslink with the polymer via the hydroxyl groups.

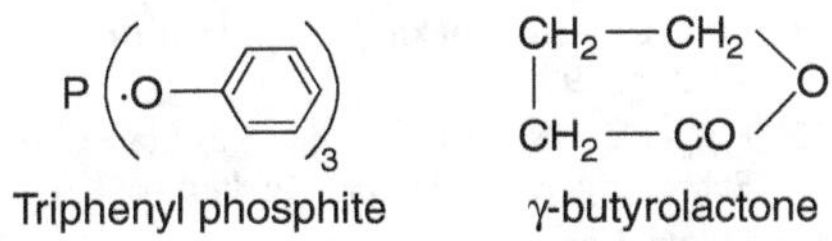

FIGURE 6.7 Chemical structures of triphenyl phosphite and γ-butyrolactone.

Other nonepoxy-containing reactive diluents may be co-curing agents, unsaturated molecules, active hydrogen-containing molecules, materials that are capable of promoting transesterification with available hydroxyl groups, and materials capable of reacting with the curing agents. These will affect the physical properties by influencing the crosslink density and system functionality.

It should be pointed out that diluents are not the only way to lower the viscosity of filled epoxy resins systems. Surface active agents can also be added to the system. They provide better wetting of the filler by the epoxy resin matrix. This can lead to substantial viscosity reduction for systems having equivalent filler concentration. The surface active agent, in turn, could also be used to produce formulations with higher filler loading at equivalent viscosity. These surface active agents are discussed in Chap. 10.

FIGURE 6.8 Reaction of γ-butyrolactone with amine.[13]

REFERENCES

1. Shell Chemical Company, Technical Literature. Also see Lee, H., and Neville, K., *Handbook of Epoxy Resins*, McGraw-Hill, New York, 1967, p. 24.30.
2. Alter and Sooler, "Molecular Structure of Epoxy Resin Polymers as a Basis for Adhesion," *ACS Symposium*, New York, September 1957.
3. Lee and Neville, *Handbook of Epoxy Resins*, p. 24–32.
4. Dannenberg, H., et al., "Infrared Spectroscopy of Surface Coatings in Reflected Light," *Analytical Chemistry*, March 1960.
5. Bauer, R. S., "Formulating Weatherable Epoxy Resin for Maximum Performance," *Waterborne and Higher Solids Coating Symposium*, New Orleans, 1982.
6. Behm, D. T., and Gannon, J., "Epoxies," in *Adhesives and Sealants*, vol. 3, Engineered Materials Handbook, ASM International, Materials Park, OH, 1990.
7. Potter, W. G., *Epoxide Resins*, Springer-Verlag, New York, 1970, p. 110.
8. Lee and Neville, *Handbook of Epoxy Resins*, p. 13.6.
9. Russel, D. H., Epoxy Resin Composition, U.S. Patent 3,050,474, 1962.
10. Meath, A. R., "Epoxy Resin Adhesives," in *Handbook of Adhesives*, Skeist, I., ed., 3d ed., van Nostrand Reinhold, New York, 1990, p. 355.
11. May, C., and Tanaka, Y., *Epoxy Resins: Chemistry and Technology*, Marcel Dekker, New York, 1973, p. 299.
12. Bolger, J., "Structural Adhesives State of the Art," Chapter 7 in *Adhesives in Manufacturing*, G. L. Schneberger, ed., Marcel Dekker, New York, 1983, p. 172.
13. Potter, *Epoxide Resins*, p. 115.

CHAPTER 7
HYBRID RESINS

7.1 INTRODUCTION

A variety of polymers, both thermosets as well as thermoplastics, can be blended and coreacted with epoxy resins to provide for a specific set of desired properties. The most common of these are nitrile, phenolic, nylon, polysulfide, and polyurethane resins. At high levels of additions these additives result in hybrid or alloyed systems with epoxy resins rather than just modifiers. They differ from reactive diluents in that they are higher-molecular weight-materials, are used at higher concentrations, and generally have less deleterious effect on the cured properties of the epoxy resin.

These hybrid epoxy adhesives are generally used for demanding structural applications such as in the aerospace industry where the optimal properties from each component are desired. For example, epoxy is generally used to provide good adhesion and processing characteristics. They are blended with the following resins to provide additional improvements in the properties noted.

Phenolic—high-temperature resistance

Nylon—toughness and peel strength

Polysulfide—elongation

Vinyl—high resiliency and peel strength

Nitrile—toughness and peel strength, chemical resistance

Polyurethane—peel and impact strength

Hybrid resins have been used to improve the flexibility, thermal shock resistance, elongation, heat distortion temperature, and impact strength of unmodified epoxy adhesives. However, there can also be some sacrifice in certain physical properties due to the characteristics of the additive. These alloys result in a balance of properties, and they almost never result in the combination of only the beneficial properties from each component without carrying along some of their downside.

These blends can take a number of different forms. The added resin may be reacted with the epoxy resin, or it may be included as an unreacted modifier. The modifier may be blended into a continuous phase with the epoxy resin (epoxy alloys) or precipitated out as a discrete phase within the epoxy resin matrix (as is generally done in the case of toughening modifiers). Epoxy hybrid adhesives are often used as film (supported and unsupported) or tape because of the ease with which formulated systems can be dissolved into solvent and applied to a carrier or deposited as a freestanding film. Some systems, notably epoxy-urethanes and epoxy-polysulfides, can be employed as a liquid or paste formulation because of the low-viscosity characteristics of the components.

TABLE 7.1 Common Epoxy Hybrid Adhesive Systems

Hybrid resin	Characteristics
Epoxy-nylon	Films are blends containing 30 to 50% by weight of epoxy resin. The nylon constituent provides high tensile shear strength as well as high peel strength. Suitable catalysts are dicyandiamide and aromatic polyamines. These adhesives have useful properties at low temperatures but have only moderately good high-temperature properties.
Epoxy-phenolic	Epoxy-phenolic adhesives are generally used in aerospace applications requiring high shear strength at temperatures in excess of 150°C. Usually the phenolics are a resole-type, and often the epoxy is a minor component. These adhesives are relatively brittle and have low peel and impact strengths.
Epoxy-nitrile	Nitrile-epoxy adhesives are composed of solid epoxy resin modified with *carboxyl-terminated butadiene nitrile* (CTBN) copolymer. The CBTN is introduced into the epoxy resin at elevated temperatures. The modification provides toughness and high peel strength without sacrificing heat and chemical resistance. The film adhesives are widely used in the aerospace industry in the construction of jetliners.
Epoxy-vinyl	Vinyl-epoxy adhesives have moderate strength at temperatures up to 150°C. Oxidation stability is excellent. Vinyl constituents increase the toughness and peel strength. The adhesive is often used for bonding safety glass, aerospace engine structures, and structural panels.
Epoxy-polysulfide	Epoxy-polysulfide adhesives and sealants have a very high degree of elongation but poor tensile strength compared to unmodified epoxies. They are used primarily as sealants or coatings in the building and construction industries.
Epoxy-polyurethane	Epoxy-polyurethane adhesives have a great deal of toughness. They have high elongation and very good peel strength and impact properties. They also have a relatively high degree of tensile strength, given their elongation. These adhesives bond well to flexible substrates such as plastics. They also bond well to oily surfaces. The main market for epoxy-polyurethane adhesives is automotive.

TABLE 7. 2 Range of Bond Strengths (Measured at Room Temperature) of Various Hybrid Adhesives

Hybrid adhesive	Tensile shear strength, psi	T-peel strength, lb/in
Vinyl-phenolic	3000–4500	15–35
Nitrile-phenolic	3000–4500	15–60
Epoxy-phenolic	2000–3200	6–12
Epoxy-nylon	5500–7200	80–130
Epoxy-nitrile	3700–6000	22–90

The most common types of epoxy hybrid adhesives are

1. Epoxy resins that are toughened with elastomeric and nonepoxy resins
2. Alloyed blends consisting of epoxy-phenolic, epoxy-nylon, and epoxy-polysulfide adhesives

The first group, resins that are used primarily to toughen epoxy adhesive systems, is described in Chap. 8. This chapter focuses on the resinous modifiers that are used as alloy blends. Characteristics of commercially available epoxy alloy adhesives are presented in Table 7.1. Tensile shear and peel strengths that are typical of these hybrid adhesives are compared in Table 7.2.

7.2 EPOXY-NITRILE (SINGLE-PHASE)

This section focuses on the modification of epoxy resins by blending with acrylonitrile butadiene (nitrile) resins. These are true alloyed blends since the nitrile rubber usually contains no groups that are normally reactive with epoxy groups. The nitrile molecules and the epoxy molecules intermingle as a blend to provide a single-phase alloy. If a large elastomer concentration is used, no phase separation will occur to form precipitates.

These single-phase hybrids are very different from the two-phase toughened epoxy-nitrile adhesives that are discussed in Chap. 8. These two-phase adhesives have redefined structural adhesives to a great extent and have opened the door to many applications that were previously not possible because of the epoxy resin's inherent rigidity. The polymer mixtures that exist as separate phases provide significant increases in toughness but have only a small improvement in elongation at typical use levels.

The single-phase epoxy-nitrile adhesive is made by choosing an epoxy resin–curing agent combination in which the nitrile rubber is soluble. Therefore, after cure the adhesive is a clear, rubbery, single-phase resin. The nitrile component is introduced into the epoxy resin at elevated temperatures. The blend is cured with standard epoxy curing agents and catalysts at elevated temperatures. Nitrogen-containing curing agents are generally used with these modifiers because they catalyze the reaction of the carboxyl group with the epoxy functionality.

The final product is generally an adhesive solution or an adhesive film. The film is widely used in the construction of commercial jetliners where a high degree of peel strength and good resistance to moisture is required. The cured hybrid generally has a Shore A hardness of 75 and an elongation of over 100 percent.

The epoxy-nitrile adhesives were introduced commercially in the late 1960s. They consisted primarily of DGEBA epoxy resin modified with carboxyl-terminated butadiene nitrile (CTBN) rubber. These first nitrile copolymers were available from B.F. Goodrich under the trade name of Hycar. The most convenient form of epoxy nitrile adhesive, especially when one is bonding large parts (aircraft structures), is a supported film. However, solvent solutions of epoxy-nitrile adhesives have also been commercially available.

Table 7.3 shows the effect of adding a nitrile rubber to a rigid epoxy formulation. There is significant improvement in toughness and elasticity in the cured bond line, which translates into high tensile shear and peel strength. As a result, nitrile-modified epoxy adhesives have found application in bonding airframe structures and other applications where a combination of high tensile strength with peel and impact resistance is required.

The improvements in peel strength compare favorably with the improvements provided by epoxy resins that are blended with nylon (discussed below). One advantage of the nitrile-epoxy

TABLE 7.3 Comparison of Bond Strengths of Epoxy Adhesive Formulations with and without Nitrile Rubber Addition[1]

	Nitrile-modified epoxy adhesive		Unmodified epoxy adhesives	
Test temperature, °C	Tensile shear strength, psi	T-peel strength, lb/in	Tensile shear strength, psi	T-peel strength, lb/in
–40	4900	48	2100	2
–18	5100	47	2500	3
–4	5200	44	2700	3
23	3600	22	2200	4
38	2800	21	3000	5

formulations is that they provide a higher degree of peel strength retention at subzero temperatures. The nitrile-epoxy adhesives also provide better resistance to moisture.

However, the single-phase epoxy nitrile adhesive achieves its high peel strength by bulk elongation. Other properties, such as heat and chemical resistance, are generally degraded as they are when other flexibilizers or plasticizers are added to the epoxy. The two-phase version of this adhesive hybrid solves many of these problems.

7.3 EPOXY-PHENOLIC

Epoxy-phenolic adhesives are made by blending epoxy resins with phenolic resins to improve the high-temperature capabilities of the standard epoxy resins. Developed in the early 1950s, they were the first "high-temperature" epoxy adhesives to become commercially available.[2,3]

Epoxy-phenolic adhesives were developed primarily for bonding metal joints in high-temperature applications. Their first major application was to join major aircraft components. They are also commonly used for bonding glass, ceramics, and phenolic composites. Because of their relatively good flow properties, epoxy phenolics are also used for bonding honeycomb sandwich composites.

These adhesives are generally based on blends of solid epoxy resins with resole-type phenolic resin. The epoxy resin component is often not the predominant component in the blend, depending on the end properties required. Phenolics are compatible with epoxy resins and will react through the phenolic hydroxyl group. The amount of phenolic resin used is generally much greater than that required to crosslink with the epoxy, so one can debate whether (1) the epoxy toughens the phenolic adhesive or (2) the phenolic increases the heat resistance of the epoxy.

Adhesives based on epoxy-phenolic blends are good for continuous high-temperature service in the 175°C range or intermittent service as high as 260°C. They retain their properties over an extended temperature range, as shown in Fig. 7.1. Shear strengths of up to 3000 psi at room temperature and 1000 to 2000 psi at 260°C are possible. Resistance to weathering, oil, solvents, and moisture is very good. However, because of the rigid nature of the constituent materials, epoxy-phenolic adhesives have relatively low peel and impact strength and limited thermal-shock resistance.

These adhesives are available as pastes, solvent solutions, and film supported on glass fabric. The adhesive films generally give better strengths than do liquid systems. Cure requires a temperature of 175°C for 1 h under 15- to 50-psi pressure. Most adhesives of this

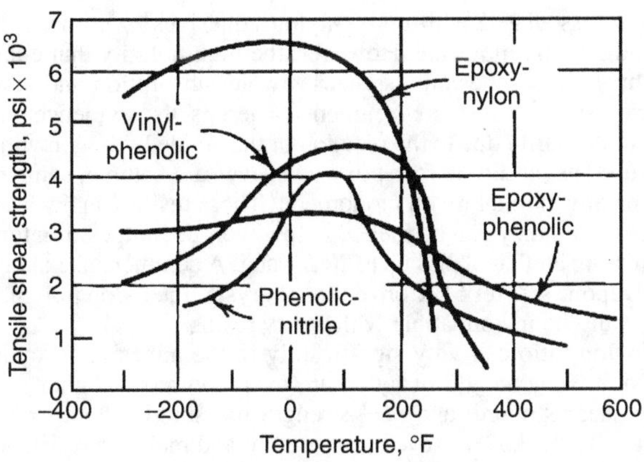

FIGURE 7.1 The effect of temperature on the tensile shear strength of modified epoxy-phenolic compared to other hybrid adhesives (substrate is aluminum).[4]

type have limited storage life at room temperature and are generally stored under refrigerated conditions.

Typical compositions of epoxy-phenolic adhesives that have been commercially available over the years are shown in Table 7.4. Note that additives such as quinolate or gallate act as chelating agents to inhibit the reaction with iron in metallic substrates, such as stainless steel, at high temperatures. The current trend in formulating epoxy-phenolic adhesive is to increase the epoxy content for greater toughness and toward using acid accelerators for faster cure. The epoxy-phenolic adhesives are composed with other high-temperature epoxy adhesives in Chap. 15.

7.4 EPOXY-NYLON

Epoxy-nylon adhesives, introduced in the 1960s, were one of the first structural adhesives designed specifically to have high shear strength and extremely high peel strength. These characteristics are achieved by blending epoxy resins with copolymers from the polyamide (or nylon) resin family which is noted for toughness and tensile properties.

TABLE 7.4 Epoxy-Phenolic Adhesive Compositions of Commercially Available Types[5]

FPL 878		EPON 422J		EPON 422	
Component	pbw	Component	pbw	Component	pbw
EPON 1007 DGEBA	20	EPON 1001 DGEBA	33	EPON 1004 DGEBA	75
Resole phenolic	160	Resole phenolic	67	Resole phenolic	25
Solvent	20	Aluminum powder	100	H_3PO_4	1
Hydroxy-napthanoic acid	3	Copper quinolate	1		
n-Propyl gallate	1.5				

The toughness associated with epoxy-nylon adhesives is primarily due to hydrogen bonding within the nylon molecule. However, the high polarity that contributes to a high degree of toughness and adhesion also causes water absorption and a significant loss of strength from exposure to moist environments. There is also evidence that the toughening action of the nylon is partly due to the reaction of the amide hydrogen with the epoxy ring.[6]

The nylons used as modifiers for epoxy adhesives are soluble, semicrystalline copolymers made from conventional nylon monomers.[7] These resins can be dissolved in alcohols and other solvents, and they can be melted below their decomposition temperature. A preferred solvent mixture is ethanol plus up to 20% water. A commercial example of an adhesive-grade nylon is Dupont's Zytel 61. Conventional crystalline nylon polymers, such as nylon 6 or nylon 66, would be incompatible with epoxy resins.

The epoxy/nylon ratio can vary significantly in the adhesive formulation, but blends containing 30 to 50% by weight of nylon are most common. Higher ratios of nylon will increase tensile shear strength and peel strength measured at temperatures from –55 to 80°C, as shown in Table 7.5. However, temperature and moisture resistance will eventually be degraded by high nylon concentration.

Epoxy-nylon resins are one of the best materials to use in film and tape adhesives for applications where the service environment is not severe. Their main advantages are better flexibility and a large increase in peel strength compared with unmodified epoxy adhesives. Epoxy-nylon adhesives offer both high tensile shear and peel strengths. Epoxy-nylon adhesives can provide tensile strength greater than 5000 psi and peel strength greater than 50 lb/in. In addition, epoxy-nylon adhesives have good fatigue and impact resistance. They maintain their tensile lap shear properties at cryogenic temperatures, but the peel strength is poor at low temperatures.

Epoxy-nylon adhesives are limited to a maximum service temperature of 85°C, and they exhibit poor creep resistance. Possibly their most serious limitation is poor moisture resistance because of the hydrophilic nylon (polyamide) constituent.[9] The degradation by exposure to moisture occurs with both the cured and uncured adhesives.

Not only is low tensile shear strength noticed on moisture aging, but also the mode of failure changes from one of cohesion to adhesion. Table 7.6 shows the effect of humidity and water immersion on an epoxy-nylon adhesive compared to a nitrile-phenolic adhesive. Substrate primers have been used with epoxy-nylon adhesives to provide improved moisture

TABLE 7.5 Epoxy-Nylon Adhesive Composition[8]

Adhesive composition			Tensile shear strength, psi, at temperature			Peel strength, MIL-A-25463, lb/in, at temperature		
Nylon	Epoxy	Triazine	–55°C	23°C	83°C	–55°C	23°C	83°C
0	100	24	2690	2855	3500	3.0	3.0	2.0
20	80	19	2690	2950	3400	3.7	4.5	6.0
40	60	15	4850	4800	4200	14	49	71
50	50	12	6100	5700	4900	10	78	102
70	30	7.3	7240	6420	5200	26	135	119
85	15	3.6	8250	5740	4850	37	180	143
95	5	1.2	7560	4125	2700	18	145	82

Adhesive was prepared by dissolving nylon (Zytel 61), epoxy resin (ERL 2774), and curing agent (2,4-dihydrozino-6-methamino-S-triazine) in methanol/water/furfural alcohol solvent solution. The adhesive was applied to acid-etched aluminum, the solvent was allowed to evaporate, and cure was 60 min at 175°C.

TABLE 7.6 Effect of Humidity and Water Immersion on the Tensile Shear Strength of Epoxy-Nylon and Nitrile-Phenolic Adhesives*,[10]

	Epoxy-nylon adhesive		Nitrile-phenolic adhesive	
Exposure time, months	Humidity cycle†	Water immersion	Humidity cycle†	Water immersion
0	4370	4370	3052	3052
2	1170	2890	2180	2740
6	950	1700	2370	2280
12	795	500	2380	2380
18	1025	200	2350	2640
24	850	120	2440	2390

*Substrate is etched aluminum.
†Humidity cycle of 93% RH between 65 and 30°C with a cycle time of 48 h.

resistance. However, they do not achieve the moisture resistance associated with unmodified epoxy or nitrile-epoxy adhesives.

Epoxy-nylon adhesives are generally available as unsupported film or as solvent solutions. Adhesive films are made either by solution casting or by dry mixing and calendering. Common solvent solutions are blends of methanol, water, and furfural alcohol. A curing agent is 2,4-dihydazino-6-methylamino-S-trazine. Other suitable catalysts are dicyandiamide and aromatic polyamines. Typical formulations are shown in Table 7.7.[11]

A pressure of 25 psi and temperature of 175°C are required for 1 h to cure the epoxy-nylon adhesive. Because of their excellent filleting properties and high peel strength, epoxy-nylon adhesives are often used to bond aluminum skins to honeycomb core in aircraft structures. In these applications, climbing drum peel strengths in excess of 150 lb/in have been achieved.

TABLE 7.7 Effect of LP-3 Polymer on the Physical Properties of Liquid Epoxy Resin[13]

	Formulation, pbw						
Component	A	B	C	D	E	F	G
Epoxy resin (EEW 175 to 210)	100	100	100	100	100	100	100
LP-3	—	25	33	50	75	100	200
DMP-30	10	10	10	10	10	10	10
Physical properties after curing 7 days at 25°C							
Tensile, psi	3500	5500	6500	7200	2075	2350	150
Elongation, %	0	1	2	5	7	10	300
Shore D hardness	80	80	80	80	76	76	15
Coef. of thermal expansion, in/(in · °C) × 10^5	4.5	5.5	6.0	7.5	10.0	13.5	15.0
Impact resistance, ft · lb	2	1	3	5	27	70	100

7.5 EPOXY-POLYSULFIDE

Polysulfide compounds that are compatible with epoxy resins are liquid elastomers at room temperature. The most significant commercial resin of this type is LP-3 from Toray. The predominant product is a mercaptan-terminated liquid polymer (LP) that contains approximately 37% bound sulfur (see Fig. 7.2). It is the high concentration of sulfur linkages that provides these products with their unique chemical properties. A sulfur odor is noticeable during processing, making ventilation important.

Polysulfides can be cured by themselves with an oxidizing agent as a catalyst. They can also be used to cure epoxy resins (see Chap. 5); however, the rate of cure is very slow for a practical adhesive. Thus, polysulfide resins are generally added into epoxy formulations as a modifier to increase flexibility. In applications where maximum flexibility is required, the level of polysulfide may be greater than the epoxy resin present in the formulation.

For adhesive systems, the liquid epoxy resins most widely used with LP-3 polymers are liquid unmodified and diluent-modified bisphenol A resins and liquid blends of bisphenol A and bisphenol F resins. Solid bisphenol A, multifunctional, and aliphatic diepoxy resins have also been used. Ratios of liquid polysulfide polymer to epoxy are in the range of 1:2 to 2:1. The effect of various degrees of polysulfide on cure properties of a DGEBA epoxy is shown in Table 7.7. An increase in elongation and impact strength is the result of increased amounts of the liquid polysulfide polymer.

The addition of the polysulfide resin acts as both an elastomeric modifier and a diluent for the epoxy resin. The low viscosity of LP-3, for example, can drastically reduce the viscosity of the overall formulation. This provides greater ease of mixing and application and the ability to be applied and sprayed without a solvent.

Suitable curatives for the polysulfide-epoxy reaction include liquid aliphatic amines, liquid aliphatic amine adducts, solid amine adducts, liquid cycloaliphatic amines, liquid amide-amines, liquid aromatic amines, polyamides, and tertiary amines. Primary and secondary amines are preferred for thermal stability and low-temperature performance. Not all amines are completely compatible with polysulfide resins. The incompatible amines may require a three-part adhesive system. The liquid polysulfides are generally added to the liquid epoxy resin component because of possible compatibility problems. Optimum elevated-temperature performance is obtained with either an elevated-temperature cure or a postcure.

Polysulfide resins combine with epoxy resins to provide adhesives and sealants with excellent flexibility and chemical resistance. These adhesives bond well to many different substrates. Tensile shear strength and elevated-temperature properties are low. However, resistance to peel forces and low temperatures is very good. Epoxy polysulfides have good adhesive properties down to –100°C, and they stay flexible to –65°C. The maximum service temperature is about 50 to 85°C depending on the epoxy concentration in the formulation. Temperature resistance increases with the epoxy content of the system. Resistance to solvents, oil and grease, and exterior weathering and aging is superior to that of most thermoplastic elastomers.

Liquid polymer:

$$HC(C_2H_4OCH_2H_4SS)_xC_2H_4OCH_2OC_2H_4SH$$

Curing with epoxy resin:

$$2\,R-\underset{\diagdown O \diagup}{CH-CH_2} + HS-R'-SH = R-\underset{\displaystyle OH}{CH}-CH_2-S-R'-CH_2-\underset{\displaystyle OH}{CH}-R$$

FIGURE 7.2 Conventional polysulfide structure and curing mechanism.[12]

TABLE 7.8 Typical Formulations and Properties of Polysulfide-Epoxy Adhesives[14]

	Formulation, pbw		
Component	A	B	C
DGEBA epoxy resin (EPON 828)	100	—	100
ZL-1612 Epoxy-terminated polysulfide	—	100	—
LP-3 Liquid polysulfide	—	—	75
Tertiary amine	10	7	—
Cycloaliphatic amine	—	—	46
Property			
Tensile strength, psi	7230	6980	4530
Elongation, %	3	4	10
Hardness, Shore D	82	82	74

The epoxy-polysulfide adhesive is usually supplied as a two-part, flowable paste that cures to a rubbery solid at room temperature. These systems can be heavily filled without adversely affecting their properties. Epoxy polysulfides are often used in applications requiring a high degree of elongation. Typical formulations are shown in Table 7.8 for adhesive systems formulated with polysulfide as a modifier and for a system where an epoxy-terminated polysulfide is employed. These adhesives are generally used to bond concrete in floors, roadways, and airport runways. Other principal uses include sealing applications, bonding of glass, potting, and bonding of rubber to metal.

7.6 EPOXY-VINYL

Epoxy resins may be blended with certain vinyl polymers to improve the impact strength and peel strength of the adhesive. Polyvinyl acetals, such as polyvinyl butyral and polyvinyl formal, and polyvinyl esters are compatible with DGEBA epoxy resins when added at concentrations of 10 to 20% by weight. The addition improves the resulting impact resistance and peel strength of the cured adhesive. However, temperature and chemical resistance are sacrificed by the addition of the low-glass-transition-temperature vinyl resins.

The solubility of the vinyl polymer in the epoxy resins will depend on the molecular weight and grade of vinyl used. Polyvinyl esters can be dissolved in liquid epoxy resins that are heated to about 100°C or in melted solid epoxy resins. However, most epoxy-vinyl blends are normally prepared via solvent solution.

The formulated adhesives are generally available as films or solvent solutions. They are commonly used as laminating adhesives for film or metallic foil because of their high peel strength. A composition consisting of a plasticized polyvinyl chloride copolymer and an epoxy resin can be cured with an aliphatic polyamine, which will crosslink by reacting with both resins. This adhesive possesses excellent adhesion to metals.

7.7 EPOXY-URETHANE

Epoxy-urethane adhesives provide properties when cured that are similar to those of epoxy-nylon adhesives except they offer a major improvement in moisture resistance. Isocyanate monomers and prepolymer react with the hydroxyl groups on epoxy resins to give tough, flexible hybrids. These adhesives have been noted for their excellent adhesion to many high-energy plastic substrates and for their propensity to bond to oily metal substrates.

The reactivities of isocyanates and epoxy resins are similar in that both can be cured with amines or other hydrogen-containing compounds. There have been many attempts to marry the rigidity, heat, and chemical resistance of the epoxies with the toughness and peel strength of the polyurethanes. However, these attempts have often resulted in adhesive systems that display the worst properties of the parent resins, rather than the best properties. This is due primarily to the significant different reaction rates between the –NCO groups on the isocyanate and the epoxy groups on the epoxy molecule with common curing agents. The common reaction mechanism is shown in Fig. 7.3.

The isocyanate reaction is several orders of magnitude faster than the epoxy reaction. Hence, attempts to cure mixtures of epoxy and isocyanate resins must provide for essentially complete consumption of all –NCO groups before any of the epoxy rings can react.

Another toughened epoxy formulation comes from mixtures of epoxies and urethane oligomers with pendant epoxy groups.[15] Curing of the pendant epoxy groups unites the urethane and nonurethane components through conventional epoxy reactions to give tough, durable films. Tensile elongation of the films can vary from 15 percent to more than 100 percent compared to about 2 percent for most unmodified cured epoxies. This approach, however, results in a decrease in high-temperature strength and chemical resistance although 500- to 1000-psi shear strength is achievable at temperatures in the 80 to 100°C range. Straight polyurethane adhesives show much poorer strength at temperatures in this range.

These reactive hybrid epoxy-urethane adhesives were developed initially for bonding to oily cold-rolled steel, but they have given good results on other substrates as well. It is believed that the oil contaminant on the substrate is adsorbed into the uncured adhesive and acts as a plasticizer. Table 7.9 shows tensile shear values obtained after curing 20 min at 177°C on various untreated substrates. The epoxy-urethane adhesives resist hydrolysis and give useful bond strengths up to about 100°C.

Another approach for formulating a high-peel-strength epoxy-urethane adhesive has been developed by Hawkins.[17] Toluene diisocyanate is first reacted with a mixture of long-chain polyether triols plus short-chain diols to obtain a viscous liquid resin terminated with hydroxyl groups. This resin is then blended with a diepoxy resin, dicyandiamide curing agent, fillers, and diluents to give a one-component adhesive. When cured for 60 min at 180°C, the dicyandiamide reacts with the epoxy and catalyzes the reaction between the epoxy and the –OH groups on the modifying resin. T-peel strength on aluminum is claimed to increase from 5 lb/in to over 50 lb/in via the addition of the urethane modifier. The increase in peel strength is accompanied by only a minor decrease in heat distortion temperature.

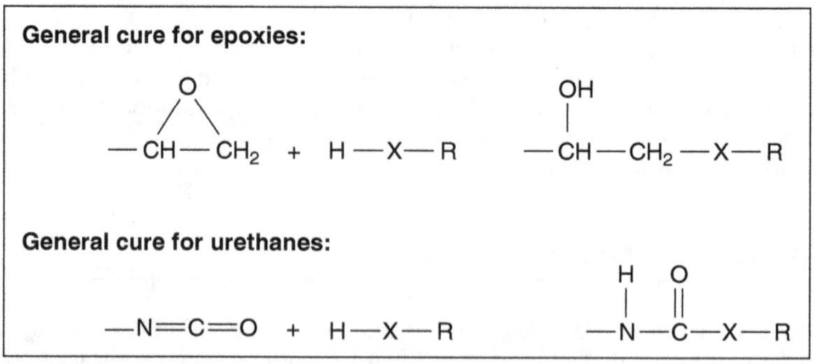

FIGURE 7.3 Common amine reaction mechanisms for epoxy and polyurethane resins.

TABLE 7.9 Cured Tensile Shear Strength of Uncleaned Substrates Bonded with Epoxy-Urethane Hybrid System[16]

Substrate	Substrate thickness, mils	Tensile shear strength, psi
Oily steel	58	3400
Galvanized steel	75	4770
Polyester painted steel	18	1440
Painted steel/galvanized steel	18/75	1760
Electrogalvanized steel	28	1930
Organic coated steel	33	2160
Sheet molding compound (polyester/glass)	100	660*
Aluminum	65	3340

*Failure of the substrate.

Some more recent examples of epoxy-urethane hybrids in one- and two-component structural adhesives are noted below. These have had success in the automotive industry where high impact strength, fatigue resistance, and high bond strength to a variety of substrates are valued characteristics.

Blocked isocyanate prepolymers have been mixed with epoxy resins and cured with amines.[18,19] These blocked prepolymers will initially react with the amines to form amine-terminated prepolymers that crosslink the epoxy resin. Urethane amines are also offered commercially for use with epoxy resins to develop hybrid adhesive systems.[20]

Novel, toughened one-component epoxy structural adhesives based on epoxy-terminated polyurethane prepolymer incorporating an oxolidone structure were developed to provide improved toughness, fracture resistance and adhesive properties with good chemical and moisture resistance.[21] The hybrid resin cures with a standard latent curing agent/accelerator.

An acrylate-terminated polyurethane modified epoxy compound and a polyethylene polyamine homologue and fatty acid combination were formulated into a two-component adhesive system. The adhesive is useful for bonding various thermoplastic resins such as ABS, PC, PBT-PC blends, and PPO.[22]

A latent curable, urethane modified epoxy resin was found to provide high shear and peel strengths on metals. The modified epoxy is first reacted with an isocyanate-terminated urethane prepolymer and an acidic phosphorus compound.[23]

7.8 OTHER HYBRIDS

The number of possible hybrid systems that can be manufactured with epoxy resins is nearly infinite, and many adhesive formulations have been attempted in a quest to improve the main disadvantages of a cured epoxy: brittleness and rigidity.

Functionalized, liquid polybutadiene derivatives have also been developed as hybrid flexiblizers for epoxy resins. Carboxyl-terminated butadiene/acrylonitrile polymers, butadiene homopolymers, and maleic anhydride–amino acid grafted butadiene homopolymers have been used as flexibilizers to impart good low-temperature strength and water resistance to DGEBA-based epoxy adhesives. An epoxy system toughened by polybutadiene with maleic anhydride is claimed to provide a hydrophobic backbone, low viscosity, softness, and high tensile strength and adhesion (Table 7.10).

TABLE 7.10 Polybutadiene-Toughened Epoxy System[24]

	Component	Grade, supplier	Parts by weight*,†
Part A	Polybutadiene	Ricon 130MA20, Sartomer	83.5
	Maleic anhydride	Lindride 12, Lindau Corp.	16.1
	Catalyst	Byk A500, Byk Chemie	0.2
	Catalyst	Byk A515, Byk Chemie	0.2
Part B	Epoxy resin	DER 331, Dow	100
Part C	Dialkymethylamine (accelerator)	DAMA 1010	1.3% of total

*The material is mixed 2:1, for part A : part B, and then the DAMA 1010 is mixed into the formulation.
†Properties:
Hardness = 75 Shore D
Tensile strength = 2650 psi
Lap Shear strength = 9350 psi
Glass transition temperature = 119–127°C

Epoxy-PVC plastisols are a type of PVC plastisol adhesive that is used in large quantities in the automobile industry. It is used to bond sheet steel to inner stiffener panels and to seal around the crimped panel edges. These adhesives are formulated as high-solids, thixotropic pastes and are applied as discrete dots or droplets to the stiffener surface or panel edge before joining or crimping. These adhesives are called *Hershey drops* in the trade because of the characteristic geometry of the droplets.

The epoxy-PVC plastisol type is a mixture of a plastisol-grade PVC powder, primary PVC plasticizers (e.g., dioctyl phthalate), a liquid DGEBA epoxy resin, thickeners, stabilizers, surfactants, and other additives. The epoxy serves as a secondary plasticizer, acts as a stabilizer (acid scavenger), and helps to fortify the plastisol by crosslinking during cure.

These adhesives require relatively high cure temperatures (above 130°C). In the past this had not been a problem, because the adhesive cured in the paint bake cycle. However, with newer paint systems and efforts to reduce energy consumption, lower-cure-temperature adhesives are needed, and epoxy-urethane adhesives and other flexibilized adhesives are beginning to replace the plastisols.

The epoxy-PVC plastisol adhesives are soft and flexible after cure. They provide 50 to 100 percent elongation and tensile shear strength of about 400 to 700 psi on steel. Higher bond strengths are generally not needed in the auto applications because of the large bond area. The important properties in addition to low cost are long-term toughness and adhesion as well as adhesion to oily steel.

Fluoroepoxies have gained interest because of the unique adhesion properties that can be provided by the fluorine groups. There have been several attempts to marry the properties of epoxy resins with those of fluorocarbon resins. In general these have focused on adhesive systems that (1) have a lower surface tension than unmodified epoxy or (2) have significant hydrophobicity to resist exposures in moist environments.

A fluoroepoxy compound was made from a fluorodiepoxy resin cured with an amine adduct. The use of fluoroepoxies is most advantageous in bonding fluorocarbons such as polytetrafluoroethylene. Good bond strengths are achieved on fluorocarbons without the need to surface-treat the substrate because the surface tension of the fluoroepoxy adhesive is reduced from about 45 to 33 dyn/cm.[25]

Fluoroepoxy resins have also been developed that when cured with silicone amines or fluoroanhydrides show substantial hydrophobicity. As the resin's fluorocarbon content is increased, so is the hydrophobicity. Adhesives made with these compounds showed much less degradation in glass transition temperature as a function of exposure to high-moisture environments.

Epoxy-silicone hybrid resins (Fig. 7.4) have generally been developed for use in the molding of microelectronic packages. These resins have moderate strength but exhibit a degree of elongation of about 60 percent at break.

Certain adhesive formulations have been developed by reaction of silicone polymers containing amine termination with epoxy resins. The resulting products exhibit good flexibility while still maintaining good chemical and thermal stability.[27]

Epoxy-acrylic hybrid systems have been developed with major applications in the automotive industry. The system may be a one-component or two-component adhesive, or it may be a one-component adhesive taking the form of a film or paste. In one instance, the hybrid takes the form of a film in which the acrylic (the monomethacrylate of DGEBA) has been polymerized by uv light. The epoxy is a DGEBA-dicyandiamide latent curing system containing a polyvinyl phenol–amine complex. After irradiation, the film has good tack and will adhere to oily steel. It can be cured at 180°C for 30 min to achieve final strength.[28]

An *interpenetrating polymer network* (IPN) consisting of an epoxy and an elastomer has been developed by Isayama.[29] This is a two-component adhesive-sealant where the components are simultaneously polymerized. It consists of the MS polymer, developed in Japan by Kanegafuchi and commonly used in sealant formulations, with the homopolymerization of DGEBA using a phenol catalyst and a small amount of silane as a graft site to connect the MS polymer and epoxy homopolymer networks.

Special monomers can be added to epoxies to reduce shrinkage during cure.[30] These monomers are cyclic compounds such as cyclosporanes, for which bonds are broken during curing. The bond-breaking process increases the distance between atoms and offsets shrinkage, which would otherwise result when bonds form as molecules link together during the curing process. Of the various compounds developed, dinorborene (dinorborene-spiroorthocarbonate, DNSOC) had the most promise.

The use of 11.7% DNSOC in a low-viscosity DGEBA epoxy cured at 180°C was able to reduce the shrinkage from 5.4 percent to zero. Although there was great hope for these materials as adhesives, a commercial product is no longer for sale. However, other cyclic compounds have been synthesized and are being studied.

Other resins that are used to modify epoxy formulations are shown below with their reactive functionality:

- Polyesters (terminal carboxyl groups)
- Aniline-formaldehyde (terminal amine groups)
- Urea-formaldehyde (methylol groups)
- Furfural resins (methylol groups)

These hybrids generally yield improved flexibility and toughness, but they are more commonly used in epoxy coating formulations than in adhesive systems.

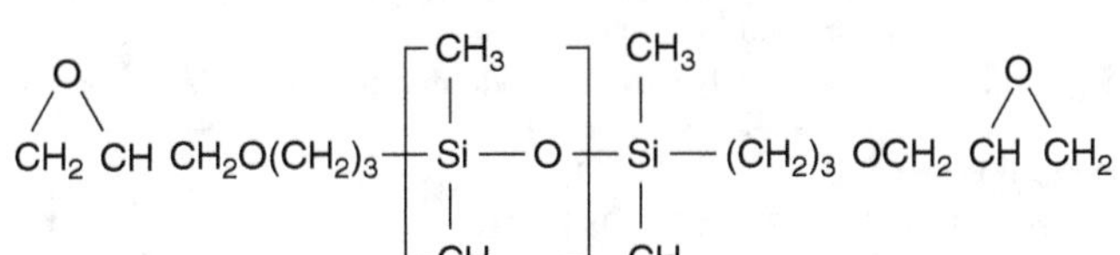

FIGURE 7.4 Chemical structure of epoxy-silicone hybrid resin.[26]

REFERENCES

1. Lewis, A. F., and Saxon, R., "Epoxy Resin Adhesives," Chapter 10 in *Epoxy Resins*, H. Kakuichi, ed., Marcel Dekker, New York, 1969.
2. Black, J. M., and Bloomquist, R. F., "Development of Metal Bonding Adhesives with Improved Heat Resistance," *Report FPL-710*, Forest Products Laboratory, Madison, WI, May 1974.
3. Black, J. M., and Bloomquist, R. F., "Polymer Structure and the Thermal Deterioration of Adhesives in Metal Joints," *Adhesives Age*, vol. 5, nos. 2 and 3, 1962.
4. Burgman, H. A., "Selecting Structural Adhesive Materials," *Electrotechnology*, June 1965.
5. Bulletin SC-54-57, Shell Chemical Company; also in Bolger, J., "Structural Adhesives State of the Art," Chapter 7 in *Adhesives in Manufacturing*, G. L. Schneberger, ed., Marcel Dekker, New York, 1983, p. 155.
6. Gorton, B. S., "Interaction of Nylon Polymers with Epoxy Resins in Adhesive Blends," *Journal of Applied Polymer Science*, vol. 8, 1964, p. 1287.
7. Smith, M. B., and Sussman, S. E., "Development of Adhesives for Very Low Temperature Applications," Narmco Research and Development, NASA Contract NAS-8-1565, May 1963.
8. Frigstad, R. A., U. S. Patent to 3M Co., 3,449,280, 1970.
9. DeLollis, N. J., and Montoya, O., "Mode of Failure in Structural Adhesive Bonds," *Journal of Applied Polymer Science*, vol. 11, 1967, pp. 983–989.
10. DeLollis and Montoya, "Mode of Failure in Structural Adhesive Bonds."
11. Frigstad, R. A., U.S. Patent No. 3,449,280.
12. Amstock, J. S., "Polysulfide and LP Polymers," Chapter 11 in *Handbook of Adhesives and Sealants in Construction*, McGraw-Hill, New York, 2001.
13. Panek, J. R., "Polysulfide Sealants and Adhesives," in *Handbook of Adhesives*, 3d ed., I. Skeist, ed., van Nostrand Reinhold, New York, 1992.
14. Peterson, E. A., "Polysulfides," in *Adhesives and Sealants*, vol. 3, Engineered Materials Handbook, AMS International, Materials Park, OH, 1990.
15. Guthrie, J. L., and Ching Lin, S., "One-Part Modified Epoxies for Unprimed Metal, Plastics," *Adhesives Age*, July 1985.
16. Guthrie, and Lin, "One Part Modified Epoxies for Unprimed Metal, Plastics," pp. 23–35.
17. Hawkins, J. M., U.S. Patent No. 3,525,779, 1970.
18. Grieves, R., U.S. Patent 4,623,702, 1986.
19. Lay, D. G., and Millard, T. G., Paper presented at ASC Meeting, St. Louis, MO, April 14–17, 1991.
20. Durig, J., et al., "High Performance Flexible Epoxy System for Civil Engineering," Paper presented at the SPI-Epoxy Resin Formulators Meeting, San Francisco, February 20–22, 1991.
21. Air Products Patent, EP 703277-A2.
22. Shah, D., and Dawdy, T. H., Lord Corporation, U.S. Patent, no. 5,232,996, 1993.
23. Asahi Chemical Co. Patent, EP 703259-A2.
24. "Ricon Toughened Epoxy System," *Sartomer Application Bulletin*, Sartomer Company, Exton, PA, 2004.
25. Sheng, Yen Le, "The Use of Fluoroepoxy Compounds as Adhesives to Bond Fluoroplastics Without Surface Treatment," *SAMPE Quarterly*, vol. 19, no. 2, 1988.
26. Goodman, S. H., "Epoxy Resins," in *Handbook of Thermoset Plastics*, 2d ed., S. H. Goodman, ed., Noyes Publishing, Westwood, NJ, 2000, p. 207.
27. Rifle, J. S., et al., "Elastomeric Polysiloxane Modifiers for Epoxy Networks," *ACS Symposium Series*, vol. 221, 1983, pp. 21–54.
28. Strasila, D., U.S. Patent No. 5,151,318, 1992.
29. Isayama, K., et al., U.S. Patent No. 4,657,986, 1987.
30. Sadhir, R. K., and Luck, R. M., *Expanding Monomers: Synthesis, Characterization, and Applications*, CRC Press, Boca Raton, FL, 1992.

CHAPTER 8
FLEXIBILIZERS AND TOUGHENERS

8.1 INTRODUCTION

Polymerized epoxy adhesives are amorphous and highly crosslinked materials. This microstructure results in many useful properties such as high modulus and failure strength, low creep, and good chemical and heat resistance. However, the structure of epoxy resins also leads to one undesirable property—they are relatively brittle materials. As such, epoxy adhesives tend to have poor resistance to crack initiation and growth, which results in poor impact and peel properties. In sealant formulations, epoxy resins do not often provide the degree of elongation or movement that is required for many applications.

Thus, two qualities are often required for adhesives or sealants that unmodified epoxy resins lack—flexibility and toughness. Formulators have overcome these problems by either flexibilizing the epoxy structure or incorporating modifiers into the adhesive. It must be recognized, however, that flexibility and toughness have different meanings in adhesive technology, and flexibilizers and tougheners operate by different mechanisms.

Flexibility is primarily characterized by a material's elongation. Flexibilizers in epoxy systems work by allowing the material to deform under stress. In this way stresses on the joint are distributed over a larger area. The flexibilized epoxy resin has the capability of compensating for differences in thermal expansion or elastic moduli of the substrates.

Flexibilizers improve peel and impact strength generally by allowing the adhesive to deform under the application of stress. Figure 8.1 shows that an adhesive that has greater elongation distributes peel stress over a much larger area than do more rigid adhesives. The adhesive with higher elongation generally provides a higher peel strength as measured in pounds per inch of bond width.

Flexibilized adhesives reduce mechanical damage by lowering the modulus and increasing elongation via plasticization, and this allows deformation of the adhesive.[2,3] This increase in flexibility is mainly due to decreasing the crosslink density of the cured adhesive. Decreasing the crosslink density, unfortunately, may also result in the lowering of certain properties such as tensile strength and glass transition temperature.

Toughness, on the other hand, is characterized by a material having both high elongation *and* tensile strength (i.e., maximization of the area under the stress-strain curve). Whereas flexibilizers work by deforming, tougheners improve properties because elastomeric inclusions in the epoxy matrix absorb energy and stop a crack from propagating throughout the bond line. Toughened adhesives tolerate damage by preventing crack growth and, thus, limiting the damage area. This results in enhanced fracture, impact, and stress resistance with minimal change in the crosslink density. Tougheners enable formulators

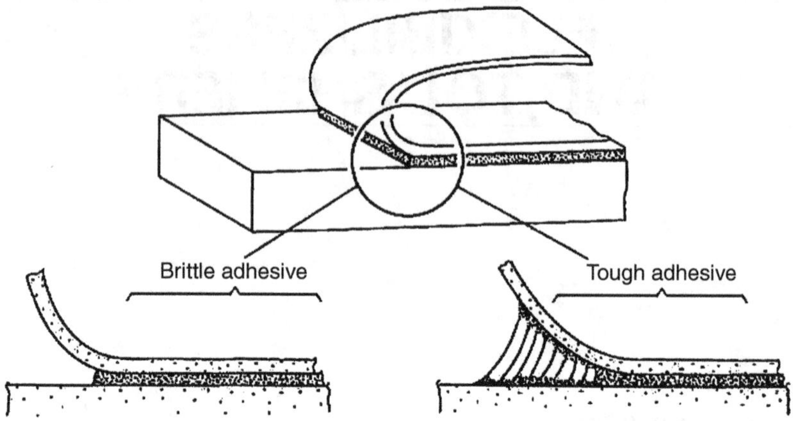

FIGURE 8.1 Tough, flexible adhesive distributes peel stress over a larger area.[1]

and end users to have the best of two worlds—resistance to peel, impact, and fatigue *and* a high degree of temperature and chemical resistance.

8.2 IMPROVEMENTS IN FLEXIBILITY

Epoxy adhesive formulators have generally addressed the problem of improving flexibility in several ways:

1. By adding flexible chemical groups to the cured epoxy structure, via either the epoxy base resin or the curing agent
2. By adding nonepoxy flexibilizing resins to the formulation to create an epoxy-hybrid adhesive system
3. By adding flexible diluents to the base epoxy resin formulation

Flexibilizers have been previously discussed as resins (Chap. 4), curing agents (Chap. 5), and hybrid modifiers (Chap. 7). This chapter primarily discusses flexibilizers that are used as additives to rigid epoxy resins either during the adhesive formulation stage or during the epoxy resin manufacturing process. They are generally reacted into the epoxy resin or have reactivity with various curing agents so that the cured properties are not significantly reduced due to the addition of the flexibilizer.

8.2.1 Flexibility through Resin and Curing Agent

Flexibility can be provided through the resin or hardener constituents by incorporating large groups in the molecular chain, which increases the distance between crosslinks. This has previously been illustrated in Figs. 3.10 through 3.12. Table 8.1 summarizes the tradeoffs in the properties of the cured adhesive that can occur. The reduction in crosslink density increases the flexibility (elongation) of the resulting molecule, but at the expense of a lowering of the glass transition temperature. This, in turn, generally results in a decrease in tensile and shear strength as well as a decrease in other performance properties, such as chemical and heat resistance.

TABLE 8.1 Tradeoffs in Properties That Occur due to Flexibilizing the Adhesive by Reducing Crosslink Density

Decrease in crosslink density will generally cause		Increase in crosslink density will generally cause	
Increase in	Decrease in	Increase in	Decrease in
Peel strength	Internal stresses	Internal stresses	Peel strength
Impact strength	Glass transition temperature	Glass transition temperature	Impact strength
Fatigue properties	Heat resistance	Heat resistance	Fatigue properties
Thermal shock resistance	Chemical resistance	Chemical resistance	Thermal shock resistance
Elongation	Hardness	Hardness	Elongation
Coefficient of thermal expansion	Modulus	Modulus	Coefficient of thermal expansion
	Tensile strength	Tensile strength	

Flexibility can be achieved within the cured epoxy molecule in a number of ways, most notably by

1. Using a higher-molecular-weight, more flexible resin
2. Using a higher-molecular-weight, more flexible curing agent
3. Decreasing the potential number of crosslinking sites by either choosing resin and curing agents with lower functionality or varying the resin–curing agent stoichiometry.

The main reason for the greater flexibility is due to long-chain difunctional materials that upon cure become part of the epoxy matrix. The result is a single-phase, flexible system. The disadvantage of this approach is a reduction in the crosslink density and consequently reductions in glass transition temperatures as well as the heat and chemical resistance of the system.

Both the epoxy resin and the curing agent components can be made more flexible by increasing their molecular weight. Polysulfides, polyamides, and *n*-aminoethylpiperazine curing agents are examples of long-chain difunctional curing agents that offer improved flexibility in epoxy adhesive formulations. There are many other examples of incorporating flexibility into an epoxy adhesive by appropriate selection of the curing agent. For example, by changing from hexahydrophthalic anhydride to hexamethylenediamine, one can double the impact resistance of a resin system and slightly increase its tensile elongation at break.[4]

Gross changes in the flexibility of a resin system can also be obtained through a major change in the structure of the cured resin. Such changes include shifting from an aromatic structure to a more aliphatic hydrocarbon or reducing curing agent functionality. Insertion of long hydrocarbon side chains also can provide additional flexibility to the epoxy molecule. The effect on a typical epoxy formulation, using Araldite 508 flexible epoxy resin, is shown in Table 8.2. Cured epoxies without the addition of flexible resins show elongation less than 2 percent.

8.2.2 Flexibility through Hybrid Formulation

Another common method of flexibilizing epoxy adhesives is by blending the primary epoxy resin with other, more elastic polymers. Epoxy-nylon, epoxy-polysulfide, and to a certain extent epoxy-urethane hybrids use such a mechanism to provide flexibility. These flexibilizers are important additives for epoxy adhesives even though they may reduce certain

TABLE 8.2 Typical Flexible Epoxy Formulations and Their Effect on Properties[5]

	Formulations		
Epoxy adhesive components	A	B	C
Araldite 6010, pbw	50	33	33
Araldite 508, pbw	50	67	67
n-Aminoethylpiperazine, pbw	17	14	14
Properties			
Cure schedule	7 days at 25°C	7 days at 25°C	Gel at 25°C plus 2 h at 100°C
Tensile strength, 10^3 psi	3.34	2.23	2.51
Elongation at failure, percent	88	107	91
Modulus of elasticity, 10^6 psi	0.1	0.013	0.029
Hardness, Barcol	70	38	56

properties. Flexible hybrid resins (Chap. 7) improve adhesive properties, especially low-temperature performance and adhesion to flexible substrates. They also relieve stresses within the adhesive by allowing chains to relax after cure, as well as any stresses that occur in the bonded joint during use.

A good example of a flexible epoxy hybrid adhesive is the epoxy polysulfides. Figure 8.2 shows the improved impact resistance that can be realized by using increasing amounts of

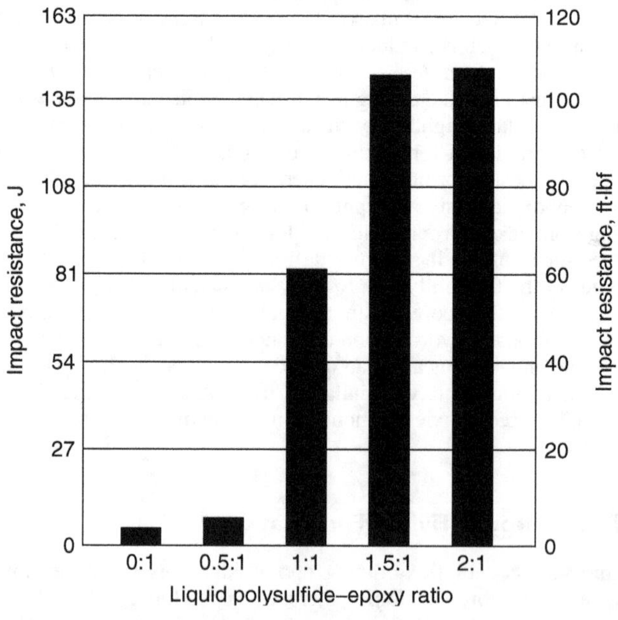

FIGURE 8.2 Impact resistance of varying concentrations of polysulfide resin in an epoxy adhesive formulation.[6]

polysulfide in epoxy blends. The elongation properties of the cured epoxy hybrid are similarly increased.

Soluble thermoplastic polymer additions (e.g., epoxy-nylon and epoxy-vinyl) provide very tough and somewhat flexible systems, but they are limited because of their high viscosity and high raw materials costs. Almost all these systems need to be handled at elevated temperatures or as solvent solutions. They require elevated temperature to achieve wetting and cure.

8.2.3 Flexibility through Diluents

Diluents are primarily used in epoxy adhesive formulations to reduce viscosity. However, diluents also provide a degree of flexibility by two mechanisms:

1. By introducing long-chain, flexible molecules into the epoxy structure that remain unreacted after cure
2. By introducing long-chain, flexible molecules into the epoxy structure that react during cure

The first category consists of unreactive diluents, and they may be considered mainly as plasticizers. Generally they do not increase flexibility, and often they result in the overall degradation of physical properties. The cured resin has reduced strength and hardness and could even exhibit a "cheeselike" structure. The second category consists of reactive diluents, and they can be considered as true flexibilizers.

Of the two groups, the reactive diluents are generally preferred for structural adhesives applications because they do not degrade end properties as much as the unreactive diluents do. Note that diluents, either reactive or unreactive, do not necessarily result in an increase in elongation of the cured epoxy. They may only reduce modulus and hardness. Their enhancement regarding flexibility will depend on the component's functionality and molecular nature.

Flexibilizers of these types do not convert a rigid epoxy structure to one that has the elongation and hardness of synthetic elastomers. Rather, they provide a degree of resiliency to a normally rigid system, thereby improving properties such as peel strength, impact resistance, and thermal shock. These flexibilizers are also valued in adhesive systems because they reduce internal stresses that could occur within the joint during cure or on thermal cycling. For example, they provide a degree of resiliency that is better suited for counteracting differing thermal expansion rates between the adhesive and the substrate, and they also reduce the amount of stress caused by shrinkage during cure.

Nonreactive diluents for epoxy resins are discussed in Chap. 6. Of the various plasticizers for epoxy that have been investigated, dibutyl phthalate is perhaps the most popular. However, the use of unreactive diluents or plasticizers to develop highly flexible epoxy adhesives has generally been unsatisfactory. Plasticizers do not provide sufficiently high levels of flexibility to offset large reductions in tear resistance, chemical and heat resistance, and adhesion properties. Few of the typical plasticizers are fully compatible, separating from the resin during or after cure. Liquid hydrocarbon resins and benzyl alcohol have been used with certain epoxy-amine systems, but at only low concentrations (5 to 15% by weight).

Reactive diluents, on the other hand, have one or two reactive epoxy groups that allow the modifier to be compatible and nonmigrating. Thus, they are more effective than nonreactive diluents for adhesive systems.

Monofunctional epoxy diluents are long-chain molecules that have one functional group that can react with the epoxy system. The linkage prevents the diluent from migrating out of the cured epoxy during aging. The long chains force the molecules in the cured structure to be farther apart, and this permits the crosslinked molecule freer movement under stress. However, it also results in decreased crosslink density and thus in degradation of thermal and chemical resistance.

In adhesive formulations, glycidyl ethers are the most common monofunctional diluent. These include butyl glycidyl ether and allyl glycidyl ether. Other effective monofunctional

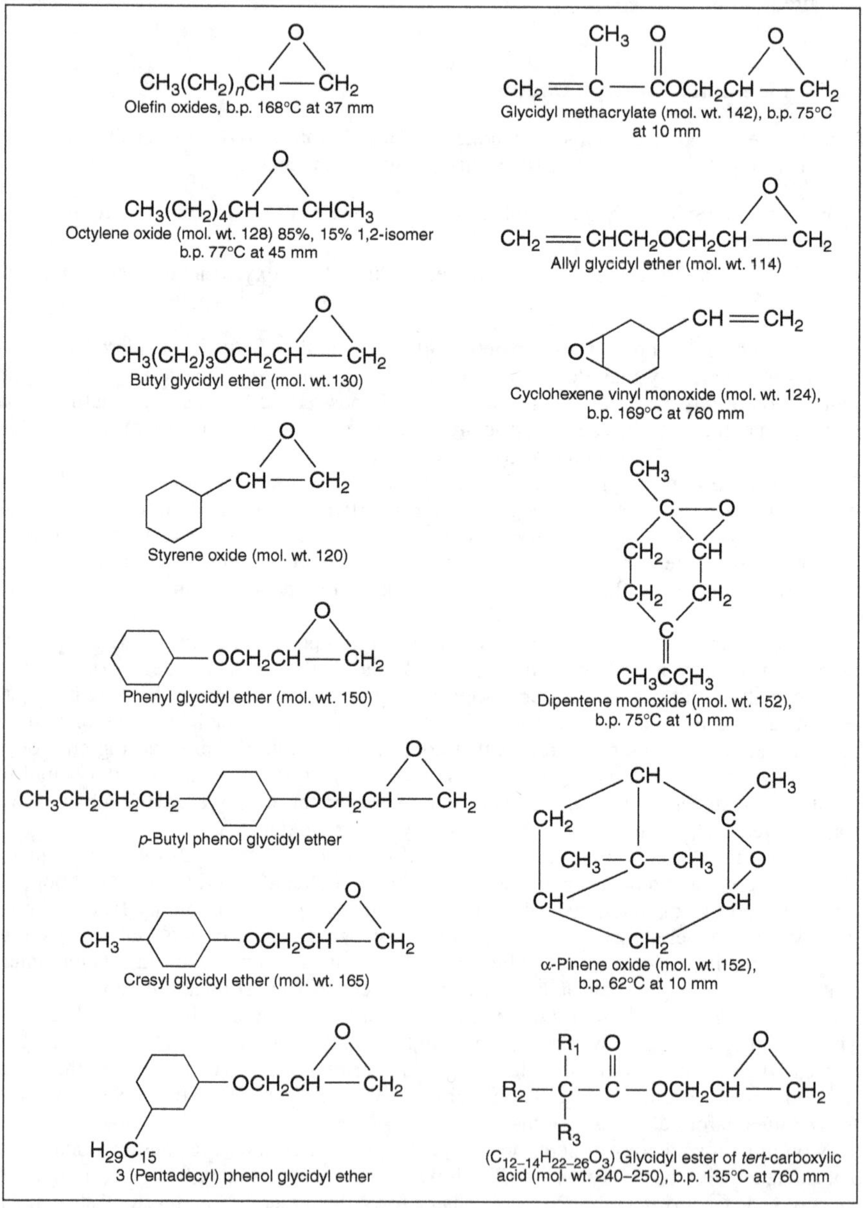

FIGURE 8.3 Commercial epoxy-containing monofunctional reactive diluents.[7]

TABLE 8.3 Characteristics of Reactive Diluents Commonly Used as Flexibilizers in Epoxy Systems

Diluent type	Examples	Characteristics
Epoxy resins from polyols	Epoxidized polyether glycol	Varies in molecular weight from $n = 2$ to 7. Higher-molecular-weight types provide greater flexibility.
Epoxy resins from aliphatic acids	Cyclic diglycidyl ester	When used with primary amine curing agents, reactivity is reduced.
Hydroxyl-terminated compounds	Hydroxyl-terminated polyesters and polyols without ester linkages	Used with cycloaliphthalic epoxies that do not contain hydroxyl groups.

diluents are epoxidized vegetable oils, olefin oxides, styrene oxide, and pinene oxide. Figure 8.3 shows the chemical structure of several commercial epoxy-containing monofunctional reactive diluents.

Di- or polyfunctional epoxy resins may be considered reactive diluents that provide a degree of resiliency, yet most preserve the inherent properties of the base epoxy resin. When added to an epoxy formulation, they will not reduce functionality, and in some cases an actual increase in crosslink density is noticed. Typical resins of these types are shown in Fig. 8.4. As flexibilizers, these diluents are often blended with the more reactive epoxy resins. They, of course, can also be used as epoxy resins in themselves.

The commercial flexibilizing reactive diluents are predominantly glycidyl derivatives of glycols, dimerized acids, or reaction products of dicarboxylic acids with epoxy resins. The properties of certain polyepoxy diluents are summarized in Table 8.3. Some of, but not all, these materials provide good reactivity with aliphatic primary amines at room temperature.

The above-mentioned methods used to achieve flexibility are widely recognized and practiced in commercial epoxy formulations. The following discussion summarizes some of the newer methods that have been developed.

Modern flexibilizers include isocyanate monomers and prepolymers. These react with the hydroxyl groups on epoxy resins to give tough, flexible hybrids. Silicone polymer containing amine termination will also react with epoxy groups to produce more flexible epoxies, while still maintaining good chemical and thermal stability. Acrylic polymers with carboxyl termination have also been investigated as epoxy flexibilizers.

Silyl-terminated polyether can be blended with epoxy resins to form elastic adhesives at room temperature.[9] In this system, small rigid epoxy particles are chemically linked by an elastic polyether phase. A typical formulation is shown in Table 8.4. Elongation and peel

TABLE 8.4 Nonfiller Silylated Polyether/Epoxy Blend[10]

	Component	Grade, supplier	Parts by weight
Part A	Silyl	SAX 350, Kaneka	100
	Epoxide hardener	Tertiary amine	5
	Compatibilizer	Alkyl amino silane	2
Part B	Epoxy	Epon 828, Resolution Performance Products	50
	Silyl hardener	Dialkyltincarboxylate	2
	Demineralized water		0.5

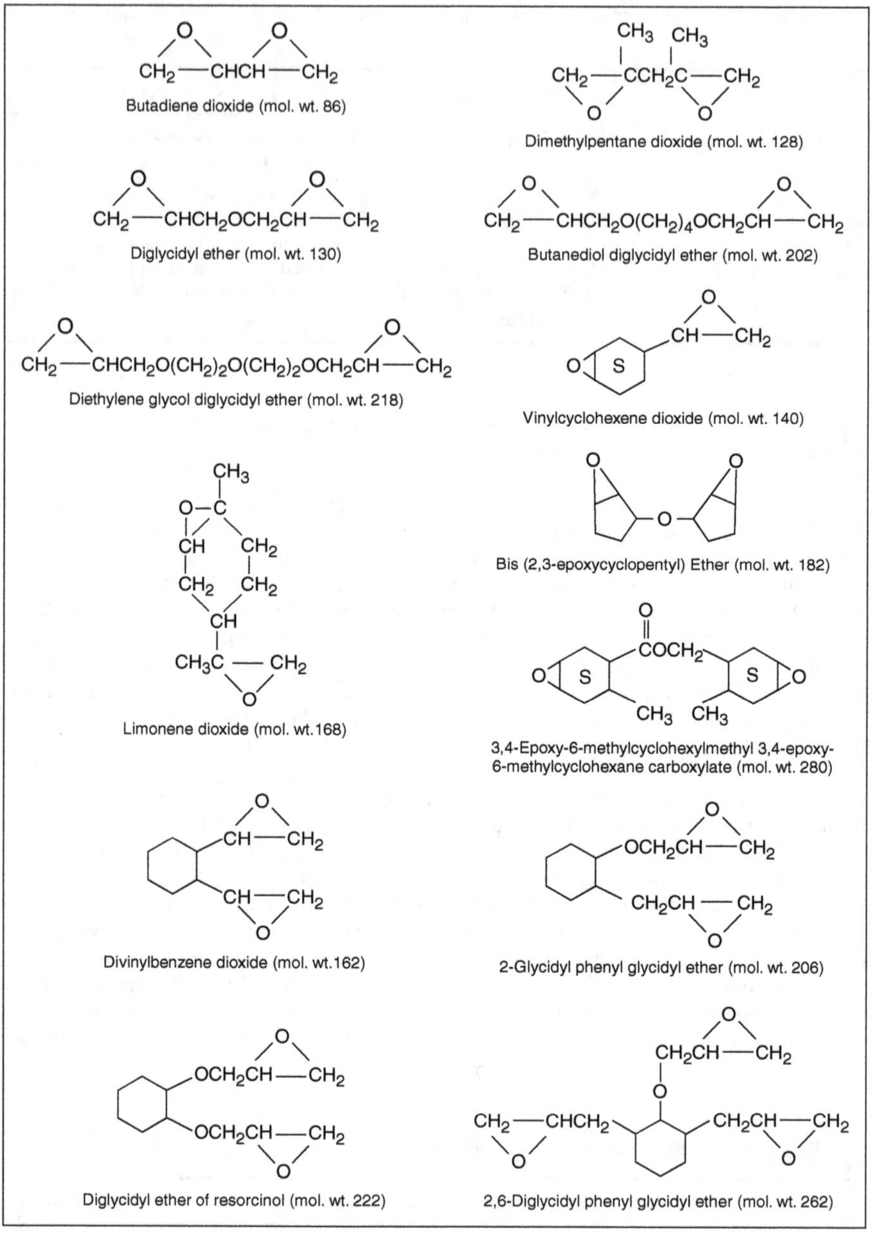

FIGURE 8.4 Typical di- and polyfunctional epoxy resin diluents.[8]

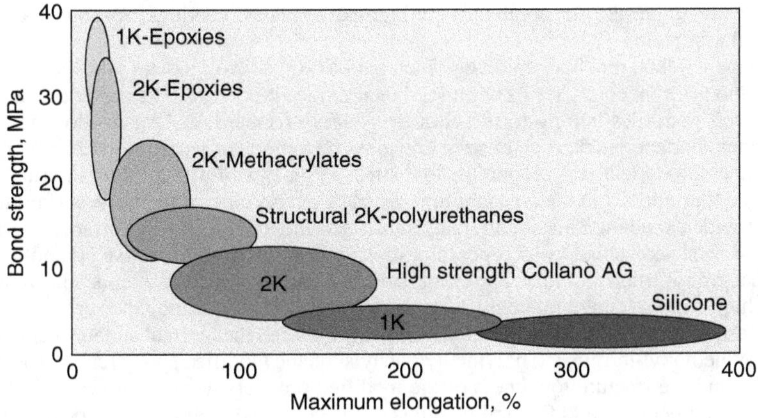

FIGURE 8.5 Bond strengths and elongation for typical structural adhesives.[11]

strength (20 to 22 piw*) are significantly increased with a moderate decrease in tensile properties (tensile shear strengths are 900 to 1500 psi). Tuning of the mechanical properties (i.e., strength from the epoxy and high impact resistance from the polyether) can be achieved by adjusting the ratio and selection of the silylated polyether and the epoxy resin. Maximum strength (both t-peel and shear strength) properties are achieved at a silylated polyether/epoxy ratio of approximately 100/40.

Commercial silyl-epoxy hybrid adhesives have been developed by Collano AG. These are polymer alloys consisting of a matrix of silyl reactive polymers (elastic phase), which host domains of epoxy reactive polymers (hard domains). With a different choice of components, mixing ratios, and the size of respective domains, the overall properties of the bond line can be tailored for specific purposes.

Figure 8.5 depicts representative ranges of bond strength and elongations for typical structural adhesives such as epoxies, methacrylate, polyurethanes, and silicones. The silyl-epoxy hybrids conveniently fill the gap between the silicones and polyurethanes.

Flexibilizers based on functionalized, liquid polybutadiene derivatives have also been developed for epoxy resins. Carboxy-terminated butadiene-acrylonitrile polymers, butadiene homopolymers, and maleic anhydride–amino acid grafted butadiene homopolymers have been used as flexibilizers to impart good low-temperature strength and water resistance to DGEBA-based epoxy adhesives. An epoxy system toughened by polybutadiene with maleic anhydride is claimed to provide a hydrophobic backbone, low viscosity, and softness, along with high tensile strength and adhesion (Table 7.10).

Polyoxyalkylene ethers based on propylene oxide or propylene oxide/ethylene oxide copolymers have also been described as useful flexibilizers. DGEBA-terminated ethylene oxide/propylene oxide polyethers (30/70) have been used to modify DGEBA-based adhesives by improving room temperature shear and peel strengths.

Diglycidyl ether of ethoxylated resorcinol (DGER), a resorcinol-based epoxy monomer, can be cured with various epoxy curing agents to yield a balance of flexibility, low viscosity, and fast cure capability.[12] The molecular flexibility of DGER is claimed to improve the impact strength and fracture toughness of the cured epoxy system. These flexibilized

*piw = pounds per inch width.

systems are suggested for application in high-performance coatings, adhesives, and composite matrix resins.

In recent years, much attention has been paid to siloxane oligomers as a specific type of rubber modifier for epoxy resins because of their unique properties. These liquid functionally terminated siloxanes belong to the elastomer class of materials. Organosiloxanes exhibit important characteristics, such as very low glass transition temperature (−120°C), moisture resistance, good electrical properties, low stress, high flexibility, good weathering ability, and good thermal and oxidative stabilities. In addition, because of their low surface tension, they provide excellent adhesion to many substrates and a hydrophobic surface.

It has been shown that by incorporating siloxane oligomers with reactive organofunctional terminal groups, such as amine, epoxy, or carboxyl, into the structure of epoxy networks, one can achieve improvement in the fracture toughness, water absorption, and surface properties of the resultant polymer.[13] Yayun[14] and Menghuo[15] studied the thermal and mechanical properties of methoxyl-terminated polymethylphenylsiloxane modified epoxy resins. The thermal stability and the fracture toughness of the modified products were improved, and the glass transition temperature and flexural modulus of the crosslinked epoxy network were not significantly lowered.

8.3 IMPROVEMENTS IN TOUGHNESS

Flexibilizers, such as those mentioned above that work by reducing crosslink density, provide a major dilemma for both the epoxy adhesive formulator and the end user. Although flexibility is somewhat improved, other valuable properties are reduced. Reductions in tensile strength and heat and chemical resistance are characteristic of many flexibilized epoxy systems. Until the development of toughened epoxy formulations, it was difficult, if not impossible, to provide an epoxy adhesive having high impact and peel strength *and* high heat resistance.

Toughened structural adhesives generally have two distinct phases: The larger phase is the base resin, and the minor phase consists of small (on the order of 1 μm in diameter), distributed elastomeric entities. The addition of the second phase significantly improves fracture toughness by providing crack pinning and stress distribution mechanisms within the material. The elastomeric, particulate phase is often thought of as a "crack stopper" for improving fracture toughness.

A variety of toughening agents have been used to modify epoxy adhesives to improve peel strength and fracture toughness without significantly affecting other properties of the epoxy base resin. Generally these modifiers can be classified into three types:

1. Reactive liquid rubbers
2. Functionally terminated engineering thermoplastics
3. Inorganic and hybrid particles

8.3.1 Reactive Liquid Rubber

Early structural epoxy adhesives were based on modification of epoxy resins by the use of liquid butadiene-acrylonitrile polymers with terminal carboxyl groups. By using low levels of these liquid polymers, the fracture energy of cured DGEBA-type epoxy resins was increased by a factor of 15. Since that time, considerable work has been done on developing toughening agents for epoxy resins using liquid polymers with reactive terminal groups.

Formulations have been developed where small rubber domains of a definite size and shape are formed in situ during cure of the epoxy matrix. The domains cease growing at gelation. After cure is complete, the adhesive consists of an epoxy matrix with embedded rubber particles. The formation of a fully dispersed phase depends on a delicate balance between the miscibility of the elastomer, or its adduct with the resin, with the resin-hardener mixture and appropriate precipitation during the crosslinking reaction.

The advantage of this toughening process is that the high modulus and T_g of the epoxy resin are not sacrificed. The T_g of the product is indicative of the epoxy resin used as the matrix, although the precipitated domain has a T_g of about –40 to –50°C.

The butadiene-acrylonitrile tougheners that have been especially successful in both DGEBA and DGEBF types of epoxy resins are

- Carboxyl-terminated butadiene-acrylonitrile (CTBN)
- Amine-terminated butadiene-acrylonitrile (ATBN)
- Vinyl-terminated butadiene-acrylonitrile (VTBN)

CTBN and ATBN are the most commonly used in structural epoxy adhesive formulations. CTBN (Fig. 8.6) is generally the elastomer of choice because of its miscibility in many epoxy resins. These tougheners were originally developed by BF Goodrich (now Noveon, Inc.) under the tradename Hycar.

The degree of toughness is determined by the crosslink density of the matrix, the elastomer particle size and size distribution, the volume fraction of the elastomeric phase, and the degree of adhesion between the epoxy matrix and the particle. The formulating procedure was found to have as strong an effect on the fracture toughness as the materials themselves.[16]

In general, the elastomer must be prereacted (adducted) with the epoxy for the toughening effect to take place. Adducts reduce the likelihood of early phase separation and maintain the solubility of the elastomer in the uncured resin system. For CTBN the reaction is carried out at high temperatures (150 to 160°C) and usually in the presence of a catalyst, such as tris-dimethylamino phenol or piperidine. The resulting epoxy-CTBN adducts are available from several suppliers, and they can be easily formulated into epoxy adhesives.

Table 8.5 shows a comparison of adhesives formulated with and without a CTBN adduct. When compared to the control epoxy, the toughened formulation exhibits significantly higher peel strength and moderately higher tensile shear strength. CTBN modified epoxy adhesives are generally one-part systems, cured with dicyandiamide at elevated temperature. However, two-part, room or mildly elevated-temperature curing systems are also possible and provide similar improvements in properties (Table 8.6).

One of the disadvantages of CTBN-epoxy adhesives has been their high viscosity, which limits additional formulation options. Recently new adducts, such as EPON 58003 and RSM-2577 from Resolution Performance Products LLC, have been introduced which have significantly lower viscosities.[19] In addition, lower concentrations of these new CBTN-epoxy adducts are generally required to achieve equivalent adhesive performance.

ATBN liquid toughening agents are synthesized by reacting a CTBN polymer with an amine. For example, *n*-aminoethylpiperazine (AEP) will give a low-molecular-weight,

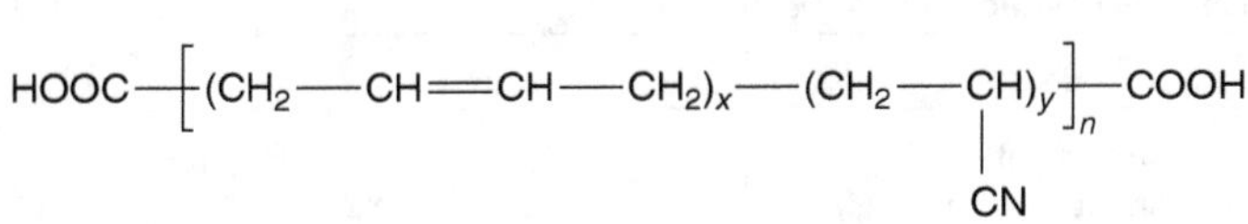

FIGURE 8.6 Chemical structure of carboxyl-terminated butadiene acrylonitrile (CTBN).

TABLE 8.5 Properties of a CTBN Modified Epoxy Adhesive[17]

Formulation components	Unmodified epoxy	CTBN modified epoxy
DGEBA liquid epoxy (EEW: 190), pbw	100	77.5
CTBN–DGEBA epoxy adduct (EEW: 140), pbw	—	37.5
Tabular alumina, pbw	40	40
Fumed silica, pbw	3.5	3.5
Dicyandiamide, pbw	6	6
3-Phenyl-1,1-dimethyl urea, pbw	2	2
Properties on specimen cured 1 h at 171°C		
Tensile shear at room temperature, kpsi	1.1	1.78
T-peel at room temperature, kg/cm	2.5	6.98
Glass transition temperature, °C	125	113

amine-terminated product at a reaction temperature of 130°C. The terminal secondary amine groups undergo the typical reaction chemistry of secondary amines. As a result, ATBN liquid polymers usually are not mixed directly into the epoxy resin component of a two-part adhesive or in a one-part adhesive since crosslinking and shortened shelf life will result. ATBN adducts are, therefore, often mixed with the curing agent component of two-component epoxy adhesives.[20] A typical two-part ATBN modified epoxy adhesive is shown in Table 8.7 and compared to a CTBN modified epoxy adhesive in Table 8.8.

ATBN tougheners also increase the ultimate tensile shear and T-peel strengths and environmental durability using conventional room temperature curing agents such as fatty polyamides, amidoamines, and amine ethers. However, the improvement in toughness, as measured by peel tests, becomes less significant with the faster-curing hardeners.

8.3.2 Thermoplastic Additives

Various kinds of engineering thermoplastics such as polyether sulfone, polyether imide, polyaryl ether ketone, and polyphenylene oxide have been studied as toughening agents for

TABLE 8.6 Formulas for Two-Part CTBN Modified Epoxy Adhesive System Cured Two Week at 25°C[18]

Components	Control epoxy	CTBN-epoxy
DGEBA epoxy resin (EEW: 190), pbw	100	77.5
$CaCO_3$ filler, pbw	30	30
DGEBA-CTBN adduct (Hycar 1300X8), pbw	—	37.5
Amine hardener (Ancamine AD), pbw	60	59.3
Properties		
Lap shear strength at –40°C, MPa	6.9	22.7
Lap shear strength at 25°C, MPa	8.6	16.9
Lap shear strength at 75°C, MPa	4.4	5.5
180° T-peel strength at 25°C, kN/m	0.5	5.0

TABLE 8.7 Two-Part ATBN Modified Epoxy Adhesive[21]

Components	Parts by weight
Part A	
DEGEBA epoxy resin (EEW: 190)	60
Diethylene glycol	1
Aluminum (Toyal 101)	37
Fumed silica (Cab-O-Sil TS720)	2
Part B	
Polyamide (Ancamide 2482)	38
ATBN (Hycar 1300x16)	12
Aluminum (Toyal 101)	22
Talc (Microtuff 325F)	27
Fumed silica (Cab-O-Sil TS720)	1
Properties	
Mix ratio, by volume, part A : part B	1:1
Cure schedule	7 days at 25°C or 30 min at 125°C
Tensile shear strength, psi	
• On aluminum after curing at RT	2500
• On aluminum after curing at ET	2750
• On ABS after curing at RT	550
T-peel strength, lb/in	
• On aluminum after curing at RT	22
• On aluminum after curing at ET	20

TABLE 8.8 Typical Two-Part CTBN and ATBN Toughened Epoxy Adhesives[22]

	CTBN modified epoxy		ATBN modified epoxy	
Components	Control	Modified	Control	Modified
Part A				
DGEBA epoxy resin, pbw	100	77.5	100	100
$CaCO_3$ filler (Atomite whiting), pbw	30	30	30	30
DGEBA-CTBN adduct,* pbw		37.5		
Part B				
ATBN (Hycar 1300x16), pbw				30
Polyamine (Ancamine AD, Air Products and Chemicals), pbw	60	59.3		
Amidoamine (Ancamide 501, Air Products and Chemicals), pbw			35	32.8
Properties				
Cure schedule	2 weeks at 25°C			
Tensile shear strength, MPa, on aluminum measured at				
• –40°C	6.9	22.7	8.6	15.5
• 25°C	8.6	17.9	9.6	17.2
• 75°C	4.4	5.5	13.8	11.7
T-peel strength, kN/m, on aluminum measured at 25°C	0.5	5.0	0.5	5.3

*Adduct contains 40% CTBN (Hycar 1300x13).

epoxy systems. Unlike early epoxy-hybrid adhesives (e.g., epoxy nylons, epoxy polysulfides), these modern attempts use *engineering thermoplastics*—tough, ductile, chemically and thermally stable polymers having a high T_g. These are chosen so as to improve toughness and not to negatively affect high-temperature performance. Due to the high modulus and T_g of the thermoplastic modifier, the modulus and T_g of thermoplastic modified epoxy resins will reach or even exceed the corresponding values of unmodified resin. In addition to improved fracture resistance, this modification leads to reduced shrinkage on cure.

The thermoplastic resins are usually blended with the epoxy resin in a solvent solution. Early researchers realized that to make this approach effective, it was necessary to increase the compatibility and interfacial adhesion of the thermoplastic modifier and the epoxy resins. The problem of poor miscibility of the thermoplastic resins and poor processability of the final product are the main reasons that these materials have not achieved commercial success.

Several research efforts have been directed at making a more miscible system. In an effort to toughen an epoxy resin by incorporating engineering thermoplastic units into the main chain, Kun-Soo Lee and coworkers developed epoxy resins containing ether ether ketone units.[23] These resins provided toughness without phase separation at elevated temperatures. Low-molecular-weight polyphenylene ether (PPE) resins have also been developed that, unlike their high-molecular-weight counterparts, are readily miscible with epoxy resins and form low-viscosity blends with them.[24] The cured thermoset product exhibits improved thermal properties as well as enhanced fracture toughness.

Compared to the carboxylated nitrile elastomer additives, the use of thermoplastics has primarily been focused on the aerospace industry. On a cost per pound basis, the two-phase nitrile additives offer the best combination of property improvement without negative impact. The thermoplastic additives, however, may offer better high-temperature performance, but they are more difficult to formulate and to process as adhesives. As a result, the cost of these adhesives is generally much higher than that of other toughened epoxy mechanisms.

8.3.3 Inorganic Particles and Preformed Modifiers

Some inorganic particulate fillers have also been considered as toughening agents for epoxy materials. Glass beads, fly ash, alumina trihydrate, and silica were used early on to improve the toughness of filled epoxy resins. Various studies, however, have demonstrated that the fracture energy of filled epoxies reaches a maximum at a specific filler concentration.

Brittle, hollow glass beads have been found to be both an extender and an impact modifier for coatings and adhesives. The extender particles have to be designed in such a way that they actively dissipate deformation energy and hence reduce the likelihood of loss of adhesion between the epoxy and the substrate.[25]

In recent years, preformed particles such as thermoplastic powders or core-shell polymers made from elastomeric latexes are increasingly being used as modifiers to improve mechanical and thermal properties of polymer systems. However, these modifiers have yet to gain widespread use in commercial epoxy adhesive systems. Figure 8.7 shows an idealized view of an epoxy matrix toughened with a core-shell polymer. There are several advantages from using preformed particles as toughening agents:

- They can be manufactured relatively easily into particles of different sizes.
- They maximize the volume fraction of the toughening phase.
- They enhance toughness without any loss in thermomechanical properties.

A surface activated rubber particle, Vistamer, was recently introduced as a low-cost toughener for epoxy and polyurethane adhesives. The source of the rubber is reprocessed

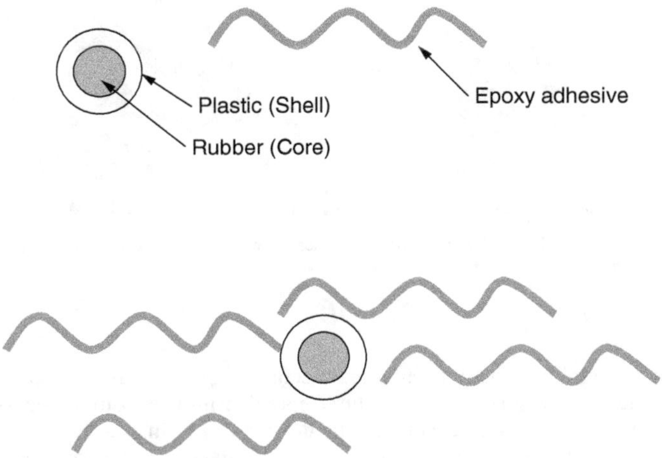

FIGURE 8.7 Morphology of a core-shell toughened epoxy adhesive.

tire treads or whole tires. Following the grinding process, the rubber particles are exposed to a reactive oxidizing gas, which creates polar groups on the particle surface. The resulting higher-surface-energy particles are easily wetted by polar polymers.[26]

Multiwalled carbon nanotubes were recently discovered as a toughener for epoxy adhesives used to bond graphite fiber–epoxy composite substrates.[27] Significant enhancements of the bonding performance were observed as the weight fraction of the carbon nanotubes was increased. Nanomaterials are materials whose size is on the order of angstroms. Nanomaterial particles dispersed in epoxy resin hold the promise for producing toughened adhesives with properties that are even more remarkable than those present in today's adhesive products.

Recently it has also been shown that modified spherical nanosilica particles can be used to toughen epoxy resins without the loss of other properties, such as glass transition temperature or modulus.[28] When such surface modified nanoparticles are added to CTBN toughened epoxy resins, the performance of both one- and two-component epoxy adhesives was greatly improved.

8.3.4 Interpenetrating Polymer Network (IPN) Tougheners

A series of adhesives has also been modified by creating interpenetrating networks, allowing for extensive control of adhesive properties without sacrificing thermal stability.[29] These adhesives are based on resins containing epoxy groups. Additionally, they contain monomers that can create a network of their own using a different chemical mechanism, without reacting with the epoxy monomer. During the curing two independent networks are created, penetrating each other but not being covalently bonded (Fig. 8.8).

Such a mixture of networks is called an *interpenetrating polymer network* (IPN). Much higher shear strength and impact resistance are claimed for IPN epoxy systems than for standalone single networks, and at the same time the high-temperature strength is improved.

Core-shell structures (Fig. 8.7) have also been suggested as a method of toughening epoxy resin formulations. The polymeric shell, which surrounds an elastomeric core, provides good adhesion to the epoxy matrix. Hyperbranched polymers with hydroxyl

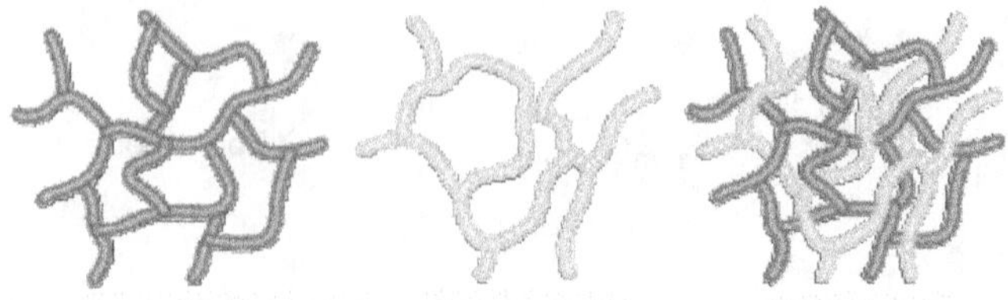

FIGURE 8.8 Interpenetrating polymer networks. (*Courtesy: Polymerics GmBH*)

functionality, resembling the molecular architecture of dendrimers, give core-shell structures in curable epoxy resins.[30] Nanostructured epoxy resins using diblock copolymers give an intermittent swollen mesophase in the gel/cure of epoxy resin. This results in a core-shell rubber structure as a nanoregion reinforcement of the glassy, crosslinked epoxy.[31]

REFERENCES

1. Rider, D. K., "Which Adhesives for Bonded Metal Assembly," *Production Engineering*, May 25, 1964.
2. Farrakhov, A. G., and Khozin, V. G., "Mechanical Behavior of Plasticized Epoxy Polymers," *Crosslinked Epoxies*, B. Sadlacek and J. Kahovec, eds., Walter de Gruyter and Co., Berlin, 1987, pp. 585–597.
3. Misra, S. C., et al., "Effect of Crosslink Density Distribution on the Engineering Behavior of Epoxies," *Epoxy Resin Chemistry*, ACS Symposium Series 114, R. S. Bauer, ed., American Chemical Society, 1979, pp. 137–156.
4. *EPON Resin Structural Reference Manual—Additive Selection*, Resolution Performance Products, Houston, TX, 2001, p. 8.
5. Buckley, W. O., and Schroeder, K. J., "Adhesive Modifiers," in *Adhesives and Sealants*, vol. 3, *Engineering Materials Handbook*, ASM International, Materials Park, OH, 1990.
6. Buckley and Schroeder, "Adhesive Modifiers."
7. Lee, H., and Neville, K., *Handbook of Epoxy Resins*, McGraw-Hill, New York, 1968, p. 13.10.
8. Lee and Neville, *Handbook of Epoxy Resins*, p. 13.16.
9. Devroey, D. R. E., and Homma, M., "Blends of Silyl-Terminated Polyethers and Epoxides as Elastic Adhesives," *International Journal of Adhesion and Adhesives*, vol. 21, 2001, pp. 275–280.
10. Devroey and Homma, "Blends of Silyl-Terminated Polyethers and Epoxides as Elastic Adhesives."
11. Ferrand, D., and Schwotzer, W., "Elastic Structural Adhesives by Hybrid Technology," *Adhesives and Sealants Industry*, April 2004.
12. Press Release, "INDSPEC Chemical Corporation Formulates New Epoxy Monomer," INDSPEC Chemical Corporation, Pittsburgh, PA, March 5, 2003.
13. Ho, T., et al., "Modification of Epoxy Resins with Polysiloxane TPU for Electronic Encapsulation," *Journal of Applied Polymer Science*, vol. 60, 1996, p. 1097.
14. Yayun, L, Master's thesis, Syracuse University, NY, 1999.
15. Menghuo, C., Master's thesis, Syracuse University, NY, 2001.
16. Klug, J. H., "Model High Performance Adhesive Systems," *Journal of Applied Polymer Science*, vol. 66, 1997, pp. 1953–1963.
17. Buckley and Schroeder, "Adhesive Modifiers," p. 185.

18. Technical Bulletin, "Hycar CTBN Modified Epoxy Adhesives," B. F. Goodrich Company, AB-8, Cleveland, OH, 1990. (Hycar products now manufactured by Noveon, Inc.)
19. Farris, R. D., and Steward, S. L., "New Epoxy Tougheners Widen the Adhesive Formulation Window," *Adhesives and Sealants Industry*, January 2002.
20. Petrie, E. M., "Improving the Toughnes of Structural Adhesives," at www.SpecialChem4Adhesives.com, April 7, 2004.
21. "Two Component Adhesive Starting Formulation," General Purpose Adhesive, Air Products and Chemicals Company, Allentown, PA, 2001.
22. "Two Component Adhesive Starting Formulation," General Purpose Adhesive, Air Products and Chemicals Company.
23. Lee, Kun-Soo, et al., "Preparation of Epoxy Resins Containing Ether Ether Ketone Unit and Their Thermal Properties," *Bulletin of Korean Chemical Society*, vol. 22, no. 4, 2001.
24. Yeager, G. W., et al., "Use of Low Molecular Poly(phenylene ether) Resin in Epoxy Thermosets," *International SAMPE Symposium and Exhibit (Proceedings)*, vol. 44, pt. 1, 1999, pp. 1075–1089.
25. Rosler, M., and Kinke, E., "Hollow Glass Beads as Impact Absorbing Extenders," *Farbe and Lack*, vol. 103, no. 1, 1997, pp. 50–54.
26. Gerace, M. J., and Gerace, J. M., "Surface Activated Rubber Particles Improve Structural Adhesives," *Adhesives Age*, December 1995, pp. 26–31.
27. Hsiao, Kuang-Ting, et al., "Use of Epoxy/Multiwalled Carbon Nanotubes as Adhesives to Join Graphite Fiber Reinforced Polymer Composites," *Nanotechnology*, vol. 24, July 2003, pp. 791–793.
28. Bowtell, M., "Novel Adhesive Solutions," Review of 17th International Swissbonding Symposium, Zurich, *European Adhesives & Sealants*, December 2003.
29. Technical Bulletin, "IPN Technology," Polymerics GmBH, Berlin, Germany, 2004.
30. Boogh, L., et al., "Novel Tougheners for Epoxy Based Composites," *SAMPE Conference Proceedings*, vol. 28, 1996.
31. Hillmyer, M., et al., "Nanostructured Thermosets," *Journal of American Chemical Society*, vol. 119, 1997, p. 2749.

CHAPTER 9
FILLERS AND EXTENDERS

9.1 INTRODUCTION

Fillers and extenders are used in epoxy adhesive formulations to improve properties and to lower cost. Properties that can be selectively improved include both the processing properties of the adhesive as well as its performance properties in a cured joint. However, the use of fillers can also impair certain properties. Typically, the formulator has to balance the improvements against property decline. The advantages and disadvantages of filler addition in epoxy formulations are listed in Table 9.1. Common fillers used in epoxy formulations and the properties that they are used to modify are shown in Table 9.2.

Fillers generally represent one of the major components by weight in an adhesive formulation. However, their concentration is quite often limited by viscosity constraints, cost, and negative effects on certain properties. The degree of improvement provided by a filler in an epoxy formulation will heavily depend on the type of filler and its properties (particle size, shape, size distribution, and concentration), surface chemistry, dispersion characteristics, dryness, and compatibility with the other components in the formulation. Table 9.3 summarizes the properties of selected fillers.

A significant number of fillers can be used to modify the properties of an adhesive. They can be either organic or inorganic. The most common types of fillers used in epoxy adhesive formulation are inorganic solids. They are commercially available in particle sizes greater than 0.1 μm and in various grades. Naturally occurring organic materials such as cellulose, cotton fiber, etc., are also sometimes used as fillers and extenders in epoxy formulations. Synthetic resins such as thermoplastic powders may be used as fillers; however, they are more often considered as a resinous modifier of toughening agent.

9.2 FORMULATING WITH FILLERS

Selection of the proper filler is based on a number of factors. The most important, of course, is the improvements in processing and end properties that it will provide. However, other important considerations when one is selecting a filler include availability, cost, surface chemistry, water adsorption, oil adsorption, density, and other physical or chemical properties of the filler grade itself.

A filler should be dry, nonreactive with the uncured resin, and of a neutral or only slightly basic pH. Adsorbed water, which is present in some degree in most fillers, inhibits dispersion. Thus, most fillers must be dried before they are added to the adhesive formulation. The drying process will drive off adsorbed moisture and gases from the surface of the filler. The filler should generally be nonreactive with the base resins or curing agents that

TABLE 9.1 Advantages and Disadvantages of Fillers in Epoxy Formulations (Depending on Type of Fillers Employed)[1]

Advantages	Disadvantages	Advantage or disadvantage (depending on application)
Lower cost of product	Increased weight	Increased viscosity
Reduced shrinkage on curing	Increased water absorption (depending on filler)	Increased thermal and electrical conductivity
Decreased exothermic temperature on curing	Loss of transparency	Reduced thermal expansion coefficient
Improved tensile shear strength	Difficulty in machining hard fillers	
Increased surface hardness		
Improved abrasion resistance		
Improved heat-aging properties		
Increased compressive strength		
Increased electrical strength		
Improved toughness if fibrous fillers are used		

are used in the formulation. If there is reactivity, stoichiometric considerations must be observed. Some hydroxyl-bearing fillers are reactive and can be used advantageously since they provide crosslinks to the epoxy system.

Certain types of fillers, even though unreactive, will affect the pot life and exotherm of the adhesive system. Generally, modifications of cure or reactivity are not the prime functions of

TABLE 9.2 Fillers for Common Epoxy Adhesive Formulations[2]

Filler	Function
Aluminum	Machineability
Alumina	Abrasion resistance, electrical
Aluminum silicate	Extender
Aluminum trioxide	Flame-retardant
Barium sulfate	Extender
Calcium carbonate	Extender
Calcium sulfate	Extender
Carbon black	Pigment, reinforcement
Copper	Machineability, electrical conductivity
Glass fiber	Reinforcement
Graphite	Lubricity
Iron	Abrasion resistance
Kaolin clay	Extender
Lead	Radiation-shielding
Mica	Electrical resistance
Phenolic or glass microspheres	Decreased density
Silica sand	Abrasion resistance, electrical properties
Silicon carbide	Abrasion resistance
Silver	Electrical conductivity
Titanium dioxide	Pigment
Zinc	Adhesion, corrosion resistance
Zirconium silicate	Arc resistance

TABLE 9.3 Selected Properties of Fillers Used in Epoxy Adhesive Formulations[3]

Property	Calcium carbonate	Kaolin	Talc	Mica	Glass microballoons	Hydrous alumina	Silica	Wood flour
Thermal conductivity, W/(m · K)	2.3	2.0	2.1	2.5	0.008	0.08	2.9	0.3
Coefficient of thermal expansion, 10^{-6}/K	10	8	8	8	8.8	4.4	10	5–50
Hardness, mohs	2.5–3	2	1	2.5–3	5	2	6.5–7	2
Density, g/cm^3	2.71	2.58	2.8	2.82	0.15–0.3	2.4–2.42	2.65	0.5–0.7
Dielectric constant	6.14	2.6	5.5–7.5	2–2.6	1.5	7	4.3	5

the filler. However, these effects need to be considered, especially when the curing process is critical. The effects of fillers on working life and exotherm are described in Sec. 9.3.12.

The filler concentration in any epoxy adhesive formulation will depend primarily on the following three factors:

1. Handling and viscosity characteristics of the formulated adhesive
2. Wetting characteristics of the filler with regard to other components in the formulation
3. Ultimate properties desired

The particle size, density, and oil absorptivity will determine the maximum weight loading that is achievable in a specific formulation. The high-oil-absorption lightweight fillers, such as diatomaceous silicas and chopped glass, can greatly increase the viscosity at lower filler loadings, even at concentrations of several parts per hundred. Medium-weight granular filler, such as talc, powdered aluminum, and alumina, can be used at loadings up to 200 pph. The heavier, nonporous fillers, such as aluminum oxide, silica, and the calcium carbonates, can be used at levels as high as 700 to 800 pph without causing the viscosity to be unworkable. Fillers with fine particle size will tend to settle out less than those with larger filler sizes.

Particle interactions resulting in aggregates of particles will adversely affect dispersion. Special surface treatments are provided to reduce these aggregation forces and achieve higher loadings and better suspension stability with less effect on viscosity. These surface treatments can be applied directly to the filler, and many grades of treated fillers are commercially available.

The filler surface treatment can also be applied as an additive to the adhesive formulation. Coupling agents, such as organotitanates, zircoaluminates, or organosilanes, are added to the formulation so that they preferentially find their way to the filler's surface and provide for optimal surface chemistry and adhesion between the filler and the matrix resin. These coupling agents are discussed in Chap. 10.

Particle size and shape will affect the degree of mixing required. Particles with large aspect ratios, such as fibrous fillers, and particles of large size are typically more difficult to disperse. In addition, high filler loading makes sufficient wetting of the filler more difficult because of the increased viscosities encountered.

In certain circumstances, blends of different-size fillers are utilized. With heavier or larger fillers, there is a greater tendency of the filler to settle, unless lightweight secondary fillers are employed as antisettling agents, or unless the filler is thoroughly milled into the system. A broad size distribution of a singular type of filler is an effective way of achieving antisettling and good dispersion properties.

To obtain the best results with fillers, they should be heated before being incorporated into the epoxy formulation. This will drive off any moisture or adsorbed gases from the surface of the fillers. Once thoroughly dry, the fillers may be stirred directly into the resin.

High-shear mixing is generally required for compounding of most formulations. The mixing vessel must sometimes be equipped with a vacuum to eliminate the possibility of air being beaten into the adhesive formulation. Often the formulation is warmed to a moderately elevated temperature to facilitate mixing. For very viscous formulations, roll-milling equipment may be required to achieve efficient mixing and dispersion of the filler. In some cases it may be necessary to employ a grinding device to accomplish the mixing with sufficient thoroughness. Unless complete and efficient mixing is accomplished, the final formulation will not possess the desired properties.

Solvent addition or blending the epoxy resin with low-molecular-weight diluents is another method of lowering the viscosity so that fillers can be efficiently added to the epoxy adhesive formulation. However, in these cases the formulator must address the high vapor pressures of the solvent or diluent (as well as various health, safety, and environmental issues). In the case of diluent addition, the reduction in crosslinking density and thermal or

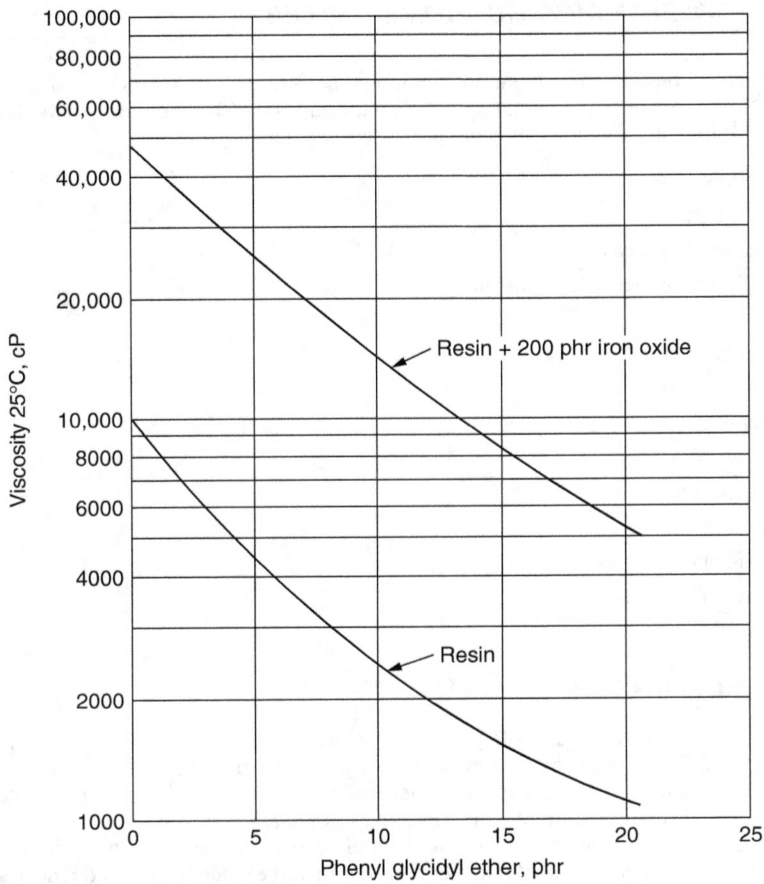

FIGURE 9.1 Effect of reactive diluent concentration on the viscosity of a filled epoxy resin.[4]

chemical properties must be considered. Figure 9.1 illustrates the viscosity of a filled and unfilled resin as a function of percentage of the diluent phenyl glycidyl ether.

In two-part epoxy adhesive systems, the filler can generally be incorporated into either the epoxy resin or the curing agent component. Prevailing factors will be the characteristics mentioned above (viscosity of each component, dispersion characteristics, etc.). However, the effect on the mix ratio and on the ease of mixing the two components prior to application must also be considered. Fillers are often used to provide practical mix ratios for two-component adhesives.

Fillers are available in many forms including particles, fibers, and mats or fabrics. Table 9.4 lists common forms of fillers and reinforcements that are used with epoxy adhesives. Note that epoxy adhesive film carriers such as fabrics or mats can be considered as a type of filler.

TABLE 9.4 Forms of Fillers and Reinforcing Materials Commonly Used with Epoxy Resins[5]

1. Powders	9. Continuous strand
2. Spheres	10. Yarn
3. Granules	11. Spun roving
4. Fibers	12. Fabric
5. Shreds	13. Woven roving
6. Whiskers	14. Chopped strands
7. Needles	15. Reinforced mats
8. Flakes	

9.3 PROPERTY MODIFICATION BY FILLERS

This section considers only the more common types of fillers for epoxy adhesive formulations. They are categorized by the specific contributions that they provide to adhesive properties. Modifications to the following properties are discussed:

- Materials cost
- Flow properties
- Bond line thickness
- Coefficient of thermal expansion
- Shrinkage
- Conductivity
- Electrical properties
- Specific gravity
- Mechanical strength
- Heat and chemical resistance
- Adhesion
- Pot life and exotherm
- Fire resistance
- Color

9.3.1 Materials Cost

Fillers are often used for the sole purpose of reducing the raw materials cost of the adhesive system. They do this by replacing relatively expensive synthetic organic components with inexpensive inorganic components, generally naturally occurring minerals. Fillers that are used for this purpose are often referred to as *extenders*.

Extenders may also provide one or several secondary functions in addition to cost reduction. Depending on the type and concentration of the extender, it could reduce shrinkage that occurs on polymerization or during aging, lower the coefficient of thermal expansion of the adhesive formulation, provide a higher viscosity to avoid adhesive starvation in the joint, or perform a number of other property improvements.

Extenders have two very broad classifications: inorganic and organic. Inorganic extenders are primarily naturally occurring minerals, and these are often used in epoxy formulations. Organic extenders are naturally occurring resins or fibrous particles. In either case, the most distinguishing features of these extenders are their low cost and availability.

Inorganic Extenders. There are many mineral extenders that are commonly used in epoxy adhesive formulations. The more important ones are summarized below.

Calcium carbonate is the most commonly used extender. It is widely available and low in cost, and it provides for improvements in certain performance properties. The material is a mineral that is mined throughout the world. Common forms of calcium carbonate include limestone, marble, calcite, chalk, and dolomite. It is manufactured by precipitation processes and is commercially available from a number of sources. Calcium carbonate is available in many different particle sizes and in various grades. To improve dispersion in certain resins, the filler is often coated with calcium stearate or stearic acid.

Silica is also often used as an extender in adhesive formulations. Similar to calcium carbonate, silica is an abundant mineral found in crystalline form (quartz) and amorphous form (diatomaceous silica). Diatomaceous silica is used more extensively than quartz because it is a softer material providing fewer machining and abrasive problems. There is also concern over respiratory problems possibly associated with inhalation of finely divided quartz. Although it is an excellent additive for increasing viscosity, diatomaceous silica has a low oil adsorption and very large surface area so that it is not a particularly good extender, since it cannot be easily incorporated into the formulation in high concentrations.

Kaolin, commonly called clay, is another naturally occurring mineral that is often used as an extender in adhesive formulations. It is a hydrated aluminum silicate with a hexagonal platelike crystal structure. In addition to reducing cost, kaolin clay provides shrinkage control, dimensional stability, viscosity increase, and pigmentation. Certain clays, such as bentonite, are highly alkaline and can accelerate or even catalyze the epoxy curing reactions. Therefore, their reactivity must be taken into consideration when one is calculating the stoichiometric mix ratio of resin and curing agent.

Talc is a hydrated magnesium silicate that is composed of thin platelets primarily white in color. Talc is useful for lowering the cost of the formulation with minimal effect on physical properties. Because of its platy structure and aspect ratio, these extenders are considered reinforcement. Polymers filled with platy talc exhibit higher stiffness, tensile strength, and creep resistance, at ambient as well as elevated temperatures, than do polymers filled with particulate fillers. Talc is inert to most chemical reagents and acids. The actual chemical composition for commercial talc varies and is highly dependent on the location of its mining site.

Organic Extenders. Organic extenders are primarily of two types: (1) fillers derived from organic materials and (2) low-cost, naturally occurring or synthetic resins. Of the first type, wood flour, shell flour, and other cellulosic fillers are the most common. They also provide a margin of mechanical property reinforcement because of their relatively high aspect ratio. Of the resinous types these are petroleum-based derivatives as well as soluble lignin and scrap synthetic resins.

Coal tar pitch is the most widely used resinous extender for epoxy resins. It is primarily used in surface coating formulations, but can also be used as a cost reducer and flexibilizer in epoxy adhesives and sealants. In addition to the increase in flexibility (and reduction in thermal and chemical resistance), coal tar pitch extenders provide excellent water resistance. Their primary applications, therefore, are often in the marine, pipe, tank, and general industrial maintenance areas.

Petroleum-derived bitumens are also used as extenders in epoxy formulations. High-boiling-point petroleum distillates can serve as low-cost extenders, but a compatibilizer such as an alkylphenol must be present in the mix to achieve compatibility between the resin and the extender. Furfural resins can also be used with epoxies, to reduce cost and to obtain increased resistance to acids.

9.3.2 Flow Properties

Controlling flow is an important part of the adhesive formulation process. Control of flow is important for several reasons.

1. It allows easy and reproducible metering and mixing of two-component systems prior to application.
2. It provides for certain application characteristics of the adhesive (brush, spray, trowel, penetration, etc.).
3. It can provide sag resistance (thixotropy) for adhesives that are applied to vertical surfaces.
4. It can provide for a practical and reproducible bond line thickness in the final joint.

The first three factors are generally controlled by the rheological properties of the liquid adhesive through the application of fillers in the formulation. The final factor can be controlled through the viscosity; however, other methods are also possible to control the bond line thickness.

Control of Viscosity. To ensure that adhesives and sealants function well during their application and end use, the formulator must be able to control the flow properties of the product. The challenge that faces the formulator is that the adhesive or sealant may need different flow characteristics at different times. For example, adhesives must flow readily so that they can be evenly applied to a substrate and wet out the surface. Yet, there should not be an excess of penetration into porous substrates, nor should the adhesive "run" or "bleed" to create a starved joint. Certain adhesives and sealants must be capable of convenient flow application by trowel or extrusion, but they must also exhibit sag and slump resistance, once applied. Therefore, the flow properties, or rheology, of the material must fit the desired method of application.

Adhesive and sealant manufacturers employ rheological additives for thickening and to prevent sag of their products. In practice, rheological additives may provide benefits in addition to viscosity or flow control. When properly formulated into adhesives and sealants, rheological additives can

- Lower energy needed to pump, mix, extrude, spray, etc.
- Prevent phase separation
- Improve suspension characteristics
- Increase cohesive strength through reinforcement of the base polymer
- Control bond line thickness
- Improve surface texture
- Reduce materials cost

In practice, formulators have a variety of additives available to manipulate these properties.

The primary rheological property of concern is viscosity, which can be increased by the addition of fillers. Fibrous fillers cause a larger viscosity increase than do particulate fillers. Finer filler particle size having a larger surface area will generally but not always result in higher viscosity than an equal concentration of larger particle sizes. Increased viscosity provides a method to reduce the flow characteristics of the adhesive. However, too high a viscosity can yield undesirable processing properties.

The maximum filler loading for any system is frequently set by the maximum viscosity allowable for its method of application. Table 9.5 shows the maximum amount of certain fillers that can be tolerated in a liquid epoxy for pouring. Figure 9.2 illustrates the degree of viscosity increase that can be obtained with various weight percentages of different fillers in a DGEBA epoxy diluted with dibutyl phthalate.

Control of Thixotropy. Often the adhesive application will require that the product be fluid for mixing and application, but it must not flow or sag once applied. For example, ASTM C920 defines a nonsag sealant as "one that permits application in joints on vertical surfaces without sagging or slumping." This property is called *thixotropy*. Thixotropic materials, such as tomato catsup, toothpaste, etc., undergo a decrease in viscosity when subject to shearing.

Although most fillers provide epoxy systems with viscosities that are unaffected by the shear rate, certain fillers can provide thixotropy that results in an adhesive that will not flow under low levels of stress (e.g., under its own weight when applied to vertical surfaces). Yet the product will exhibit lower viscosities when under higher levels of stress, such as when being dispensed or applied to a substrate.

TABLE 9.5 Maximum Amount of Filler in DGEBA Epoxy Resin (EEW = 180) for a Pourable Mixture[6]

Filler	Concentration, pph, at maximum pourable viscosity
Black iron oxide	300
Tabular aluminum oxide	200
Atomized aluminum	150
Graphite	50
Titanium dioxide	50
Calcium carbonate	30
Silica flour	30

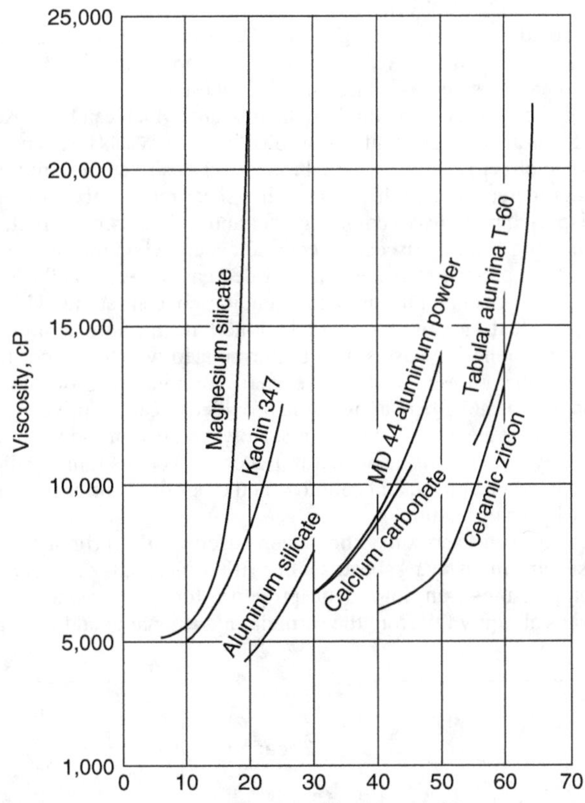

FIGURE 9.2 Viscosity of DGEBA epoxy resin diluted with dibutyl phthalate as a function of filler loading volume.[7]

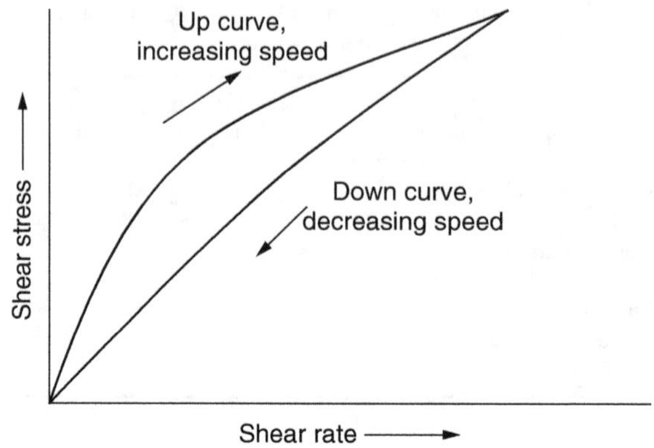

FIGURE 9.3 Hysteresis due to thixotropy.

Thixotropy provides a shear thinning effect; that is, viscosity decreases as the shear rate increases, and vice versa. This not only allows easy pumping, dispensing, and mixing of the adhesive, but also provides sag resistance once the adhesive is applied. The thixotropic fillers work by forming a temporary "structure" in the mixture, which can be broken down at high rates of shear. This structure is generally the result of van der Waals forces between molecules.

The thixotropic effect is shown in Fig. 9.3. The viscosity curves are for a material that is exposed to first increasing and then decreasing shear rates. Thixotropy results from the ability of the dispersed particles to come together and form network structures when at rest or under low shearing forces. Viscosity decrease occurs when this structure breaks down due to shearing stress and the resistance to flow decreases. (See Fig. 9.4.)

Thixotropy can be obtained at fairly low loading concentrations with colloidal silica, bentonite, metallic leafing powders, and hydrated magnesium aluminum silicates. If required, thixotropic adhesive pastes may be formulated which will not flow during cure even at elevated temperatures and which are useful for bonding loose-fitting joints.

The addition of asbestos fibers at one time provided excellent thixotropic adhesive formulations, especially at elevated temperatures. However, health and environmental regulations have severely limited the use of this material. Today, fumed silica, precipitated calcium carbonate, certain clays, and cellulose and other fibers offer thixotropic properties at relatively low levels of loading.

Fibers. At one time asbestos fibers were used as the primary thixotrope in the adhesive and sealant industry. Asbestos fiber filled materials are very resistant to sag under high temperatures—an important property for many adhesives and sealants. However, asbestos abruptly fell from the formulator's arsenal of additives primarily due to

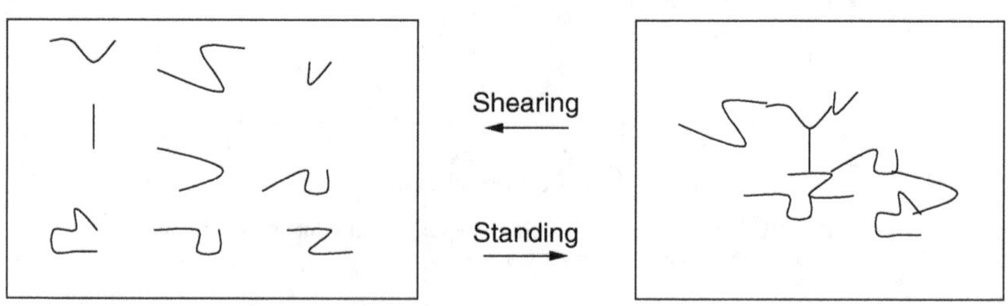

FIGURE 9.4 Thixotropic structures.

health and environmental factors. Asbestos fillers also stiffen and harden the resulting product; so it is not the ideal additive when adhesives and sealants need to remain flexible. Although no single material is a direct replacement for asbestos, the need was eventually filled by a number of materials.

Several types of polymeric fibers are used as thixotropes in both adhesive and sealant formulations. These are cellulose, polyolefin, and aramid fibers. A fiber that exhibits good thixotropy usually has several key characteristics including

- The ability to interact with the polymer matrix via dipolar and/or dispersion interactions
- The ability to achieve polymeric entanglement
- The presence of branchlike fibrils on the fiber which increase the surface area and promote rheological effects
- A high surface area with surface activity

It is interesting to note that these characteristics also are necessary to impart physical reinforcement and resulting cohesive strength improvement to the adhesive or sealant.

Cellulose fibers (e.g., Interfibe RT Cellulose Fiber from Interfibe) are perhaps the best accepted thixotrope of this group. This material provides good thixotropy, reinforcement of the bulk polymer, minimal surface texture, and low cost. Cellulose fibers are claimed to provide greater formulation latitude than the other thixotropes at reduced cost.

Studies have been conducted on adhesive and sealant systems using cellulose fibers as a thixotrope in epoxy adhesives. Table 9.6 shows thixotropic epoxy adhesive formulations and resulting properties using fumed silica and *reinforcing thixotropic* (RT) cellulose fiber additives. Note that a certain percentage of fumed silica is still required in the cellulose fiber filled formulations for optimal nonsag properties. The cellulose fibers' low specific gravity (1.1) relative to other filler materials also contributes to weight savings and further improves the sag resistance of these systems.

Polyolefin (Spectra from Honeywell-Allied Signal) and aramid (Kevlar from DuPont) fibers have much lower surface areas and polarities than asbestos fibers and do not bond

TABLE 9.6 Cellulose Fibers in an Epoxy Adhesive[8]

Formulation	A (control)	B
Epoxy resin 1, pbw	47	40
Epoxy resin 2, pbw	22	18.7
Dicyandiamide, pbw	3.7	3.1
Cure accelerator, pbw	1.3	1.1
Ground limestone, pbw	20	28.4
Fumed silica, treated, pbw	6	3
Cellulose fibers, pbw	0	5.6
Properties		
Viscosity, P		
Initial	86	91
Aged	105	120
Slump resistance ($^1/_2$ in × $^1/_2$ in × 4 in)	No movement	No movement
Tensile shear strength, psi		
Initial	1770	1801
250-h humidity	1579	1727
250-h salt spray	1288	1592

well with all polymeric matrix resins. However, aramid fibers have been proposed as a replacement for asbestos or fumed silica.[9] They are claimed to provide better sag resistance than fumed silica. Disadvantages are that aramid fibers are very expensive, and they increase the modulus as much as 2.5 times that of fumed silica.

Fumed Silica. Today colloidal silica (fumed silica) is the most common thixotropic agent in epoxy resins. Fumed silica, an amorphous silicon dioxide, is a versatile, efficient additive used for flow control and thixotropy. Fumed silica has long been the dominant thixotrope employed in the adhesive and sealant industry.

Fumed silica is generally incorporated at concentrations of less than 10 pph. Epoxy systems based on lower-viscosity resins generally tend to hold thixotropic action better at elevated temperature than systems based on the higher-viscosity resins.

Fumed silica is typically available with sizes of 7 to 40 nm and surface areas ranging from 50 to 380 m^2/g. Unlike precipitated silica, fumed silica has no internal surface area. The specific gravity of fumed silica is approximately 2.2. Because of its high surface area to weight ratio, formulations generally require only a minor amount of fumed silica (1 to 5 percent by weight) to achieve thixotropic properties.

The surface chemistry of fumed silica is extremely important because of its influence on the rheological behavior of the formulation. Three types of chemical groups can be formed on the surface of the particle depending on the processing procedures:

1. Isolated hydroxyl
2. Hydrogen bonded hydroxyl
3. Siloxane

The isolated and the hydrogen bonded hydroxyl groups are hydrophilic sites, whereas the siloxane is a hydrophobic site. Thus, fumed silica grades are generally characterized by their surface area and whether they are hydrophilic (standard grade) or hydrophobic. The hydrophilic silica is most effective in nonpolar and medium polar media. The performance of hydrophilic fumed silica is often improved by adding a polar substance such as ethylene glycol, glycerin, or some secondary amines to the formulation. In medium polar to polar media, hydrophobic fumed silica is a more efficient thickening agent and generally preferred. Hydrophobic fumed silica is also noted for providing superior moisture resistance in adhesives, sealants, and coatings.[10]

The thixotropic characteristics provided by fumed silica are due to its ability to develop a loosely woven, latticelike network by hydrogen bonding between particles. This network raises the apparent viscosity of the system, increases the cohesive forces, and contributes to the suspension of the solid. Because the hydrogen bonds themselves are relatively weak, they are easily disrupted through the action of an applied stress or shearing force and quickly reform when the stress or shearing force is removed.

Parameters that are important to the performance of fumed silica systems include

- The nature of the resin system (polarity)
- Concentration of the silica
- Grade of silica used (particle size, surface area, density, surface chemistry, etc.)
- Degree of dispersion
- Presence of additives in the formulation other than the fumed silica

At times the results of the formulation are less than expected because these factors are not considered or understood relative to the final rheological properties. For example, proper dispersion can maximize the efficiency of both hydrophilic and hydrophobic fumed silica. To ensure proper dispersion, the addition of silica to the formulation in the right sequence and the effectiveness of the dispersion equipment become very important. High-shear mixing

TABLE 9.7 Comparative Sag Resistance Properties of Fumed Silicas[11]

Thixotrope	Type surface	Maximum nonsag film thickness, mils; aging period at 60°C			
		Initial	1 Week	2 Weeks	4 Weeks
Cab-O-Sil PTG	Hydrophilic	60	60	45	26
Aerosil 200 / Aluminum Oxide C	Hydrophilic	60	40	25	13
Aerosil COK 84	Hydrophilic	60	60	40	16
Cab-O-Sil N70-TS	Hydrophobic	60	60	60	60
Aerosil R972	Hydrophobic	60	30	25	16

equipment will improve the efficiency of fumed silica in the formation. In most cases, silica should be added to the resin directly or to the more viscous part of the formulation with as little solvent or diluent as possible. Incorporating the silica before adding any fillers or pigments ensures homogeneous distribution. Dispersing the silica in a concentrated base, or master batching, provides optimum efficiency and stability.

Fumed silica is often used in 100 percent solids, liquid polymers. With epoxy adhesives and sealants only a few percent by weight of the additive will eliminate common problems such as slumping and separation. The fumed silica also raises the effective viscosity of the base resin to prevent other components from settling while the extrudability or spreadability is unaffected. Also note that fumed silica provides a surface that is free of texture. This is important in architectural-grade paints and sealants.

Comparative sag resistance properties of various commercial types of fumed silica are shown in Table 9.7. In most liquid epoxy resin systems, fumed silica is only employed at a 1 to 3 wt percent basis to provide thixotropic characteristics.

Precipitated Calcium Carbonate and Other Minerals. Precipitated calcium carbonate ($CaCO_3$) functions as a thixotrope in sealant and adhesive formulations as well as being a low-cost extender (often used in the 40 to 50 percent by weight range) and reinforcement. Ultrafine (<100 nm) precipitated calcium carbonate provides the greatest efficiency. These have surface areas from 15 to 30 m^2/g. $CaCO_3$ is hydrophilic, and as a result it has a tendency to agglomerate in organic polymers and plasticizers.

Precipitated calcium carbonates are generally surface-treated to render them hydrophobic and to improve their dispensability in hydrophobic systems. Conventional surface treatment of $CaCO_3$ is with fatty acids. However, the use of different types of acids yields subtle differences in filler wettability and polymer compatibility.

Other mineral additives have been used for many years as functional extenders and fillers in adhesives and sealants. These include kaolin (hydrated aluminum silicate), bentonite (hectorite clay), talc (magnesium silicate), and attapulgite (hydrated magnesium aluminum silicate) additives.[12,13] Kaolin and talc are considered to be viscosity thickeners whereas attapulgite is more of a conventional thixotrope. They are considered to be very cost-effective rheological additives.

Kaolin is a common inexpensive filler used primarily as an extender in adhesive and sealant formulations. The cost of the adhesive is reduced because the kaolin addition increases the product's volume. Depending on the grade, kaolin can also

- Control viscosity to prevent drip or sag
- Improve surface smoothness
- Provide reinforcement
- Reduce shrinkage

The finer particle sizes build viscosity more efficiently and provide better reinforcement than do their coarser counterparts. Fine grades of kaolin have particle size of 0.5 μm or less. Kaolin has a hexagonal particle with the finer particles occurring as platelets and coarser particles forming stacks or books of platelets. It is relatively soft and is a nonabrasive filler. As with other solid rheological additives, the key parameters in choosing a particular grade are particle size, shape, and particle distribution. Kaolin is also available in hydrous and calcinated (where dehydroxylation is used to remove water) grades. The calcinated grades have a more porous particle shape.

Special surface modifications are available to further improve reinforcement. The objective of the surface treatment is to increase filler loading and/or improve physical properties without loss of rheological characteristics. A variety of surface-modified kaolins have been introduced including clays treated with silane, titanate, polyester, and metal hydroxide. Silane-treated kaolin is used in applications requiring maximum aging characteristics in the service environment.

Bentonite is a colloidal clay that is both hydrophilic and organophilic. It is water-swelling with some types of clay absorbing as much as 5 times their own weight in water. It is used in emulsions, adhesives, and sealants. It is a gritty, abrasive white particle filler. A macroscopic particle of bentonite is composed of many thousands of stacked and/or overlapped submicroscopic flakes.

Talc is also often used as an extender in epoxy adhesives and sealants, but it also has flow control properties. Talc is used in higher-solids, high-viscosity applications such as caulking compounds, automotive putties, mastics, and sealants. Talc is a hydrophobic and organophilic material.

Talcs are either platy or acicular in particle shape. Thin platelet particles have aspect ratios varying from 20 : 1 to 5 : 1. Coarse particle sizes (10 to 75 μm) are commonly used in these applications at loading levels of 5 to 30 percent. Fine talcs (1 to 10 μm) are more expensive and require intensive dispersion processes. Platy grades enhance barrier properties and air, water, and chemical resistance.

Polymers filled with platy talc exhibit higher stiffness, tensile strength, and creep resistance than do polymers filled with standard particulate fillers. These properties are maintained at both ambient and elevated temperatures. Surface treatments for talc particles include magnesium and zinc stearates, silanes, and titanates.

Attapulgite is a very cost-effective thixotrope.[14] However, unlike kaolin or talc, it is not used at high loadings and thus is not considered as an extender. Attapulgite consists of needlelike particles with a size of about 0.1 μm. Thickening occurs upon the mutual separation of individual particles as the porous mineral surface is exposed to the surrounding medium and forms stable colloids. Cationic surfactants often enhance the performance of attapulgite thixotropes in organic systems.

Typical usage levels range from 2 to 8 percent by weight. Attapulgite particles are easy to disperse and commonly added to the formulation with other dry components. At levels ranging from 10 to 20 percent attapulgite is considered to be a partial asbestos replacement in polymeric-based sealants for the construction and automotive industries.

Other Thixotropic Additives. Microcellular fillers such as glass and plastic are also used to provide nonsag properties to adhesives and sealants. However, their mode of operation is very much different from that of the rheological modifiers mentioned above. Microcellular fillers work because they drastically reduce the specific gravity of the product. Thus, because the adhesive or sealant is lighter, there is less stress on the adhesive or sealant due to the forces of gravity.

Microcellular fillers such as glass microballoons have been around for some time and are commonly used in syntactic foam. However, they are expensive raw materials. Sibrico Corporation, of Hodgkins, IL, has recently introduced a new, relatively inexpensive filler called Sil-Cell. This is a microsized, multiwall hollow glass filler. Sil-Cells range in size from 1 to 150 μm. The particles have both spherical and irregular geometries, which is claimed to provide advantages in improving tensile strength. The resin matrix mechanically keys to the microglass bubble whereas spherical bubbles must rely on coupling agents for maximum interfacial strength.

9.3.3 Control of Bond Line Thickness

If the adhesive has a propensity to flow easily before and during cure, then one risks the possibility of a final joint that is starved of adhesive material. If the adhesive flows only with the application of a great amount of external pressure, then one risks the possibility of entrapping air at the interface and too thick of a bond line. These factors could result in localized high-stress areas within the joint and reduction of the ultimate joint strength.

Flow characteristics can be regulated by the incorporation of fillers, by the use of scrims or woven tapes as "internal shims" within the adhesive itself, or by the careful regulation of the cure cycle. All these options along with a few more are available to the adhesive formulator and end user. Generally, fillers are incorporated to control the viscosity of the adhesives as well as other properties such as thixotropy and sag resistance.

The type and amount of fillers are chosen so that a practical bond line thickness will result after application of the necessary pressure (usually only contact pressure, approximately 5 psi) during cure. Ordinarily, the objective is a bond line thickness of 2 to 10 mils. Consideration, of course, must be given to the curing temperature. Viscosity of the formulation could drastically be reduced at elevated temperatures, and unless there is a furrow designed into the joint to contain the adhesive, much of the adhesive could flow out of the joint area before the adhesive is completely cured.

Glass, nylon, polyester, and cotton fabric or mat are often used as a carrier in tape or film adhesive systems. In addition to being a carrier and a reinforcement, the strands of the fabric offer an "internal shim" so that the bond line cannot be thinner than the thickness of these strands. Sufficient pressure need only be applied to cause the adhesive to flow so that the shims meet the substrate surfaces to provide a positive stop. Paper, mat, and other carrier materials may also be used for this purpose.

Glass or polymeric microballoons, incorporated directly into the adhesive formulation, can also provide the shimming function. Here the diameters of the microballoons are the positive stops that will prevent too thin of a bond line.

Another option is to design mechanical shims into the joint itself. The parts to be assembled are designed with lips or stops so that the adhesive cannot flow out of the joint area or so that a certain predetermined adhesive thickness is always maintained.

9.3.4 Coefficient of Thermal Expansion

Depending on the substrate, the curing temperatures, and the service temperatures that are expected, the adhesive formulator may want to adjust the coefficient of thermal expansion of the adhesive system. This will lessen internal stresses that occur due to differences in thermal expansion between the substrate and the adhesive. These stresses act to degrade the joint strength.

There are several occasions when the difference in coefficient of thermal expansion between the substrate and adhesive will result in internal stresses in the joint. Common occurrences are (1) when the cured joint is taken to a temperature that is different from the curing temperature and (2) when the joint is exposed to thermal cycling.

When a liquid adhesive solidifies, the theoretical strength of the joint is reduced because of internal stresses and stress concentrations that usually develop. The most common cause of internal stress is due to the difference in the thermal expansion coefficients of the adhesive and the adherends. These stresses must be considered when the adhesive or sealant solidifies at a temperature that is different from the normal temperature that it will be exposed to in service. Figure 3.7 shows thermal expansion coefficients for some common adhesives and substrates. The significant difference in thermal expansion coefficient between epoxy adhesives and some substrate materials means that the bulk adhesive will move more than 10 times as far as the substrate when the temperature changes, thereby causing stress at the interface.

There are several possible solutions to the expansion mismatch problem. One is to use a resilient adhesive that deforms with the substrate during temperature change. The penalty here is possible creep of the adhesives, and highly deformable adhesives usually have low cohesive strength. Another approach is to adjust the expansion coefficient of the adhesive to a value that is nearer to that of the substrate. This is generally accomplished by formulating the adhesive with specific fillers to "tailor" the thermal expansion coefficient.

The general effect of most fillers is to reduce the coefficient of thermal expansion of the cured epoxy resin in proportion to the degree of filler loading. Certain fillers, such as zirconium silicate and carbon fiber, have a negative coefficient of thermal expansion. These are very effective in lowering the expansion rate of the epoxy, especially at elevated temperatures.

Ideally, the coefficient of thermal expansion should be lowered (or raised) to match that of the material being bonded. With two different substrate materials, the adhesive's coefficient of thermal expansion should be adjusted to a value between those of the two substrates. This is generally done by using fillers, as shown in Fig. 9.5.

It is usually not possible to employ a sufficiently large filler loading to accomplish the degree of thermal expansion modification required to match the substrate. High loading volumes increase viscosity to the point where the adhesive cannot be applied or cannot wet the substrate. For some applications and with some fillers, loading volumes up to 200 pph may be employed, but optimum cohesive strength values are usually obtained with lesser amounts.

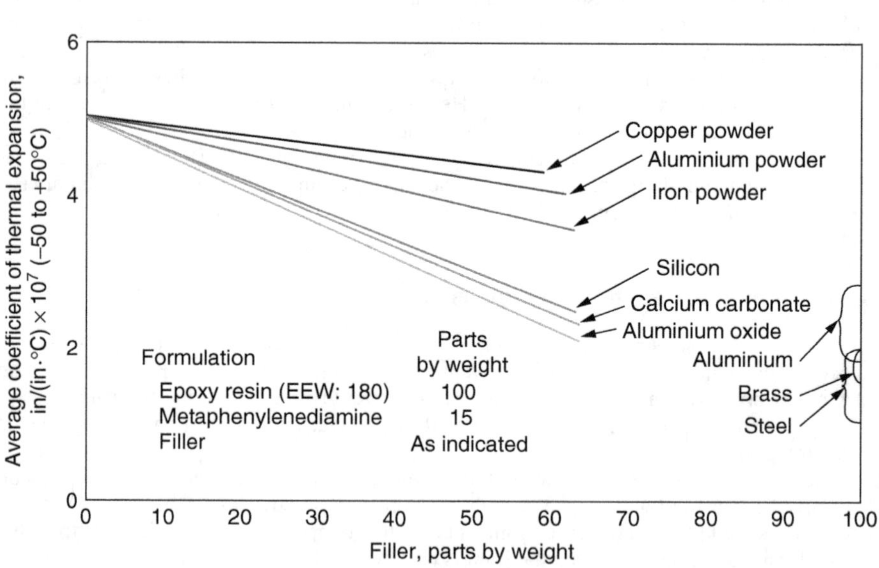

FIGURE 9.5 The coefficients of thermal expansion of filled epoxy resins compared with those of common metals.[15]

9.3.5 Shrinkage

Nearly all polymeric materials (including adhesives and sealants) shrink during solidification. Sometimes they shrink because of escaping solvent, leaving less mass in the bond line. Even 100 percent reactive adhesives, such as epoxies and urethanes, experience some shrinkage because their solid polymerized mass occupies less volume than the liquid reactants. Table 3.6 shows typical percentage cure shrinkage for various reactive adhesive systems.

The result of such shrinkage is internal stresses at the adhesive-substrate interface and the possible formation of cracks and voids within the bond line itself. Depending on the primary base resin, the adhesive formulator may need to reduce the amount of shrinkage when the adhesive hardens. This can be accomplished in several ways.

Elastic adhesives deform when exposed to such internal stress and are less affected by shrinkage. Fillers also reduce the rate of shrinkage by bulk displacement of the resin in the adhesive formulation. This results in an increase in the inherent bond strength of the adhesive. Fillers may improve operational bond strength by 50 to 100 percent although all of this improvement is not due to reduced shrinkage (see Sec. 9.3.11) .

9.3.6 Electrical and Thermal Conductivity

In certain applications, such as in the electrical and electronic industries, adhesive systems must have a degree of electrical and/or thermal conductivity. Electrical conductivity is, of course, important in adhesives that must make electrical interconnection between components and in adhesives that must provide *electromagnetic* or *radio-frequency interference* (EMI and RFI) functions.

Thermal conductivity is also important in highly integrated electronic applications where the heat generated by components must be transferred to a heat pipe or by some other means outside the electronic package. Thermal conductivity within adhesive systems is also a means of reducing exotherm and stresses that could develop during the curing cycle or other excursions to elevated temperatures.

Increased electrical or thermal conductivity is generally accomplished by adding conductive fillers to the epoxy adhesive formulations. For optimum electrical or thermal conductivity, the particles must be so concentrated that they come into contact with one another. This generally requires such a high level of filler loading that other properties such as flexibility and tensile shear strength may be significantly degraded.

Electrical Conductivity. Unmodified polymeric resins are natural insulators and do not exhibit electrical conductivity. There are certain applications, however, where electrically conductive adhesives provide a significant value. One such application is the use of conductive adhesives as an alternative for wire or circuit board soldering. Another application, with less of a requirement for conductivity, is the assembly of components that are shields or protection from electromagnetic interference.

Appropriate fillers have been used to produce adhesives with high electrical conductivity. Note that, regardless of the adhesive system itself, electrical conductivity is improved by minimizing the adhesive bond line and by minimizing the organic or nonconductive part of the adhesive.

Electrically conductive adhesives owe their conductivity as well as their high cost to the incorporation of high loadings of metal powders or other special fillers of the types shown in Table 9.8. If enough metal particles are added to form a network within the polymer matrix, electrons can flow across the particle contact points, making the mixture electrically conductive. Virtually all high-performance conductive products today are based on flake or powdered silver. Silver offers an advantage in conductivity stability that cannot

TABLE 9.8 Volume Resistivities of Metals, Conductive Plastics, and Various Insulation Materials at 25°C[16]

Material	Specific gravity, g/cm^3	Volume resistivity, $\Omega \cdot$cm
Silver	10.5	1.6×10^{-6}
Copper	8.9	1.8×10^{-6}
Gold	19.3	2.3×10^{-6}
Aluminum	2.7	2.9×10^{-6}
Best silver filled inks and coatings		1×10^{-4}
Best silver filled epoxy adhesives		1×10^{-3}
Unfilled epoxy adhesives	1.1	10^{14}–10^{15}

be matched by copper or other lower-cost metal powders. Conductive carbon (amorphous carbon or fine graphite) can also be used in conductive adhesive formulations if the degree of conductivity can be sacrificed for a lower-cost adhesive.

Conductive adhesives are generally formulated from base polymers that are low-viscosity, thermosetting resins such as epoxies. Where elastomeric properties are required, silver-filled flexible epoxy and silver-filled silicone rubber systems are commercially available.

Generally, the adhesive electrical conductivity varies according to the direction in which the conductivity is measured. Often, applied thin films show isotropic characteristics due to the shearing forces that occur during the application process. This tends to align the filler particles and create greater conductivity in the direction of the application stress. However, certain adhesives require directional conductivity. For example, conductive pressure-sensitive adhesive tapes in certain applications may require conductivity in the *z* axis and not necessarily the *x* or *y* axis. EMI applications generally require the same degree of conductivity in all directions. There are several methods of orienting the conductive filler particles in the desired direction. New anisotropic conductive films have been developed with arrayed conductive particles.[17]

Thermal Conductivity. Many of today's electronic products feature miniaturization. In these applications, higher thermal conductivity and better dimensional stability are required from adhesive systems. Thermal management has become a significant area of development, and thermally conductive adhesives provide a way to transfer heat away from sensitive electronic components.

Most unmodified polymeric resins have a very low thermal conductivity. There are certain applications where high levels of thermal conductivity are required. For example, power electronic devices and other heat-generating components are bonded to heat sinks and other metal sources. Metal powder filled adhesives, such as those described above for electrically conductive adhesives, can conduct both heat and electricity.

Some applications, however, must conduct heat but not electricity. In these applications the adhesive must permit high transfer of heat plus a degree of electrical insulation. Fillers used for achieving thermal conductivity alone include aluminum oxide, beryllium oxide, boron nitride, and silica. Table 9.9 lists thermal conductivity values for several metals as well as for beryllium oxide, aluminum oxide, and several filled and unfilled resins.

Theoretically, boron nitride is an optimum filler for thermally conductive adhesives. However, it is difficult to fill systems greater than 40 percent by weight with boron nitride. Beryllium oxide is high in cost, and its thermal conductivity drops drastically when it is mixed with organic resins. Therefore, aluminum, aluminum oxide, and copper fillers are commonly used in thermally conductive adhesive systems.

TABLE 9.9 Thermal Conductivity of Metals, Oxides, and Conductive Adhesives at 25°C[18]

Material	Thermal conductivity, Btu/(h · °F · ft²/ft)
Silver	240
Copper	220
Beryllium oxide	130
Aluminum	110
Aluminum oxide	20
Best silver filled epoxy adhesives	1–4
Aluminum filled epoxy (50%)	1–2
Unfilled epoxies	0.1–0.15

Titanium or beryllium oxide also provides a degree of improvement in thermal conductivity to epoxy systems. Magnesium oxide and aluminum oxide have also been commonly used for this purpose, although the degree of improvement is not as great as with the fillers discussed above. The effect of various fillers on the thermal conductivity of cured adhesive is shown in Fig. 9.6. The incorporation of metal fibers with metal powders has been shown to provide synergistic improvement to the thermal conductivity of adhesive systems,

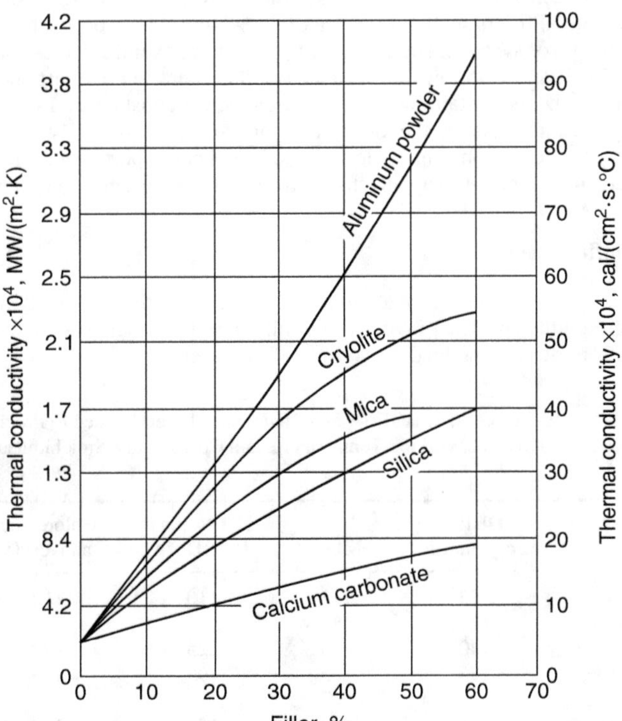

FIGURE 9.6 Effect of various fillers on adhesive thermal conductivity.[21]

probably due to the effect of the fibers physically connecting the particulate filler in the system.[19,20]

Thermal conductivity also helps to improve heat transfer during cure. This reduces exothermic temperatures and extends pot life, particularly at high filler loadings. Shrinkage during cure is also reduced, as explained in the sections above.

9.3.7 Electrical Properties

Nonconductive fillers are employed with electrical-grade epoxy adhesive formulations to provide assembled components with specific electrical properties. Metallic fillers generally degrade electrical resistance values, although they could be used to provide a degree of conductivity as discussed above.

The effect of electrical-grade fillers (e.g., silica) on the electrical properties of the adhesive is usually marginal. Generally fillers are not used to improve electrical resistance characteristics such as dielectric strength. The unfilled epoxy is usually optimal as an insulator. Also under conditions of high humidity, fillers may tend to wick moisture and considerably degrade the electrical resistance properties of the adhesive.

The one exception where certain fillers can provide electrical property improvement is in arc resistance. Here hydrated aluminum oxide and hydrated calcium sulfates will improve arc resistance if cure is sufficiently low to prevent dehydration of the filler particles. Electrical-grade fillers generally improve the arc resistance of cured epoxy systems, as indicated in Table 9.10.

Mineral fillers generally result in increased dielectric constants and dissipation (or loss) factor. Some mineral fillers such as titanium dioxide have been reported to lower the dissipation factor of epoxy systems. Dielectric constants of epoxy adhesive can significantly be reduced by using fillers with low dielectric constant (generally polymers such as HDPE or PTFE). Hollow spheres containing air (with a dielectric constant of 1.0) can be used as fillers to significantly reduce the dielectric constant of epoxy. By use of inorganic microballoons, dielectric constants as low as 1.4 have been reported at a filler loading volume of 20 percent compared to a dielectric constant of 4.4 for unfilled systems.

9.3.8 Specific Gravity

The majority of mineral fillers have a higher specific gravity than do epoxy resins and will, therefore, increase the specific gravity of the fully cured product. The increase in specific gravity is in proportion to the loading volume of the filler.

TABLE 9.10 Arc Resistance, s, of DGEBA as a Function of Filler, Loaded to Produce an Initial Viscosity of 4000 to 5000 cP at Working Temperature, and Cured with a Stoichiometric Amount of Listed Curing Agent[22]

Filler	DETA/ acrylonitrile	MDA	HHPA plus 1% BDMA	Chlorendic anhydride	Methane diamine
No filler	85	115	110	110	95
Mica, 1000-mesh	125	180			140
Silica, 325-mesh	140	185	225	190	185
Zirconium silicate, <1.5 μm	125	155	195	140	
Hydrated alumina, 0.08–0.12 μm	135	150	150	130	155

Fillers with a density lower than that of epoxy can be used to provide reduced specific gravity in cured products. These are usually gas-filled microballoons. Although they generally bring about a significant increase in viscosity, the microballoon filled epoxies (sometimes called *syntactic foam adhesives*) are often used in marine applications where low density and buoyancy are important criteria.

Properties of an epoxy system filled with glass microballoons are shown in Table 9.11. Other low-density microballoons, based on phenolic and other materials, have been developed for use in epoxy adhesive formulations.

9.3.9 Cohesive Mechanical Properties

Only certain fillers can be used to increase the cohesive strength of the cured epoxy adhesive formulation. Generally fibrous or flake fillers, such as talc, glass fiber, or mica, will bring about a certain degree of increase in tensile and flexural strength. However, to notice a significant increase in physical strength, these fillers must be used at a relatively high concentration. Particulate fillers, on the other hand, usually reduce room temperature tensile and flexural strength, although some improvement may be experienced at elevated temperatures.

Tensile strength and flexural modulus are generally increased and ultimate elongation is reduced in proportion to the amount of filler, when tested at room temperature. The reverse is generally true when the cured epoxy is tested at elevated temperatures. Impact strength is generally adversely affected by particulate fillers.

The addition of particulate fillers generally decreases compression fatigue, but increases ultimate compressive modulus and compressive yield strength, because of a stiffening effect. Compressive strength as a function of filler loading is shown for a cured epoxy formulation in Fig. 9.7.

The reinforcement of adhesives has typically been achieved through the use of fibrous and flake fillers. Asbestos historically has been one of the most widely used reinforcing fibers in adhesive systems due to its needlelike shape. However, it is now seldom used because of health concerns.

One material that has been used to replace asbestos in certain applications is wollastonite or calcium metasilicate. This is also a fibrous mineral filler but with a lower aspect ratio than asbestos. Surface-treated versions are available to improve adhesion to the epoxy matrix. It can be applied at relatively high loading levels to provide for high strength and improvements in moisture resistance.[25]

Talc or hydrated magnesium silicate is another mineral that is used to reinforce epoxy adhesives. It has a platelike structure that provides good stiffness and creep resistance at elevated temperatures. It also provides good electrical and chemical resistance characteristics.[26] It is relatively inexpensive and disperses well in the resin.

TABLE 9.11 Properties of DGEBA Filled with Glass Microballoons Compared with Resin Filled with Mica (Epoxy Cured with 12 pph Diethanolamine)[23]

Property	Glass microballoon, 43 pph	1000-Mesh mica, 100 pph
Density, pcf	53	102
Tensile strength, psi	4660	7700
Tensile modulus, psi x 106	0.44	1.2
Compressive strength, psi	10,650	17,970
Compressive modulus, psi X 106	0.42	1.0

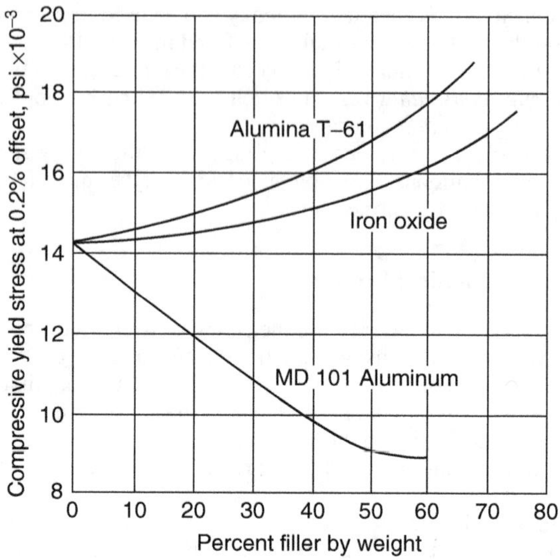

FIGURE 9.7 Effect of filler concentration on compressive strength of DGEBA cured with 8 pph of DETA.[24]

Calcium carbonate filler is a particulate filler that is generally employed to impart thixotropic properties and as an extender. However, improvements in physical properties such as tensile strength, tear, and elongation have been noticed as a function of surface area. This effect is noted more so with the coated $CaCO_3$ products.

Surface treatment also improves adhesion to the polymer matrix and resulting physical properties. The combination of particle size and surface treatment is critical in the selection of precipitated calcium carbonate fillers to obtain desired properties. Often graded combinations of ultrafine precipitated $CaCO_3$ and larger $CaCO_3$ particles are used for optimum properties and value.

The improvements in adhesive strength of cured epoxy joints that are attributable to fillers are not as much related to the improved cohesive characteristics of the adhesive as to the reduction in internal stress due to modification of the coefficient of thermal expansion, shrinkage, etc.

9.3.10 Heat and Chemical Resistance

Fillers improve the thermal properties of the cured epoxy formulation by bulk displacement of organic components. This reduces long-term shrinkage and thermally induced weight loss. However, the glass transition temperature is not significantly affected and in some cases may be lowered.

The main effect of fillers on thermal properties is achieved through improvement in thermal shock resistance. This is achieved via modification of the thermal expansion coefficient. Differences in particulate fillers have been noticed with regard to thermal shock resistance. Silica, for example, provides relatively poor thermal shock improvement. Mica and fibrous fillers used in relatively high loading provide very good improvement. However, their use is often precluded due to their handling difficulties. Fillers having high thermal conductivity will also increase the heat transfer in the cured epoxy system. The improved heat dissipation will result in improved thermal cycling properties.

Aluminum powder, in particular, is frequently employed at relatively high concentrations in high-temperature epoxy adhesive formulations. The filler provides improvement in both tensile strength and heat resistance, and it increases the thermal conductivity of the adhesive. Aluminum powder fillers also reduce undercut corrosion and, hence, improve adhesion and durability of epoxy adhesive between bare steel substrates. It is believed that this is accomplished by the aluminum filler providing a sacrificial electrochemical mechanism.[27]

Fillers often have a significant effect on the moisture resistance, the moisture-vapor transmission rate, and the solvent and chemical resistance of the cured adhesive bond. The effect, however, can be in either direction. Some fillers such as calcium carbonate tend to lower acid resistance, whereas others, such as silica or aluminum, may tend to lower alkali resistance.

Many fibrous fillers exhibit a wicking action for moisture, and this is particularly true of high-surface-energy glass fibers and especially when the fibers are exposed, such as when the cured epoxy is machined or a crack is formed in the adhesive resin matrix. For this reason glass fibers that are used in adhesives are often treated with a coupling agent such as an organosilane to improve the bond between the fiber and the epoxy matrix.

Particulate fillers, on the other hand, are believed to extend the pathway along which the water must diffuse, resulting in a reduction in the rate of water absorption. Silica flour (1 to 75 μm) has been reported as improving the boiling water resistance of epoxy films. Figure 9.8 shows the water absorption rate at 40°C of water in DGEBA epoxy filled with various types of materials.

9.3.11 Adhesive Properties

Fillers have very little effect on the actual adhesion mechanism, since most of this is controlled by intimate contact of the liquid epoxy and the substrate. However, as mentioned

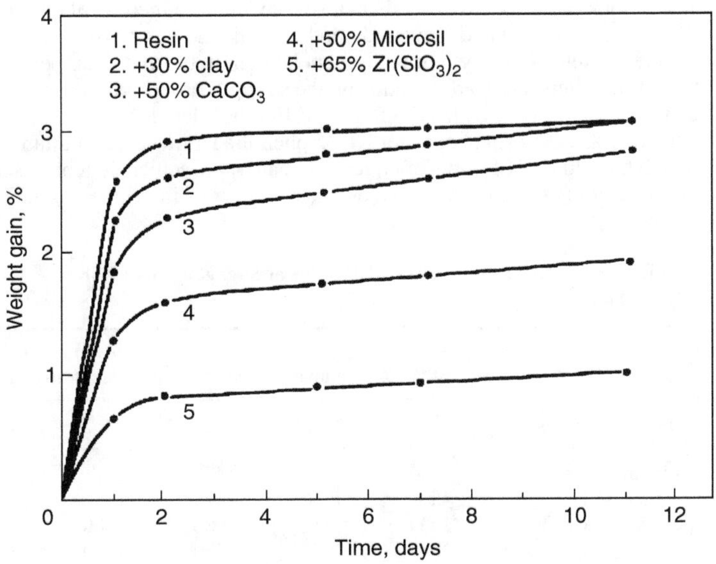

FIGURE 9.8 Water absorption versus filler for cured DGEBA.[28]

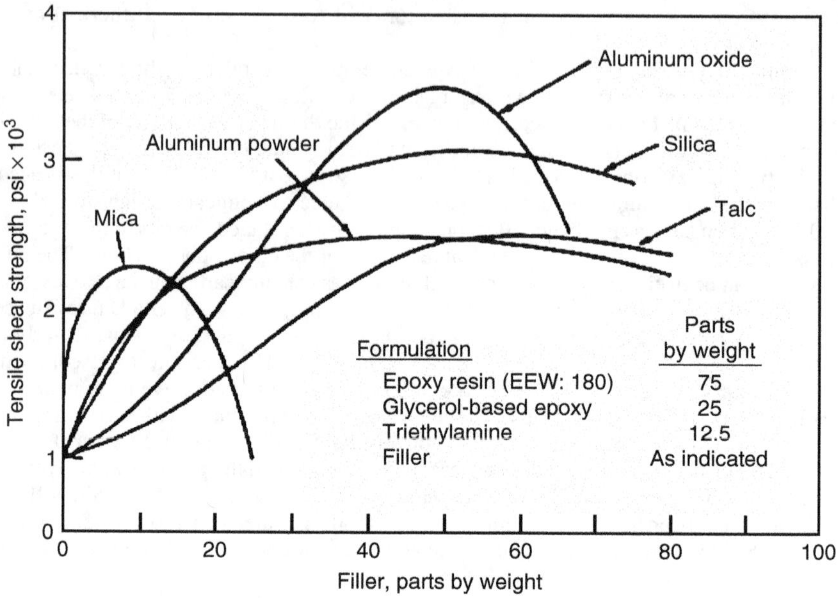

FIGURE 9.9 The effect of mineral extenders on the tensile shear strength of epoxy adhesive (the substrate is aluminum).[29]

above, fillers improve the measured adhesive strength of cured epoxy joints by decreasing internal stresses related to shrinkage and differences in the coefficient of thermal expansion. Because of this, one can generally observe a significant improvement in tensile shear strength with the addition of fillers. Fillers also provide for a method of controlling viscosity so that a predetermined and regulated bond line thickness is possible.

The degree of tensile strength improvement is often in the 50 to 100 percent range. The effect of various fillers and loading ratios on the strength properties of epoxy adhesive formulation is indicated in Fig. 9.9. The effect of different fillers loaded at a constant 100 pph is indicated in Table 9.12 for shear strength on phenolic laminate and aluminum substrates.

Although the addition of fillers often results in an improvement in tensile shear strength, as noted above, it also often results in a decrease in peel or impact strength. This is due to

TABLE 9.12 Effect of Fillers on Block Shear Strength, psi, of Adhesive Formulation[30]

	Phenolic laminate tested at			Aluminum block tested at	
Filler at 100 pph	23°C	75°C	90°C	23°C	105°C
Aluminum powder	2790		1470		
Ignited Al_2O_3	4600	1360	1195	1650	1150
Carbon black	2000	555	980	2000	1170
Silica	2840	1600	1250	1530	600
Zinc dust	2510	600	300	4080	3865

the reduction in elongation and toughness that is often accompanied by high filler loadings in epoxy adhesive systems. When an adhesive joint is stressed in peel or cleavage, the more rigid bond line results in all the stress being focused at a very small region near the edge of the bond. Adhesives with a higher degree of elongation distribute the stress over a large area, resulting in higher peel and cleavage strength.

9.3.12 Working Life and Exotherm

Fillers generally lower the curing reaction rate and reduce the degree of exotherm. This is primarily due to their diluting effect and the resulting increase in thermal conductivity. The effect of various fillers on working life and peak exotherm is presented in Table 9.13. The more filler added, the less will be the heat evolved during cure. However, even with highly filled epoxy systems, the filled resin is still a relatively poor conductor of heat.

9.3.13 Fire Resistance

Flame-retardant additives work by acting chemically and/or physically in the condensed phase or gas phase. The types of flame-retardant additives and their operating characteristics are described below.

Char formers: Usually phosphorus compounds, which remove the carbon fuel source and provide an insulation layer against the fire's heat

Heat absorbers: Usually metal hydrates such as aluminum trihydrate (ATH) or magnesium hydroxide, which remove heat by using it to evaporate water in their structure

Flame quenchers: Usually bromine- or chlorine-based halogen systems that interfere with the reactions in a flame

Synergists: Usually antimony compounds, which enhance performance of the flame quencher

Flame retardants for polymers can also be classified as reactive (the flame-retarding agent reacts chemically with the polymer to become an integral part of the molecule) or additive (nonreactive agents that are simply blended or mixed into the compound). Flame retardants can also be classified by their major chemical group, as shown in Table 9.14.

There are many families of flame retardants, each with advantages and disadvantages. It is common to formulate polymers with multiple flame retardants, typically a primary

TABLE 9.13 Effect of 50 percent Filler Loading on Viscosity, Exotherm, Modulus of Rupture, and Shrinkage of an Epoxy Resin[31]

Filler	Viscosity, cP, at 25°C	Peak exotherm, °C	Working life, min	Modulus of rupture, psi	Shrinkage, percent
None	1,100	223	48	18,174	0.91
Silica	51,500	53	95	16,875	0.77
Mica	54,500	51	94	12,358	0.66
Limestone	10,000	59	90	12,433	0.47
Atomized aluminum 101	4,600	47	100	13,710	0.8
Barytes	4,200	83	84	17,417	0.71

TABLE 9.14 Major Types of Flame Retardants: Typical Products, Applications, and Suppliers*

Type	Typical products	Typical applications	Typical suppliers
Alumina trihydrate	Available in various particle-size grades and in surface-treated grades	Polyesters, phenolics, and epoxies. Also polypropylene and ethylene/rubber compounds	Albemarle-Martinswerk, Alcan Chemicals, Alcoa, Aluchem, Amspec Chemical, DuPont Polymer Modifiers, J.M. Huber, Nabaltec
Magnesium	Magnesium hydroxide Magnesium carbonate	Synergists with ATH for smoke reduction	Addenda, Albemarle-Martinswerk, Amspec Chemical, Dead Sea Periclase, DSBG, J.M. Huber, Kyowa Chemical Industry, Sumitomo Chemical, Witttaker
Chlorines	Chlorinated paraffins Tris (dichloropropyl) phosphate, methyl pentachlorostearate, and other chlorinated phosphates Cycloaliphatic chlorine Chlorendic anhydride	LDPE film and flexible PVC Urethane foam and topical fabric Polypropylene and nylon Reactive intermediate in making polyester and epoxy flame-retardant resins	Amspec Chemical, Asahi Denka Kogyo, ICC Industries-Dover, Kettlitz Chemie, Lehmann & Voss, OxyChem, Polytechs
Bromines	Aromatic bromines (e.g., decarbromophenyl oxide ether, tetrabromo-bisphenol A, pentabromodiphenyl oxide) Aliphatic bromines Ionic bromines Benzene-ethyltribromo derivatives Polypentabromobenzyl acrylate	Polyolefins, ABS, polyesters Thermosets and thermoplastics Urethane foams and polyesters Plastic foams and polyester fibers Thermoplastics usually as a synergist Engineering thermoplastics Glass-reinforced nylon and PBT	Albemarle-Martinswerk, Amspec Chemical, Constab Additive Polymers, DSBG, Great Lakes, Heveatex, ICC Industries-Dover, Nyacol, O'Neil Color & Compounding, OxyChem, Polytechs, Solvay, Teijin Chemicals, Tiarco Chemical, Unitex Chemicals
Phosphorus	Phosphate esters and others (halogenated and nonhalogenated)	Polyurethane foams, polyesters, and thermoplastics such as flexible PVC, modified PPO, and cellulosics Also polyethylene, polypropylene, polystyrene, and ethylene/propylene copolymers	Akzo Nobel, Albemarle, Amfine Chemical Corp., Amspec Chemical, Bayer, Ciba Specialty Chemical-Melapur, Clariant, Cytec, Daihachi Chemical Industry, Great Lakes, Italmatch Chemicals, Nitroil, Rhodia

Antimony oxide	Dusting and nondusting grades available in various particle sizes	Works synergistically with reactive or additive halogenated compounds ABS, polyethylene, polypropylene, polystyrene, thermoplastic polyester, unsaturated polyesters	Addenda, Amspec Chemical, Anzon, Champine, Chemisphere, Frilvam, Great Lakes, Nyacol, OxyChem, Polytechs, Tiarco Chemical
Borates	Zinc borate, barium metaborates, ammonium fluoroborate, boric acid	Flexible PVC, polyolefins, unsaturated polyesters, thermoplastic polyesters, epoxies, nylons, urethanes, and phenolics	Alcan, Amspec Chemical, Asahi Denka Kogyo, Borax, Buckman Laboratories, Great Lakes
Silicone	Polydimethylsiloxane	Works by producing a char surface Polyolefins, polyolefin blends, EVA, polycarbonate, polyurethane	Dow Corning, GE Silicones
Melamine	Pure melamine, melamine derivatives (salts and inorganic acids), melamine homologues	Polyurethane, polyamide, polyolefins, thermoplastic polyesters	Akzo Nobel, Budenheim, Ciba Specialty Chemicals-Melapur

*For additional information see the Flame Retardant Center at SpecialChem4Polymers.com.

flame retardant plus a synergist such as antimony oxide, to enhance overall flame resistance at the lowest cost. Several hundred different flame-retardant systems are used by the polymer industry because of these formulation practices.

Selecting a flame retardant for an adhesive system has many ramifications, depending on the formulation being modified, the end use, how it will be processed, and the cost/performance ratio. When one is choosing a flame retardant, characteristics such as water extraction, particle size, viscosity, toxicity, dusting, uniformity, as well as economics must be considered. The materials chosen to perform the function of flame retardation must not interfere with the final product's performance. The major problem with incorporating flame retardants in adhesives is that very often a significant amount is required, and they interfere with the other properties of the adhesive and contribute to the cost. This is why bromo bisphenol epoxy resins are often employed in flame-retardant epoxy adhesives.

The criteria in selecting and using the right flame-retardant system for a specific adhesive formulation can be broad and somewhat complex. Care must also be taken with regard to choosing the proper test method and determining the parameters of the test. Consultation with the flame-retardant suppliers is generally necessary to optimize this process.

9.3.14 Color

Fillers can be used as pigments to provide color to the epoxy formulation. Adhesive coloring is used to match the color of substrates. Different coloring is often added to each component in a two-part epoxy as a method for determining efficient mixing of the components. The resultant color from the mixture helps the user know that the mix ratios are close and that the degree of mixing is sufficient.

Both pigments and dyes have been used successfully as colorants. With liquid epoxy resins, inorganic pigments seem to provide optimal properties. Organic pigments are generally less effective with epoxy resins.

Pigments are generally dispersed into the resin formulation at relativity low percentages (1 to 3 percent) to provide the required color. Titanium dioxide is frequently employed in connection with the colorant to provide a whiting agent with the necessary hiding power suitable for tinting. The pigments are generally dispersed directly into the epoxy formulation. However, concentrated dispersions in liquid DGEBA and in diluents, such as dibutyl phthalate, are commercially available and may provide for easier incorporation of the pigment.

REFERENCES

1. Potter, W. G., *Epoxide Resins*, Springer-Verlag, New York, 1970, pp. 115–116.
2. *Materials Engineering*, May 1976, p. 73.
3. Katz, H. S., and Milewski, J. V., eds., *Handbook of Fillers for Plastics*, van Nostrand Reinhold, New York, 1987.
4. Lee, H., and Neville, K., *Epoxy Resins*, McGraw-Hill, New York, 1957, p. 148.
5. Magee, J. E., "Formulating Techniques and Evaluations," in *Epoxy Resin Technology*, P. F. Bruins, ed., John Wiley & Sons, New York, 1968, p. 120.
6. Potter, *Epoxide Resins*, p. 118.
7. Lee, H., and Neville, K., *Handbook of Adhesive Resins*, McGraw-Hill, New York, 1982, p. 14.7.
8. Grace, M. J., and Sullivan, C., "Cellulose Fiber as a Reinforcing Thixotrope in Adhesive and Sealants," *Adhesives Age*, July 1995.
9. Technical Bulletin, "Kevlar: Applications," Sealants and Adhesive Group, DuPont Company, Wilmington, DE, 2001.

10. Nargiello, M., "Hydrophobic Silicas Contribute to Consistency in Formulating," *Adhesives Age*, July 1989, pp. 30–31.
11. Katz and Milewski, *Handbook of Fillers for Plastics*, p. 180.
12. Ralston, H. P., "Kaolin, Talc, and Attapulgite Extend and Thicken Adhesives," *Adhesives Age*, July 1989, pp. 32–34.
13. *Additives for Adhesives and Sealants*, Elementis Specialties, Hightstown, NJ, 2002.
14. Carr, J. B., "Attapulgite-Based Thixotropes Offer Performance and Flexibility," *American Paint and Coatings Journal*, November 1997.
15. Lee and Neville, *Epoxy Resins*, p. 151.
16. Conn, R. C., "Special Effects," *Adhesives Age*, February 2002, pp. 47–48.
17. Ishibashi, K., and Kimura, J., "A New Anisotropic Conductive Film with Arrayed Conductive Particles," *AMP Journal of Technology*, vol. 5, June 1996, p. 2430.
18. Bolger, J. C., and Morano, S. L., "Conductive Adhesives: How and Why They Work," *Adhesives Age*, June 1984.
19. Katz and Milewski, *Handbook of Fillers for Plastics*.
20. Lee and Neville, *Handbook of Adhesive Resins*.
21. Buckley, W. O., and Schroeder, K. J., "Adhesive Modifiers," in *Adhesives and Sealants*, Advanced Materials Handbook, ASM International, Materials Park, OH, 1990, p. 178.
22. Lee and Neville, *Handbook of Adhesive Resins*, p. 14.35.
23. Quaint, A., "Low Density Potting Compound," Sandia Corp., Preprint SCR-417, 1961.
24. Technical Bulletin, *The Effect of Fillers on the Physical Properties of Epoxy Resins*, Shell Chemical Corp., Houston, TX, 1960.
25. Seymour, R. B., ed., *Additives for Plastics*, Academic Press, New York, 1978.
26. Katz and Milewski, *Handbook of Fillers for Plastics*.
27. Wright, C. D., and Muggee, J. M., "Epoxy Structural Adhesives," Chapter 3 in *Structural Adhesives*, S. R. Hartshorn, ed., Plenum Press, New York, 1986.
28. Formo, J., and Bolstad, L., "Where and How to Use Epoxies," *Modern Plastics*, July 1955.
29. Burgman, H. A., "Selecting Structural Adhesive Materials," *Electrotechnology*, June 1965.
30. Lee and Neville, "Epoxy Adhesives," Chapter 21 in *Handbook of Adhesive Resins*.
31. Lee and Neville, *Epoxy Resins*, p. 150.

CHAPTER 10
ADHESION PROMOTERS AND PRIMERS

10.1 INTRODUCTION

Some adhesives and sealants may provide only marginal adhesion to certain substrates. This could be due to the low surface energy of the substrate relative to the adhesive (e.g., epoxy bonding polyethylene) or to a boundary layer that is cohesively weak (e.g., powdery surface on concrete). The substrate may also be pervious, allowing moisture and environmental chemicals to easily pass through the substrate to the adhesive interface, thereby degrading the bond's permanence. Generally, attempts are made to overcome these problems through adhesive formulation and by substrate surface treatment. When these approaches do not work, additional bond strength and permanence may possibly be provided by the use of (1) primers or (2) adhesion promoters.

Primers and adhesion promoters work in a similar fashion to improve adhesion. They add a new, usually organic, layer at the interface. The new layer can be bifunctional and bond well to both the substrate and the adhesive or sealant. The new layer is very thin so that it provides improved interfacial bonding characteristics, yet it is not so thick that its bulk properties significantly affect the overall properties of the bond.

Both primers and adhesion promoters are strongly adsorbed onto the surface of the substrate. The adsorption may be so strong that instead of merely being physical adsorption, it has the nature of a chemical bond. Such adsorption is referred to as *chemisorption* to distinguish it from reversible physical adsorption.

The main difference between primers and adhesion promoters is that primers are liquids that are applied to the substrate as a relatively heavy surface coating prior to application of the adhesive. Adhesion promoters, on the other hand, are liquids that form a very thin (usually monomolecular) layer between the substrate and the adhesive. Usually chemical bonds are formed (1) between the adhesion promoter and the adhesive and (2) between the adhesion promoter and the substrate surface. These bonds are stronger than the internal chemical bonds within the adhesive. These new bonds also provide an interface region that is more resistant to chemical attack from the environment. Adhesion promoters are also sometimes referred to as *coupling agents.*

Adhesion promoters can be applied either by incorporating them directly into the adhesive formulation or by applying them directly to a substrate, similar to a primer. When applied in situ, through the adhesive formulation, the adhesion promoter migrates to the interface region and attaches itself between the adhesive molecule and the substrate before the adhesive cures.

Adhesion promoters or coupling agents are also used in applications other than improving the adhesive-substrate interface. With highly filled compounds, adhesion promoters provide formulators a way of effectively improving properties and reducing the overall cost of the compound. Adhesion promoters can be applied to particulate fillers or fibers to improve their wetting characteristics. In this way, viscosity can be decreased or more filler can be added while maintaining a specified viscosity. Bulk properties of the resulting adhesive, such as tensile strength and modulus, can also be significantly improved. Thus, when one is discussing adhesion promoters, the term *substrate* takes on the added possibilities of fillers, reinforcements, etc., as well as conventional adherends.

10.2 ADHESION PROMOTERS

Adhesion promoters are a group of specialty bifunctional compounds that can react chemically with both the substrate and the adhesive. The adhesion promoter forms covalent bonds across the interface that are both strong and durable. Adhesion promoters can be applied directly to the substrate, similar to primers, or they can be mixed with the adhesive itself. When mixed into the adhesive formulation, adhesion promoters are sometimes referred to as coupling agents because they can interact with both the substrate and the other fillers in the formulation.

When mixed with the adhesive, the adhesion promoter is capable of migrating to the interface and reacting with the substrate surface as the adhesive cures. When applied directly to the substrate, adhesion promoters are used as a very thin coating that ideally is only one molecular layer thick.

Adhesion promoters can also be applied directly to mineral fillers or fibers prior to their addition to a formulation. The adhesion promoter provides for a formulation that has several unique benefits:

1. Enhanced processability, shown by reduced viscosity, easier flow, and lower energy required to compound and mix
2. Increased tensile, flexural, and impact strength
3. Better retention of these properties under adverse temperature and humidity conditions over long periods of time

Adhesion promoters usually consist of molecules with short organic chains having different chemical compositions on either end of the chain. On one end is an organofunctional group that is particularly compatible with the given adhesive material. At the other end of the chain is an inorganic functionality that is especially compatible with a given substrate. The adhesion promoter, therefore, acts as a chemical bridge between the adhesive and the substrate.

Adhesion promoters were first used to treat glass fibers and other fillers before they are incorporated into liquid resin to make composite materials. In the fiber industry, adhesion promoters are also known as finishes. Certain finishes have been specially developed to match a fiber with a resin matrix. Without adhesion promoters, the interfacial resin–glass fiber adhesion is weak, and water can diffuse along the interface with catastrophic results on the end properties of the composite.

10.2.1 Organosilane Adhesion Promoters

Silanes are the most common commercial adhesion promoter. They are commonly used to enhance adhesion between polymeric and inorganic materials.[1,2] They usually have the form $X_3Si–R$, where X is typically a chlorine or alkoxy group and R is the organofunctionality.

The organofunctional portion bonds with the resin in the adhesive or the organic medium, and the silane portion bonds to the inorganic or substrate surface. Silane coupling agents are commonly used between the adhesive and the adherend, between resin matrix and reinforcing fibers in composites, and between resin matrix and mineral fillers in compounds. The resulting interface provides

- A chemical bridge generated between the surface and organic polymer
- A barrier to prevent moisture penetration along the interface
- Transfer of stress from the resin to the substrate or inorganic filler component, thereby improving joint strength or bulk properties
- Effective dispersion of fillers to lower the apparent viscosity of the system

These chemicals are usually applied to fillers or to the substrate surface as an aqueous solution, but it can also be applied via solvent solution. The solutions usually are very dilute, only 0.01 to 2 percent by weight of silane to keep the highly reactive hydrolyzed molecules from reacting with one another. The bond strength enhancement increases with silane concentration up to a maximum of about 2 percent, and then the enhancement falls off gradually with additional concentration. Silane coupling agents react with water in aqueous solutions to form hydrolyzed silanes, which then further react with the surface of the inorganic substrate. The bound silanes then polymerize, building up layers outward from the substrate with the organic functionality oriented toward the adhesive. This process is shown in Fig. 10.1.

Silanes form strongly adsorbed polysiloxane films on ceramic and metal surfaces. The chemical and mechanical integrity of these films are highly dependent on application parameters such as solution concentration, solution pH, and drying time and temperature. The character of the substrate may also influence the polysiloxane film structure.[4] Silanes are applied as primers to the surface of the substrate by wiping, spraying, brushing, or dipping. The film thickness is less than 0.1 mil. The solvent in the silane system is removed by drying at 50 to 60°C for 10 min. The advantage of the primer application method is that it efficiently utilizes the silane material, and there are minimal stability problems. The disadvantages are that it is a two-step process and that it is difficult to see the clear silane coating unless it is pigmented. Therefore, complete surface coverage may be in doubt.

Tests have shown that silanes arrange themselves in layers with a high degree of order, influenced to a great extent by the surface of the substrate. The molecules order themselves virtually perpendicular to the surface to which they attach, and subsequent layers arrange themselves steplike in a head-to-head fashion. A rough surface can break up the first ordered layer, preventing the formation of the second. The thickness of the silane interphase has important effects on mechanical performance. Thinner, but continuous, layers seem to give stronger and more durable adhesive bonds.

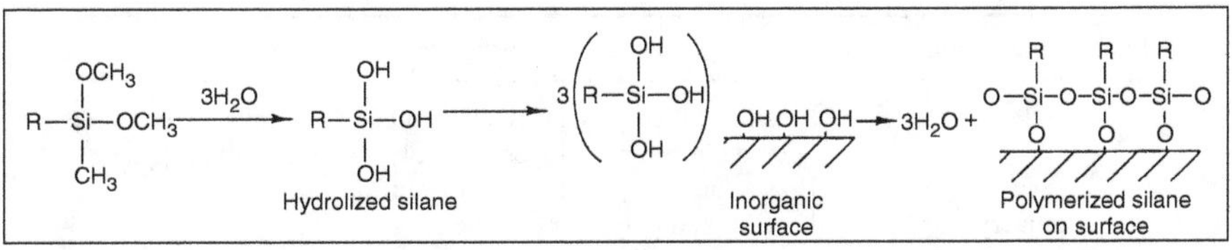

FIGURE 10.1 Organosilane coupling agents react with water in aqueous solutions to form hydrolyzed silanes that react with the surface of inorganic fillers or the adherend. The bound silane polymerizes, building up layers outward, with the organic functionality oriented to the resin matrix.[3]

Three primary mechanisms have been suggested for enhanced adhesion via silane coupling agents.[5] The classical explanation is that the functional group on the silane molecule reacts with the adhesive resin. Another possibility is that the polysiloxane surface layer has an open porous structure. The liquid adhesive penetrates the porosity and then hardens to form an interpenetrating interphase region. The third mechanism applies only to polymeric adherends. It is possible that the solvent used to dilute and apply the silane adhesion promoter opens the molecular structure on the substrate surface, allowing the silane to penetrate and diffuse into the adherend.

The interphase provided by the adhesion promoter may be hard or soft and could affect the mechanical properties. A soft interphase, for example, can significantly improve fatigue and other properties. A soft interphase will reduce stress concentrations. A rigid interphase improves stress transfer of resin to the filler or adherend and improves interfacial shear strength. Adhesion promoters generally increase adhesion between the resin matrix and substrate, thus raising the fracture energy required to initiate a crack.

There are a number of silane adhesion promoters available, and they differ in the nature of their reactivity to the resin or adhesive. Silanes may be produced with amine, epoxy, mercaptan, and other functionalities. Some examples are given in Table 10.1.

Many mineral fillers are also commercially available with an organofunctional silane surface treatment. Suppliers of treated mineral fillers are shown in Table 10.2. These fillers are used in a variety of applications having critical property requirements that must be protected from moisture. Surface treatment of the mineral fillers also provides the formulator with a tool to reduce the viscosity of highly filled systems.

Glass fibers and fabrics are commercially available with silane treatments. The resulting physical properties and the resistance of the composite to deterioration by water immersion are greatly enhanced by the addition of the coupling agent. Table 10.3 shows the effect of silanes on the dry and wet flexural strength of several composite materials.

The moisture resistance properties of filled molding compounds are enhanced by the treatment of the fillers with silane adhesion promoters prior to compounding. Silane promoters on wollastonite fillers in thermoplastic polyester molding compounds (50 percent filled) will improve the flexural strength after 16 h in 50°C water by as much as 40 percent. Silane-treated silica fillers have been found to significantly increase the moisture resistance of epoxy adhesives used in the electronics industry for chip, surface-mounted, and printed-circuit processes.[8]

Silanes are generally effective in improving adhesion to metals, including aluminum, steel, cadmium, copper, and zinc. Figure 10.2 shows the relative influence of the type of substrate on the effectiveness of the silane coupling agent in improving adhesion. Note that smooth, high-energy substrates are excellent substrates for silane attachment, and rough, discontinuous substrates show very little benefit.

TABLE 10.1 Recommended Silane Coupling Agents for Various Resins

Silane functionality	Applications
Vinyl	Free radical cure systems: crosslinked polyethylene, peroxide cured elastomers, polyesters. Polyethylene. Polypropylene.
Epoxy	Epoxy, acrylics, urethanes, polysulfide
Methacryl	Unsaturated polyester, acrylic
Amino	Epoxy, phenolic, melamine, urethane, butyl rubber
Mercapto	Epoxy, sulfur cure rubbers, urethane, polysulfide
Ureido	Phenolic, urethane

TABLE 10.2 Commercial Sources of Organofunctional Silane Treated Fillers[6]

Filler	Organic functionality	Treated filler	Supplier
Alumina trihydrate	Amino Vinyl Methacryloxy Epoxy	Hyflex	Solem Industries
Calcinated clay	Vinyl	Burgess KE	Burgess Pigment
	Vinyl	Translink 37	Freeport-Kaolin
	Amino	Translink 445	
Hydrous clay	Mercapto	Nucap 200	J. M. Hubber, Clay Division
	Amino	Nulok 321	
Silica	Amino Vinyl	Insil series	Illinois Minerals
	Amino Epoxy Vinyl Methacryloxy	Novakup series	Malvern Minerals
Wollastonite (calcium silicate)	Amino Vinyl Methacryloxy Epoxy	Wollastokup series	NYCO
Glass spheres (solid)	Amino Methacryloxy		Potters Industries
	Amino		Ferro Corp., Cataphote Division
Glass spheres (hollow)	Amino Methacryloxy		PQ Corp.

TABLE 10.3 Effect of Various Silanes on Glass Reinforced Thermoset Resins[7]

		Flexural strength, psi	
Resin system	Silane	Dry	Wet
Polyester	Control	60,000	35,000
	A-174	87,700	79,000
Epoxy	Control	78,000	29,000
	A-186	101,000	66,000
Melamine	Control	42,000	17,000
	A-187	91,000	86,000
Phenolic	Control	42,000	
	A-1100	91,000	

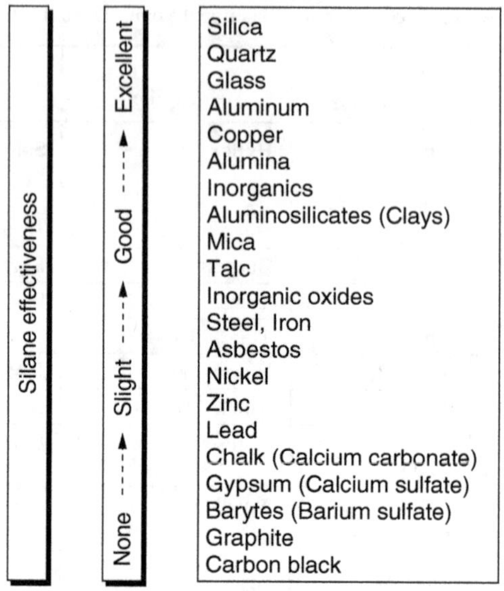

FIGURE 10.2 Effect of substrate type on silane adhesion promotion.[9]

Silane adhesion promoters increase the initial bond strength and also stabilize the surface to increase the durability and permanence of the bonded joint in moist aging environments.[10,11] The lap shear values in Table 10.4 show the improvement in initial bond strength when silane coupling agents are incorporated into the adhesive formulation. Silane-based coupling agents are capable of increasing the environmental resistance of aluminum,[12] titanium,[13] stainless steel,[14] and other metal joints. The effect of a silane adhesion promoter on the durability of mild steel joint bonded with an epoxy adhesive is shown in Fig. 10.3.

Although the best results can be obtained in using silanes as substrate primers, they can also be added to the adhesive with some effect. The integral blend method of applying the silane involves adding 0.1 to 2.0 percent by weight of the silane to the polymer matrix prior to application. The advantage of this method is that it does not require a separate substrate

TABLE 10.4 Tensile Shear Strengths of Several Structural Adhesives Incorporating Silane Adhesion Promoters[15]

Adhesive type	Adhesion promoter	Percentage of promoter	Lap shear, psi
Nitrile phenolic	None	—	1450
Nitrile phenolic	Gamma-aminopropyltriethoxysilane	1	2350
Nitrile phenolic	Gamma-mercaptopropyltrimethoxysilane	1	3120
Epoxy	None	—	1410
Epoxy	Gamma-aminopropyltrimethoxysilane	1	1675
Epoxy	Gamma- mercaptopropyltrimethoxysilane	1	1580

2024T3 aluminum, 0.063-in thickness, $\frac{1}{2}$-in overlap. Tests run at 23°C; appropriate cure schedule used.

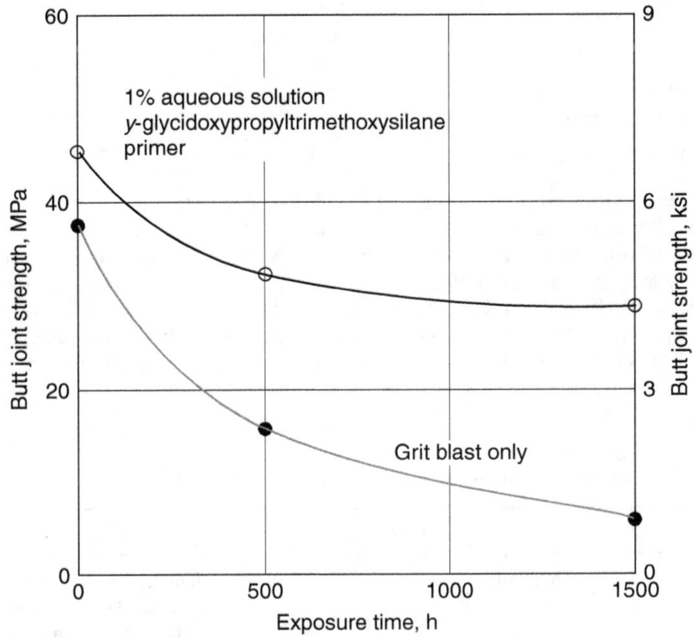

FIGURE 10.3 Effect of silane adhesion promoter on the durability of a mild steel joint bonded with epoxy adhesives.[16]

coating step. The disadvantage is that there could be stability problems in storage due to hydrolysis and polymerization. Also, the efficiency of promoter coverage at the interface is reduced due to the migration process.

Conventional silanes are limited to high-solids and solvent adhesive and sealant applications in which moisture is not encountered until use. However, water-stable epoxy silane promoters have recently been developed to enhance the wet strength of waterborne adhesives and sealants.[17]

10.2.2 Organometallic Adhesion Promoters

While silanes predominate as adhesion promoters, the uses of titanate, zirconate, and other agents are growing. These coupling agents also offer increased impact strength and other physical properties when used with fillers or applied at the interface. Like the silanes, they react with surface hydroxyl groups, but there is no condensation polymerization to produce a polymer network at the interface. Similar to the silanes, organometallic adhesion promoters typically provide a dual function of improving processing and improving adhesion.

Organic titanates and zirconates have been found to provide several important functions as additives for organic adhesives and sealants. At least eight primary functions have been proposed.[18] Organic titanates and zirconates can be used as

1. Catalysts for manufacture of adhesive and sealant prepolymers
2. Adhesion promoters
3. Wetting agents

4. Surface protection
5. Water scavengers
6. Crosslinkers
7. Catalyst for crosslinking
8. Thixotropic agents

The mechanisms and processes by which the organic titanates and zirconates provide these important functions are summarized in Table 10.5.

The use of organic titanates and zirconates for surface modification is based on their ability to hydrolyze to a coating that is very thin, amorphous, and primarily inorganic. The properties of this film depend on the type and amount of organometallic coupling agent used, the chemistry of the organometallic, and the processing properties used to apply the coating. These coatings modify the surface of the filler or substrate to provide the following unique properties.

TABLE 10.5 Common Functions of Organic Titanates and Zirconates[19]

Function	Effect	Use
Catalysis	Act as Lewis acids to catalyze reactions	Manufacture of acrylates, methacrylates, polyamides, polyester, polyurethanes, epoxies, etc.
Adhesion promotion	Improved adhesion between unreactive inorganic or organic substrates and resin matrix	Adhesion promoter in form of a primer solution or as a polymer additive. Sometimes used with organosilanes
Wetting agent	Surface tension can be reduced, and a better wetting of the polymer is achieved	Used as a primer or as a polymer additive
Dispersing agent	Improved dispersion of pigments or fillers, viscosity reduction, increase of filler content, better end properties	Pretreatment of fillers or pigments or as a polymer additive
Surface protection	A thin protective metal oxide layer can be formed on the surface of organic or inorganic substrates	Priming of the substrate by solutions of titanium/zirconate or by partial hydrolyzing and heating or chemical vapor deposition
Water scavenger	Acts as a water scavenger or drying agent by forming polymeric titanium oxide hydrate or other compounds	Used as an additive in silicone or polyurethane sealants (0.5–5% by weight)
Crosslinking	Can react with the OH or COOH groups of polymer, resulting in crosslinking	Curing, improvement in temperature and chemical resistance, gel formation
Crosslinking catalyst	Through choice of the appropriate metal (Ti or Zr) and chelating agent reactivity can be controlled	Increased reaction rate in silicone, polyurethane, and epoxy systems
Thixotropic agent	Undergo reversible crosslinking reactions via hydrogen bridge bonds, forming a gel structure	Added as the last ingredient in the formulation. Mainly used for water-based acrylics

- They promote adhesion of adhesives and coatings to glass, metal, and plastics.
- The organometallic interface improves dispersibility of pigments and fillers in aqueous and nonaqueous systems and reduces viscosity.
- It can provide scratch-resistant and reflective properties to glass.
- It can modify frictional characteristics of the substrate.

Titanates The titanate structure may be tailored to provide desired properties through the functionalities on the basic structure shown below.

$$(RO)_m\text{–Ti–}(O\text{–}X\text{–}R_2\text{–}Y)_n$$

The $(RO)_m$ is the hydrolyzable proton that attaches to the inorganic substrate, as shown in Fig. 10.4. It also controls dispersion, adhesion, viscosity, and hydrophobicity. The X group enhances corrosion protection and acid resistance and may provide antioxidant effects, depending on the chemistry. The R_2 provides entanglements with long hydrocarbon chains and bonding via van der Waals forces. The Y group provides thermoset reactivity, chemically bonding the filler to the polymer. It can be one of any number of chemical groups reactive with a number of different matrices.

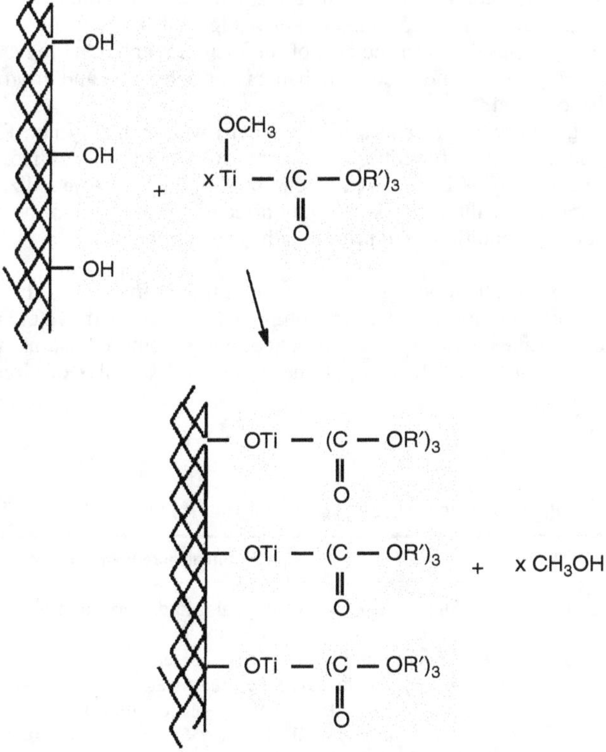

FIGURE 10.4 Titanate condensation on a hydroxyl-containing surface.[20]

Like the silanes, the organic titanates react with surface hydroxyl groups, but there is no condensation polymerization to produce a polymer network at the interface. Titanate coupling agents are unique in that their reaction with free protons on the substrate surface results in a monomolecular layer whether it is a filler or substrate. Thus, one of the most common problems in using organic titanates is overconcentration. Since excess titanate (amount greater than necessary to form a monolayer) does not result in a polymer network, it is suspected that it can form a weak boundary layer, resulting in degraded properties. Thus, the amount of titanate that is used is an important parameter.

Typically, titanate-treated inorganic fillers or reinforcements are hydrophobic, organophilic, and organofunctional and, therefore, exhibit enhanced dispersibility and bonding with the polymer matrix. When used in filled polymer systems, titanates claim to improve impact strength, exhibit lower viscosity, and enhance the maintenance of mechanical properties during aging.

Major suppliers of titanate coupling agents include DuPont (Tyzor) and Kenrich Petrochemicals (Ken-React). Typical titanate coupling agents are shown in Table 10.6. There are many types and chemistries available. The manufacturer of the titanate is an excellent source for information on the appropriate type, dosage, and processing requirements in specific applications. Titanates can be made available as liquids (solutions and water-soluble salts). New neoalkoxy technology allows powdered and pellet coupling agents to be processed as concentrates without pretreatment steps.[21]

Adhesion promoters based on organotitanium have been used to improve particulate filler adhesion and to pretreat aluminum alloy for adhesive bonding.[22] Titanates have been used predominantly to modify the viscosity of filled systems.[23] It has been shown that a small percentage of titanate in a heavily filled resin system can reduce the viscosity significantly. Thus, titanate adhesion promoters allow higher fillings of particulate matter to either improve properties or lower the cost of the system without having a negative effect on the viscosity. Improved bond strength to halocarbon substrates and improved hydrolytic stability are also claimed.[24]

Typical dosage of organotitanates is 0.2 percent by weight of polymer. They can also be used to treat aluminum alloy for adhesive bonding. Researchers have investigated titanate as a primer for adhesion of 1-2, polybutadiene modified epoxy to anodized aluminum.[25] The breaking strength of the joints is strongly affected by the amount of titanate used as well as by the drying condition used prior to adhesive application.

Zirconates Zirconate coupling agents have very similar structure to the titanates. They also perform similar functions. Zirconium compounds exist in both water and organic solvent soluble forms. Like the titanates, zirconate coupling agents are useful in improving the dispersion characteristics of fillers in polymer systems. Examples of zirconate coupling

TABLE 10.6 Common Titanate Coupling Agents and Their Applications

Titanate	Application/advantages
Monoalkoxy titanate	Stearic acid functionality; aids in dispersion of mineral fillers in polyolefins
Chelate titanate	Greater stability in wet environments
Quat titanate	Water-soluble, aids adhesion of water-soluble coatings and adhesives
Neoalkoxy titanate	Eliminates pretreatment associated with fillers, can be used as a concentrated solid additive
Cycloheteroatom titanate	Ultrahigh thermal properties for specialty applications

TABLE 10.7 Common Zirconate Coupling Agents and Their Applications

Zirconate	Application/advantage
Coordinate zirconate	Phosphite functionality; reduces epoxy viscosity without accelerating cure
Neoalkoxy zirconate	Accelerates peroxide and air-based cures (e.g., polyester SMC/BMC); adhesion promoter and primer for organic substrates
Zirconium propionate	Adhesion promoter for printing inks on treated polyolefin films
Zircoaluminates	Comparable to organosilanes at lower cost
Zirconium acetylacetonate, zirconium methacrylate	Adhesion promoters and primers for treated polyolefins

agents and their applications are shown in Table 10.7. Manufacturers of organic zirconate coupling agents include DuPont and Kenrich Petrochemical.

The aqueous chemistry of zirconium is complex and dominated by hydrolysis. One aspect is that polymerization takes place when salt solutions are diluted. The polymeric species can be cationic, anionic, or neutral. Polymers that are formed include ammonium zirconium carbonate, zirconium acetate, and zirconium oxychloride.

Zircoaluminates, introduced in 1983, claim performance at least comparable to that of silanes at substantial cost savings. Several functionalities are available. They are stable and soluble in an aqueous environment and do not require the presence of water to function. The surface reaction is irreversible. Among the fillers treated successfully are silica, clay, calcium carbonate, alumina trihydrate, and titanium dioxide.

10.2.3 Other Organometallic Adhesion Promoters

Chrome complexes have been developed as adhesion promoters by the reaction of chromium chloride with methacrylic acid. The chromium oxide portion of the adhesion promoter reacts with a substrate while the methacrylic portion reacts with a free radical curing outer layer. Chrome-based adhesion promoters are commonly used as a primer for aluminum foil to increase the strength and durability of aluminum/polyethylene interfaces.[26]

Other types of coupling agents include 1,2-diketones for steel,[27] nitrogen heterocyclic compounds such as benzotriazole for copper,[28,29] and some cobalt compounds for the adhesion of brass-plated tire cords to rubber.[30]

10.3 PRIMERS

Primers are resinous liquids that are applied to a substrate prior to application of an adhesive or a sealant. They provide a thicker coating at the interface than do adhesion promoters, and therefore often affect the bulk properties of the joint. The reasons for their use are varied and may include, either singularly or in combination, the following:

- Protection of surfaces after treatment (primers can be used to extend the time between preparing the adherend surface and bonding)
- Adjustment of the free surface energy by providing a surface that is more easily wettable than the substrate

- Dissolving of low levels of organic contamination that otherwise would remain at the interface as a weak boundary layer
- Promotion of the chemical reaction between adhesive and adherend
- Inhibition of corrosion of the substrate during service
- An intermediate layer to enhance the physical properties of the joint and improve bond strength (e.g., adjustment of the rheological properties at the interface or strengthening weak substrate regions)

Being lower in viscosity than the adhesive or sealant, primers can be used to penetrate porous or rough surfaces to provide better mechanical interlocking and for sealing such surfaces from the environment. Primers are often applied and appear as protective surface coatings. Application processes and equipment for applying primers are similar to those used in applying a paint coating to a substrate.

The application of a primer is an additional step in the bonding process, and it comes with associated costs and quality control requirements. Therefore, primers should be used only when justified. The most likely occasions for a primer to be used are when (1) the adhesive or sealant cannot be applied immediately after surface preparation, (2) the substrate surface is weak or porous, or (3) the adhesive-adherend interface requires additional protection from service environments such as moisture.

Epoxy-based primers are commonly used in the aerospace and automotive industries. These primers have good chemical resistance and provide corrosion resistance to aluminum and other common metals. Primer base resins, curing agents, and additives are much like adhesive or sealant formulations except for the addition of solvents or low-viscosity resins to provide a high degree of flow.

10.3.1 Application and Use of Primers

Unlike substrate surface treatments, primers always add a new organic layer to the surface and two new interfaces to the joint structure. Most primers are developed for specific adhesives, and many are developed for specific adhesive/substrate combinations.

Primers are applied quickly after surface preparation and result in a dry or slightly tacky film. It is generally recommended that they have a dried coating thickness range of tenths of a mil to approximately 2 mils. It is necessary to control the primer thickness, since if the primer layer becomes too thick, its bulk properties may predominate, and the primer could become the weakest part of the joint.

Primers usually require solvent evaporation and several curing steps before the adhesive or sealant can be applied. Adhesive primers are usually not fully cured during their initial application. They are dried at room temperature, and some are forced-air dried for 30 to 60 min at 65°C. This provides a dry, nontacky surface that can be protected from contamination and physical damage by good housekeeping practices until the substrate is ready to be bonded with an adhesive. Full primer cure is generally achieved during the cure of the adhesive.

Primers developed to protect treated surfaces prior to bonding are generally proprietary formulations manufactured by the adhesive producer to match the chemistry of the adhesive. These usually consist of a diluted solution (approximately 10 percent by weight) of the base adhesive in an organic solvent. Like the adhesive formulation, the primer may also contain wetting agents, flow control agents, and toughening compounds. If the primer is for a metal surface, corrosion inhibitors such as zinc and strontium chromate and other inorganic chromate salts may also be added to the primer formulation.

The application of corrosion-resistant primers has become standard practice for the structural bonding of aluminum in the automotive and aerospace industries. The adhesive-primer combinations are chosen to provide both maximum durability in severe environments and higher initial joint strength. Improved service life is typically achieved by establishing strong and moisture-resistant interfacial bonds and protecting the substrates' surface region from hydration and corrosion.

Primers can be used to protect both treated metal and nonmetal substrates after surface treatment. The use of a primer as a shop protectant may increase production costs, but it may also provide enhanced and more consistent adhesive strength. The use of a primer greatly increases production flexibility in bonding operations. Usually primer application can be incorporated as the final step in the surface preparation process. The primer is applied as soon as possible after surface preparation and usually no more than a few hours later. The actual application of the adhesive may then be delayed significantly.

With aluminum, for example, the maximum safe *surface exposure time* (SET) interval between surface preparation and bonding is 12 h. Many other substrates have maximum SETs that are less than this. By utilizing an adhesive primer, the SET may be extended to days and even months depending on the particular adhesive-primer system used and the storage conditions prior to bonding. This process allows a shop to prepare the surfaces of a large number of parts, prime them, and store them for relatively long periods prior to bonding. It also enables an assembly shop to outsource the more hazardous surface preparation processes. The primer provides protection of the treated joints during transportation between the treating shop and the bonding shop. With primers, scheduling of the entire assembly operation is not dependent on the nature of the surface preparation.

As with metallic substrates, primers may be used to protect treated nonmetallic substrates. After surface treatment, a high-energy substrate has an active surface that will readily adsorb atmospheric contamination. The primer protects the treated surface until the time when the adhesive or sealant is applied. Primers are especially useful in this way for the protection of polymeric parts that are treated by flame or corona discharge. Primers also find benefit on polymeric substrates in that their solvents will soften the surface, and some of the primer resin will diffuse into the bulk of the substrate, thereby increasing the adhesive strength by molecular diffusion.

Another production-related reason for priming is associated with bonded assemblies having many subsections. With primers, individual subsections can be treated, primed, and then fit into place before the bonding step without regard to time. This then allows the entire assembly to be bonded at one time.

Sometimes primers can take the place of surface treatments. Two examples are with porous substrates and with certain plastic substrates. With weak porous substrates, such as wood, cement, or porous stone, the primer can be formulated to penetrate and bind weakly adhering material to provide a new, tightly anchored surface for the adhesive. Chlorinated polyolefin primers will increase the adhesion of coatings and adhesives to polypropylene and to thermoplastic olefins. The chlorine atoms in the outer surface of the primer increase surface energy and enhance adhesion of adhesives, sealants, and paints.

Low-viscosity primers can also easily fill the irregularities on the substrate surface and displace air and fill hollows. This can improve the wetting properties of the adhesive "system." For example, if the adhesive is a hot melt and it is applied to a bare metallic surface, the adhesive will gel before it gets a chance to efficiently wet the surface and mechanically interact with any surface roughness. However, if a dilute primer were first applied to the substrate and dried, the hot-melt adhesive could bond directly to the primer that in turn has bonded to the interstices of the substrate, thus providing excellent adhesion.

10.3.2 Primers for Corrosion Protection

When a corrosive medium contacts the edge of a bonded joint and finds an extremely active surface, such as that produced by a fresh acid treatment of the metal substrate to improve adhesion, corrosion at the metal-adhesives interface can occur. This initial corrosion and the subsequent penetration can take several forms.

Some primers will inhibit the corrosion of metal adherends during service. By protecting the substrates' surface area from hydration and corrosion, these primers suppress the formation of weak boundary layers that could develop during exposure to wet environments. Primers that contain film-forming resins are sometimes considered interfacial water barriers. They keep water out of the joint interface area and prevent corrosion of the metal surfaces. By establishing strong, moisture-resistant bonds, the primer protects the adhesive-adherend interface and lengthens the service life of the bonded joint. However, moisture can diffuse through any polymeric primer, and eventually it will reach the interface area of the joint. Therefore, the onset of corrosion and other degradation reactions can only be delayed by the application of a primer, unless the primer contains corrosion inhibitors or it chemically reacts with the substrate to provide a completely new surface layer that provides additional protection.

Representative data are shown in Fig. 10.5 for aluminum joints bonded with an epoxy film adhesive and a standard chromate-containing primer. Up until recently, standard corrosion-resistant primers contained high levels of solvent, contributing to high levels of volatile organic compounds (VOCs) and chromium compounds, which are considered to be carcinogens. As a result, development programs have been conducted on waterborne adhesive primers that contain low VOC levels and little or no chrome. Data are presented on several of these primers in Tables 10.8 and 10.9.

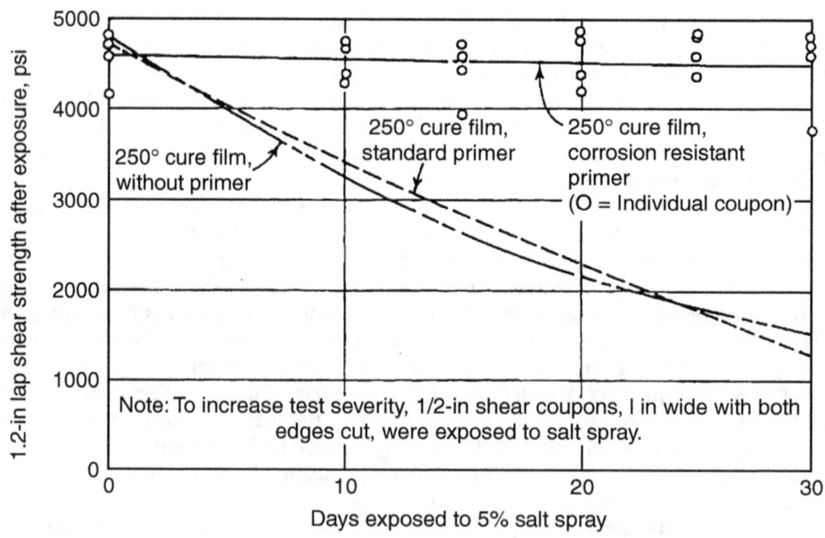

FIGURE 10.5 Effect of primer on lap shear strength of aluminum joints exposed to 5% salt spray.[31]

TABLE 10.8 Examples of Commercial Primers for Structural Adhesives[32]

Primer (source)	Adhesive	Thickness	Application process	Properties
BR 127 (*a*)	FM 73, 94, 903 toughened epoxy adhesives	2.5–5.0 μm	Air drying 30 min; curing at 120°C for 30 min	Corrosion-inhibiting; protects metal oxide from hydrolysis
EA 9203 (*b*)	EA 9602 and other epoxy adhesives	5–10 μm	Air drying 60 min	Corrosion-inhibiting for use with room temperature curing adhesives
Primer 2000 (*b*)	Thermosetting acrylic adhesives	5–10 μm	Air drying in 3–5 min at room temperature	Cleans and etches aluminum and stainless steel Provides improved durability to salt water
EC-2320 (*c*)	AF-111, AF-126, and AF-125-2 film epoxy adhesives	1.3–5.1 μm	Air drying up to 120°C	Improves shear and peel strengths as well as environmental resistance
Nitrilotrismethylene phosphoric acid (NTMP)	Epoxy adhesives	Single molecular layer	Application in aqueous solutions of 1–1000 ppm	Improves durability of aluminum joints in wet conditions
Chlorinated polyolefin: Unistole (*d*), Trapylen (*e*)	Paint, adhesives, sealants	Thin, uniform coating	Application in organic solvent	Chlorine atoms in primer increase polarity and enhance paint adhesion to molded PP, TPO, TPE, EPDM

Source:
(*a*) Cytec Industries
(*b*) Hysol – Loctite Div. of Henkel
(*c*) 3M Company
(*d*) Mitsui Chemicals
(*e*) Tramaco GmbH

TABLE 10.9 Tensile Lap Shear Strengths Using Environmentally Acceptable Adhesive Primers for 121°C Curing Adhesives[33]

		Tensile lap shear strength, psi				
Primer (source)	Adhesive	22°C	82°C*	82°C wet†	30-Day salt fog‡	60-Day salt fog‡
BR 127 (*a*)	FM 73 (*a*)	6005	4312	2622	6105	5846
	AF163-2K (*c*)	6421	4617	3499	6505	6400
BR 6747-1 (*a*)	FM73, FM 94, FM903 (*a*)					
BR 6750 (*a*)	Most 177°C curing epoxy adhesives					
XEA 9290 (*b*)	FM 73 (*a*)	5255	—	—	4791	—
EA 9257 (*b*)	EA 9657 (*b*)					
EA 9296 (*b*)	EA 9696 (*b*)					
EC 3982 (*c*)	FM 73 (*a*)	6182	4395	2891	6427	6051
	AF163-2K (*c*)	6733	4271	3633	6711	6573
EC 2320 (*c*)	AF111, AF126, and AF125-2 (*c*)					

*Heat soaked at 82°C for 10 min.
†Wet specimens conditioned at 60°C and 95–100% RH for 60 days. Heat soaked at 82°C for 4 min.
‡50% salt fog at 35°C.
Source:
(*a*) Cytec Industries
(*b*) Hysol – Loctite Div. of Henkel
(*c*) 3M Company

REFERENCES

1. Plueddemann, E. P., *Silane Coupling Agents*, Plenum Press, New York, 1991.
2. Mittal, K. L., *Silane and Other Coupling Agents*, VSP, Utrecht, The Netherlands, 1992.
3. English, L. K., "Fabrication of the Future with Composite Materials, Part IV: The Interface," *Materials Engineering*, March 1987.
4. Bascom, W. D., "Primers and Coupling Agents," *Adhesives and Sealants*, vol. 5, *Engineered Materials Handbook*, ASM International, Materials Park, OH, 1990.
5. Kinloch, A. J., *Adhesion and Adhesives*, Chapman and Hall, London, 1987, pp. 156–157.
6. Berger, S. E., "Organofunctional Silanes," in *Handbook of Fillers for Plastics*, H. S. Katz and J. V. Milewski, eds., van Nostrand Reinhold, New York, 1987.
7. Marsden, J. G., and Sterman, S., "Organofunctional Silane Coupling Agents," in *Handbook of Adhesives*, 3d ed., I. Skeist, ed., van Nostrand Reinhold, New York, 1990.
8. Conn, R. C., "How Silane Fumed Silica Influences Adhesion Properties of a Model Epoxy Adhesive System after Water Immersion," *Adhesives Age*, February 2002.
9. Waldman, B. A., "Organofunctional Silanes as Adhesion Promoters and Crosslinkers," Kent State University, Ohio, Adhesion Principles and Practice Course, May 1997.
10. Plueddemann, E. P., *Silane Coupling Agents*.
11. Bascom, "Primers and Coupling Agents."
12. Patrick, R. L., et al., *Applied Polymer Symposium*, vol. 16, 1981, p. 87.

13. Boerio, F. J., and Dillingham, R. G., in *Adhesive Joints*, K. Mittal, ed., Plenum Press, New York, 1984, p. 541.
14. Chovelon, J. M., et al., "Silanization of Stainless Steel Surfaces: Influence of Application Parameters," *Journal of Adhesion*, vol. 50, 1995, pp. 43–58.
15. Kothari, V. M., "Specialty Adhesives Can Be Compounded for Applications in Many Market Niches," *Elastomerics*, October 1989.
16. Kinloch, A. J., "Predicting and Increasing the Durability of Structural Adhesive Joints," in *Adhesion*, vol. 3, K. W. Allen, ed., Elsevier Science Publisher, New York, 1978.
17. Huang, M. W., and Waldman, B. A., "Water Stable Epoxysilanes Enhance Wet Strength of Sealants and Adhesives," *Adhesives and Sealants Industry*, Oct./Nov. 1998.
18. Technical Note, *Adhesives and Sealants*, DuPont Tyzor Organic Titanates, Wilmington, DE, 2001.
19. Technical Note, *Adhesives and Sealants*.
20. Pocius, A.V., *Adhesion and Adhesives Technology*, Hanser Publishers, New York, 1997, p. 139.
21. Monte, S. J., and Sugerman, G., "New Neoalkoxy Titanate Coupling Agents Designed to Eliminate Particulate Pretreatment," 125th Meeting of the Rubber Division, Indianapolis, IN, May 8–11, 1984.
22. Calvert, P. D., et al., "Interfacial Coupling by Alkoxytitanium and Zirconium Tricarboxylate," in *Adhesion Aspects of Polymer Coatings*, K. L. Mittal, ed., Plenum Press, New York, 1983, p. 457.
23. Waddington, S., and Briggs, D., "Adhesion Mechanism between Polymer Coatings and Polypropylene Studies by XPS and SIMS," *Polymer Communications*, vol. 32, 1991, p. 506.
24. Monte, S. J., "Titanates," in *Modern Plastics Encyclopedia*, McGraw-Hill, New York, 1989, p. 177.
25. Monte, S. J., "Titanates," Chapter 4 in *Handbook of Fillers for Plastics*, H. S. Katz and J. V. Milewski, eds., van Nostrand Reinhold, New York, 2000, pp. 88–90.
26. Kinloch, *Adhesion and Adhesives*, pp. 156–157.
27. De Nicola, A. J., and Bell, J. P., "Synthesis and Testing of β-Diketone Coupling Agents for Improved Durability of Epoxy Adhesion to Steel," in *Adhesion Aspects of Polymeric Coatings*, K. L. Mittal, ed., Plenum Press, New York, 1983, p. 443.
28. Yoshida, S., and Ishida, H., "A FT-IR Reflection Absorption Spectroscopic Study of an Epoxy Coating on Imidazole Treated Copper," *Journal of Adhesion*, vol. 16, 1984, p. 217.
29. Cotton, J. B., et al., U.S. Patent 3,837,964, 1974.
30. van Ooij, W. J., and Biemond, M. E. F., "A Novel Class of Rubber to Steel Tire Cord Adhesion Promoters," *Rubber Chemistry and Technology*, vol. 57, 1984, p. 688.
31. Krieger, R. B., "Advances in Corrosion Resistance of Bonded Structures," in *Proceedings of the National SAMPE Technology Conference*, vol. 2, *Aerospace Adhesives and Elastomers*, 1970.
32. Comyn, J., *Adhesion Science*, Royal Society of Chemistry, Cambridge, England, 1997, pp. 27–28.
33. Kuhbander, R. J., and Mazza, J. J., "Evaluations of Low VOC Chromated and Nonchromated Primers for Adhesive Bonding," *38th Annual SAMPE Symposium*, May 1993, pp. 785–795.

CHAPTER 11
ROOM TEMPERATURE CURING EPOXY ADHESIVES

11.1 INTRODUCTION

One of the distinct advantages of epoxy adhesives is that they can be cured at room temperature or even at lower temperatures. Epoxy adhesives are often divided into room temperature curing types and elevated-temperature curing types. This chapter discusses room temperature epoxy formulations. The major advantages and disadvantages of room temperature curing epoxy adhesives are shown in Table 11.1.

Room temperature curing epoxy adhesives are *two-component* adhesives, indicating that the curing agent portion and the epoxy resin portion are packaged separately. These are often referred to as "2K" adhesives, with the K derived from the German spelling of *komponent.*

One of the two components is commonly referred to as a *base*, and it contains the epoxy resin including additives. The second component is commonly referred to as the *curing agent*, hardener, catalyst, activator, or accelerator, and it contains the triggering agent for the crosslinking reaction. The two components are also sometimes referred to as part A (base) and part B (curing agent).

The two parts are mixed just prior to application. Metering occurs before the mixing operation at the time of assembly, or the two components are packaged in preweighted containers of various types and sizes.

There are also one-component epoxy systems that do not require mixing and cure at room temperature. These, however, need an energy source such as ultraviolet light or electron beam for the crosslinking reaction to proceed. There are also two-component epoxy formulations that are premixed by the supplier, immediately frozen, and shipped to the user in the frozen state. These, of course, must be kept in the frozen state prior to use, and even in this situation the shelf life is relatively limited. However, to the user, this type of adhesive has many of the application advantages of a one-component adhesive. These uv cure epoxy adhesives and innovative packaging technologies are discussed in upcoming chapters.

Once mixed, the two-part room temperature curing epoxy adhesives are designed to react at ambient conditions and temperatures near room temperature (20 to 25°C) or lower. In many cases the cure time may be accelerated with heat, but this is not a requirement. Once the adhesive is mixed, the crosslinking reaction begins immediately, and the resulting working life is limited. The cure rate and working life are dependent on the specific formulation as well as the ambient temperature. Although the adhesive hardens to a handling strength in usually a short time, it continues to develop strength for a much longer period.

TABLE 11.1 Advantages and Disadvantages of Room Temperature Curing Epoxy Adhesives

Advantages	Disadvantages
Very long shelf life of stored components at room temperature	Short working life of mixed adhesive that can result in waste, difficult application
No thermal energy required to heat either the adhesive or the parts being bonded	Components that must be metered accurately and mixed thoroughly
Can be accelerated by exposure to elevated temperatures	Tensile strength, heat and chemical resistance not as high as when cured at elevated temperatures
Can be cured at room temperature to a handling strength and then exposed to elevated temperatures for final cure	Long cure times that may limit production and require fixturing
Less shrinkage and internal stress resulting from thermal expansion differences	
Moderate strength as well as heat and chemical resistance	
Good toughness and moderate peel and impact strength	
Lower cost due to less energy and no heating equipment	
Less hazardous (less vapors, no high-temperature equipment)	

With room temperature curing epoxy adhesives, very reactive curing agents are required. An exothermic temperature increase is generated during the cure reaction, the degree of which depends on the specific epoxy adhesive system and the mass of adhesive that is mixed. This exotherm may limit the working time of the adhesive before it is applied to the substrate. However, any heat that is produced by the exothermic reaction is quickly dissipated due to the thin film of the adhesive and the thermal conductivity of the adjoining substrates.

Depending on the type of curing agent chosen, epoxy adhesive systems can reach full strength in minutes to several days. The most common types of room temperature curing epoxy adhesives reach handling strength in several hours and full cure in 5 to 7 days. The faster-curing agents can reach handling strength in as little as 5 min. It is generally good practice to allow all room temperature epoxy adhesives to cure for 5 to 7 days for the full strength to develop.

The most widely used epoxy adhesive of this type employs a polyamide or amidoamine curing agent. Even the simplest of such systems, consisting of equal parts of a liquid DGEBA epoxy resin and a polyamide hardener, provide good bond strength and durability in most environments. However, other curing agents may also be used depending on the application and performance properties required. These include crosslinking aliphatic amines, such as DETA and triethylenetetramine; catalytic amines, such as DMP-30; sulfur curing agents such as the mercaptans; and amine adducts. The curing mechanisms relative to these various room temperature curing agents are presented in Chap. 5.

Two-part, room temperature curing adhesives are generally preferred for large parts due to the energy required to heat the substrates. Two-part room temperature curing epoxies eliminate ovens and heaters while offering a reasonably short time to reach handling strength. If a large part is to be assembled or a large amount of adhesive is to be applied, a two-component adhesive with a short pot life would be inappropriate. However, automatic metering, mixing, and dispensing equipment is available to minimize the time for adhesive preparation and application. As a convenient alternative to bulk containers, many two-part epoxy adhesives are available in dual-chamber cartridges for economical handheld applications, which simultaneously meter, mix, and dispense the epoxy through disposable static mixing nozzles.

Another characteristic difference between a two-component, room temperature curing epoxy adhesive system and a one-component, heat curing system is the shelf life. The shelf life of a one-component epoxy can be 3 to 6 months when stored at room temperature and may even require refrigeration or freezing for a practical shelf life. The shelf life of each component in a two-component system is generally 6 months to 1 year.

Two-part, room temperature curing epoxy adhesives are generally more flexible than one-part epoxy adhesives. The curing mechanism is much different, and the resulting properties are more dependent on the type of curing agent used than are those of the one-component epoxy adhesive systems. As a result of the greater flexibility, these adhesives have generally higher bond strength to flexible substrates (e.g., thermoplastics) or substrates that tend to change dimensions with the environment (e.g., wood, cardboard). The higher degree of flexibility results in improved low-temperature performance and better thermal shock resistance. The greater flexibility also provides these adhesives with a higher degree of peel and cleavage strength and resistance to impact-type forces.

Table 11.2 shows typical physical property and performance data for several two-component epoxy adhesives. As can be seen from the property data, the adhesives can be

TABLE 11.2 Properties of Two-Component Epoxy Adhesives[1]

Property	General-purpose	Fast-setting	High-performance
Viscosity, Pa · s			
• Resin	50	260	91
• Hardener	35	160	103
• Mixed	45	250	54
Mix ratio, weight	100 resin/ 80 hardener	100 resin/ 100 hardener	100 resin/ 71 hardener
Specific gravity			
• Resin	1.17	1.48	1.36
• Hardener	0.92	1.44	0.97
Gel time	2 h	4 min	1 h
Cure schedule	24 h at 25°C or 30 min at 100°C	4 h at 25°C	5 days at 25°C or 2 h at 88°C
Elongation at break, percent	—	—	11
Aluminum tensile shear strength, psi, at			
• –60°C	2900	1500	4200
• 25°C	2600	2900	4500
• 82°C	300	1200	2600
• 121°C	—	—	1000
Tensile shear after			
• 30 days at 60°C	2200	3000	3800
• 30 days after humidity aging	1700 (40°C, 92% RH)	1000 (54°C, 95% RH)	2900 (54°C, 95% RH)
T-peel strength, lb/in	—	—	11

TABLE 11.3 Properties of One-Component Epoxy Adhesives[2]

Property	Epoxy-film	Epoxy paste
Appearance	Black film on carrier	Red-brown paste
Viscosity	—	300 Pa · s
Specific gravity	1.14	1.44
Shelf life	3 months at 25°C	3 months at 25°C
Cure schedule	Various elevated-temperature schedules	Various elevated-temperature schedules
Elongation at break, %	—	7
Cold-rolled steel tensile shear strength, psi, at		
• −30°C	3000	3300
• 25°C	2300	3200
• 82°C	2200	2800
Salt spray endurance (500 h), psi	2200	2600

formulated to meet a broad range of application and performance requirements. Comparison data for several one-component, heat-curing epoxy adhesives are shown in Table 11.3.

The most significant differences in performance properties between a two-component, room temperature curing epoxy adhesive and a one-component, heat-curing type are the heat and chemical resistance. These differences are due primarily to the lower crosslink density or glass transition temperature of the room temperature curing types.

With most room temperature curing epoxy adhesives, a significant drop in tensile shear strength occurs between 80 and 120°C. The exact temperature where the drop in strength occurs will depend very much on the specific curing agent used, the fillers or modifiers

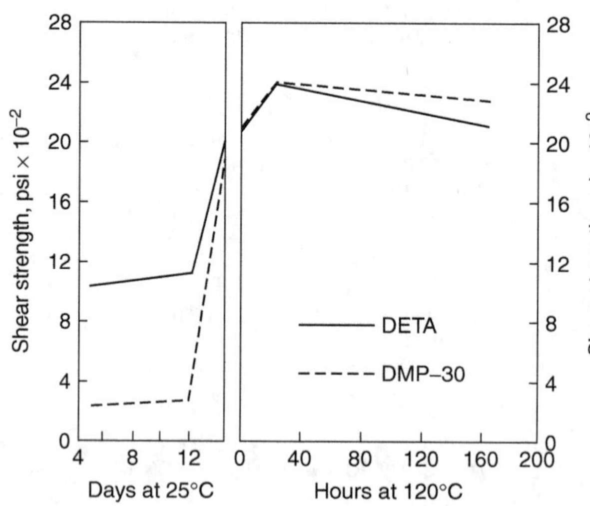

FIGURE 11.1 Effect of cure on shear strength of DGEBA formulations.[3]

used in the formulation, and the length of cure. When room temperature curing epoxy adhesives are cured at elevated temperatures or are given a postcure at elevated temperatures, the heat resistance is somewhat improved. However, the heat resistance of epoxies that are fully cured at elevated temperatures is generally far superior. The effect of cure temperature and time on typical room temperature curing rigid epoxy adhesives is shown in Fig. 11.1.

Chemical resistance is similarly affected by curing agent type and the temperature of the cure. Moisture, solvent, and general chemical resistance are usually superior for epoxy adhesives that are cured at elevated temperatures. Room temperature epoxy adhesives, having a lower glass transition temperature, are more severely affected by environmental exposures.

Room temperature curing epoxy adhesives provide widely varying application and performance properties depending on the formulation employed. The following sections highlight certain formulations and commercial products that fall under this popular classification of epoxy adhesives.

11.2 GENERAL-PURPOSE ADHESIVES

Many curing agents can be used in two-part, room temperature curable epoxy adhesives. However, the most useful curing agent families are polyamides, amidoamines, and aliphatic amines. These curing agents react directly with the epoxy and become part of the crosslinked epoxy structure, once cured. The polyamides and amidoamine curing agents provide flexibility and a degree of toughness to the cured epoxy. The aliphatic amines are somewhat more reactive and provide moderate temperature performance and good chemical resistance. However, they are generally more brittle than the polyamides or amidoamines. Table 11.4 shows the effect of each of these curing agents on selected properties of an epoxy adhesive.

11.2.1 Polyamides and Amidoamines

Polyamides and amidoamines are perhaps the most popular type of curing agent used in general-purpose epoxy adhesives. They are either used as the sole curing agent or blended with other curing agents or accelerators in consumer epoxy adhesive products, such as two-tube, ready-to-mix systems. They provide low toxicity, excellent flexibility, good toughness,

TABLE 11.4 Curing Agent Selection for General-Purpose Room Temperature Curing Epoxy Adhesives

Property	Polyamine	Amine adduct	Polyamide	Amidoamine
Tensile strength	+	+	–	+
Peel, impact strength			+	+
Heat resistance	+	+	–	–
Chemical resistance	+	+	–	–
Water resistance	+	+	–	+
Safety, hazardous vapors	–	+	+	+

+ Relatively good.
– Relatively poor.

and good water resistance. Older types of polyamide curing agents have high viscosity that presents compatibility problems with certain liquid epoxy resins and long cure times that could be degraded by moisture in the ambient environment. Newer polyamide and amidoamine products, however, have lower viscosity and better cure properties under adverse conditions.

The polyamide and amidoamine curing agents react directly with the epoxy resin and become an integral part of the cured product. Their flexibility helps to distribute the stress and impact loads throughout the adhesive joint. A typical liquid DGEBA cured with a polyamide will give peel strengths of about 16 lb/in on metal. Peel strengths of only 4 to 5 lb/in are common for the more rigid epoxy adhesives.

Higher percentages of polyamide or amidoamine curing agents in the epoxy adhesive will improve peel strength; lower percentages will provide higher-temperature and chemical resistance at the cost of reduced peel and impact properties. The epoxy–curing agent ratio for these systems can range from 1 : 2 to 2 : 1.

The pot life of polyamide or amidoamine cured epoxy adhesives is generally on the order of hours at room temperature. Full cure is achieved in 5 to 7 days at room temperature, and handling strength is achieved in about 16 to 24 h. A faster cure can be achieved in 20 min to 4 h by heating to 60 to 150°C. When room temperature cures are required, an accelerator such as an amine is often added to the formulation.

Table 11.5 shows a typical formulation for an accelerated and unaccelerated epoxy-amidoamine adhesive system. The accelerated formulation provides a faster-curing adhesive, but it is not as flexible. Common fillers such as alumina, talc, and silica can be incorporated up to a level of 100 pph before any deleterious effect on properties is observed.

A typical epoxy-polyamide adhesive that is useful for general metal bonding applications is described in Table 11.6. The effect of varying the filler type and concentration on lap shear strength at room and elevated temperatures and resistance to impact is also shown. Some fillers increase the tensile shear strength of epoxy adhesives that are cured with polyamide or amidoamine curing agents, and other fillers will reduce the strength.

As a family of curing agents for epoxy resins, the amidoamines are lower in viscosity than the polyamides. They exhibit very good adhesive properties due to their chemical structure and easy penetration. Amidoamine cured epoxy adhesives have shown very good properties on concrete and other porous substrates. They cure extremely well under humid conditions. In fact amidoamine cured epoxy formulations have been used to cure underwater in certain applications. A typical general-purpose room temperature curing epoxy-amidoamine system is described in Table 11.7. This adhesive is used as a general-purpose metal-to-metal adhesive and body solder in the automotive industry.

11.2.2 Aliphatic Amines

Aliphatic amines typically provide two-component, room temperature curing epoxy formulations with fast reactivity. They are often used as accelerators for polyamide and

TABLE 11.5 General-Purpose Epoxy Adhesive with Amidoamine Curing Agents[4]

Component	Parts by weight	
Epoxy resin (EEW: 180–200)	100	100
Amidoamine (e.g., Versamid 115 or equivalent)	70	35
Tertiary amine (e.g., DMP-30)	—	5
Filler or reinforcement	As desired	As desired

TABLE 11.6 Effect of Filler on Tensile Shear Strength of Polyamide Cured Epoxy Adhesives[5]

Type of filler	Filler, pph	Tensile shear strength, psi: At 23°C	At 120°C	After mechanical shock of 80 in · lb
Tabular aluminum	20	3200	2560	2940
	60	3080	—	1060
	100	2960	2070	440
Calcium carbonate	20	2730	1620	2240
	60	3200	—	300
	100	3400	1430	130

Adhesive formulation:

DGEBA epoxy (EEW: 180–200)	45 pbw
Fatty polyamide (amine value: 210–230)	50 pbw
Butyl glycidyl ether	5 pbw
Filler	As specified

This formulation with aluminum filler will provide tensile shear strengths as follows on various substrates:

Aluminum	3200 psi
Cold-rolled steel	2800 psi
Naval brass	2100 psi
Hard cooper	1400 psi

amidoamine curing agents. When compared to polyamide and amidoamine curing agents, aliphatic amines provide high shear strength; moisture insensitivity; good chemical resistance, particularly to solvents; and somewhat higher operating temperatures, especially when cured at elevated temperatures. Typical of the aliphatic amines are diethylene tetramine (DETA), triethylenetetramine (TETA), and catalytic amines such as DMP-30.

Table 11.8 presents a typical triethylenetetramine cured epoxy adhesive formulated with selected fillers. In this formulation the use of aluminum powder and alumina increases substantially the resistance of the adhesive to boiling water.[7] This is also true when DETA is used as the curing agent.[8] A typical room temperature cured aliphatic amine cured epoxy adhesive for general-purpose use is shown in Table 11.9. This shows the difference that is achieved in shear strength by curing at elevated temperatures versus room temperature.

As with amidoamine and polyamide cured adhesives, epoxy resins cured with aliphatic amines exhibit tensile shear strength that is dependent on the type of filler and concentration. Table 11.10 shows the effect of filler loading on strength of a simple general-purpose, room temperature curing epoxy adhesive composed of liquid DGEBA epoxy mixed with 10 pph of a tertiary amine.

TABLE 11.7 Typical Formulation for an Epoxy Adhesive Cured with an Amidoamine[6]

Component	Parts by weight
DGEBA (EEW: 180–200)	100
Amidoamine	85
Butyl glycidyl ether	15
Aluminum flakes	15
Colloidal silica	12.6

TABLE 11.8 Typical Epoxy Adhesive Cured with Triethylenetetramine[9]

Component	Parts by weight
DGEBA (molecular weight: 450)	100
Polyglycidyl ether of glycerol	25
Allyl glycidyl ether	10
Filler	100
Triethylenetetramine	12.5

TABLE 11.9 Room Temperature Cure, General-Purpose Epoxy Adhesive Cured with Triethylenetetramine[10]

Component	Parts by weight	
Part A		
Liquid epoxy resin (EEW: 190)	70	
Aliphatic epoxy resin (e.g., DER 732, Dow)	30	
Alumina T-60	50	
Part B		
Triethylenetetramine	13	
Cure conditions	7 days at 25°C	2 h at 95°C
Tensile shear strength, psi		
Aluminum, 16-gauge	1150	1600
Stainless steel, 16-gauge	1400	1730

TABLE 11.10 Effect of Filler Loading on Tensile Shear Strength of DGEBA Adhesive Formulation Cured with 10 pph Tertiary Amine[11]

		Tensile shear strength on aluminum, psi	
Filler	Filler, pph	At 23°C	At 60°C
None	—	1030	1205
Aluminum	40	2525	3725
	60	2525	3580
	80	2390	2620
Aluminum oxide	30	2505	2805
	50	3615	3600
	70	2555	3556
Mica	6	2210	1810
	14	2360	2815
	26	1055	1210
Silica	40	600	610
	90	1460	1225
Talc	30	1870	2360
	60	2650	2655
	80	2520	2400

11.3 FAST-CURING SYSTEMS

Highly reactive two-component epoxy adhesives have several advantages. These adhesives can come to a practical handling strength in minutes at room temperatures and can develop useful bond strengths at ambient temperatures as low as –20°C. The various uses for these adhesives include

1. Bonding difficult-to-clamp fixtures such as surface mounting electrical receptacles, brackets, and racks to masonry walls
2. Rapid production line assembling, such as bonding together sections of reinforced plastic car bodies
3. Bonding traffic buttons to highways when minimum traffic delay is important
4. Emergency repair of leaks
5. Cold weather construction uses such as pressure grouting, pipe joint sealing, and bridge support structure repair[12]

These fast-curing systems also have difficulties, however, that are directly related to their reactivity. Generally, the disadvantages must be balanced against the relative merits of using these adhesives. The major disadvantages of using fast-curing epoxy adhesives include the following:

1. Shrinkage on polymerization is greater than with slower-curing adhesives, and the increased shrinkage can result in increased stresses within the joint.
2. Highly reactive epoxy adhesives are generally more brittle than slower-reacting systems, and this results in lower peel and impact strengths.
3. Highly reactive adhesives are more difficult to mix and apply because of their short working life, resulting in inaccuracies and material waste. Automated metering, mixing, and dispensing equipment may be required.
4. There may be insufficient time to properly align the substrates, and misaligned parts can result in waste.
5. Many of the more reactive curing agents and accelerators cause skin irritation and toxicological problems.

Thus, fast cure is often looked upon as a liability rather than a benefit. Because the fast-acting catalysts react so quickly with epoxy resins and because they are used in very small proportions, homogeneous mixing is difficult and variation in crosslinking density can occur throughout the adhesive. Another consequence of high reactivity is a high exotherm. A high exothermic temperature can be degrading to the final epoxy adhesive properties and even dangerous if large masses of mixed adhesives overheat.

The curing agents used with fast-setting epoxy adhesives are typically accelerated aliphatic amines and accelerated polymercaptan-terminated curatives. Lewis acids, such as BF_3, can also provide exceptionally fast cure, frequently within seconds of introducing the catalyst, although they are not generally used in practical adhesive systems. Figure 11.2 compares the pot lives and bond line gel times of these curing agents.

Many fast-reacting epoxy adhesives utilize a polymercaptan and an accelerator as the main curing agent. The mixing ratio provided by the mercaptan is very practical. The polymercaptan-epoxy formulations bond to most substrates except low-surface-energy plastics. They are especially effective on alkaline surfaces such as concrete, gypsum board, and rubber containing amine antioxidants. Epoxy cured with mercaptan-terminated curatives

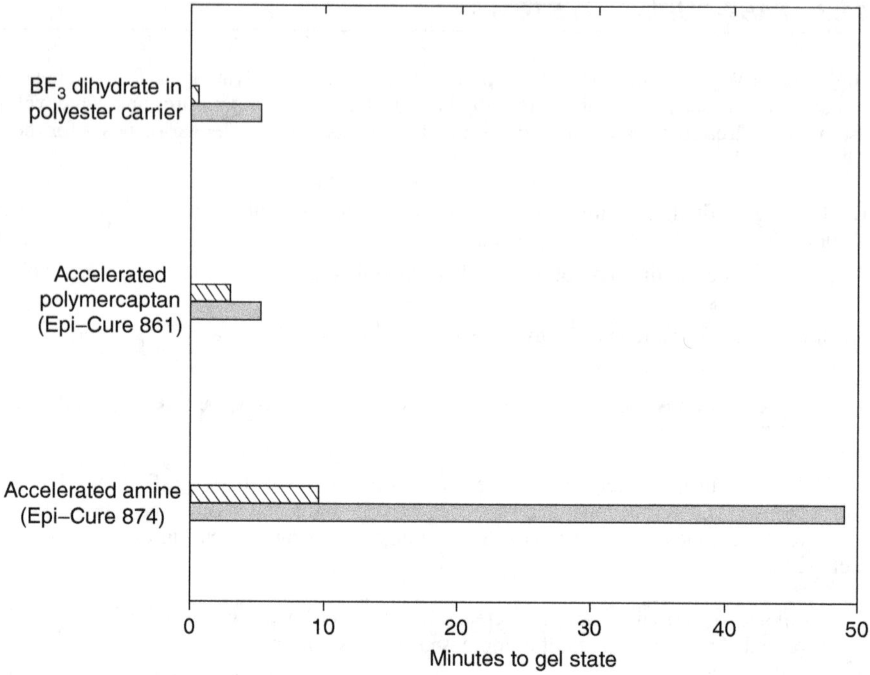

FIGURE 11.2 Gel time relationship of highly reactive adhesives (100 g mixing mass at 25°C; 0.006-in-thick adhesive film at 25°C).[13]

TABLE 11.11 Quick-Curing Epoxy Adhesive Cured with a Polymercaptan[14]

Component	Parts by weight
Part A	
Epoxy resin (EEW: 190–210)	60
Silica flour	60
Carbon black	0.1
Fibrous filler or colloidal silica	3
Part B	
Polymercaptan (e.g., Dion 3-800LC)	75
Polyamide	12
Tertiary amine	8
Silica flour	50
Titanium dioxide	10
Fibrous filler or colloidal silica	4

Gel time: 8 min at 23°C. Shear strength (aluminum-to-aluminum): 2270 psi at 23°C.

and accelerated with amine catalysts is the basis of the "5-min epoxies" often found in hardware stores.

A fast-curing polymercaptan-epoxy adhesive formulation is shown in Table 11.11. The gel time for the mixed adhesive is about 8 min at room temperature. Tensile shear strength on aluminum is reported to be 2270 psi at 23°C. Note that the different color pigments used in parts A (carbon black) and B (titanium dioxide) are to assist by visually indicating when the two components are well mixed.

The primary disadvantages of polymercaptans over other fast-reacting curing systems are (1) skin irritation caused by aromatic mercaptans (aliphatic mercaptans are less dermatitic than common polyamides) and (2) an objectionable sulfur odor, especially when the vapor pressure of the mercaptan is relatively high.

Figures 11.3 and 11.4 show the tensile shear strength development at 23 and −18°C, respectively, of mercaptan cured epoxy formulations relative to accelerated amine and amidoamine curing systems. The practical cure times at −18°C illustrate why mercaptan cured epoxy adhesive systems are popular in the building and construction industries where the adhesive can be applied at cold, outdoor temperatures.

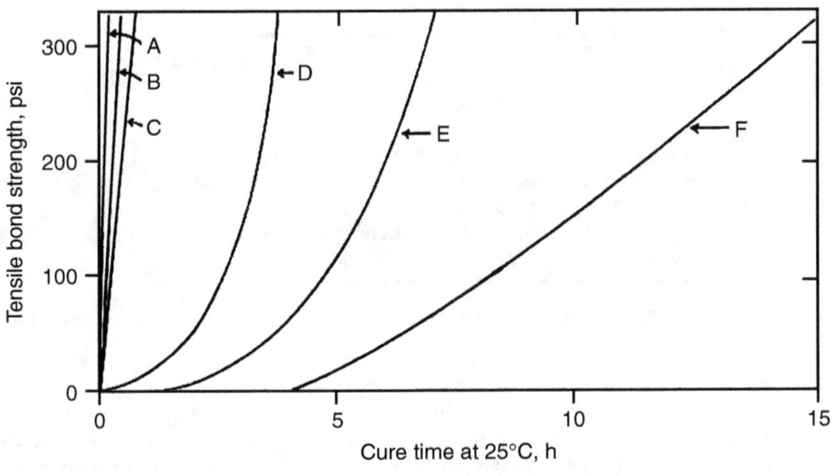

Adhesive formulations

Component	Parts by weight					
	A	B	C	D	E	F
Epoxy novolac resin (Epi-Rez 5159)	100		100			
DGEBA resin (Epi-Rez 510)		60		60	60	100
Aliphatic triglycidyl ether (Epi-Rez 5044)		40		40	40	
Accelerated polymercaptan (Epi-Cure 861)	92	99			46	
Accelerated polymercaptan (Epi-Cure 862)			87			
Accelerated aliphatic amine (Epi-Cure 874)				19		
Amidoamine (Epi-Cure 855)					25	
Accelerated aromatic amine (Epi-Cure 8491)						65

FIGURE 11.3 Tensile strength development at 25°C of various adhesives (steel-to-concrete).[15]

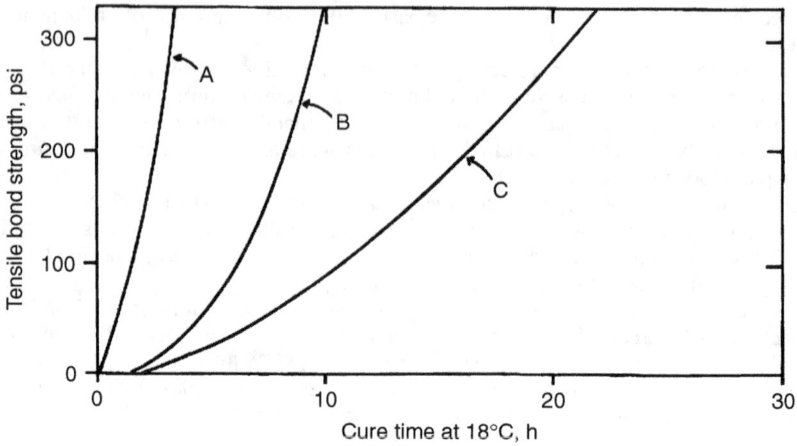

Adhesive formulations

Component	Parts by weight A	B	C
Epoxy novolac resin (Epi-Rez 5159)	100		100
DGEBA resin (Epi-Rez 510)		60	
Aliphatic triglycidyl ether (Epi-Rez 5044)		40	
Accelerated polymercaptan (Epi-Cure 861)	92	91	
Accelerated polymercaptan (Epi-Cure 862)			87

FIGURE 11.4 Tensile bond strength development at −18°C of various adhesives (steel-to-concrete).[16]

Exceptionally fast cures can also be achieved with an epoxy acrylate resin. These resins provide very high reactivity when combined with aliphatic amine curing agents. These can be used alone or as a highly reactive modifier in fast-setting adhesives or in adhesives that must set at low temperatures. Substituting conventional DGEBA epoxy for the epoxy acrylate can decelerate the cure speed. This allows the ability to "dial in" a preferred rate of gelation and cure speed.

Table 11.12 shows two rapid-setting, room temperature cure epoxy adhesives based on epoxy acrylate resins with aliphatic amine curing agents. These adhesives have gel times of less than 5 min for a 100-g mass. The bond strength development is rapid with handling strength occurring in about 1 h at room temperature.

11.4 IMPROVING FLEXIBILITY

The room temperature cured epoxy adhesives discussed thus far exhibit a general lack of flexibility, especially when considered next to elastomeric sealants. Flexibility is generally desired when the performance requirements include high peel strength, impact strength, and resistance to thermal shock or thermal cycling.

TABLE 11.12 Starting Formulations for Rapid-Setting, Room Temperature Curing Epoxy Adhesives[17]

Component	Parts by weight	
	Formulation A	Formulation B
Part A		
Epoxy acrylate resin (EPON 8111, Resolution Performance Products)	100	100
Fumed silica (Cab-O-Sil M5 Silica, Cabot Corp.)	5	5
Ground quartz (Novacite 1250, Malvern Minerals Co.)	150	75
Part B		
Aliphatic amine (Epi-Cure 3271, Resolution Performance Products)	22.1	5
Aliphatic amine (Epi-Cure 3270, Resolution Performance Products)		80
Fumed silica (Cab-O-Sil M5 Silica, Cabot Corp.)		5
Ground quartz (Novacite 1250, Malvern Minerals Co.)		77
Property		
Mix ratio by weight (part A : part B)	11.5 : 1	100 : 93
Gel time, s, at 25°C, 100-g mass	105	225
Bond line gel time, min	7	45
Tensile shear strength, psi, on etched aluminum		
• 15 min at 25°C	180	—
• 1 h at 25°C	340	100
• 24 h at 25°C	2600	2800
• 1 day at 25°C followed by 2 h at 100°C	2400	4000

Several methods can be used by the formulator to improve the flexibility of room temperature curing epoxy resin adhesive formulations:

1. High-molecular-weight resins to lower crosslink density
2. Polymerized fatty acid modification of resins and/or curing agents (e.g., polyamides, amidoamines)
3. Polyalkylene glycol modification of resins and/or curing agents
4. Curing agents that provide low crosslink density

The most common method is by using long-chain, flexible curing agents or epoxy resins (items 1 and 2 above). Flexible polyamide and amidoamine curing agents have been used as described to provide less brittleness in epoxy adhesives. Table 11.13 is a starting formulation for a flexible, high-peel-strength epoxy adhesive using both polyamide curing agents and reactive diluent as flexibilizers for the base epoxy molecule. This can be used as a general purpose adhesive on metal and plastic substrates.

Figure 11.5 shows 90° peel strength and tensile shear strength for epoxy adhesives consisting of various amounts of a flexibilized epoxy resin (Epi-Rez 505) in a mix with a standard DGEBA liquid epoxy resin. The peel strength increased dramatically with the concentration of flexibilized resin. The accompanying increase in tensile shear strength is because of better stress distribution within the lap shear joint due to the greater flexibility rather than an increase in cohesive strength of the adhesive.

TABLE 11.13 Starting Formulation for a Flexible Epoxy Adhesive Containing Polyamide Curing Agent and Reactive Diluent[18]

Component	Parts by weight
Part A	
DGEBA epoxy resin (EEW : 190)	50.5
Epodil 748 diluent (Air Products and Chemicals Inc.)	7.0
Talc (Microtuff 325F, Barretts Minerals, Inc.)	40.0
Fumed silica (Cab-O-Sil TS-720, Cabot Corp.)	2.5
Part B	
Polyamide (Ancamide 910, Air Products and Chemicals, Inc.)	38.0
Modified polyamide (Ancamide 2482, Air Products and Chemicals, Inc.)	15.0
Talc (Microtuff 325F, Barretts Minerals, Inc.)	25.0
Fumed silica (Cab-O-Sil TS-720, Cabot Corp.)	2.0
Property	
Mix ratio (part A : part B)	
• By weight	1 : 1
• By volume	1 : 1
Cure schedule	7 days at 25°C
Bond line set time, h	8
Tensile shear strength, psi	
• Aluminum, ambient cure	1770
• Aluminum, heat cure (30 min at 120°C)	2040
• ABS, ambient cure	1020
T-peel strength, lb/in	
• Aluminum, ambient cure	11
• Aluminum, heat cure (30 min at 120°C)	14

Flexibilized reactive epoxy compounds have also been used for providing stress relief and properties of high peel and impact strength to an epoxy adhesive. A starting formulation for such an adhesive is shown in Table 11.14. This formulation incorporates a modifier (HELOXY 505, Resolution Performance Products) that is a polyglycidyl ether of castor oil. It provides the formulation with good flexibility and relatively low viscosity. This can be used as a general-purpose adhesive where good peel and impact strength and resistance to thermal cycling are required.

The remainder of this section explores room temperature curing epoxy formulations where a great deal more flexibility is required. The most important application for this type of epoxy system is as a sealant or potting compound.

Polysulfides provide substantially greater flexibility than do the polyamide or amidoamine curing agents. They generally also provide lower working viscosities and are widely used for concrete bonding. Liquid polysulfides can be utilized over a rather wide mixing range with liquid DGEBA epoxy resins to provide controlled amounts of flexibility to adhesive joints. Two typical formulations for a two-component, epoxy-polysulfide adhesive are described in Table 11.15.

The mixed components provide a pot life of about 20 min at room temperature. The adhesive will reach a handling strength overnight. After 7 to 14 days of curing at room temperature or 2 h at 70°C, the bond will develop a tensile shear strength of about 500 psi. Epoxy-polysulfide adhesives are noted by their rather low degree of cohesive strength, but high degree of elongation. By varying the concentration of DMP-30, the cure speed can be varied according to need.

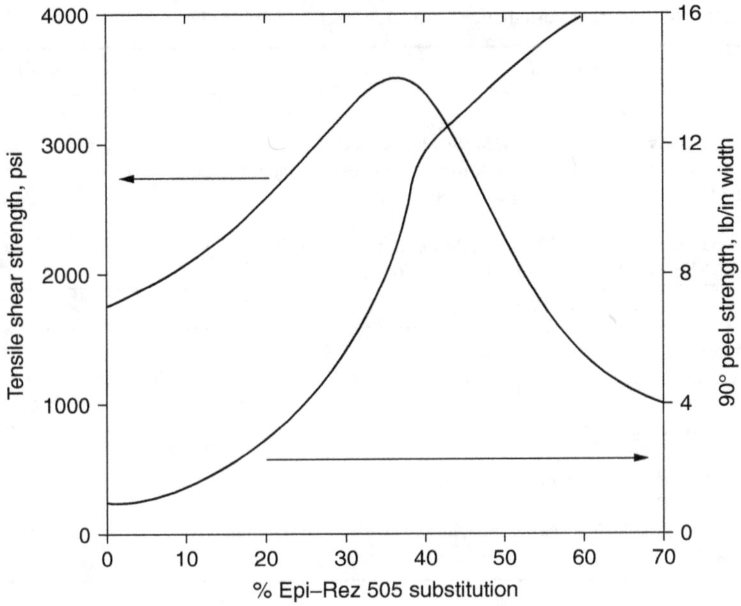

Adhesive formulation

DGEBA epoxy resin (e.g., Epi-Rez 510)	100 pbw
Amidoamine (e.g., Epi-Cure 8525)	100 pbw
Aluminum T-60	20 pbw
Thixotrope	As required

FIGURE 11.5 Effect of a flexibilized epoxy on tensile shear and peel strength of epoxy adhesive formulation.[19]

The polysulfide additive provides excellent adhesion to concrete because of its chemical nature and low viscosity, resulting in good penetration. Table 11.16 describes a formulation for a flexible epoxy-polysulfide adhesive. This adhesive is claimed to provide excellent adhesion between new and old concrete surfaces. Table 11.17 describes several epoxy-polysulfide formulations for general-purpose adhesives.

Epoxy-polysulfide systems are used in the construction, electrical, and transportation industries because of their unique combination of cured flexibility, adhesion to many substrates, and chemical resistance. They are typically used as adhesives for

- Steel, aluminum, ceramics, wood, and glass
- Crack repair for concrete
- Patch repair for concrete and mortar
- Grouting compounds
- Lead-free automotive body solder

Because of their excellent flexibility, epoxy-polysulfide adhesives are commonly used for bonding plastic substrates where differences in thermal expansion are a concern.

TABLE 11.14 Starting Formulation for a High-Peel-Strength Adhesive[20]

Component	Parts by weight
Part A	
DGEBA epoxy resin (EPON 828, Resolution Performance Products)	75
PGE of castor oil (HELOXY 505, Resolution Performance Products)	25
Fumed silica (Cab-O-Sil TS-720, Cabot Corp.)	2
Part B	
Amidoamine (Epi-Cure 3046, Resolution Performance Products)	42
Fumed silica (Cab-O-Sil TS-720, Cabot Corp.)	0.6
Property	
Mix ratio (part A : part B)	
• By weight	100 : 42
• By volume	2 : 1
Viscosity, cP, at 25°C	10,000
Working life, h, 100 g at 25°C	1.75
Tensile shear strength, psi, on acid-etched aluminum cured 3 days at 25°C	2450
90° Peel strength, lb/in of width, on acid-etched aluminum cured 14 days at 25°C	9

Table 11.18 shows the effect of increasing the polysulfide-epoxy ratio on impact properties; the effect at concentrations of polysulfide polymer greater than 1 : 1 is especially significant. The combination of increased flexibility, tensile strength, and elongation is very desirable in certain adhesive applications.

Polysulfide-epoxy adhesives can be used in various chemical and solvent environments, even under immersion conditions. The systems exhibit low permeability to water and water

TABLE 11.15 Epoxy-Polysulfide Adhesive Formulation[21,22]

Component	Parts by weight	
Part A		
DGEBA epoxy resin (Bakelite ERL-2774, Bakelite AG)	100	
DGEBA epoxy resin (EPON 828, Resolution Performance Products)		50
Surfex M.M. (Diamond Shamrock Corp.)		50
Part B		
Polysulfide LP-3 (Toray Chemicals)	100	25
Silica	80	
Tertiary amine (e.g., DMP-30)	20	
Aliphatic amine (Epi-Cure 3253, Resolution Performance Products)		5
Property		
Viscosity	Paste	Paste
Working life, 1 lb at 25°C		30 min
Tensile shear strength, psi		
• At 25°C		1480
• At 93°C		1464
• At 100°C		1000

TABLE 11.16 Formulation for a Flexible Epoxy-Polysulfide Adhesive[23]

Component	Parts by weight
Part A	
LP-3 Polysulfide	100
Hydrite Clay 121	140
Trimethylaminomethlyphenol	20
Toluene	65
Part B	
Epoxy resin (DER 331, Dow)	200
Hydrite Clay 121	105
Toluene	5
Property*	
Tensile strength, psi	450
Flexural strength, psi	345
Compressive strength, psi	4350

*New-to-old concrete after 7-day cure at room temperature.

vapor, which provides improved water and corrosion resistance. Epoxy-polysulfide adhesives also cure with very low shrinkage, in contrast to unmodified epoxy resins, which may exhibit volume shrinkage as high as 5 percent. Epoxy-polysulfide adhesives provide resistance to thermal shock and cycling as well as resistance to low-temperature environments.

Typical properties of several epoxy-polysulfide adhesives are shown in Table 11.18. Room temperature tensile shear strength ranges from about 500 to 3000 psi, depending on the substrate and the adhesive formulation. Generally, tensile strength properties reach a maximum at about 10 to 30 percent polysulfide concentration. Peel strength can be as high as 20 lb/in.

The heat distortion temperature is only slightly affected by the incorporation of 20 percent polysulfide polymer, but at a 1 : 1 ratio the drop becomes significant. This property prevents the use of epoxy-polysulfide adhesives at elevated temperatures.

TABLE 11.17 Epoxy-Polysulfide Adhesive Formulations[24]

	Parts by weight		
Components	Tough, general-purpose adhesive	Tough, general-purpose adhesive	Concrete bonding adhesive
Part A			
Epoxy resin (Araldite 6020)	100	100	
Epoxy resin (Epotuf 6140)			100
Part B			
Polysulfide LP-3	50	50	50
Dimethyl aminopropylamine (DMAPA)	10		
DMP-30		6	7.5
Calcium carbonate			120

TABLE 11.18 The Effect of LP-3 Polymer on Physical Properties of Liquid Epoxy Resin[25]

	Parts by weight						
Component	A	B	C	D	E	F	G
Epoxy resin (EEW : 175–210)	100	100	100	100	100	100	100
Polysulfide LP-3	—	25	33	50	75	100	200
DMP-30	10	10	10	10	10	10	10
Property							
Tensile, psi	3500	5500	6500	7200	2075	2350	150
Elongation, percent	0	1	2	5	7	10	300
Shore D hardness	80	80	80	80	76	76	15
Coefficient of thermal expansion, in/(in · °C) × 10^5	4.5	5.5	6.0	7.5	10.0	13.5	15.0
Impact resistance, ft · lb	2	1	3	5	27	70	100

*After curing 7 days at 25°C.

Epoxy-polysulfide systems can be formulated either as a liquid DGEBA epoxy mixed with liquid polysulfide polymer or as an epoxy-terminated polysulfide polymer; either may be cured with a tertiary amine such as DMP-30. Table 11.19 describes the formulation and shows the physical properties of these epoxy-polysulfide adhesives compared to an unmodified epoxy adhesive.

11.5 IMPROVING TOUGHNESS

Two-component, room temperature curing epoxy adhesives have evolved over three distinct generations, from brittle to flexible to toughened. The quest for toughening is due to the inherent lack of peel and impact strength properties in brittle epoxy adhesives and

TABLE 11.19 Typical Formulations and Properties of Epoxy-Terminated and Nonterminated Polysulfide-Epoxy Adhesives[26]

	Parts by weight		
Component	A	B	C
DGEBA epoxy resin (EPON 828)	100	—	100
ZL-1612 epoxy-terminated polysulfide	—	100	—
LP-3 liquid polysulfide	—	—	75
Tertiary amine	10	7	—
Cycloaliphatic amine	—	—	46
Property			
Tensile-strength, psi	7230	6980	4530
Elongation, percent	3	4	10
Hardness, Shore D	82	82	74

the reduced cohesive strength and heat and chemical resistance in flexibilized epoxy adhesives. The "toughened" epoxy adhesive claims to have the best of several worlds: improved peel strength, greater resistance to impact and thermal shock, and the ability to absorb internal stresses in the joint without sacrificing high shear strength or temperature and chemical resistance. Recent advances in toughened two-part epoxy adhesives have resulted in systems that can cure at room temperature in practical times.

The toughening mechanism of elastomer modified epoxy systems is different from that of flexibilized epoxy systems and can be used in combination with them. Flexibilized epoxy systems reduce mechanical damage by a reduction in modulus or plasticization of the adhesive. This allows stress to be relieved through distortion of the adhesive, but it also generally results in a lowering of the adhesive's glass transition temperature with an accompanying reduction in heat and chemical resistance.

By contrast, tougheners work by relieving stress through the prevention of crack growth by incorporating elastomeric particles in an epoxy resin matrix. These adhesives generally maintain a large percentage of the modulus and temperature resistance of the unmodified resin system. The primary elastomers used in these toughened epoxy systems have been functionalized butadiene acrylonitrile copolymers that are one of the following:

- Epoxy-terminated butadiene nitrile (ETBN)
- Carboxy-terminated butadiene nitrile (CTBN)
- Amine-terminated butadiene nitrile (ATBN)
- Vinyl-terminated butadiene nitrile (VTBN)

Carboxy-terminated curative, such as CTBN, provides excellent toughening in part due to its miscibility in many epoxy resins. Phase separation during cure is required to obtain toughening, and generally the phase separation requires an elevated-temperature cure. However, by prereacting the CTBN with a portion of the epoxy to obtain an adduct, a room temperature curing toughened epoxy is possible. Adduction reduces the likelihood of early phase separation and maintains the solubility of the elastomer in the uncured resin system.

CTBN modified epoxy adhesives are generally one-part systems, cured with dicyandiamide at elevated temperature. However, two-part, room or mildly elevated-temperature curing systems are also possible and provide similar improvements in properties (Table 11.20). Common room temperature curing agents used with CTBN adduct–epoxy resin blends can be polyamides or amidoamines. The CBTN modified epoxy adhesives have good shear and peel strength and are useful in general-purpose applications requiring high-performance adhesion to metals, ceramics, wood, glass, and polar thermoplastics such as polystyrene, ABS, and polycarbonate.

One of the disadvantages of CTBN epoxy adhesives has been their high viscosity, which limits formulation options. Recently new adducts, such as EPON 58003 and RSM-2577 from Resolution Performance Products, have been introduced that have significantly lower viscosities.[28] In addition, lower concentrations of these new CTBN epoxy adducts are generally required to achieve equivalent adhesive performance. New lower-viscosity CTBN adducts have also resulted in formulations where a greater concentration of cost-reducing filler, such as calcium carbonate, can be used.

CTBN epoxy adducts can be of significant value in modifying certain adhesive formulations that are relatively rigid because of high reactivity and crosslink density. For example, Table 11.21 shows the starting formulations for several fast-setting, general-purpose adhesives based on epoxy acrylate resins and aliphatic amine curing agents. The CTBN modified epoxy resin (EPON 58034, Resolution Performance Products) is used to improve the toughness and adhesive properties. These adhesives can be used for rapid-setting, general-purpose applications. Fillers and pigments can be incorporated according to need.

TABLE 11.20 Formulation for a Two-Part CTBN Modified Epoxy Adhesive System[27]

Component	Parts by weight
Part A	
DGEBA epoxy resin (EPON 828, Resolution Performance Products)	80
CTBN modified epoxy resin (EPON 58034, Resolution Performance Products)	20
Part B	
Amidoamine (Epi-Cure 3072, Resolution Performance Products)	32
Property	
Mix ratio, part A : part B	
• By Weight	3.1 : 1
• By Volume	2.7 : 1
Viscosity, cP, at 25°C	4600
Tensile shear strength, psi, measured at 25°C	
• Aluminum cured 7 days at 25°C	3500
• Aluminum cured 15 min at 118°C	4075
• Epoxy laminate cured 7 days at 25°C	4000
• Epoxy laminate cured 15 min at 118°C	4300
90° Peel strength, lb/in of width, on aluminum cured at 25°C and tested at 25°C	15–18

ATBN liquid toughening agents are synthesized by reacting a CTBN polymer with an amine. For example, *n*-aminoethylpiperazine (AEP) will give a low-molecular-mass amine-terminated product at a reaction temperature of 130°C. The terminal secondary amine groups undergo the typical reaction chemistry of secondary amines. Therefore,

TABLE 11.21 Starting Formulation for a Fast-Setting, Toughened Epoxy Adhesive[29]

	Parts by weight		
Component	A	B	C
Part A			
Epoxy acrylate resin (EPON 8121, Resolution Performance Products)	80	60	40
DGEBA epoxy resin (EPON 828)	0	20	40
CTBN modified epoxy resin (EPON 58304)	20	20	20
Part B			
Aliphatic amine (Epi-Cure 3273, Resolution Performance Products)	86	86	86
Property			
Mix ratio (part A : part B)			
• By weight	100 : 86	100 : 86	100 : 86
• By volume	1 : 1	1 : 1	1 : 1
Gel time, min, at 25°C, 50 g	2	3	5.5
Cure rate			
• Handling strength	2 h at 25°C		
• Full strength	8–10 h at 25°C		

ATBN liquid polymers cannot be mixed directly into the epoxy resin component of a two-part epoxy adhesive or in a one-part adhesive since crosslinking and shortened shelf life will result. ATBN adducts are, therefore, mixed with the curing agent component of two-component epoxy adhesives.[30]

ATBN tougheners increase the ultimate tensile shear and T-peel strengths and environmental durability of epoxy resins with conventional room temperature curing agents such as fatty polyamides, amidoamines, and amine ethers. However, the improvement in toughness, as measured by peel tests, becomes less significant with the faster-curing hardeners. Formulations for typical ATBN two-part epoxy adhesives are given in Tables 8.7 and 8.8.

11.6 IMPROVING ENVIRONMENTAL RESISTANCE

There are several ways by which the formulator can moderately improve the heat or chemical resistance of room temperature curing epoxy adhesives. Using an elevated-temperature cure or a postcure will, of course, improve the temperature resistance by virtue of improved crosslink density. However, this section describes formulations that have been developed for moderately improved heat resistance after only a cure at room temperature. Optimal (heat-curing) high-temperature and chemically resistant epoxy adhesives are discussed in Chap. 15.

11.6.1 High-Temperature Resistance

Generally aliphatic amines will provide greater heat resistance than polyamide or amidoamine curing agents. Higher-functionality epoxy resins such as epoxy-novolacs will also provide better heat resistance. Table 11.22 provides a formulation for a room temperature curing, two-component epoxy adhesive that maintains high shear strength when tested at −55 to +150°C.

Cycloaliphatic amines can also provide high heat resistance when cured with a liquid DGEBA epoxy at room temperature (although better high-temperature characteristics can be achieved with an elevated-temperature cure). A range of cure schedules and working

TABLE 11.22 High-Temperature Epoxy Adhesive That Cures at Room Temperature[31]

Component	Parts by weight
Part A	
Epoxy novolac resin (e.g., DEN 438, Dow)	40
Epoxy resin (EEW: 180–200)	40
Fibrous filler or colloidal silica	20
Part B	
Triethylenetetramine (TETA)	12
Property	
Tensile shear strength, psi, on aluminum when tested at	
• −55°C	1900
• 150°C	1850

TABLE 11.23 H_2SO_4 Acid Resistance of Epoxy Adhesive Formulations Cured 1 Week at 25°C[32]

Component	Parts by weight			
	A	B	C	D
Part A				
DGEBA epoxy resin (e.g., Epi-Rez 510)	100	100	100	100
Part B				
Accelerated aliphatic amine (e.g., Epi-Cure 874)	18	18	18	—
Petroleum extender		73		
Coal tar extender			81	
Accelerated aromatic amine (e.g., Epi-Cure 8494)				45
Weight gain, %, after immersion in 33% H_2SO_4 for 2 weeks at 65°C	2.9	3.6	3.9	1.5

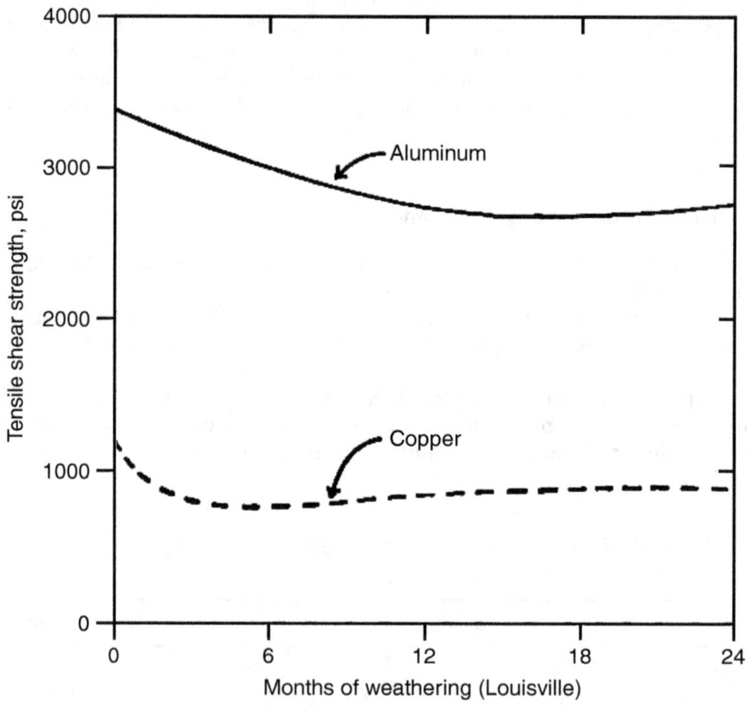

Adhesive formulation

Epoxy resin (EEW: 180–200), e.g., Epi-Rez 510	100 pbw
Amidoamine curing agent, e.g., Epi-Cure 8525	100 pbw
Alumina T-60	20 pbw
Thixotrope	20 pbw

FIGURE 11.6 Tensile shear strength retention of weathered aluminum and copper test specimens.[23]

lives are possible with these curing agents. Cycloaliphatic amines provide good adhesion and very good chemical resistance. These curing agents have been discussed in Sec. 5.2.1.

11.6.2 Chemical Resistance

Generally, those factors tending to promote thermal stability also tend to improve chemical resistance. A relatively high degree of chemical resistance can be obtained with aliphatic amine or cycloaliphatic amine cured epoxy adhesives. Accelerated aliphatic polyamines are frequently employed as curing agents in battery adhesive formulations. However, these do not provide the higher degree of chemical resistance of aromatic amines.

Systems cured with DETA and TETA generally have excellent resistance to aqueous sodium hydroxide even at 50% concentration and 80°C. Their resistance to 25 percent sulfuric acid, 23 percent hydrochloric acid, and 25 percent chromic acid is also very good up to 80°C. Strong organic acids such as acetic acid generally attack DETA and TETA cured epoxy resins as well as strong (40%) nitric acid. Resistance to intermittent solvent exposure is excellent; however, long-term immersion in any aliphatic solvent such as kerosene is not recommended.

Table 11.23 presents H_2SO_4 acid resistance data for epoxy adhesive formulations consisting of accelerated aliphatic amines and accelerated aromatic amines. Generally a weight gain of less than 4% is indicative of moderate chemical resistance. Aliphatic and aromatic amines as well as polymercaptan curing agents are often used in bonding reinforced plastic pipe in the chemical industry. The combination of a high degree of chemical resistance along with rapid, on-site cure makes these adhesive systems especially valuable.

Although most epoxy adhesives have good weather resistance, optimum properties are generally achieved when the adhesive has a combination of good water resistance and thermal shock resistance. Figure 11.6 illustrates the retention of tensile shear strength of copper and aluminum strips bonded with an amidoamine cured epoxy after 2 years of weathering in a temperate climate.

Great caution must be exercised in exposing any adhesive joint to the simultaneous effects of environment and stress. The stress can act to accelerate the degradation caused by the environment, and vice versa. Joints that will be exposed to both high-humidity environments and high load at the same time are especially vulnerable, and prototype specimens need to be tested. This degradation mechanism and the performance of several epoxy adhesive systems to combined environmental stress conditions are discussed in Chap. 15.

REFERENCES

1. Behm, D. T., and Gannon, J., "Epoxies," in *Adhesives and Sealants*, vol. 3, *Engineered Materials Handbook*, ASM International, Materials Park, OH, 1990, p. 101.
2. Behm, D. T., and Gannon, J.,"Epoxies," p. 101.
3. Petronio, M. et al., "Glass to Metal Bonding," *Adhesives Age*, January 1959.
4. Savia, M., "Epoxy Resin Adhesives," Chapter 26 in *Handbook of Adhesives*, 2d ed., I. Skeist, ed., van Nostrand Reinhold, New York, 1977.
5. Lee, H., and Neville, K., *Handbook of Epoxy Resins*, McGraw-Hill, New York, 1967, p. 21.35.
6. Lee and Neville, *Handbook of Epoxy Resins*, p. 21.35.
7. Lee and Neville, *Handbook of Epoxy Resins*, p. 21.38.
8. Wiles, Q. T., U.S. Patents 2,528,934 and 2,528,933, 1950.
9. Lee and Neville, *Handbook of Epoxy Resins*, p. 21.38.

10. Meath, A. R., "Epoxy Resin Adhesives," Chapter 19 in *Handbook of Adhesives*, 3d ed., I. Skeist, ed., van Nostrand Reinhold, New York, 1990, p. 356.
11. Lee and Neville, *Handbook of Epoxy Resins*, p. 21.39.
12. Bruins, P. F., *Epoxy Resin Technology*, Interscience Publishers, New York, 1968, p. 166.
13. Bruins, *Epoxy Resin Technology*, p. 168.
14. Savia, M., "*Epoxy Resin Adhesives*," p. 443.
15. Bruins, *Epoxy Resin Technology*, p. 169.
16. Bruins, *Epoxy Resin Technology*, p. 171.
17. Resolution Performance Products, Starting Formulations 4000 and 4001, "Rapid Setting Room Temperature Cure Adhesive," Houston, TX, August 2003.
18. Air Products and Chemicals, Inc., "Flexible Adhesive," Allentown, PA, 2001.
19. Bruins, *Epoxy Resin Technology*, p. 165.
20. Resolution Performance Products, Starting Formulation 4007, "High Peel Strength Adhesive," Houston, TX, August 2003.
21. Dannenberg, H., and May, C. A., "Epoxide Adhesives," Chapter 2 in *Treatise on Adhesion and Adhesives*, R. L. Patrick, ed., Marcel Dekker, New York, 1969.
22. Resolution Performance Products, Starting Formulation 4026, "Metal Adhesive," Houston, TX, August 2003.
23. Meath, "Epoxy Resin Adhesives," p. 356.
24. Amstock, J. S., *Handbook of Adhesives and Sealants in Construction*, McGraw-Hill, New York, 2002, p. 2.11.
25. Panek, J. R., "Polysulfide Sealants and Adhesives," in *Handbook of Adhesives*, 3d ed., I. Skeist, ed., van Nostrand Reinhold, New York, 1992.
26. Peterson, E. A., "Polysulfides," in *Adhesives and Sealants*, vol. 3, *Engineered Materials Handbook*, AMS International, Materials Park, OH, 1990.
27. Resolution Performance Products, Starting Formulation 4018, "High Strength Room Temperature or Heat Cure Adhesive," Houston, Tx, August 2003.
28. Farris, R. D., and Steward, S. L., "New Epoxy Tougheners Widen the Adhesive Formulation Window," *Adhesives and Sealants Industry*, January 2002.
29. Resolution Performance Products, Starting Formulation 4008, "Adhesive for Thermoplastic Substrate," Houston, TX, August 2003.
30. Petrie, E. M., "Improving the Toughness of Structural Adhesives," at www.SpecialChem4Adhesives.com, April 7, 2004.
31. Savia, "Epoxy Resin Adhesives," p. 443.
32. Bruins, *Epoxy Resin Technology*, p. 162
33. Bruins, *Epoxy Resin Technology*, p. 161.

CHAPTER 12
ELEVATED-TEMPERATURE CURING LIQUID AND PASTE EPOXY ADHESIVES

12.1 INTRODUCTION

Although room temperature curing epoxy adhesives are very convenient to use in many applications, the short working life, long curing time, and inherently suboptimal physical properties of these systems are often a serious disadvantage to their use. In these cases, elevated-temperature curing epoxy adhesives are required. The major advantages and disadvantages of elevated-temperature curing epoxy adhesives are shown in Table 12.1.

Heat curing epoxy systems may consist of one or two components. The nomenclature is similar to that employed with the room temperature curing adhesives (e.g., *resin component*, *curing agent component*, *2K system*, etc.). Many heat curing epoxy adhesives systems are liquids or pastes. However, heat curing systems also can be processed into solid adhesive forms (e.g., powders and films). This chapter describes the paste and liquid formulations; solid adhesive forms are considered separately in Chap. 13.

Two-component, elevated-temperature curing epoxy adhesives, once mixed, are designed to cure only at elevated temperatures. However, these systems generally will have a finite working life at room temperature, although it is much longer than that of room temperature curing epoxy adhesives. Generally, a minimum temperature of around 80°C is required to fully cure these systems. Various cure temperatures and cure schedules are often provided to offer the user flexibility in processing.

One-component (or "1K"), elevated-temperature curing epoxy adhesives have the catalyst or curing agent incorporated directly with the epoxy resin. Thus, no metering or mixing is required. However, since the curative is integrated with the resin, these systems have limited shelf life, and special storage conditions such as refrigeration may be required. Elevated-temperature curing epoxy adhesives can be desirable when the short working life of a room temperature curing adhesive cannot be tolerated. One-component epoxy adhesives are advantageous when it is necessary to eliminate measuring or mixing errors or when dispensing equipment cannot handle multicomponent systems.

Several important factors must be considered relative to the "elevated temperatures" that are required to cure these adhesives. The curing temperature and time specified represent the temperature and time that the adhesive film actually experiences. For example, if a cure of 60 min at 150°C is recommended, this does not mean that the assembly should be simply placed in a 150°C oven for 60 min. The temperature is to be measured at the adhesive within the joint. A large assembly will act as a heat sink and may require substantial time for an adhesive within the joint to reach the necessary temperature. In this example,

TABLE 12.1 Advantages and Disadvantages of Elevated-Temperature Curing Epoxy Adhesives

Advantages	Disadvantages
Fast cure time results in higher production speed and less tie-up of fixturing equipment	Greater internal stresses than in room temperature curing adhesive because of • Higher shrinkage on polymerization • Thermal expansion coefficient differences
Long working life	More brittle than room temperature curing adhesives (poorer peel and impact properties)
High-temperature resistance	Safety and hazardous nature of high temperatures
Good chemical resistance	One-component systems have shorter shelf life; may require refrigeration
One-component systems possible (no need to meter and mix)	Solid systems easier to apply, less of a health hazard than liquids or pastes
Long shelf life for two-component systems	Energy consumption
Solid systems (film, preforms, powders) possible	Viscosity of adhesive decrease at elevated temperature could result in starved joint
Viscosity decrease at elevated temperature provides more efficient wetting of substrate	

the total oven time would be 60 min in addition to the time that it takes to bring the adhesive film up to 150°C.

Bond line temperatures are best measured by thermocouples placed very close to the adhesive. In some cases it may be desirable to place the thermocouple directly in the adhesive for the first few assemblies being cured.

Oven heating is the most common source of heat for curing epoxy adhesives that require an elevated-temperature cure. Good air circulation within the oven is mandatory for uniform heating. Temperature distribution within an oven should always be checked before items are placed in the oven. Many ovens will have significant temperature distribution variation and dead spaces, especially in corners where air circulation is not uniform. The number of items and their arrangement in the oven also affect the time for the joint to reach the proper temperature. The geometry and size of the part may affect air circulation and cause variations in temperature distribution.

Heated-platen presses are also commonly used for bonding flat or moderately contoured panels. Autoclaves or vacuum bagging is used for large parts such as in the aerospace industry. In addition to direct heating, epoxy adhesives can be cured by induction and dielectric heating as well as other indirect heating sources. Formulations specifically developed for indirect heating are discussed in Chap. 14. Chapter 17 provides additional information regarding the processing and equipment used to cure epoxy adhesive at elevated temperature.

Regardless of the heating source, elevated-temperature curing epoxy adhesives necessitate that the parts be held with clamps, pressure pads, or fixturing methods to apply constant and uniform pressure throughout the curing cycle. Reduction in viscosity due to elevated temperatures could result in excessive adhesive flow during cure. This could result in joints that are starved of adhesive or movement of the parts, if not firmly fixtured.

Although room temperature curing epoxy adhesive can also be cured at elevated temperatures, there are important differences in morphology and internal stresses between a room temperature cured adhesive and a heat cured epoxy adhesive (see Chap. 3). One should not assume that the properties would be equivalent or better if the adhesive were cured at elevated temperature rather than room temperature. A number of factors need to be considered in such a deliberation.

Whereas most room temperature curing epoxy adhesives are cured with aliphatic amines, polyamides, or amidoamines, most elevated-temperature curing epoxy adhesives are cured with aromatic amines, modified aliphatic amines, alcoholic and phenolic hydroxyls, acid anhydrides, Lewis acids, and a host of other curatives. Latent curing agents, such as dicyandiamide and imidazoles, are typically used in one-component epoxy adhesives systems.

12.2 TWO-COMPONENT ADHESIVE FORMULATIONS

The primary advantage of a two-component, elevated-temperature curing epoxy adhesive over a one-component system is that the curing agent typically used provides a tougher bond. This is primarily due to the polyaddition reactions that take place with two-part systems and the homopolymerization reactions that generally occur with one-part epoxy formulations. The two-component curatives also give the formulator and user greater latitude in processing conditions. Two-component epoxy adhesives often have lower cure temperatures and/or less cycle time than their one-component counterparts. This is because one-component formulations require a lower reactivity to preserve shelf life.

The primary curing agents used in two-component, elevated-temperature curing epoxy adhesives are modified aliphatic amines and aromatic amines. Diethylaminopropylamine (DEAPA) and dimethylaminopropylamine (DMAPA), both aliphatic amines, are widely used as curing agents. They provide properties similar to those of DETA but have better strength retention at elevated temperatures and a working life of about 4 h at room temperature. The DMAPA is more active than the DEAPA, but both are used at concentrations of about 10 pph.

Table 12.2 describes an epoxy formulation cured with DEAPA. Cure conditions of only 45 min at 100°C are adequate to obtain high bond strength on aluminum and moderately better performance at elevated temperatures than that for systems cured at room temperature. In this formulation the low-molecular-weight, high-epoxy-content resin (EPON 1009)

TABLE 12.2 General-Purpose Two-Component Epoxy Adhesive Cured with DEAPA[1]

Component	Parts by weight
Part A	
Liquid epoxy resin (e.g., EPON 828)	100
Epoxy resin (e.g., EPON 1009)	10
Epoxidized soybean oil	12
Thixotrope	As required
Part B	
Diethylaminopropylamine (DEAPA)	11
Property	
Tensile shear strength, psi, on aluminum at	
• −57°C	2620
• 25°C	3390
• 83°C	3445

*Cure conditions: 2 h at 150°C.

is added to the more conventional liquid epoxy resin to improve high-temperature performance. The epoxidized soybean oil is used to reduce viscosity.

Cycloaliphatic amines, such as aminoethylpiperizine (AEP), isophoronediamine (IPDA), and bis(*p*-aminocycloethyl)methane (PACM), provide longer pot life than the aliphatic amines along with greater temperature resistance and toughness when cured. AEP is used to provide higher impact strength and toughness than aliphatic amines; however, it will not cure at room temperature in thin sections. IPDA and PACM have been considered as alternative to aromatic amines. They provide shorter pot life than the aromatic amines, but they are colorless and non-staining to the skin. IPDA and PACM also provide a high heat distortion temperature.

General-purpose rigid and flexible two-component, elevated-temperature cure epoxy adhesives can easily be formulated from cycloaliphatic amine (rigid) or polyamide (flexible) curing agents. Table 12.3 describes two such formulations. Note that the flexible adhesive system exhibits a T-peel strength on aluminum of 14 lb/in compared to only 4 lb/in for the rigid formulation. However, the rigid formulation has a higher tensile shear strength and better heat resistance.

Temperature-resistant two-part, elevated-temperature curing epoxy adhesives can be formulated with aromatic amines, such as metaphenylenediamine (MPDA), methylene dianiline (MDA), or a eutectic blend of the two. These adhesives will provide relatively high temperature strength, but they are generally brittle. When mixed with epoxy resin at concentrations of about 15 pph for MPDA and 26 pph for MDA, they provide complete cure in about 30 min at 175°C. The aromatic amines also provide a working life of several hours at room temperature. Starting formulations for aromatic amine cured epoxy adhesives are shown in Table 12.4.

MDA cured epoxy adhesive will provide shear strength over 4 times that of DETA at temperatures of 100 to 150°C. Table 12.5 indicates the high-temperature performance that

TABLE 12.3 Elevated-Temperature Curing Epoxy Adhesive Formulations

Component	Parts by weight	
	Rigid	Flexible
Part A		
DGEBA epoxy resin (EEW: 190)	68	50.5
Diluent (Epodil 748)		7
Talc (Microtuff 325F)	29.5	40
Fumed silica	2.5	2.5
Part B		
Ancamine 2264 (Air Products and Chemicals Inc.)	40	
Ancamide 910 (Air Products and Chemicals Inc.)		38
Ancamide 2482 (Air Products and Chemicals Inc.)		15
Aluminum (Toyal 101)	24	20
Talc (Microtuff 325F)	24	25
Fumed silica	2	2
Property		
Mix ratio, by weight, part A : part B	2 : 1	1 : 1
Cure schedule	30 min at 150°C	30 min at 120°C
Tensile shear strength, psi, on aluminum measured at		
• 25°C	1734	1640
• 120°C	396	175
T-peel strength, lb/in, on aluminum measured at 25°C	4	14

TABLE 12.4 Starting Formulations for Epoxy Adhesives Cured with MPDA[2]

Component	Parts by weight
Part A	
Epoxy resin (EPON 828)	100
Allyl glycidyl ether	2
Polyvinyl acetate	6
Thixotrope	As required
Part B	
Metaphenylenediamine (MPDA)	15.2
Property	
Tensile shear strength, psi, on aluminum measured at	
• 23°C	3160
• 83°C	3560
• 126°C	3250
• 150°C	2140

can be achieved with several aromatic amines. The aromatic amines, however, generally provide poorer bond strength at lower temperature due to their relatively brittle nature and high degree of shrinkage on polymerization. Aromatic amines can be used in one-component epoxy adhesives although their shelf life is relatively short when compared to that of the more commonly used latent curatives.

Low-melting-point blends of aromatic amines (e.g., Shell Chemical Curing Agent Z or Celanese Epi-Cure 841) can be used to formulate two-component epoxies, which cure satisfactorily at temperatures below 100°C. These adhesives are often used to bond heat-sensitive

TABLE 12.5 Effect of Aromatic Amines on Tensile Shear Strength of DGEBA Adhesive Supported on Glass Cloth Carrier[3]

		Bonding conditions			Tensile shear strength, psi, at			
Curing agent	pph	°C	psi	min	23°C	127°C	177°C	204°C
Metaphenylenediamine (MPDA)	15	150	10	45	3100	3275	736	507
		150	100	45	2935	3510	790	660
		177	100	45	3330	2978	830	645
Methylene dianiline (MDA)	24	150	10	45	4200	3170	930	575
		150	100	45	3705	3470	870	590
		177	100	45	3700	3150	955	615
	27	150	10	45	3220	3150	700	550
		150	100	45	3940	4160	920	610
		177	100	45	4050	4630	960	755
	29.6	150	10	45	4970	4350	800	565
		150	100	45	4630	3800	740	590
		177	100	45	4310	3610	710	530

adherends. The aromatic amines also provide B-stage cure characteristics that prove useful in formulating solid adhesive forms.

Diaminodiphenyl sulfone (DADPS) cured epoxy adhesive also exhibits very good elevated-temperature properties. With this curing agent and resin blends of DGEBA epoxy and polyglycidyl of tetraphenolethane, adhesive strengths of greater than 1000 psi have been realized on aluminium substrates after aging for 400 h at 260°C.[4] The DADPS is generally mixed with the epoxy resin at a concentration of about 30 pph. Another formulation consisting of 100 parts EPON 1031 (a tetrafunctional solid epoxy resin), 100 parts aluminum powder, and 30 parts DADPS was found to yield bond strength in excess of 1000 psi when cured 30 min at 175°C and tested at 260°C.[5]

Anhydride cured epoxy adhesives are sometimes used where high-temperature performance or chemical resistance is desired. These curing agents are not widely used in adhesive formulations, however, because the adhesive properties are inferior to those realized by the amine curing agents and because anhydrides require rather long cure cycles. Anhydride curing agents are used where exceptionally long working life is required. The exotherm that is generated with epoxy-anhydride systems is low, and the reaction rate is slow. Elevated-temperature cures (greater than 150°C) are necessary, and long postcures are required to develop ultimate properties. Tertiary amines are the most widely used accelerator for this system. Anhydride cured epoxy resins exhibit very low viscosities.

The physical and electrical characteristics of the anhydride cured systems are very good over a wide temperature range. Compared to amines, anhydride cured epoxies exhibit better chemical resistance to aqueous acids, but less chemical resistance to some other reagents. When epoxy resins are cured with anhydrides, the product has relatively high heat distortion temperature and low moisture sensitivity.

Phthalic anhydride (PA) and its derivatives are among the most common anhydride curing agents. Table 12.6 describes an anhydride cured epoxy adhesive that is accelerated with

TABLE 12.6 Formulation for an Anhydride Cured Epoxy Adhesive[7]

	Parts by weight	
Component	A	B
Part A		
Epoxy resin (EPON 828)	100	12
Condensation product of DGEBA epoxy and ethylene glycol		88
Thixotrope	As required	As required
Part B		
Phthalic anyhydride	43	43
Phthalic acid	20	
Benzyldimethylamine	0.8	
Pyromellitic dianhydride		2.5
Pyridine		0.3
Property		
Cure conditions	2 h at 150°C	15 min at 205°C
Tensile shear strength, psi, on aluminum at		
• −57°C	3040	
• 25°C	3930	
• 83°C	4840	
• 105°C	3160	

a tertiary amine. This formulation cures in 2 h at 150°C. At higher anhydride contents, lower room temperature strength results owing to a reduced hydroxyl content in the cured resin.[6]

Other anhydrides such as dodecyl succinic anhydride (DDSA) or adducts of DDSA with polyglycols, can also be used for formulating heat cured epoxy adhesives. These have excellent electrical properties and good thermal shock resistance. Anhydride cured epoxies are also useful for bonding plastics, notably polyester such as Mylar.[8]

12.3 ONE-COMPONENT ADHESIVE FORMULATIONS

In one-component epoxy adhesives, the curing agent and resin are compounded together as a single product by the adhesive formulator. The curing agent system is chosen so that it does not react with the resin until the appropriate processing conditions are applied.

One-component epoxy curing agents, such as dicyandiamide, BF_3 complexes, aromatic amines, and imidazoles, are commonly used since these compounds are stable at room temperature but cure when heated past an activation temperature. All these curing agents are slow-reacting curatives that provide a relatively long shelf life when mixed into the epoxy resin. However, the greatest success has been realized by a group of curatives known as *latent* curing agents.

Latent curing agents work by several mechanisms. The most common is due to the curing agents' insolubility in the epoxy resin and, therefore, its inactivity at room temperature. When heated, the curative melts and dissolves into the resin thereby activating the curing process. These types of curing agents are best represented by dicyandiamide and several other curatives. They represent the most common curatives used in single-component epoxy adhesives.

A second method of achieving latency is by using a curative precursor, which is chemically inactive at room temperature but then converts to an active curative at the cure temperature. Examples of this type include Monuron and di-urea adduct of toluene diisocyanate and dimethylamine. Full cure in about 1 h at 125°C is achievable with these materials.

Dicyandiamide (sometimes referred to as "dicy") or its derivatives are used in most commercial one-component epoxy adhesives. This curative is a white crystalline solid and is easy to incorporate into an epoxy formulation as a finely ground powder. When cured with epoxy resin, dicyandiamide provides an excellent set of performance properties.

Dicyandiamide has an activation temperature of about 177°C. However, this can be reduced to about 120°C by the use of accelerators. When mixed into liquid DGEBA resins, dicyandiamide has very good shelf life due to its latency. The shelf life of unaccelerated systems is about 6 months at room temperature or longer if refrigerated. The shelf life is reduced significantly with the addition of accelerators.

Dicyandiamide cured epoxies have been known to exhibit high strength and chemical resistance; however, they are relatively brittle. Through the utilization of toughening agents, very resilient and tough adhesives can be formulated without sacrificing any of the inherently good properties of the unmodified system. With toughened dicyandiamide cured epoxies, peel strengths on the order of 30 lb/in have been achieved along with tensile shear strength in the range of 3000 to 4500 psi. The toughened dicyandiamide cured epoxy adhesive also shows very good resistance to thermocycling.

A typical dicyandiamide cured epoxy adhesive formulation is described in Table 12.7. These formulations are unaccelerated so that they have a shelf life of greater than 6 months at room temperatures. The adhesives will cure in 30 to 90 min at 149 to 177°C. Tensile shear strength on aluminum is in the range of 2500 to 3000 psi.

TABLE 12.7 One-Component Epoxy Adhesive Cured with Dicyandiamide[9]

Component	Parts by weight			
	A	B	C	D
DGEBA epoxy resin (EPON 828, Resolution Performance Products)	100	100	100	100
Hycar CTBN (B.F. Goodrich)		10		
Dicyandiamide (SKW Corporation)	10	10		10
Dicyandiamide (Dyhard 100SF, Degussa Corp.)			6	
Ground tabular alumina (Alumina T60/T64, Cabot Corp.)			20	
Fumed silica (Cab-O-Sil, Cabot Corp.)			10	
Bentone 27 (Rheox, Inc.)	10	5		5
Tetramethyl ammonium chloride (Distillation Products Industries)	3			
No. 1 white calcium carbonate (Thompson, Weinman, & Co.)	50			
Karmex (DuPont Co.)		7.5		
Aluminum 120 (Reynolds Metals Co.)		20		50
Methanol (Hoechst Celanese Corp.)				1.67
Property				
Viscosity, cP, at 25°C	Paste	Thixotropic paste	Thixotropic paste	Thixotropic paste
Shelf life at 25°C, months	>3	>6	>6	12
Cure schedule	30 min at 149°C or 40 min at 135°C	30 min at 149°C or 40 min at 123°C	1–1.5 h at 177°C	1 h at 149°C
Tensile shear strength, psi, on aluminum				
• At 25°C		3500	2630	2900
• At 150°C		700	—	1050

Accelerators for dicyandiamide cured epoxy adhesive formulations include tertiary amines, modified aliphatic amines, imidiazoles, and substituted ureas. All except the substituted ureas can cure epoxy resins by themselves. All these materials provide good latency and excellent adhesive applications.

Probably the most effective accelerator for dicyandiamide systems is the substituted ureas because of their synergistic contribution to the performance properties of the adhesive and their exceptionally good latency. It has been shown that adding 10 pph of a substituted urea to 10 pph of dicyandiamide will produce an adhesive system for liquid DGEBA epoxy resins that can cure in only 90 min at 110°C. Yet this adhesive will exhibit a shelf life of 3 to 6 weeks at room temperature. Cures can be achieved at temperature even down to 85°C if longer cure times are acceptable.[10]

Table 12.8 shows the effect of three commercially available substituted ureas on shelf life, cure rate, exotherm, and glass transition temperature of a dicyandiamide cured epoxy adhesive. The accelerators are compared at use levels of 1, 3, and 5 pph in a one-component adhesive consisting of 100 pph of DGEBA epoxy, 8 pph of dicyandiamide, and 3 pph of

TABLE 12.8 Properties of DGEBA-Dicyandiamide Cured Adhesive Accelerated with Substituted Ureas[11]

	Concentration of substituted urea, pph, in DGEBA/dicyandiamide/fumed silica (100/8/3) system								
Control	0								
U-52 (CVC Specialties)*		1	3	5					
U-405 (CVC Specialties)†					1	3	5		
Diuron‡								1	3
Property									
Viscosity, P, at 25°C after									
• Initial	290	370	390	420	310	320	320	320	350
• 12 weeks	310	360	400	430	370	1900	2140	390	430
• 24 weeks	300	400	450	470	970	—	—	470	660
• 48 weeks	330	540	650	740					
Time to double viscosity, weeks	>136	60	55	57	20	10	9	30	25
Peak exotherm, °C	173	171	165	161	172	167	163	176	168
Minutes to 95% cure at 120°C	—	47	27	20	43	22	15	64	26
Glass transition temperature, °C	140	133	127	119	130	118	110	131	122

*U-52: 4,4′-methylene bis(phenyl dimethyl urea).
†U-405: phenyl dimethyl urea.
‡Diuron: 3-(3,4-dichlorophenyl)-1,1-dimethyl urea.

fumed silica. Usable shelf life is generally taken as the time for a twofold increase in viscosity to occur.

Table 12.9 shows a formulation for an accelerated general-purpose one-component, dicyandiamide cured epoxy adhesive compared to one with a modified aliphatic amine curing agent. Notice that the dicyandiamide cured system provides a higher glass transition tem-

TABLE 12.9 Formulations for a Dicyandiamide and Modified Aliphatic Amine Cured Epoxy Adhesive

Component	Parts by weight A	Parts by weight B
DGEBA epoxy resin (EEW: 190)	100	100
Dicyandiamide (Amicure DG-1200, Air Products and Chemicals Inc.)	6	
Modified aliphatic amine (Ancamine 2014AS, Air Products and Chemicals, Inc.)	5	28
Fumed silica (Cab-O-Sil TS720)	2	2
Property		
Glass transition temperature, °C	120	85
Gel time at 140°C, min	5.5	1
Shelf life (time required to double viscosity) at 42°C	>3 months	11 weeks

TABLE 12.10 Starting Formulation for an Epoxy Adhesive Cured with BF_3 Amine Catalyst[12]

Component	Parts by weight
DGEBA epoxy resin (EPON 828, Resolution Performance Products)	50
Ground calcium carbonate (ExCal W3, Excalibar Minerals Inc.)	48
Fumed silica (Cab-O-Sil TS-720, Cabot Corp.)	2
BF_3 amine catalyst (Leecure 8-239B, Leepoxy Plastics, Inc.)	1.5
Property	
Viscosity, cP, at 25°C	Thixotropic paste
Shelf life at 25°C	4 months
Cure schedule	30 min at 177°C or 2 h at 135°C
Tensile shear strength, psi, on aluminum after 7 days' cure at 93°C	875

perature by virtue of its greater crosslinking density. The dicyandiamide also has greater than 3 months' shelf stability even though it is catalyzed. The accelerated dicyandiamide cured epoxy adhesive formulation is typical of the general-purpose one-component epoxy adhesive products that are commercially available today from many adhesive suppliers.

Boron trifluoride monoethylene amine complex (BF_3-MEA) is a Lewis acid that has also been used as a latent curing agent. It is an adduct of boron trifluoride and diethylamine existing as a solid that melts at 80 to 85°C. When mixed with a DGEBA epoxy, the formulation exhibits exceptionally long shelf life of 6 months to 1 year at room temperature. There is no significant cure that takes place until the curing temperature exceeds 100 to 125°C. The cured adhesive exhibits good physical properties when tested in the 150 to 175°C temperature range; however, the peel and impact properties are somewhat poorer than those of dicyandiamide cured epoxies.

The BF3-MEA complex offers a slightly faster rate of cure and a reduced shelf live in liquid epoxy systems when compared to unaccelerated dicyandiamide cured epoxy formulations. However, when used as a sole curing agent, BF3-MEA has not had the commercial success of dicyandiamide because of their lower bond strength and brittleness. The BF3-MEA complex compounds also hydrolyze in the presence of moisture, so that mixtures with epoxy resins must be stored in tightly closed containers to maintain shelf life.

BF_3-MEA cured epoxy adhesives are very economical to produce since only a small amount of catalyst is required. These adhesives have good elevated service temperature. Table 12.10 is a starting formulation for a one-component epoxy adhesive cured with a BF_3 MEA catalyst. It has a shelf life of approximately 4 months at room temperature.

Imidazoles, such as 2-ethyl-4-methyl imidazole (EMI), represent another family of latent curing agents that can be used alone or as an accelerator for dicyandiamide. When EMI is mixed with DGEBA epoxy resins, the shelf life is greater than 6 months at 40°C, yet it will gel in a very short time (minutes) at temperature from 120 to 170°C. Cure of this adhesive at 120°C for 30 min gives strong water-resistant bonds to stainless steel. The cured adhesive exhibits a high degree of thermal and chemical resistance.

12.4 NOVEL ONE-COMPONENT ADHESIVE SYSTEMS

Several novel methods have been developed over the years to provide one-component systems with long shelf life and good performance properties when cured at a moderate time and low temperatures. These include frozen, microencapsulated, and molecular sieve catalyzed systems.

One method of formulating a single-component epoxy system is to blend a liquid epoxy resin with a conventional amine curing agent and then package and immediately flash-freeze the product to stop further reaction. In essence, the user then has a one-component adhesive that does not require metering or mixing, but which must be kept frozen until use. The adhesive formulations can be such that they cure at either room or slightly elevated temperatures. While offering some of the conveniences of a one-component adhesive, these types of adhesive systems do not provide the performance properties of conventional elevated-temperature cure adhesive. They are expensive because of the added processing steps and the cost of packaging in small containers to avoid waste.

Amines have also been microencapsulated within small cellulosic or polyelectrolyte capsules. This is a method for keeping the amine separate from the epoxy resin during storage. When the user decides to initiate cure, the capsules are broken, usually in the application process, and the amine is free to react with the epoxy resin. A successful example of this type of product is an epoxy adhesive that can be preapplied to machine screw threads. When the screw is ultimately threaded into place, the shearing action causes the capsule to break. Bond strengths are generally low for this type of adhesive, but this may not be important in certain applications.

Molecular sieves such as zeolite have also been used as carriers for low-molecular-weight amines. An amine, such as DETA, can be absorbed into the sieve prior to mixing with an epoxy resin. The sieve protects the amine from reacting with the epoxy resin, and a relatively long shelf life is possible. Release of the DETA is then initiated by heat or through displacement by atmospheric moisture.

12.5 IMPROVING PERFORMANCE PROPERTIES

Several formulations of elevated-temperature curing epoxy adhesives have been developed with improved thermal resistance and greater toughness. The next section describes the processes and materials that can be used to achieve moderately better heat resistance and toughness. However, formulations with the optimum temperature resistance are discussed in Chap. 15, and tougheners are described in Chap. 8.

12.5.1 Thermal Resistance

Epoxy-novolac resins and highly functional DGEBA epoxy resins have been used in combination with anhydride, phenolic, or aromatic amine curing agents to produce moderately high-temperature-resistant adhesives. Although useful tensile shear strength at temperatures as high as 260°C can be achieved for short periods of time (hundreds of hours), longer-term service temperatures are limited by thermal degradation processes to 150 to 175°C. Figure 12.1 shows tensile shear strength at test temperature for various different formulations.

A general-purpose two-component adhesive that will provide high tensile shear strength up to 150°C is described in Table 12.11. The base epoxy resin in this formulation is a mixture of an epoxy novolac and a liquid DGEBA epoxy resin.

Benzoquinone tetracarboxylic acid dianhydride (BTDA) has been found to provide epoxy adhesives with excellent high-temperature properties, in both the short and long terms. The formulation described in Table 12.12 provides good resistance to 260°C. This two-part adhesive can be cured 2 h at 200°C. The disadvantage of BTDA is that relatively high cure temperatures are required that result in a high degree of internal stress within the bond line.

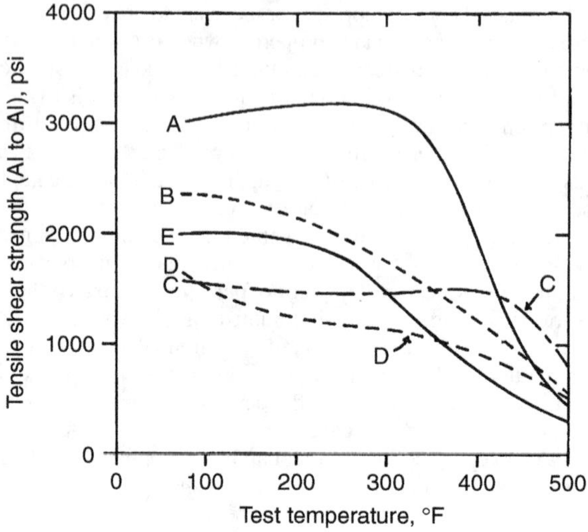

Adhesive formulations

Component	Parts by weight				
	A	B	C	D	E
Epi-Rez 508 purified DGEBA	50				
Epi-Rez 5155 epoxy novolac	50				
Epi-Rez 510 DGEBA		100			100
Epi-Rez 5108 purified DGEBA			100	100	
Cyclan 330 (trianhydride)	33				
HHPA	53				
Diethylaminoethanol	0.5				
Cyclopentanetetracarboxylic dianhydride		50			
2-Ethyl-4-methylimidazole		0.2	2	2	
Epi-Cure 841 aromatic amine					22.5
Alumina T-60			179		20
Thixotrope					20
Colloidal silica	2	2	3		
Aluminum powder	50	50			

FIGURE 12.1 High-temperature epoxy adhesives.[13]

Other curing agents that are known to provide moderately good heat resistance are pentamethyldiethylenetriamine (PMDA) and short-chain acid anhydrides, such as methyl nadic anhydride. Imidazoles, such as 2-ethyl-4-methyl imidazole, and certain aromatic amines, particularly diaminodiphenyl sulfone (DADPS), provide very good high-temperature properties in two-component epoxy systems, as noted above. Thermal stability is particularly enhanced when these curing agents are used with epoxy resins having a higher functionality than the DGEBA types. The most common epoxy resins used in high-temperature formulation, therefore, are epoxy novolacs or tetrafunctional solid epoxy.

TABLE 12.11 High-Temperature Epoxy Adhesive Utilizing Epoxy Novolac Resin[14]

Component	Parts by weight
Part A	
Epoxy novolac (DEN 438, Dow)	90
Epoxy resin (DER 736, Dow)	10
Atomized aluminum powder	40
Part B	
Cycloaliphatic amine	28

PMDA or trimellitic anhydride has also been shown to provide epoxy adhesive formulations with high-temperature properties. Table 12.13 shows the elevated-temperature tensile shear strength of an epoxy adhesive cured with 4 pph of PMDA. Another specialized formulation employing PMDA was found to provide high shear strength when tested at 260°C even after aging 1000 h at 260°C.[15]

The main difficulty lies in incorporating curing agents such as PMDA into the epoxy resin and then providing for resiliency when cured. One formulation by DuPont[16] was developed for improved toughness. It requires reacting 1 mol of dialcohol with 2 mol of PMDA to yield a more readily fusible and soluble product, which maintains two anhydride groups.

12.5.2 Toughening

For many years, the typical method of improving the toughness of high-temperature structural adhesives was to add elastomeric resins to rigid high-temperature base polymer to create a hybrid product such as epoxy-nitrile. However, the toughening of high-temperature adhesives can provide a difficult challenge, since the service temperatures usually exceed the degradation point of most rubber additives. Also, the addition of an elastomer generally resulted in lowering of the glass transition temperature of the base polymer.

TABLE 12.12 BTDA Curing Agent in High-Temperature Epoxy Adhesive

Component	Parts by weight
DGEBA epoxy resin (EEW: 190)	100
BTDA	48
Atomized aluminum	100
Fumed silica	3
Property	
Cure schedule	2 h at 200°C
Tensile shear strength, psi, on etched aluminum at	
• 23°C (initial)	2480
• 150°C (initial)	1600
• 260°C (initial)	1220
• 250°C (after aging 1000 h at 260°C)	1040

TABLE 12.13 Tensile Shear Strength of Epoxy Adhesive Cured with PMDA[17]

Test temperature, °C	Tensile shear strength, psi
–58	3240
26	3110
121	2950
149	1040
204	570
260	380

However, newer adhesives systems having moderate temperature resistance have been developed with improved toughness but without sacrificing other properties. When cured, these structural adhesives have discrete elastomeric particles embedded in the matrix. The most common toughened hybrids using this concept are acrylic and epoxy systems. The elastomer is generally a amine- or carboxyl-terminated acrylonitrile butadiene copolymer (ATBN and CTBN).

Formulations have been developed in which small rubber domains of a definite size and shape are formed in situ during cure of the epoxy matrix. The domains cease growing at gelation. After cure is complete, the adhesive consists of an epoxy matrix with embedded glass rubber particles. The formation of a disperse phase depends on a delicate balance between the miscibility of the rubber with the resin, with the resin-hardener mixture, and appropriate precipitation during the crosslinking reactions.

Table 12.14 shows a comparison of dicyandiamide cured epoxy adhesives formulated with and without a CTBN adduct. When compared to the control epoxy, the toughened formulation exhibits significantly higher peel strength and moderately higher tensile shear strength. CTBN modified epoxy adhesives are generally one-part systems, cured with dicyandiamide at elevated temperature.

TABLE 12.14 CTBN Toughened Adhesive Formulation

	Parts by weight	
Component	Formula A control	Formula B toughened
DGEBA epoxy resin (EEW: 190)	100	75
DGEBA/CTBN adduct (Hycar 1300 × 13)*		25
Tubular aluminum	40	40
Cab-O-Sil	5	5
Dicyandiamide	6	6
Melamine (accelerator)	2	2
Property		
Cure schedule	1 h at 175°C	
Tensile shear strength, MPa, on aluminum at 25°C	18.5	20.5
T-peel strength, kN/m, on aluminum at 25°C	1.1	5.5
Glass transition temperature, °C	130	129

*Adduct contains 40% CTBN (Hycar 1300 × 13).

ATBN tougheners are generally used in room temperature formulations (see Chap. 11). ATBN liquid polymers cannot be mixed directly into the epoxy resin component of a two-part adhesive or in a one-part adhesive, since crosslinking and shortened shelf life will result. ATBN adducts are, therefore, mixed with the curing agent component of two-component epoxy adhesives.

Within the past several years, improvements in the toughening of high-temperature epoxies and other reactive thermosets, such as cyanate esters and bismaleimides, have been accomplished through the incorporation of engineering thermoplastics. Additions of poly(arylene ether ketone) or PEK and poly(aryl ether sulfone) or PES have been found to improve fracture toughness. Direct addition of these thermoplastics generally improves fracture toughness but results in decreased tensile properties and reduced chemical resistance.

REFERENCES

1. Naps, M., U.S. Patent 2,682,515, Shell Development Co.
2. May, C. A., and Nixon, A. C., "Reactive Diluents for Epoxy Adhesives," *Journal of Chemical Engineering Data*. vol. 6, 1960, p. 290.
3. Bandaruk, W., "Aromatic Amines as Curing Agents," *Plastics World*, November 1957.
4. Hopper, F. C., and Naps, M., Epoxy Resin Adhesive Compositions, Their Preparation, and Tape Containing the Same, U.S. Patent 2,915,490, 1959.
5. Hopper, F. C., and Naps, M., U.S. Patent 2,915,490.
6. Dannenberg, H., and May, C., "Epoxide Adhesives," in *Treatise on Adhesion and Adhesives*, vol. 2, R. L. Patrick, ed., Marcel Dekker, New York, 1969.
7. May, C. A., "Physical Significance of Acid Anhydride Cure on Epoxy Adhesive Properties," *SPE Transactions*, vol. 3, 1963, p. 251; and Savia, M., "Epoxy Resin Adhesive," in *Handbook of Adhesives*, 2d ed., I. Skeist, ed., van Nostrand Reinhold, New York, 1977.
8. Bolger, J. C., "Structural Adhesives for Metal Bonding," in *Treatise on Adhesion and Adhesives*, vol. 3, R. L. Patrick, ed., Marcel Dekker, New York, 1973.
9. Resolution Performance Polymers, Starting Formulations 4020, 4022, 4029, and 4031, "One-Package Adhesives," Houston, TX, 2004.
10. Norwakowski, A. C., et al., to American Cyanamid, U.S. Patent 3,391,113, 1968.
11. CVC Specialty Chemicals, Inc., "Advantages of Omicure V-52 as an Accelerator for Dry Cured Epoxy Resin System," TSR 020909, Moorestown NJ, December 2004.
12. Resolution Performance Products, Starting Formulation 4021, "High Strength Adhesives for Elevated Temperature," Houston, TX, 2004.
13. Shimp, D. A., "Epoxy Adhesives," in *Epoxy Resin Technology*, P. F. Bruins, ed., Interscience Publishers, New York, 1968, p. 164.
14. Meath, A. R., "Epoxy Resin Adhesives," in *Handbook of Adhesives*, 3d ed., I. Skeist, ed., van Nostrand Reinhold, New York, 1990.
15. Epoxylite Corporation, Booklet L-800, St. Louis, MO.
16. Hyde, T. J., "The Epoxy Resin PMDA Glycol System," *ACS Division of Paint, Plastics, Printing Ink Chemistry, Preprints*, vol. 19, no. 2, September 1959.
17. Black, J. M., and Bloomquist, R. F., "Metal Bonding Adhesives for High Temperature Service," *Modern Plastics*, June 1956.

CHAPTER 13
SOLID EPOXY ADHESIVE SYSTEMS

13.1 INTRODUCTION

Epoxy adhesives are most commonly used as liquids or pastes. However, certain types of epoxies can be employed in the form of a solid. The components in these adhesives are mixed and processed to a stage where the resulting adhesive product is in a solid but still fusible (uncrosslinked) state. When the applied solid adhesive is heated, it melts, flows, and wets the substrate. Additional heating time then causes the adhesive to cure completely into a strong, thermosetting structure.

Solid epoxy adhesives can be formulated in various ways. The more common methods are described below.

1. Latent curing agents such as dicyandiamide are dissolved into solvent solutions of solid epoxy resins. This is then followed by evaporation of the solvent.
2. Soluble curing agents are added into liquid epoxy resins and cured until a B-stage condition is reached. The B stage is a solid, thermoplastic stage. When given additional heat, the B-stage epoxy will flow and continue to cure to a crosslinked condition or *C stage*.
3. Reactive powdered resins and curing agents can be combined by dry blending. These powder blends can then be preformed into various shapes by dry compression pressing.

The main mechanism in all these methods is the physical separation and restriction of molecular mobility of the epoxy resin and the curing agent that are imposed by the solid state of the product. These adhesive systems generally provide a shelf life of up to 6 months at room temperature depending on the reactivity of the curing agent and resin. All these products require elevated temperatures to liquefy and crosslink.

The most widely recognized types of solid epoxy adhesives are tapes or films that are commonly used to bond large substrates such as honeycomb skins and structural paneling. These are generally manufactured using the first or second technique described above. Solvent solutions that are used to manufacture tape or film can also be applied directly to the substrate. Once the solvent evaporates, not only is the adhesive an integral part of the substrate, but also it protects the surface during storage and handling.

Although tapes, films, or coatings applied from solvent solution make up the majority of the solid epoxy adhesives, formulations in the form of a powder or shaped solids can also be employed in certain applications. These are manufactured using the second or third technique described above.

The main advantages of solid adhesives are that they are single-component (i.e., metering and mixing are not required) and that they can be applied uniformly to a substrate with

TABLE 13.1 Advantages and Disadvantages of Solid Epoxy Adhesives over Liquids and Pastes

Advantages	Disadvantages
Elimination of plural component dispensing and metering equipment and associated labor and spare parts	More expensive by weight than liquid or paste adhesives
Reduction of labor required to oversee, maintain, and clean equipment and containers in contact with liquids or pastes	Not effective in filling a deep volume
Improved raw materials procurement and flow by replacing two containers associated with dual-component adhesive with a single-component system	Not cost-effective generally for small-volume production processes
Elimination of pot life consideration and line downtime resulting from adhesive advancement	A more limited number of formulations available than with liquids or pastes
Elimination of mix ratio tolerance concerns with resulting improvement in product consistency	Parts that must generally be designed to hold the melt before it cures to a solid
Elimination of variations from the coating process with resulting improvement in product consistency	Difficulties that can occur in application of solids to vertical or contoured surfaces
Improved worker safety and reduced hazardous materials exposure to liquid and paste systems	Heat required to make the solid epoxy adhesives flow and cure

little or no waste. Being solid, these adhesives also avoid the mess, cleanup requirements, and health hazards often associated with liquid adhesives. As a result, solid epoxy adhesives have gained a certain degree of popularity in several high-volume production applications including the assembly of electronic, automotive, and aerospace components. Table 13.1 summarizes the advantages and disadvantages of using solid epoxy adhesives over more conventional liquid or paste forms.

A significant advantage of tape and film adhesives is the greater toughness that is available compared to other adhesive types. This is primarily due to the ease with which resinous modifiers can be added to the formulation via solvent solution. Thus, hybrid epoxy adhesives such as epoxy-nylon, epoxy-phenolic, etc., are often found in tape or film form.

Solid epoxy adhesive formulations can be processed to either a thermoplastic or a thermoset state. Solid epoxy resins of exceptionally high molecular weight (e.g., phenoxy) can be used without any degree of cure as a hot-melt type of adhesive. However, fully crosslinked, thermoset systems are generally employed in structural applications.

13.2 SOLID ADHESIVE MANUFACTURING PROCESSES

Single-component, solid epoxy adhesives are made in several ways. Generally all methods consist of completely formulating the adhesive system, including resins, fillers, curing agents, etc., in the liquid state and then converting it to a thermoplastic solid. This conversion can be done through the cooling of a melt, or by the removal of a solvent from a solution, or by partial curing—a process sometimes known as *B-staging*.

A B-stage resin is one in which a limited reaction between resin and curing agent is allowed to take place. The reaction is arrested while the product is still fusible and soluble, although having a higher softening point and melt viscosity than originally. The B-stageable formulation contains sufficient curing agent to effect crosslinking on subsequent heating.

For tape or film adhesives, the curing agent is usually incorporated into either a liquid epoxy resin or a high-molecular-weight solid epoxy resin solution. The hardening process occurs by B-staging (in the case when a liquid epoxy resin is used), solvent evaporation, or both. The hardening process is usually accompanied by extruding, calendering, or casting the adhesive into thin films that are typically 5 to 15 mils thick.

The product is processed to a condition where it is either tack-free or slightly tacky to aid in application to vertical or contoured substrates. These films may be in the form of unsupported sheet, or they may be reinforced with glass fabric, paper, or another reinforcing medium. These supported films are sometimes referred to as *prepreg*.

For powder or preformed adhesives, the incorporation of curing agent can be done via various processes.

1. When high-molecular-weight solid epoxy resins are employed, they may be reduced to powder and then combined with a powdered curing agent.
2. Alternatively, the high-molecular-weight solid epoxy may be melted, the curing agent added, and the mixture cooled and pulverized.
3. With liquid epoxy resin, the procedure is to react the resin with the curing agent and advance the mixture to a B stage.

Solid adhesive made by the third process can be (1) cast directly into a solid usable shape or, more commonly, (2) cast and then ground to a powder form. This grinding operation is conducted with standard pulverizing equipment. With lower-melting-point blends, cooling may be required to prevent softening and blocking of the resin during the pulverizing operation.

The powder can be applied directly to the substrate surface by electrostatic coating processes or "dusting" onto a warmed surface. It can also be formed into shapes or preforms by the application of pressure in a die mold. This process is similar to how pharmaceutical tablets are made. In this way shaped preforms can be made that will conform to a specific joint geometry. There are a wide variety of applications to which epoxy preforms can be adapted. They can be used to

- Bond plastics, metals, composites, ceramics, and dissimilar materials
- Ruggedize fragile leads and active semiconductors prior to molding
- Seal terminals and leads into plastic molding
- Encapsulate discrete components

The incorporation of particulate fillers into the liquid epoxy formulation can be achieved with any of the processes mentioned above by the use of roll mills or screw-type kneaders. If the formulation is solid, fillers may be blended into the product by the use of a pebble mill or the like. After blending, they may be rolled under pressure and then ground if desired.

In general, solid epoxy adhesives will have a somewhat limited shelf life, and to ensure reproducible wetting and flow, it is advisable that they be stored in a refrigerated condition until their use. Depending on the chemistry and nature of the curing agents that are used, these storage conditions could require temperatures of 5°C or less. With the more reactive systems, flow will decrease on aging even with refrigeration. Thus, the storage life of the product must be rigidly controlled and verified.

13.3 CHEMISTRY

Solid epoxy adhesives generally rely on high-molecular-weight epoxy resin for the solid appearance of the uncured adhesive. This epoxy resin is generally formulated with either:

1. A latent curing agent
2. A curing agent capable of developing a B stage

The cure system must be slow at room temperature to prevent storage life and flow problems, but sufficiently rapid at elevated temperatures to permit reasonably short curing times. Even the most latent curing system in use today does not completely eliminate room temperature reaction; thus shelf life and storage requirements must be critically controlled.

A wide range of epoxy resins as well as a wide range of curing agents and catalysts are available for formulating solid epoxy adhesives. Resins with different viscosities, amounts of reactive groups, and structures are available. Additives that change the uncured resin viscosity, reduce brittleness, or impart some other property are also available.

Epoxy resins with aromatic backbones and high functionality give a strong, high-temperature, highly crosslinked matrix. However, the resulting adhesive is usually brittle. Aliphatic epoxies with lower functionality usually result in matrices with higher elongation and toughness but lower temperature capability.

Latent curing agents, such as dicyandiamide or 4, 4′-diaminodiphenyl sulfone (DADPS) are commonly used in producing solid epoxy adhesives. Being latent, these curing agents react with epoxies only on heating. They are relatively insoluble in the epoxy resin at room temperature, yet melt and become soluble at an activation temperature. Once activated, they cure relatively quickly at elevated temperatures. Both dicyandiamide and DADPS provide excellent high-temperature properties.

Dicyandiamide is a true latent catalyst for epoxy resin curing. It is also considered to be the workhorse of one-component adhesives due to its ease of use, excellent performance properties, long shelf stability, and low toxicity. In certain admixtures with DGEBA, it has demonstrated a room temperature storage life in excess of 4 years. Dicyandiamide is usually added to the solid epoxy resin in concentrations of about 3 to 6 pph. It melts at about 150°C. Cures can be conducted in the range of 120 to 175°C but are very slow at the lower temperatures. As a result, it is common practice to add accelerators such as benzyldimethylamine (BDMA) and mono- or dichlorophenyl substituted ureas to these systems.

DADPS also provides excellent high-temperature properties and chemical resistance. Of the amine curing agents, DADPS provides the best retention of strength after prolonged exposure to elevated temperatures. It melts at 135°C and can be cured with epoxy resins at 20 to 30 pph with cure temperatures ranging from 115 to 150°C. Because of the low reactivity of this system, an accelerator, such as BF_3-MEA, is usually employed at about 1 pph.

Other latent curing agents that are used in solid adhesives are dihydrazides and BF_3-MEA complexes. These compositions are also stable at room temperature but cure when heated. Solid anhydrides can be used in one-component powder blends (e.g., 10 pph of trimellitic anhydride accelerated with 0.5 pph of 2-methylimidazole). Solid systems with aromatic diamines are prepared by comelting the solid epoxy with the amine. Typically 30 pph of curing agent is used.[1]

For B-stageable adhesives, aromatic amines are generally used in liquid or solid epoxy resins. They provide excellent chemical resistance and electrical properties. However, when fully cured, they are rather rigid. The rigidity results in high tensile strength but poor toughness, peel strength, and impact properties. The aromatic amines are blended with liquid epoxy resin at 10 to 30 pph. A B-stage will occur in several hours at room temperature or faster with mild heating. Crosslinking requires a sustained heat cure at higher temperatures. The temperature range for full cure is 60 to 200°C depending on the type of resin and curing agent.

Metaphenylenediamine (MPDA) is the best known of these aromatic amines. Methylene dianiline (MDA) requires somewhat longer cures and has a higher processing viscosity. Both products are solids and generally must be melted before being blended with resin. The difficulty in handling these materials as hot melts has led to the development of aromatic amine eutectics, which are liquid at room temperature.

13.4 TYPES OF SOLID EPOXY ADHESIVES

There are several common forms of solid epoxy adhesives. These include film, tape, powder, and preformed shapes. Certain formulations are better suited for specific forms. For example, casting of tape or film adhesive from solvent solutions lends itself to working with multicomponent hybrid systems, where each resin can be solubilized and blended together in a universal solvent. B-staged systems are generally more brittle and better suited for powders or preformed adhesives.

13.4.1 Tapes and Films

Tape and film are the most common forms of solid epoxy adhesives. *Tape* and *film* are terms that are used rather loosely for adhesives in a thin sheet form. The term *tape* generally refers to an adhesive that is supported on a web of paper or nonwoven fabric, or on an open-weave scrim of glass, cotton, or nylon. *Films*, on the other hand, are free of supporting material and consist only of the adhesive. Both tape and film products may be accompanied with a release liner depending on the tack or blocking characteristics of the adhesive.

Supporting fibers in epoxy adhesive tapes are useful in that they provide for a positive stop under bonding pressure. This can be used to control bond line thickness and to help distribute stresses evenly during service. The supporting fibers that are used in these adhesives are primarily for the purposes of carrying the adhesive and convenient application to the substrate. Their reinforcing function within the epoxy matrix is generally considered to be of secondary importance.

The final thickness of epoxy tape or film adhesives is on the order of 5 to 15 mils. These adhesives may be soft and tacky, or stiff and dry, depending on their formulation. The soft and tacky products are valuable in products requiring application to contoured or vertical surfaces. If the product is especially tacky, a release liner (e.g., polyethylene film, coated paper) is generally used to keep the film from blocking. The stiff and dry products are generally used for flat surfaces where speed and ease of application are required.

The main attribute of tape or film adhesives is that they are single-component systems requiring no need for metering or mixing. They can be easily cut to size and applied between the substrates to be joined without waste or mess. Although tape and film adhesives offer a uniformly thick adhesive product that is easily dispensed, they are poor gap fillers, especially if the gap between the mating parts varies significantly across the bonding surface. To provide better gap-filling characteristics, epoxy films are sometimes formulated with chemical blowing agents that are activated during the high-temperature cure. The resulting foamed epoxy expands to 2 to 3 times its original thickness to fill relatively large gaps. The slight degree of foaming has minimal effect on the performance properties of the cured adhesive.

Once the tape or film is in place between the substrates, the joint is heated under pressure so that the adhesive becomes slightly fluid, flows into the microroughness on the substrate, and wets the substrate. With additional time at the curing temperature, the adhesive completely crosslinks to a thermosetting condition.

Tape and film adhesives are most often used to bond large areas, such as for applications in the aerospace industry. For example, the joining of aluminum honeycomb structure to flat metal skins is often accomplished with thermosetting epoxy film adhesives. These films (Fig. 13.1) can easily be applied without the need to mix, meter, or apply a liquid coating.

Typically tape or film epoxy adhesives are modified with synthetic thermoplastic polymers to improve flexibility in the uncured film and toughness in the cured adhesive. Epoxy resins can also be blended with phenolic resins for higher heat resistance. The most common hybrid systems include epoxy-phenolics, epoxy-nylon, epoxy-nitrile, and epoxy-vinyl hybrids. These hybrid film adhesives are summarized in Table 13.2, and structural properties are shown in Table 13.3.

Tape and film adhesives dominate the markets where the assembly of large parts is required. They are also used in applications where greater reliability and consistency are desired. Tape and film adhesive eliminate any variation in concentration that may occur from metering and mixing operations. Also, the incorporation of a carrier, such as glass or polyester fabric, into the adhesive acts to control the bond line thickness, thus avoiding thin, adhesive-starved areas where part curvature or external pressure may be greatest. Although these adhesives were originally developed for the aerospace markets, they are now finding applications in automotive, building, and general industrial areas.

Guidelines for formulating solid epoxy adhesives are similar to those employed for liquid or paste adhesives. Table 13.4 shows starting formulations for several epoxy adhesive tapes and films.

Epoxy-phenolic adhesive was the first true high-temperature adhesive. It was developed in the early 1950s as a high-temperature aircraft adhesive. An example formulation is provided in Table 13.4, but these adhesives are discussed predominantly in Chap. 15.

Epoxy-nylon adhesives were developed in the 1960s for high-peel-strength applications. The key to their development is the use of noncrystalline nylons that are soluble in alcohols and other epoxy-compatible solvents. A commercial example of adhesive-grade nylon includes DuPont's Zytel 61. Standard nylons are not practical because of their incompatibility with most other resins.

FIGURE 13.1 Epoxy film adhesive with release sheets.

TABLE 13.2 Common Epoxy Hybrid Resins Used in the Formulation of Tape and Film Adhesives

Hybrid resin	Characteristics
Nylon-epoxy	Films are blends containing 30 to 50% by weight of epoxy resin. The nylon constituent provides high tensile shear strength as well as high peel strength. Suitable catalysts are dicyandiamide and aromatic polyamines. These adhesives have useful properties at low temperatures but have only moderately good high-temperature properties.
Epoxy-phenolic	Epoxy-phenolic adhesives are generally used in aerospace applications requiring high shear strength at temperatures in excess of 150°C. Usually the phenolics are a resole type, and often the epoxy is a minor component. These adhesives are relatively brittle and have low peel and impact strengths.
Nitrile-epoxy	Nitrile-epoxy adhesives are composed of solid epoxy resin modified with carboxyl-terminated butadiene nitrile (CTBN) copolymer. The CTBN is introduced into the epoxy resin at elevated temperatures. The modification provides toughness and high peel strength without sacrificing heat and chemical resistance. The film adhesives are widely used in the aerospace industry in the construction of jetliners.
Vinyl-epoxy	Vinyl-epoxy adhesives have moderate strength at temperatures up to 150°C. Oxidation stability is excellent. Vinyl constituents increase toughness and peel strength. The adhesive is often used for bonding safety glass, aerospace engine structures, and structural panels.

Epoxy-nylon film adhesive can be manufactured by solution casting processes. However, a more efficient and environmentally acceptable method is to calender dry blends of powdered nylon with a liquid epoxy resin with accelerators and other modifying resins directly onto a mesh support.

Epoxy-nylon adhesives show exceptionally high tensile shear and peel strengths; however, they have poor resistance to moisture and elevated temperatures. These adhesives can absorb significant amounts of water from the ambient environment before and after cure. Table 7.5 gives tensile shear and peel strengths for a series of adhesives made by dissolving various ratios of nylon and epoxy resins in a alcohol-water mixtures.

Epoxy-nitrile adhesives have also been developed for the purpose of toughening conventional epoxy adhesives. These early adhesives were made from homogeneous blends of nitrile elastomer and epoxy resin. They are not similar to the modern-day CTBN toughened epoxy adhesive where the nitrile phase consists of discrete particles. An advantage of the

TABLE 13.3 Structural Properties of Various Types of Film Adhesives[2]

	Lap shear at temperature (°C), ksi								
Adhesive	−55	23	83	121	149	177	204	260	Metal peel, lb/in of width
Nylon-epoxy	7	7	3.5	2.2	—	—	—	—	140
Epoxy-phenolic	3.2	3.5	NA	NA	2.7			2	<10
Nitrile-epoxy	3.5	3	2.8	NA	2.7	1.6	—	—	15
Vinyl-epoxy	3	3.7	4	4	2.5	1.4	—	—	10–15

TABLE 13.4 Starting Formulations for Epoxy Tape and Film Adhesives[3]

Phenolic-epoxy tape adhesive	Parts by weight
Medium-molecular-weight phenolic resin, 80% solids	125
High-molecular-weight phenolic resins, 60% solids	33
Solid epoxy resin	20
Ethyl acetate	20
1-Hydroxy-2-naphthoic acid	2.8
n-Propyl gallate	1.4
Tape adhesive is prepared by impregnating the above liquid adhesive into a glass mat and procuring at 133°C for 35 min. The adhesive can withstand temperatures up to 287°C.	
Vinyl-epoxy tape adhesive	**Parts by weight**
Polyglycidyl ether of 1,1,2,2-tetrakis-(4-hydroxyl phenyl) ethane	80
Tetrahydrofuran	55
Solid epoxy resin	20
Polyvinyl formal resin	10
Aluminum dust	100
Diaminodiphenyl sulfone	24.5
The adhesive is deposited on a glass cloth and dried at 95°C for 15 min. Cure is 30 min at 240°C. Resulting shear strength on steel is 1090 psi at 287°C after 400 h at 260°C.	
Nylon-epoxy adhesive film	**Parts by weight**
Epoxy resin	25
Nylon resin	75
Dicyandiamide	10
Cure conditions are 60 min at 163°C. Resulting adhesive will have a peel strength of 60 to 80 lb/in of width and a shear strength of 6200 to 6600 psi at room temperature.	
General-purpose epoxy adhesive tape	**Parts by weight**
Solid epoxy resin	100
Epoxy novolac resin	100
Solvent	As required
Dicyandiamide	16
BDMA	2–4
Mix resins and solvent. Ball mill curing agent into varnish. Varnish is then B-staged on desired film backing and heat cured in place.	
Unsupported general-purpose epoxy adhesive film	**Parts by weight**
Solid epoxy resin (40% solids)	212
Epoxy novolac resin (85% solids)	37.5
Dicyandiamide	12
BDMA	2.0
Solvent removal occurs at 83°C for 1.0 h. Dry film thickness is 0.002 in. Cure schedule is 1 h at 175°C and 120 psi.	

epoxy-nitrile adhesives is that they require relatively low cure temperatures in the range of 125°C.

The toughness of an epoxy-nitrile adhesive is nearly equivalent to that of an epoxy-nylon adhesive. However, the epoxy-nitrile system has much better hydrolytic stability. Also, the low-temperature properties of an epoxy-nitrile adhesive are superior to those of epoxy-nylon adhesive. Table 7.3 illustrates the effect of nitrile addition on tensile shear and peel strength.

13.4.2 Powders and Preforms

Another form of solid epoxy adhesive is powder or granules, which must be first heated to be made liquid and capable of flowing. Being a single-component system, powdered adhesives also eliminate proportioning and mixing errors. However, uniform distribution of the powder over large areas is sometimes difficult to achieve, especially if the surface is vertical or contoured. Powdered epoxy adhesives can be sifted, dusted, or sprayed onto a substrate surface, but special application processes, such as electrostatic powder coating, are often required for uniformity. Fluidized-bed coating processes have been used for applying adhesive coatings to magnetic wire and motor lamination stacks. The fundamentals of the powder coating process can be found in the Web site www.SpecialChem4Coatings.com in the Powder Coating Center.[4]

Solid epoxy adhesives may also be fused or compacted into various shapes or preforms (see Fig. 13.2) such as sticks, rings, and beads.[5] They can then be easily applied to a specifically shaped part. Similar to thermoplastic hot-melt adhesives, these thermosetting epoxy adhesives flow on the substrate by heating. However, the product will cure to an infusible thermoset stage with continued heating. Once cured, the product demonstrates properties similar to those of other structural epoxy adhesives. These solid epoxy adhesives are sometimes referred to as *thermosetting hot melts*.

Certain types of thermosetting epoxy powders can be fused into small rings that are then used in rod-and-tube types of joint design. The preformed epoxy ring can be slipped over one of the substrates before assembly. When exposed to high curing temperatures, the epoxy in the preform melts, flows, and wets the substrate. Additional heating fully cures the adhesive. In this type of application, a depression or trough must generally be designed into the joint to hold the liquid adhesive in place while it cures.

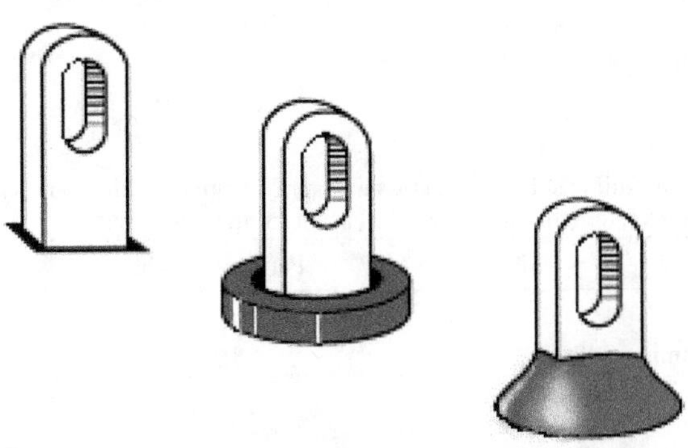

FIGURE 13.2 Solid epoxy preforms are designed to provide a simple method for sealing or bonding high-volume components. (*Courtesy: Multi-Seals, Inc., Manchester, CT.*)

Preformed sticks or tubes of solid epoxy may also be applied much as a solder is.[6] To apply, the area to be bonded is heated until sufficiently hot to cause the stick solder to flow when drawn across the surface. After application, the adhesive is cured by conventional thermal techniques. If two substrates are to be bonded, both must be preheated prior to application of the epoxy solder.

Solid shapes or powders are formulated from solid or liquid (when used as a B stage) epoxy resins and curing agent. Fillers, additives, and other modifiers are often used as they are with liquid or paste epoxy adhesive formulations. However, consideration must be given to the flow properties of the adhesive when heated as well as the application properties.

Guidelines for formulating solid epoxy adhesives are similar to those employed for liquid or paste adhesives. Table 13.5 shows starting formulations for a shaped solid and powder epoxy adhesive.

The stick solder formulation (Table 13.5) can be made in several ways. The components can be mixed as a hot melt and cast into small-diameter tubes. The components can also be ground to 60-mesh powder and then compressed into void-free rods. To apply, the area to be bonded is heated so that when the stick solder comes into contact with the surface, it melts and wets the substrate. An elevated-temperature cure is then required to crosslink the adhesive.

13.4.3 Thermoplastic Epoxy Films

Ultrahigh-molecular-weight epoxy resins can be used directly as a thermoplastic adhesive. The hydroxyl groups and good wetting characteristics provide for good adhesion to many surfaces. These adhesives are prepared in film form by solvent casting or extrusion.

The phenoxy resin has a chemical structure similar to that of epoxy resin; however, the phenoxy is a high-molecular-weight thermoplastic polymer, which needs no further conversion and has an infinite shelf life. Since the phenoxies are strongly polar polyethers, they

TABLE 13.5 Starting Formulations for Epoxy Powder and Stick Adhesives[7]

General-purpose powder adhesive	Parts by weight
Solid epoxy resin, micropulverized	100
Solid aromatic amine, micropulverized	11.5
Fumed silica	3
Red iron oxide	3

Application can be by fluidized bed or electrostatic spray. Cure conditions are 3 min at 204°C or 12 min at 149°C. Shear strength on steel is 3300 psi at room temperature.

Stick solder adhesive	Parts by weight
Solid epoxy resin	100
Metaphenylene diamine (MDA)	28.5
Powdered aluminum	60
Colloidal silica	10

The ingredients are prepared as a hot melt and cast into small-diameter tubes or are ground to 60-mesh powders and then compressed into void-free sticks.

are readily soluble in ketones, esters, and mixtures of aromatic hydrocarbons, chlorinated hydrocarbons, or alcohols with ketones.

The toughness of the resin (90 percent ultimate elongation and 9000 psi tensile strength) provides for a high degree of tensile shear strength and peel strength at room temperature. However, because the adhesive is not crosslinked, the elevated-temperature properties and creep resistance are poor. The heat deflection temperature of phenoxy is about 90°C. Applications for such adhesives include metal foil lamination, coextrusion of plastic film, and other applications where there is a need for fast processing and where elevated-temperature properties or solvent resistance is unimportant.

The phenoxy resins can form strong bonds to metals in seconds at temperatures from 315 to 340°C. At these bonding conditions, a tensile lap shear strength of 3500 to 4000 psi can be obtained on aluminum. The same bond can be made in 2 to 3 min at 260°C and in 30 min at 190°C due to the thermoplastic phenoxy's time-temperature creep characteristics and relatively low softening temperature (100°C).[8]

The phenoxy resins can also be plasticized with many of the common plasticizers and still maintain a high percentage of strength, while the bonding temperature is substantially reduced. An effective hot-melt adhesive is one in which polyvinyl ether is blended into the phenoxy resin.

Table 13.6 shows the characteristics and properties of supported and nonsupported high-molecular-weight epoxy resins cast from solvent solution. A crosslinked and uncrosslinked formulation using the same high-molecular-weight epoxy is shown for comparison.

TABLE 13.6 Unsupported Film Adhesive Cast from Solvent Solution[9]

	Parts by weight				
Component	A	B	C	D	E
Epoxy resin, mol. wt. >20,000 (e.g., Epi-Rez 2287)	100	100	374	374	374
Methyl ethyl ketone	11	11	36	36	36
Epoxy novolac resin (e.g., Epi-Rez 5155)			15	15	45
Dicyandiamide			1	1	2
Dimethylforamide			10	10	10
Benzyldimethylamine			0.05	0.05	0.05
Carrier, cloth	None	Glass	None	Glass	None
Solvent removal			1 h at 83°C		
Dry film thickness, in	0.002	0.003	0.002	0.003	0.002
Cure schedule	5 min at 170°C			1 h at 170°C	
Pressure			115 psi		
Property					
Tensile shear strength, psi, at					
• 25°C	3740	3490	4310	4165	4290
• 94°C	425	775	635	1360	2000
90° Peel strength, lb/in, at					
• 25°C	17.0	21.0	23.2	28.2	22.0
• 94°C	8.6	12.5	14.5	15.5	15.1

REFERENCES

1. Savia, M., "Epoxy Resin Adhesives," Chapter 26 in *Handbook of Adhesives*, 2d ed., I. Skeist, ed., van Nostrand Reinhold Co., New York, 1977.
2. Politi, R. E., "Structural Adhesives in the Aerospace Industry," Chapter 44 in *Handbook of Adhesives*, 3d ed., I. Skeist, ed., van Nostrand Reinhold Co., New York, 1993.
3. Savia, "Epoxy Resin Adhesives," and Politi, "Structural Adhesives in the Aerospace Industry."
4. "Powder Coating Center," www.SpecialChem4Coatings.com, 2004.
5. Harvill, K., "Finding the Cure for Epoxy Dispensing Frustrations," *ECN Magazine*, Dec. 15, 2001.
6. Paul, M. N., Process for Preparing Polyepoxide Solder, U.S. Patent 2,965,930, 1960.
7. Savia, "Epoxy Resin Adhesives," and Politi, "Structural Adhesives in the Aerospace Industry."
8. Anderson, C. C., "Adhesives," *Industrial and Engineering Chemistry*, vol. 59, no. 8, August 1967, pp. 91–94.
9. Shimp, D. A., "Epoxy Adhesives," in *Epoxy Resin Technology*, P. F. Bruins, ed., Interscience Publishers, New York, 1968, p. 175.

CHAPTER 14
UNCONVENTIONAL EPOXY ADHESIVES

14.1 INTRODUCTION

This chapter considers epoxy adhesive formulations that are rather unconventional. These include

- Epoxy adhesives that cure via radiation (ultraviolet light or electron beam energy)
- Epoxy adhesives that are available as waterborne emulsions
- Epoxy adhesives that cure via indirect thermal means (induction, dielectric, microwave, and several others)

These adhesives have found their way into several niche markets where their advantages, such as fast setting speed, are highly valued. They have not seen application in the more ordinary markets because of their materials and equipment cost. However, it is expected that these epoxy adhesive systems will grow at a faster annual rate than the average market as their advantages become more widely known.

14.2 ULTRAVIOLET AND ELECTRON BEAM CURED EPOXY ADHESIVES

Radiation curing is a production technique for polymerizing and curing of adhesives through the use of radiant energy. The source of the radiant energy is generally electron beam (EB), ultraviolet (uv) light, or visible light. These forms of radiation have the energy necessary to initiate polymerization of low-molecular-weight, unsaturated resins including various types of epoxy resins.

Polymeric inks, coatings, adhesives, or similar products can be cured with radiant energy. Examples of adhesive applications include

- Laminating and packaging adhesives
- Bonding of woven and nonwoven textiles
- Fastening (tacking and fixturing), potting, and encapsulation of electrical and electronic components
- Vehicle assembly (aerospace, automotive, recreational, mobile home, etc.)
- Glass-to-metal and glass-to-glass bonding, and jewelry assembly

- Optical-fiber connector bonding
- Plastic component bonding
- The assembly of medical components and consumer products

One of the most popular uses of radiant curing is the advancement (viscosity increase) or crosslinking of pressure-sensitive adhesives. These applications have been satisfied mostly with acrylate-based adhesive systems. With epoxy-based adhesives, the main applications are electrical and electronic components, the bonding of large aerospace structures such as composites, and the bonding of transparent substrates such as glass and plastic.

Radiation cured epoxy adhesives significantly reduce the long cure times needed for conventional adhesives. These adhesives develop excellent physical and mechanical properties in a nonthermal cure process that requires a cycle time of only seconds to minutes, versus hours for conventional thermal curing.

Radiant cured epoxy systems also provide high glass transition temperatures, low shrinkage during cure, and relatively low residual stresses. They have no volatile emissions and excellent shelf life. The good physical and adhesive properties of conventional epoxy adhesives systems are maintained with the radiant cured systems.

Radiation curing permits the use of low-viscosity, solvent-free adhesive and coating compositions. Generally, a radiant curable low-viscosity liquid resin is coated on a substrate by conventional techniques. Viscosity increase, crosslinking, and viscoelastic properties are then developed quickly by exposing the moving substrate to a radiant energy source, as illustrated in Fig. 14.1. Radiant cured pressure-sensitive adhesives have replaced solvent-based pressure-sensitive adhesives in significant quantities.

The primary advantages of radiant curing are (1) regulatory compliance through the virtual elimination of volatile organic components (VOCs) and (2) fast production speeds. Major processing advantages of radiant curing adhesives include the absence of heat, water, or solvents. Performance advantages include superior heat and chemical resistance, excellent clarity, and high shear strength when compared to many waterborne or solvent based adhesives. Some of the benefits of using radiation curable adhesives are shown in Table 14.1.

The primary disadvantages of radiant curing adhesives include relatively high material cost due to the fact that these compositions generally have no low-cost components such as solvents, extenders, or fillers. However, radiant cured adhesives are often justified on a total-cost basis when one is considering energy bills, reduced waste, labor cost, production time, and factory space availability.

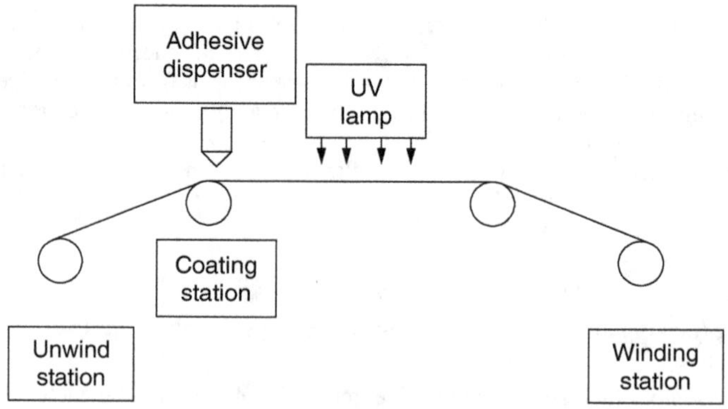

FIGURE 14.1 UV processing of a 100 percent solids liquid adhesive.

TABLE 14.1 Benefits of Radiation Curable Adhesives

Environmental/safety	Economics	Capability
No VOC emissions	More than 100% increase in line speed; web line speeds of 1000 ft/min common	Temperature, solvent, and moisture-sensitive substrates that can be coated
No regulatory reporting or permitting requirements	Energy needed to cure 1 g of pressure-sensitive adhesive about 1% that of waterborne and 4% that of solvent adhesives	Control of cure: degree of crosslinking that can be varied on the same product
Low flammability	Solvent cost savings	Multilayer composites that can be constructed
	Floor space savings	Easy to screen-print
	Savings in insurance costs	Improved adhesion to substrates due to grafting
	Reduced overall costs	Ease of coating due to lower viscosity
	Product immediately ready for testing and shipment	Radiation curing used to crosslink hot melts as well as increase molecular weight of liquids
	Single-component system	UV lamps that can usually be installed on existing production lines

Another distinct disadvantage of radiation cured adhesives is that at least one substrate must be transparent to the radiant energy. The radiant energy must also penetrate the depths of the adhesive for cure to be initiated. Complex joint geometries, thick bond lines, and shadowed areas protected from the radiant energy create problems. However, modern formulations have been specifically created to reduce these disadvantages, and partial cure by radiant energy followed by a complete cure by thermal energy is always an option.

In 1995, U.S. consumption of radiation cured products was 77 million lb, valued at $450 million.[1] Growth is forecast to average about 7 percent per year—a rate about twice that of conventional thermal cured products. Although radiation curable adhesives comprise a relatively small segment of the overall adhesive market (13 percent) and epoxy adhesives represent an even smaller component, epoxy systems are a fast-growing part of the market.

Market penetration is expected to increase further due to stricter environmental regulations and the availability of a greater variety of products. The first radiation curable adhesives were limited to acrylate and epoxy resins. Today, many different types of radiation curable adhesive systems are commercially available.

The radiation curing industry is often represented by RadTech International (www.radtech.org). RadTech is a nonprofit organization that promotes the use of radiation curing. RadTech's goal is to provide the industry with a better understanding of the science, materials, processing equipment, and applications in hopes of spreading the use of radiation curing. The organization serves as an international forum and reliable source of education and information for individuals and organizations involved with uv and EB processing.

14.2.1 Crosslinking Mechanisms

Radiation is used to crosslink (Fig. 14.2) or cure organic resins into durable coatings or adhesives having excellent physical properties with high chemical and temperature resistance. Radiation curing technology involves at least four considerations: type of radiation

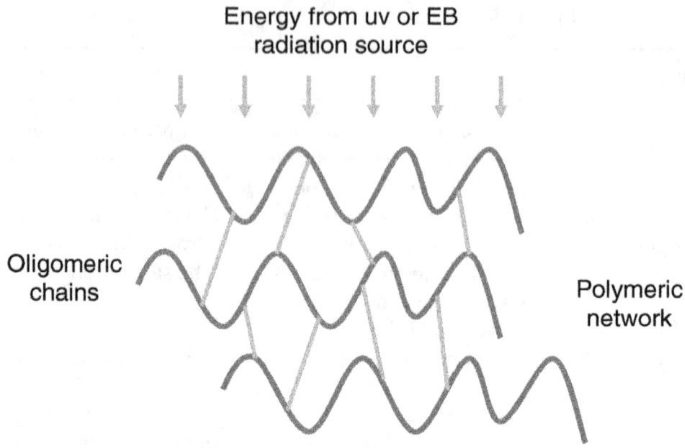

FIGURE 14.2 Radiation cured coatings react through unsaturation sites (double bonds) on oligomers and monomers.

source, organic polymer to be irradiated, mechanisms of physical and chemical interaction, and final properties associated with the cured product. Radiation cured adhesives react through unsaturation sites on oligomers and monomers. These active sites (double bonds) are capable of reacting to form larger polymers and crosslinked, three-dimensional network structures.

The effect of radiation dose on the adhesive is that as the dose increases, the molecular weight (MW) increases, resulting in a decrease of peel strength and an increase in cohesive strength. Temperature and chemical resistance are also generally increased by a greater radiation dose. The exact curing window for a product must be determined for every formulation and for each thickness. For many radiation cured systems, the processing window is narrow, making it easy to undercure or overcure the adhesive.

The main sources of energy for curing epoxy adhesives by radiation are electron beam (EB) and ultraviolet light (uv). Both provide instantaneous curing of resins that polymerize from a liquid to a solid when irradiated. The uv systems account for approximately 85 percent of the market for radiant cured adhesives, EB systems account for about 10 percent, and the remainder are chiefly adhesives that can cure by exposure to both visible and infrared light.

The type of crosslinking achieved with electron beam and uv radiation is very similar, but the way curing initiates is different. Electron beams have the higher energy, and the electron itself has sufficient energy to initiate polymerization. In uv curable materials, the polymerization reaction is not directly initiated by uv light, and a photoinitiator is required to interact with the uv radiation and produce the initiating species.

Epoxy resins that are used in uv and EB adhesive formulations are generally epoxy acrylates, and they do not have any reactive epoxy groups. Rather, they react through unsaturation sites on oligomers and monomers. These active sites (double bonds) are capable of reacting to form larger polymers and crosslinks. The mechanism of crosslinking by radiation is either free radical or ionic. If uv or visible light is utilized as the energy source, a photoinitiator is added to the mixture. The photoinitiator, when exposed to light, generates free radicals or cations, which initiate crosslinking between the unsaturation sites. In the case of EB cure, the high-energy electrons interact directly with the atoms of the unsaturated site to generate highly reactive molecules.

Many potential radiant curable epoxy systems exist, but only two have found significant commercial use in adhesives:

- Epoxy acrylates cured by free radical polymerization
- Epoxies cured by cationic polymerization

The chemistry and curing mechanisms of these systems are outlined in the following sections. Further information can be found in several textbooks on radiant cured adhesives.[2,3,4] A significant amount of information can also be acquired through RadTech, a professional society focusing on radiation curable products.[5]

UV Cure Mechanisms. There are two fundamental mechanisms for the uv curing of coatings: free radical and cationic. The more common free radical mechanism consists of a chain reaction with four primary steps:

1. Initiator radical formulation
2. Initiation
3. Propagation
4. Termination

The reaction can be illustrated by using a photoinitiator (I) and reactive monomers (R, R′), as shown in Fig. 14.3. Creating free radicals for uv curing requires the use of photoinitiators, which decompose on exposure to uv to produce initiating free radicals, which start the chain reaction. Quenching or deactivation of the photoactivated initiator by oxygen can occur, and the growing polymer radicals can also react with oxygen. This oxygen inhibition causes short polymer chains to form, resulting in tacky surfaces and poor physical properties of the coating. Fortunately with most initiator systems, the propagation rates are high, and oxygen quenching and competing reactions are minimized.

Cationic cure mechanisms are an alternative approach to uv curing. This involves the photogeneration of ions, which initiate ionic polymerization. This process is not subject to oxygen inhibition, as are some of the free radical mechanisms. Cationic cure mechanisms generally also provide less shrinkage and improved adhesion. The disadvantages are that the photoinitiators are sensitive to moisture and other basic materials. The acidic species can also promote corrosion. As a result, the vast majority of uv formulations are acrylate-based and cure by a free radical mechanism.

The energy sources of uv light cured coatings are typically medium-pressure mercury lamps, electrodeless vapor lamps, pulsed xenon lamps, or lasers. These generally emit

1. $I \longrightarrow I^*$	Initiator radical formation
2. $I^* + R \longrightarrow IR^*$	Initiation
3. $IR^* + R' \longrightarrow IRR'^*$	Propagation
4. $IRR'^* + IRR'^* \longrightarrow I_2 + RR'R'R$	Termination

FIGURE 14.3 Free radical polymerization mechanism.

electromagnetic radiation in the region of 200 to 760 nm. They also produce a certain degree of infrared radiation as heat, and this acts either to anneal the cured coating and relieve internal stresses and strains or to enhance cure rate in cationic cure systems. Pigmentation blocks or diffuses the uv radiation. Thus uv coatings that are cured with these light sources are usually clear or transparent, although thin, opaque coatings are also possible.

Electron Beam Cure Mechanisms. Electron beam curing mechanisms are similar to the uv free radical mechanism. However, the electrons are accelerated to a much higher energy state, and the electron itself has sufficient energy to initiate polymerization. The impact of these electrons is high enough to break chemical bonds and to generate ions. The ions then transform themselves into free radicals, which then initiate polymerization. Thus, the EB mechanism requires no photoinitiator. There is also greater penetration of the radiation (greater depth of cure) with less interference from pigmentation. Clear coatings of up to 20 mils and pigmented coatings of about 15 mils can be cured with EB equipment.

The absence of photoactive fragments in electron beam curable adhesives results in greater stability of both the cured and uncured adhesives. However, these advantages are offset by the relatively higher capital costs of EB curing equipment. The higher capital cost has generally restricted the use of electron beam curing to high-volume operations. EB curable adhesives are mainly used for laminating applications, such as in the production of metallized papers and plastic or foil laminates, whereas uv curable adhesives have found their way into a diversified range of industrial and commercial applications.

High-energy electron accelerators are used to cure EB adhesives. They can generate the radiant energy (150 to 500 kV) capable of curing thicker, pigmented resins, as EB energy has greater ability than uv energy to penetrate the material. The electron source is a filament that is heated inside a vacuum tube.

14.2.2 Formulation of UV and EB Epoxy Adhesives

Radiation curing adhesives are generally applied as solvent-free liquids. High-solids EB and uv curing liquid adhesives have been formulated from a variety of resins and elastomers. They include epoxy acrylates, epoxies, other acrylates, polyesters, blends of acrylate monomers with elastomers, and other compositions.

Free Radical Cure UV Adhesives. As with any adhesive, formulation variables are critical to the processing and performance characteristics. Variables such as oligomer selection, modifiers and additives, monomer structure, molecular weight, and glass transition temperature directly affect application and performance properties.

Practical radiation curable materials are composed of more than a single reactive monomer. Commercial adhesives normally contain the following:

1. Oligomers—a base resin reactive material
2. Diluents and crosslinking monomers
3. Photoinitiators (uv coatings only)
4. Stabilizers
5. Adhesion promoters
6. Surfactants, leveling agents, fillers, pigments, etc.

Examples of typical uv and EB coating compositions and their applications can be found in several sources.[6,7,8]

UV and EB formulations consist primarily of oligomers, monomers, photoinitiators (in the case of uv or light cured adhesives) and crosslinking agents or accelerators (in the case of EB cured adhesives), and the other additives that would normally be expected in adhesive formulations. However, the effect of these additives on storage and processing conditions must be determined as well as the effect on cohesive and adhesive properties. Close attention needs to be paid to the photosensitive nature of these additives and to the conditions under which they are stored and used.

Oligomers are low-molecular-weight reactive molecules that provide the backbone of the adhesive coating. The choice of oligomer has a major influence on the properties and processing characteristics of the coating. They contribute to properties such as reactivity, gloss, adhesion, chemical resistance, abrasion resistance, and nonyellowing. Oligomers also provide a bulking effect on the coating viscosity. Thus, oligomers are similar to the base resin in conventional coating formulations. Oligomers are moderately low-MW polymers, most of which are based on the acrylation of different resins. The acrylation imparts the unsaturation to the ends of the oligomer.

The two most prominent liquid radiation curable adhesives are free radical polymerization epoxy acrylates and cationic polymerization epoxies. Such adhesives are generally used as polymerizable syrups. A wide range of prepolymers can be acrylated including epoxies, urethanes, polyesters, polyethers, and rubbers. Elastomer-tackifying resin blends are often used in these formulations.

Epoxy acrylates are dominant oligomers in the radiation curable adhesives market. A bisphenol A epoxy resin is reacted with acrylic acid or methacrylate acid to provide unsaturated terminal reactive groups. The acrylic acid–epoxy reaction to make bisphenol A diacrylate destroys any free ingredients such as epichlorohydrin used to make the DGEBA epoxy starting raw material.

In most cases, epoxy acrylates do not have any free epoxy groups left from their synthesis but react through their unsaturation. Within this group of oligomers, there are several major subclassifications: aromatic difunctional epoxy acrylates, acrylated oil epoxy acrylates, novolac epoxy acrylate, aliphatic epoxy acrylate, and miscellaneous epoxy acrylates. These are described in Chap. 4.

Monomers are primarily used to lower the viscosity of the uncured material to facilitate application. The monomer must be matched with the resin to give the desired set of properties with respect to adhesion to the substrate and bulk properties such as flexibility, stiffness, cure behavior, and durability. Early radiation curable monomers had problems associated with toxicity and skin sensitivity; newly developed monomers have been significantly improved in this respect.

Monomers are usually low-MW, monofunctional materials that chemically incorporate into the cured coating rather than volatilize into the atmosphere, as is common with solvent diluents. Monomer diluents are chosen on the basis of providing good solvency, effectively reducing the viscosity of the oligomer without excessively retarding the cure rate. Certain diluents will contribute to the physical properties of the adhesive. However, generally they provide soft, thermoplastic films because of their linear and uncrosslinked nature.

Crosslinking monomers are multifunctional and contain two or more reactive sites. Thus, they crosslink the polymer chains as the film is cured, forming links between oligomer molecules and other molecules in the formulation. Monomers of this type are able to cure very rapidly.

Additives in radiation cured adhesives include stabilizers (which prevent gelation in storage and premature curing due to low levels of light exposure), color pigments, dyes, defoamers, adhesion promoters, flatting agents, and wetting agents. The curative in radiant curable adhesives is a photoinitiator for uv cured adhesives and an accelerator for EB cured adhesives. Photoinitiators are not required for EB cured systems because the electrons are able to initiate crosslinking. EB accelerators belong to a fairly wide range of chemicals, but generally they are quite polar in nature.

Other additives used to improve the performance of radiant cured adhesives are similar to those that might be found in more conventional adhesives. These include adhesion promoters, fillers, light stabilizers, antioxidants, and plasticizers.

Photoinitiators are perhaps the most important component in uv cured radiation coatings. The photoinitiator is an ingredient that absorbs light and is responsible for the production of free radicals in a free radical polymerized system or cations in a cationic photoinitiated system. The photoinitiators are usually added to the reactive coating formulations in concentration ranges from less than 1 to 20 percent by weight based on the total formulation. The absorption bands of the photoinitiators should overlap the emission spectra of the various commercial light sources.

There are two general classes of photoinitiators: (1) those that undergo direct photofragmentation on exposure to uv or visible light irradiation and produce active free radical intermediates and (2) those that undergo electron transfer followed by proton transfer to form a free radical species. The choice of photoinitiator is determined by the radiation source, the film thickness, the pigmentation, and the types of base resin employed. Examples of typical photoinitiator systems used to cure reactive resins are shown in Table 14.2. Benzophenone is perhaps one of the most common photoinitiators.

The only significant difference between uv cured adhesives and visible light (sunlight) cured adhesives is the use of a photoinitiator that absorbs the appropriate wavelength of light energy. Sunlight contains about 5 percent uv light. Adhesives developed to cure in sunlight generally do not cure when exposed solely to incandescent or fluorescent light, which contains only about 0.1 percent uv. Photoinitiators having absorption capabilities in the visible light energy range are based on dyes, quinines, diketones, and heterocyclic chemical structures.

Epoxy resins and other polymers (tetrahydrofuran, vinyl ethers, styrene, etc.) can be cured when exposed to an acid or cation intermediate species. The photoactive catalyst system commonly used to cure epoxy resins and multifunctional vinyl ether materials is composed of salts of aryldiazonium, triarylsulfonium, and diaryliodonium. These systems are commonly employed in coatings and adhesives for electronic products. The acid initiator generated from the photoinitiator continues to be active even after uv curing, and so conversion of reactants and crosslinking continue even in the absence of uv light. This phenomenon is typically referred to as *dark cure*.

TABLE 14.2 Photoinitiators Commonly Used in UV Coatings and Adhesives

Electron transfer photoinitiators	Photofragmentation photoinitiators
Benzophenone	Alkyl ethers of benzoin
Diphenoxy benzophenone	Benzil dimethyl ketal
Halogenated and amino functional benzophenones	2-Hydroxy-2-methylphenol-1-propanone
Fluorenone derivatives	2,2-Diethoxyacetophenone
Anthraquinone derivatives	2-Benzyl-2-*N*, *N*-dimethylamino-1-(4-morpholinophenyl) butanone
Zanthone derivatives	Halogenated acetophenone derivatives
Thioxanthone derivatives	Sulfonyl chlorides of aromatic compounds
Camphorquinone	Acylphosphine oxides and bis-acyl phosphine oxides
Benzil	Benzimidazoles

Visible light cured epoxy adhesives and coatings have been developed for architectural, industrial, and maintenance applications and for products difficult to heat or uv/EB cure because of their size. These are clear, one-part epoxy resins that cure by exposure to visible light for a few hours. They are formulated with cycloaliphatic epoxy compounds and a cationic photoinitiator that generates a strong acid when exposed to sunlight.

Some of the uv curable adhesives contain a combination of uv and infrared (ir) initiators to take advantage of the ir output that many uv lamps generate. At times a photoactive crosslinking agent is used to improve cohesive strength without affecting tack and peel.

Photoinitiators, sensitizers, and other radiation-sensitive agents may have an effect on the adhesive properties and especially on processing of the adhesive. Thus, all additives need to be tested with regard to storage and processing properties as well as with regard to their adhesive properties. The photoinitiator package will also need to be optimized for a given adhesive thickness and uv dosage.

Stabilizers are common additives included in uv cured adhesives to prevent premature polymerization, resulting in viscosity increase, gelation in storage, or premature curing that occurs due to low levels of light exposure. These compounds act as scavengers for free radicals. They neutralize the free radicals before they have a chance to start the chain reaction leading to polymerization. Light stabilizers, such as *hindered amine light stabilizing* (HALS) compounds, also protect the cured film from exposure to direct sunlight. These products are used to provide coatings with good outdoor stability, color retention, and nonyellowing properties.

Oxygen scavengers may be required as oxygen inhibits the curing of acrylates by quenching the photoinitiator or by scavenging free radicals. Scavenging produces stable species that slow down the cure rate but also can degrade the properties of the cured coating. Other methods of oxygen inhibition are nitrogen blanketing (a process called *inerting*) or speeding up the cure by using higher-intensity lamps and by varying the initiator type and concentration. The specific formulation determines the propensity for oxygen inhibition and the degree of sophistication needed to counteract it.

Adhesion promoters are used to provide good adhesion to such substrates as glass, hard plastics, and certain metals such as brass. These are specialized materials (e.g., organosilanes) that have the ability to promote adhesion to a substrate material and at the same time form a part of the polymer network.

Pigments generally inhibit uv curing to some degree since the pigments absorb and/or scatter uv radiation. This interferes with the ability of the photoinitiator to absorb the light energy required to initiate the polymerization reactions. Thus, the majority of commercial radiation curable adhesives are clear or contain silica.

Other additives used to improve the processing and performance of radiation cured adhesives are similar to those that might be found in more conventional formulations. These include antioxidants, defoamers, flow and wetting agents, and slip aids.

Cationic Cure Epoxies. As with free radical uv chemistry, the same principles of formulation component selection apply to cationic cured uv epoxy adhesives. Although different monomers and oligomers are normally required for this type of chemistry, the main difference lies with the photoinitiator system.

A typical cationic uv adhesive formulation contains an epoxy resin, a cure-accelerating resin, a diluent (which may or may not be reactive), and a photoinitiator. The initiation step results in the formation of a positively charged center through which an addition polymerization reaction occurs. There is no inherent termination, which may allow a significant postcure. Once the reaction is started, it continues until all the epoxy chemistry is consumed and complete cure of the resin has been achieved. Thus, these systems have been termed *living polymers*.

The cationic photoinitiator has relatively limited absorbance spectra, and the cure is more sluggish than a uv cured epoxy-acrylate. As a result, the uv may be used to develop handling strength in the epoxy adhesive, and a room temperature or elevated-temperature postcure may be necessary after the initial uv exposure for the adhesive to achieve ultimate properties.

Cationic photoinitiators are frequently found in classes of compounds such as the triaryl sulfonium, tetraaryl phosphonium, and diaryliodonium salts of large protected anions (hexafluorophosphates or antimonates). These compounds are soluble in most epoxy resins, do not activate epoxy cure until exposed to uv light, are insensitive to room lighting, and have long storage life at room temperature. Cationic photoinitiators form an acid catalyst when exposed to uv light and consequently start the cationic chemical reaction.

The cationic photoinitiators are sensitive to moisture, and the acid species formed can promote corrosion. Frequently the formulation for these curing methods contains solvent, which evaporates during curing, so that an important environmental advantage is lost. For these reasons cationic uv cure is usually preferred over free radical cure only when the higher-performance properties are justifiable.

Recently photoinitiators have been developed that can trigger photocuring when exposed to visible light.[9] As an industrial process, cure with visible light offers the advantage of increased safety because it poses less of an eye hazard than uv light. Visible light curing adhesives also provide for better curing in thick films and in shadowed areas. The visible light photoinitiators include 2,4-diiodo-6-butyoxy fluorine (DIBF). Usually a secondary photoinitiator such as (4-octyloxyphenyl)phenyliodonium hexafluoroantimonate[10] is required to trigger the cationic cure.

Cationic cured epoxies may also be crosslinked by electron beam radiation. A major application for this technology is the repair of composite aerospace structures. Direct benefits of EB processing include rapid cure, allowing completion of a permanent repair in the same or less time than a traditionally temporary repair, and ease of material handling. Other benefits include improved process control, reduced scrap due to process time limitations, reduced inventory of various repair materials, reduced spares inventory, reduced tooling costs, improved repair reliability, increased facility production, and the ability to do certain repairs not currently possible using existing repair techniques.

Recently, Acsion Industries and UCB Chemicals have undertaken a development program of cationic initiated epoxy adhesive for aluminum-to-aluminum, aluminum-to-composite, and composite-to-composite bonding. This has led to a series of bonding adhesives that have tensile shear strengths of 13 to 17 MPa on fiberglass-to-fiberglass substrates and 35 to 52 MPa on aluminum-to-aluminum substrates.[11]

In a recent study, the EB curing of epoxy resins by cationic polymerization was investigated to determine if cured materials with superior mechanical properties, high glass transition temperature, and shorter cure times could be produced.[12] Diaryliodonium salts were found to be an effective initiator for the cationic polymerization of epoxy resins when a high-energy, power electron beam accelerator was used as the source of ionizing radiation. For example, Dow Tactix 123 (a bisphenol A epoxy) containing 3 pph (4-octyloxyphenyl)phenyliodonium hexafluoroantimonate was irradiated at a total dosage of 100 kGy. Glass transition temperature of the cured material as determined by dynamic mechanical analysis was 182°C as compared to 165°C for the thermally cured material.

14.3 WATERBORNE EPOXY ADHESIVES

Waterborne epoxy dispersions have been employed effectively for many years in the coatings market primarily for the surface protection of concrete and metals. These products were developed in response to environmental regulations to reduce solvent levels in coatings.

Since these dispersions adhere well to a wide variety of substrates and provide a high degree of strength and resistance to service environments, they have also naturally found applications as adhesives.

Epoxy dispersions also can easily be blended with other waterborne polymers to make modified latex adhesives. The resulting hybrid adhesive produces performance properties and application characteristics that are superior to those of the originating latex system.

This section reviews the chemistry behind waterborne epoxy adhesives and the formulation possibilities. The characteristics of epoxy dispersions and the performance properties of cured adhesive films are addressed. The advantages and disadvantages of these adhesive systems are discussed with the focus on determining whether waterborne epoxy systems can replace traditional epoxy adhesives.

14.3.1 Background and Markets

Over the last several decades, environmental restrictions on solvents used in adhesives and coatings have become increasingly stringent. Hazardous air pollutant (HAP) limitations and the phaseout of ozone-depleting substances are only two examples of regulatory requirements that are leading to increased interest in alternatives to solvent-based systems. Another deciding factor in the development of waterborne epoxy coatings was the escalating cost of organic solvents in the mid-1970s.

Many of the attributes of solvent-borne epoxy coatings could be carried over to the waterborne epoxy coatings. These same attributes are useful in the application of waterborne epoxies as adhesive systems. They include good adhesion to a variety of substrates such as metals, wood, concrete, glass, ceramics, and many plastics; chemical resistance; low shrinkage; toughness and flexibility; and abrasion resistance.

In addition to the excellent performance properties and the reduction of solvent carriers, waterborne epoxy adhesives were found to have processing advantages. They could be easily applied by conventional coating systems (spray, roller, etc.); they were less hazardous to workers due to lower dermatitis potential and inflammability; ventilation equipment costs could be reduced; and application equipment could be easily cleaned with soap and water. In many applications, these processing advantages became the primary market drivers for waterborne epoxy adhesives as alternatives to more conventional adhesives.

However, waterborne epoxy systems are not without certain disadvantages, which have limited their application as adhesives. These disadvantages include increased use of energy to evaporate the water and dry the adhesive, lower resistance of the cured film to high-humidity environments, and storage and application limitations due to potential freezing at low temperatures.

There are many applications for polymeric waterborne adhesives. These include packaging adhesives, pressure-sensitive tape, coatings for textiles, wood adhesives, and various industrial adhesives and coatings. The potential applications for waterborne epoxy adhesives are more limited due to their lack of tack and pressure-sensitive characteristics and the time it takes for the chemical reaction to complete cure. However, waterborne epoxy systems have found significant markets in niche areas.

Waterborne epoxy coatings and adhesives have established the building and construction industry as their largest market. Commercial systems have been available for many years. The following characteristics propel their use over conventional alternatives:

- Better adhesion to damp and moist substrate including "fresh" concrete
- Good acceptance of damp fillers such as high-moisture-content sands, cements, and oxides
- Low toxicity and nonflammability; does not cause skin irritation problems

- No cost in solvent cleanup or dilution, because water is used
- Ideal properties as primers and basecoats for other adhesives and coatings

Typical applications in the building and construction industry that benefit from these characteristics include

- Use as a waterproof coating for inside or outside tanks, basements, and utility substations
- Repair of spalled concrete by use as an adhesive and protective coating
- Waterborne high- and medium-gloss wall and floor coatings to resist high-pressure cleaning in industrial plants
- Exterior textured and flexible coatings over pavement and precast concrete

Because of the environmental acceptability and economic attractiveness of waterborne epoxy adhesive, one may confidently predict increased research and development in these areas. New products and applications will continue to develop; however, the adhesive formulator must be creative in choosing waterborne raw materials and formulating products that meet both regulatory and customer requirements.

14.3.2 Preparation of Waterborne Epoxy Raw Materials

Epoxy resins are hydrophobic and consequently are not, by themselves, dispersible in water. However, water dispersibility can be conveyed to epoxy resins by two general methods:

1. Chemical modification of the epoxy resin
2. The process of emulsification

Both processes are applicable to waterborne epoxy adhesives and coatings, although the emulsification process is generally used with adhesives. Preparation of epoxy resin emulsions is covered in Chap. 4.

Several hybrid epoxy emulsions have been commercially prepared. An epoxy emulsion blended with waterborne aliphatic urethanes exhibited peel strength on aluminum of 10 lb/in—1.5 times greater than with the polyurethane itself. The optimum concentration of urethane in the final emulsion was about 50 percent by weight.[13] Epoxy-phenolic dispersions have also been developed to provide waterborne adhesive systems with high glass transition temperature and chemical resistance.

The epoxy functional waterborne dispersions can be cured with many of the same curing agents that are used with nonaqueous liquid or solvent-borne epoxy resins. The curing agents most conveniently employed are those that are water-soluble or dispersible and are stable in an aqueous medium. The hardeners commonly used in waterborne systems are polyamides or polyamidoamines. These are manufactured by dimerizing tall-oil fatty acid and then reacting the dimer acid with an aliphatic amine, typically diethylenetriamine. Dicyandiamide and substituted imidazoles have also been used for elevated-temperature, latent curing epoxy systems. Water-compatible melamine and urea-formaldehyde resins can be used to cure those epoxy dispersions, which contain sufficient hydroxyl groups.

14.3.3 Adhesive Formulations

A typical starting formulation for a two-component epoxy-polyamide emulsion is shown in Table 14.3. This formulation can be used as either a coating or an adhesive. In either application, once the mixed emulsion is applied to the substrate, the water must be evaporated. In the case of the adhesive application, this must be completed before nonporous

TABLE 14.3 Typical Two-Component Epoxy-Polyamide Emulsion Starting Point Formulation[14]

Epoxy Component	Percent by weight	Polyamide component	Percent by weight
DER 331 epoxy resin	26.25	Versamid 125	16.50
Capcure 37S	1.73	Foamaster 111	0.10
Foamaster	0.09	Titanium dioxide	45.0
Titanium dioxide	50.0	Water	38.40
Water	21.93		
Total	100.00	Total	100.00

substrates are mated. Often the drying is done in a forced-air oven at a temperature near 65°C. With the particular adhesive presented in Table 14.3, the working life is several hours at room temperature.

Table 14.4 presents formulation information for bisphenol A and polyfunctional epoxy resin emulsions that are cured with an aliphatic amidoamine curing agent. Adhesive performance data are also provided for substrates common to the automotive industry. Both formulas are based on a 1 : 1 epoxy-amine stoichiometry and they are reduced to 45 percent nonvolatiles with water. The working life of each system is several hours at room temperature.

Similar information is presented in Table 14.5 for starting adhesive formulations made from an epoxy resin emulsion and dicyandiamide latent curing agent. This adhesive has exceptionally good water resistance when cured. The adhesive was applied to the indicated substrates in a manner similar to that described above, and it was cured for 3 min at 65°C followed by 10 min at 175°C.

TABLE 14.4 Epoxy Emulsion Adhesive Formulations[15]

	Parts by weight		
Components	A	B	C
Epi-Rez 5003-W-55	100	—	100
Epi-Rez 3515-W-60	—	100	—
Epi-Kure 3046 curing agent	24.1	21.6	24.1
Tap water	50	60	50
Epoxy silane	—	—	1
Percent nonvolatile	45	45	45
Tensile shear strength, psi, on substrate*			
Aluminum	2210 c	1070 c	—
Cold-rolled steel	1480 c	780 c	1433 c
Sheet molding compound	510 c	680 c	670 c
Nylon	220 a	190 a	—
TPO	56 a	0 a	
RIM	>280 s	240 c	—

*Failure mode: a = adhesive, c = cohesive, s = substrate.

TABLE 14.5 Waterborne Epoxy-Dicyandiamide Adhesive Tensile Shear Data[16]

Formulation	Parts by weight A	Parts by weight B
Epi-Rez 3515-W-60 epoxy dispersion	100	—
Epi-Rez 3522-W-60 epoxy dispersion	—	100
Dicyandiamide	3.5	2.25
2-Methylimidazole	0.15	0.2
Tap water	20	20
Percent of nonvolatiles	50	50
Tensile shear strength, psi		
Aluminum at 25°C	3290	3560
Aluminum at 65°C	2852	1260
Aluminum at 25°C after 20 days' immersion in water	2300	—

The applications and performance characteristics of waterborne epoxy adhesives can be significantly improved by the incorporation of additives and modifiers into the adhesive formulation. Fillers such as calcium carbonate, talc, and silicas are often used to adjust the viscosity of the liquid adhesive and the thermal expansion, modulus, and strength characteristics of the cured adhesive film.

Reactive diluents can be used to reduce the modulus and increase the elongation of the cured waterborne epoxy formulations just as they are often used for 100 percent solids and solvent-borne epoxy adhesives. The reactive diluents become codispersed in the formulation with mechanical and chemical stability similar to that of the base epoxy emulsion. Polyglycidyl ether of caster oil, phenyl glycidyl ether, and diglycidyl ether of neophenyl glycol are examples of mono- and difunctional reactive diluents that have been used to improve flexibility and increase the tack-free time of waterborne epoxy adhesives.

Surfactants act as wetting agents by lowering the surface tension of the waterborne epoxy. Silanes can be used to increase adhesion to certain substrates and fillers, as shown in Table 14.4, formulation C. Water-compatible thickeners and protective colloids such as polyvinyl alcohol, substituted cellulosics and sugars, and some acrylics improve application properties and offset viscosity decrease seen with water dilution.

Some waterborne epoxy systems may contain a proportion of water-miscible cosolvent to aid in film coalescence. Its presence may allow the formulator greater latitude to control properties such as stability, drying, and particularly rheology and still meet VOC levels required by pollution legislation.

14.3.4 Blends with Other Latex Systems

Many epoxy dispersions are compatible with most types of latex emulsions including acrylic, urethane, styrene butadiene, vinyl chloride, and polyvinyl acetate. The epoxy dispersion can be used as a modifier for these emulsions to alter handling and application characteristics such as emulsion rheology, foaming tendencies, pH sensitivity, wetting properties, and coating coalescence. They can also be reacted into the latex resin either by reacting the epoxy with a functionalized latex or by use of an epoxy with a coreactant. In this way adhesive systems can be formulated that are cured at room or elevated temperatures.

Adhesives formulated with epoxy-modified latex retain the tack and conformability of the original latex but show improvements in green bond strength and fully cured bond strength. Cured epoxy latex epoxy resin systems also exhibit improved water and chemical resistance over unmodified latex systems.

In choosing an epoxy and polymeric latex, it is important that they have compatibility. Incompatibility usually occurs when the pH of the epoxy resin dispersion alters the pH of the latex into a range where the ionically stabilized latex is broken, causing agglomeration of the latex polymer. The pH of the epoxy resin's emulsion may need to be adjusted before blending with the polymeric latex.

Addition of small (10 to 20 percent by weight based on solids) amount of epoxy to a latex emulsion will generally not change the latex viscosity characteristics. Addition of the epoxy will help decrease the foaming tendencies of the system. Wetting characteristics of the latex can also be improved by using epoxy modifications since the surfactant present in the waterborne dispersion acts as a wetting agent in the latex system. This can provide a benefit in bonding low-energy substrates such as polyolefins. Also the lower molecular weight of epoxy polymers compared to latex polymers tends to make the epoxy a good plasticizer for the cured latex.

The epoxy dispersion can be reacted into the latex polymer either by reacting the epoxy with a functionalized latex (carboxyl or amino functional groups) or by use of an epoxy with an epoxy coreactant or curing agent. In the latter case, the epoxy will react with the coreactant and form a network within the latex polymer network.

It has been shown that tensile shear and peel strength for several latex polymers (ethylene vinyl acetate, polyvinyl alcohol, ethylene vinyl chloride, polyvinyl chloride, and acrylic) can be significantly increased by the addition of 10 percent by weight of an epoxy emulsion cured with a tertiary amine curing agent.[17] The epoxy modification improves the bond strength in all cases. The degree of improvement is dependent on the selection of the latex type and the chemistry of the latex polymer.

Table 14.6 illustrates typical improvements noted in epoxy hybrid formulations with vinyl chloride, acrylic, and styrene butadiene lattices. Tensile strengths of cured, latex-saturated paper substrates are listed in absolute numbers while those of latex-epoxy hybrids are listed as percent increases in tensile strength over that of the latex alone. The mechanisms believed responsible for these improvements are (1) cocuring of the epoxy group with carboxyl and amine functional groups present on the latex backbone and/or (2) homopolymerization of the epoxy catalyzed by the tertiary amine included in some hybrid formulations.

Epoxy modified polymer latex systems offer improved handling performance and moisture and chemical strength advantages over unmodified formulations. The wide range of latex polymers and the range of waterborne epoxy dispersions offer the formulator a wide latitude in performance characteristics required by specific applications.

14.4 EPOXY ADHESIVES THAT CURE BY INDIRECT HEATING

One of the major disadvantages of using structural epoxy adhesives is the cost and time required for long cure cycles. The number of fixtures, energy use, and production times (either at room temperature or in an oven) required have discouraged many from exploring epoxy adhesives as an alternative to mechanical fastening.

One approach to solve this problem has been the development of fast-reacting room temperature curing epoxy adhesive systems, as discussed in Chap. 11. However, they often result in brittle joints, poor adhesion due to reduced time to achieve substrate wetting, and

TABLE 14.6 Properties of Epoxy-Latex Hybrids[18]

Properties	Vinyl chloride-A	Acrylic-A	Styrene-Butadiene	Acrylic-B	Vinyl chloride-B	Vinyl chloride-C
Glass transition temperature, °C	73	55	–9	56	8	50
Carboxy	Yes	Yes	Yes	Yes	No	No
Amine	No	No	No	Yes	No	No
Heat-reactive	Yes	Yes	Yes	No	No	No
Dry tensile strength on paper, lb/lin · in	14.2	10.2	29.1	31.0	26.3	31.2
• Modified with Epi-Rez 3510-W-60, percent improvement	23	28	0	0	0	28
• Modified with Epi-Rez 5003-W-55, percent improvement	27	25	0	0	11	28
Wet tensile strength on paper, lb/lin · in	11.7	35.0	17.1	10.9	12.0	11.4
• Modified with Epi-Rez 3510-W-60, percent improvement	20	16	7.1	100	49	79
• Modified with Epi-Rez 5003-W-55, percent improvement	24	14	20	70	48	86

Cure: 10 min at 150°C.

waste from material that has passed its working life or parts that could not be positioned in time. Modern metering, mixing, and dispensing equipment will help to eliminate this last problem, but maintenance of the equipment, flushing of the unreacted adhesive, and the expense of the equipment result in a completely new set of problems.

An effective way of speeding the cure of conventional adhesives has resulted from innovative modifications of the heating process itself. These new processes employ various methods to "indirectly" heat the adhesive through a medium other than heated air or electrical resistance heaters that surround the joint. The processes that have been developed have been successfully employed within industries where very high production speeds and volumes are characteristic (e.g., transportation, furniture, appliance, etc.). These new heating processes include electromagnetic heating (induction and microwave), weldbonding, and embedding resistance heaters within the adhesive.

These processes have an advantage in that the heat penetrates deeply into the joint and into the epoxy material itself. With conventional thermal energy processes, the heat must be conducted into the mass of the epoxy adhesive from outside the joint. This is hindered by the presence of the substrates, the substrate geometry, and the relatively low thermal conductivity of the epoxy itself.

This section first describes some of the disadvantages of traditional heating processes that are used in adhesive bonding. Then the fundamentals of these newer indirect heating processes are discussed along with the formulation requirements of adhesives that can be used in these processes. The advantages and disadvantages of each heating method are described, as are example applications. Starting formulations that are appropriate for each process are identified.

14.4.1 Traditional Heating

Curing adhesive joints by traditional methods, such as oven heating, can be difficult. Ovens tend to be energy-inefficient. Temperature ramping times are generally quite long and depend on the type and geometry of the substrate and joint. The entire joint must be brought up to temperature and maintained at the curing temperature for the prescribed period of time. Loading and unloading specimens into the oven and into the fixturing required to hold the substrates in place during cure can be cumbersome and expensive. Temperature gradients within an oven can cause nonuniform curing, and the resulting adhesive properties can be inconsistent. Even so, hot air curing and oven curing are often used for small production runs. Large production runs often are cured in batches in large ovens, which run continuously.

Infrared radiant heaters provide an increase in the efficiency of heat transfer, exceeding that of oven heaters. Infrared heaters are useful in rapid heating of localized areas of a substrate. However, the rate of heat transfer is dependent, to some extent, on the color of the workpiece. The darker the part, the more rapid the heating.[19] Radiant heaters are also more expensive and maintenance-dependent than air circulating ovens.

The traditional thermal curing methods mainly depend on a high rate of energy transfer. The rate of heat transfer and the efficiency of the process are demonstrated in the following three modes:

- Convection heating which utilizes the flow of hot air
- Thermal conduction, which transmits the heat energy from the substrate to the adhesives via physical contact
- Radiation heating, emitting through the medium and the surroundings

The traditional heating processes used to cure epoxy adhesives are fully described in Chap. 17. These traditional curing processes include oven heating, hot presses and platens, autoclave, infrared heating, and heating by electrical resistance heaters that are in contact with or surrounding the joint area.

The primary disadvantage of these processes is that they are relatively inefficient, mainly because the entire joint must be heated to cure only several mils of epoxy adhesives. The energy consumed, the time to get up to temperature, and the time to cool down to a safe handling temperature can be prohibitive in many production applications.

14.4.2 Induction Heating

Induction heating is a form of electromagnetic heating that provides a reliable, repeatable, noncontact, and energy-efficient heat in a minimal amount of time. Experience has shown that with electromagnetic heating the resulting properties of the joint are often superior to those achieved with conventional heating.

There are two electromagnetic processes in use today: induction and dielectric. Induction is more widely used, and there are several equipment manufacturers that will provide individual heaters or complete bonding systems. Dielectric heating, primarily microwave heating, is less utilized but is finding application in certain niche markets.

The Induction Heating Process. Induction heating is a form of electromagnetic heating. Electric power is used to generate heat in conducting materials (e.g., steel or aluminum) placed in the proximity of an inductor coil through which alternating current is passed. The process is, therefore, limited to applications where at least one substrate is metal or to adhesive materials that are filled with electrically conducting powder. First developed and still used for the assembly and sealing of thermoplastics,[20] induction heating is now also used to cure thermosetting structural adhesives such as epoxies.

With induction heating, alternating current flowing through a coiled conductor produces a low-voltage, high-amperage current. This work coil is shaped to cause a heat-generating eddy current in the metallic adherend(s) or in the electromagnetic fillers (e.g., metal oxides) in the adhesive itself. The magnetic field constantly expands and contracts, so that hysteresis in the metal causes the eddy current to generate the heat necessary for curing. The work coil does not contact the workpiece.

Induction curing equipment has progressed significantly over the last decades. Improved temperature ramping cycles can be achieved with computer control of the solid-state power supply. To eliminate extra steps for loading and unloading ovens, induction heat stations can be incorporated directly into a production line. New generator designs work without water-cooling, allowing the generator to be moved as far as 100 ft from the work coils.

The induction curing process is illustrated in Fig. 14.4. Four basic components comprise the induction heating process:

- An induction generator is used to convert a 60-Hz electrical supply to 3- to 40-MHz output frequency and output power of 1 to 5 kW.
- The induction heating coil consists usually of a water-cooled copper tubing formed into hairpin-shaped loops.
- Fixturing is used to hold the parts in place.
- The bonding materials are generally (1) in the form of thermoplastic preforms that bond or seal as a hot-melt type of adhesive or (2) thermosetting paste which cures to a structural adhesive.

Speed is the most important feature of induction heating. Workpiece heating rates greater than 40°C/s are possible. One supplier of induction curing epoxy adhesives claims production speeds of 600 parts per/hour are possible.[21]

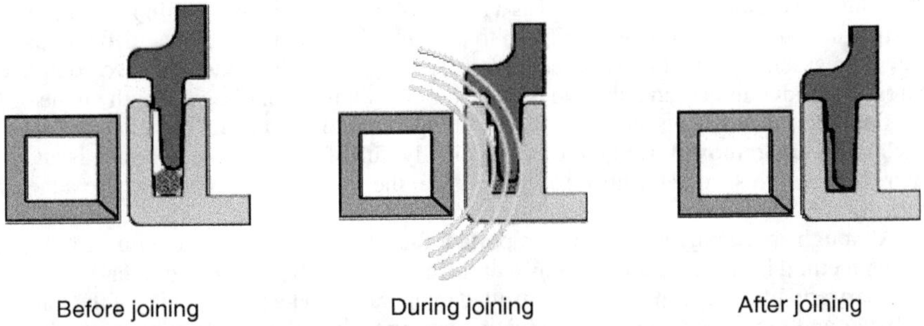

FIGURE 14.4 The induction heating process as applied to thermoplastic or thermosetting adhesives. Before joining, the adhesive is deposited in the joint, and the parts are brought together. During joining, the activated coils heat the adhesive internally (if conductive particles are present) or through metallic substrates. After joining, the adhesive has sufficient strength to be handled and moved to a postcuring process.

Frequencies used in commercial induction heating equipment range from 10 kHz to 4 Mhz. The higher the frequency, the thinner or smaller the part that can be heated. For every part size there is a critical frequency below which little heat is generated. Usually an induction power supply delivering the desired power at the lowest possible frequency is used, as this minimizes hot spots in the parts to be joined.

For high-power applications, the conductors in the workpiece may have to be cooled with water. Typical power supplies for induction curing applications range from 1 to 5 kW, depending on the parts and application requirements.

The heat induced into the workpiece will conduct instantly into the adhesive, providing the catalyst for cure. If one of the two substrates is not electrically conductive, then an adhesive can be used that includes a small percentage of metal oxide. The metal oxide particles within the adhesive become heated in the induction field and provide the source of heat to cure the matrix material in the adhesive. (See Fig. 14.5.)

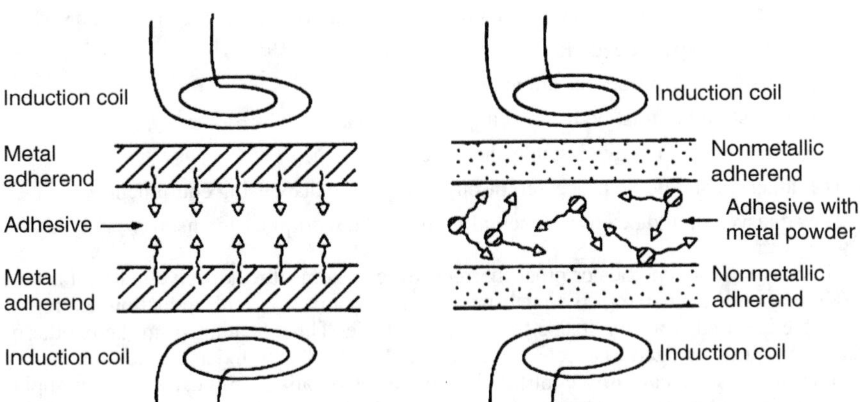

FIGURE 14.5 Left: induction heated metal parts heat thermosetting adhesive by conduction; right: parts not heated by induced electric currents are bonded with adhesive that is heated because it contains electrically conducting particles.[22]

Induction curing is generally used as a means of initiating or accelerating the initial cure of structural adhesives and getting them to a point where they can be handled and passed down the assembly line to a subprocess (e.g., paint oven) where the adhesive completely cures. The induction cured adhesive is used to rapidly attain handling strength on the order of 50 to 400 psi in a few seconds. After handling strength is attained, the substrate may be unclamped and removed from a fixture. Typically, full strength of the adhesive is attained later by way of a secondary heating source or, in the case of a two-part reactive adhesive, cure at room temperature via its natural curing process.

Although speed may be the most important factor in considering induction curing, this curing method is also characterized by a uniform heat distribution within a sharply defined boundary and by unique, precisely controlled temperature. Thus, the quality of the joint can be better and more consistent with induction curing than with other heating methods.

The advantages and disadvantages of using induction curing are shown in Table 14.7. Induction curing eliminates the biggest disadvantage of bonding with structural adhesives—the length of curing times required. Therefore, induction should be a prime consideration when the productivity gains outweigh the additional equipment costs. As a result, induction curing has been primarily developed for the demanding time constraints of the automobile industry's assembly-line environment.

The present uses of induction curing in the automobile industry include the fastening of plastic retaining clips, the bonding of manufacturer's identifying logo to the car's body, and the bonding of metallic fastening clips to the curved surfaces of glass windows and to the automobile plastic grill.[23] Also, induction curing offers a new approach to the production of hem flange automotive sheet metal joints. In this application the adhesive is applied to the upper surface of the lower steel sheet before the joint is formed. This lower sheet is then folded up and over the end of the top sheet, and the resulting joint is subjected to a 4- to 8-s period of induction heating.[24] This hem flange joining process is illustrated in Fig. 14.6.

Induction Curing Adhesives. Typically one-component epoxy adhesive systems are used for induction bonding of structural parts. However, two-component adhesives have also been used. The properly designed induction cured adhesive should provide the following application and performance properties:

- A 100% solids system is needed (no volatile organic content).
- Adhesive must have the ability to adhere to moderately contaminated or unclean surfaces.
- Ability to fill gaps is needed.
- High tensile shear properties are necessary.
- Handling strength must be reached in 1 min or less.
- The adhesive should have a stable shelf life.
- The adhesive should be processed through automatic dispensing equipment.
- The adhesive should be environmentally safe in the workplace to ensure worker safety.[26]

In conventional epoxies or other thermosetting systems, the adhesive material is cured beginning on the outside of the material and moving inward. With induction heating, the adhesive is cured from the inside toward the outside. Thus, heating from the exothermic reaction is fully utilized to accelerate the cure. However, with induction cured adhesives, exotherms must be carefully controlled to avoid overheating. In most induction applications, curing temperatures on the order of 190 to 230°C are easily achieved. This will allow many epoxy adhesive systems to reach appreciable strength in less than 1 min.

When induction curing a one-part thermosetting adhesive, one has the choice of taking the adhesive to full strength or only to partial strength in the induction field (complete cure

TABLE 14.7 Advantages and Disadvantages of Induction Heating

Advantages	Disadvantages
Accelerates the cure time of the adhesive, dramatically improving productivity	Capital costs of the induction equipment
Improves the strength of the bond by improving surface cleaning and wetting	May limit process flexibility
Process easily adaptable to many available adhesives	Risk of overcuring the adhesive especially with adhesives that have a high exotherm
Localized heat that utilizes energy only where required	Fixturing difficult relative to the working coil and geometry restraints
Very low maintenance process	

then occurs after the induction heating process). In either case, an adhesive must be chosen that has very good hot strength, since the adhesive will be at very high temperatures once cured and released from the fixturing. This will necessitate an adhesive with a high glass transition temperature and one that can compensate for the stresses caused by differences in the substrates' thermal expansion coefficients.

Overheating of the adhesive is always a critical concern with induction curing. This is especially true when the adhesive cures with a high exotherm, such as epoxy-dicyandiamide systems. Several proprietary epoxy curing agents have been developed that provide lower exotherm yet faster cure rate than typical dicyandiamide reactions.[27,28] Tertiary amines and modified polyamines[29] are often used to accelerate the cure of dicyandiamide-epoxy

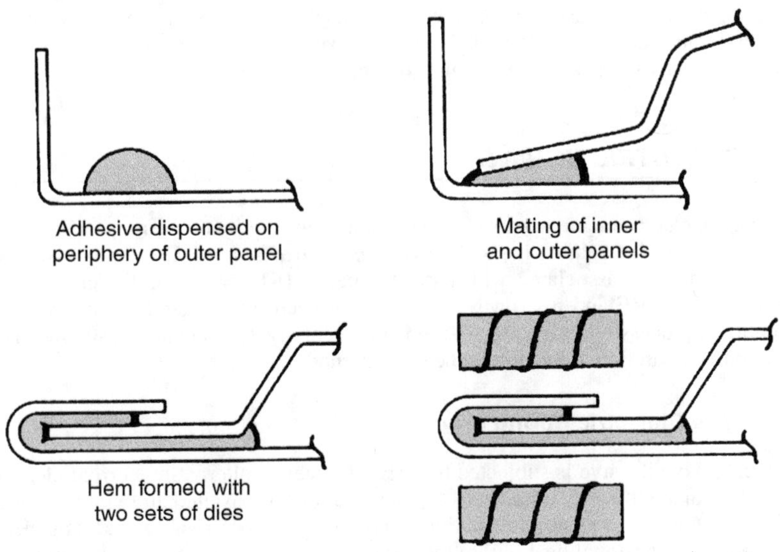

FIGURE 14.6 Hem flange joining.[25] Epoxy is applied before forming of joint and is partially cured by immediate induction heating.

adhesives and to reduce the activation temperature. They exhibit rapid green strength development at low temperatures while maintaining excellent shelf stability.

Typically a full cure will take less than 1 min and can occur with certain adhesives in 15 to 30 s. The partial cure process is usually employed on larger, more difficult to fixture parts. Heat times for sufficient handling strength can be as low as 3 to 10 s. A secondary heat source such as a final paint cure oven will then fully cure the adhesive.

Ideal induction cure epoxy adhesives can be one- or two-part and have a broad cure temperature range of 135 to 220°C. A two-part reactive epoxy adhesive is typically used where full strength must be attained and a secondary curing oven is impractical or unavailable. Although these adhesives generally have a working life of 20 min to 1 h at room temperature and achieve full cure in 24 h, induction heating will accelerate the cure time. With metal substrates and a typical two-part epoxy, handling strength can be attained in about 10 s.

One of the advantages of using a two-component room temperature curing adhesive system is that induction curing can be used to spot-cure the adhesive in a joint (similar to a spot-welding process). The adhesive that is not spot-cured by induction will then fully cure at room temperature as the part moves down the assembly line.

If both substrates to be bonded are nonconducting, then the adhesive formulation must contain a susceptor material. Susceptors can have a small percentage of magnetic iron oxide, iron filings, or carbon additives. A susceptor can also be a steel screen or perforated steel foil that is embedded in the adhesive bond line. It has been found that graphite fiber composites used in the automotive and aerospace industries are sufficiently conductive that they can be successfully heated with induction. Design considerations must be taken into account in placement of the graphite reinforcement, so that the material heats uniformly.

The induction process will then heat the adhesive directly through the susceptor additives. A typical heat cycle for this type of joint is approximately 30 to 40 s because the smaller particles are slower to react to the induction field than a metallic substrate. Induction curing offers a special advantage for bonding plastic substrates. Since the plastic substrate is not heated (assuming no metal fillers), the bulk material does not see high temperatures, which could cause degradation and warpage. Hot dies can be eliminated, and this will eliminate any surface read-through problems.

14.5 DIELECTRIC CURING

Short-length, electromagnetic waves also provide a fast method for curing structural adhesives. This method of heating is also known as *dielectric heating*. These processes use frequencies commonly associated with radio-frequency (RF) waves and microwaves. Like induction curing, RF and microwave curing have been investigated as an alternative to conventional processing. Such processing is more energy-efficient and has reduced curing time when compared to conventional heating methods.

14.5.1 The Dielectric Curing Process

When an epoxy adhesive is subjected to a high-frequency alternating current electromagnetic field, polarizable species such as dipoles and ions are excited and vibrate along with the field. The high-frequency translational motion of the polarizable species generates a large amount of frictional heat within the molecule. The heat generation, however, occurs only within those materials that have a high dielectric loss.

There are basically two forms of dielectric heating: radio-frequency and microwave. Radio-frequency heating uses a frequency (13 to 100 MHz) to generate heat in polar materials. The electrodes are generally designed into the platens of a press, and they are

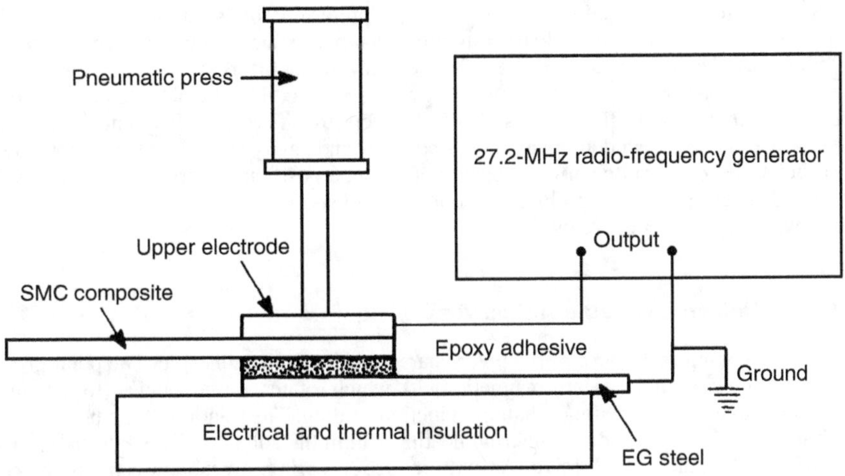

FIGURE 14.7 Illustration of dielectric heating for the bonding of an electrogalvanized steel (EGS)/SMC lap shear joint.[30]

much simpler than the shaped coils used for induction curing. This dielectric curing process is shown in Fig. 14.7. Microwave heating uses high-frequency (0.3- to 300-GHz) electromagnetic radiation to heat a polar material or a material with a susceptor located at the joint interface. The susceptor material (see below) adsorbs the microwave energy and transfers it to heat.

The Federal Communications Commission in the United States regulates microwave usage, and the frequencies for heating are designated by the International Telecommunications Union (ITU). The allowed frequencies most commonly used in the United States are 915 MHz and 2.45 GHz.

Radio-frequency heating is often used on thermoplastics with a susceptor embedded in the material and on water-based adhesives, such as acrylic, starch, and polyvinyl acetate. The RF energy is used to quickly drive off the water and dry the adhesive. Microwave heating is generally used on structural adhesives, such as the epoxies, where the heating occurs via rapid oscillation of the polar groups in the field. The feasibility of RF dielectric heating to cure one- and two-part epoxy adhesives on steel, thermoplastic, and thermoset plastic substrates has been shown.[31] Process cycle times for RF curing are about 20 to 60 s, compared with about 20 to 30 min for the same materials using conventional oven cure methods. Table 14.8 shows the bond strengths achieved on joints of SMC to steel.

TABLE 14.8 Strength of RF Dielectrically Bonded Adhesive Joints (SMC to steel)[32]

Adherends	Adhesive	Tensile shear strength, MPa	Failure mode
EGS/SMC	PG-6500	4.97	SMC fiber tear
EGS/SMC	Fusor 320/322	5.34	SMC fiber tear
EGS/SMC	PG-II	5.19	SMC fiber tear
E-EGS/SMC	PG-6500	5.12	SMC fiber tear
E-EGS/SMC	PG-II	5.74	SMC fiber tear
E-EGS/SMC	Fusor 320/322	5.59	SMC fiber tear
E-EGS/RTM	Fusor 320/322	7.42	RTM subinterfacial

Note: E-EGS is electrocoated electrogalvanized steel.

RF or microwave heating methods are not preferred for bonding metal substrates because the electromagnetic field tends to break down, causing arcing. Recently a new type of microwave heating system has been commercialized that offers significant improvement over conventional fixed-frequency microwave processing. This is the patented *variable-frequency microwave* (VFM) process.[33] VFM sweeps the microwave frequency over a predetermined range (dependent on average frequency and cavity size). In this way the power distribution becomes quite uniform because of the superposition of a great many microwave modes. This eliminates the problems of hot and cold spots and allows the use of metals in microwaves by reducing arcing.

14.5.2 Dielectric Curable Adhesives

Dielectric heating makes use of the polar characteristics of the adhesive. The polar groups become aligned with the electromagnetic field, which rotates at frequencies (for example, 2.45 GHz) that are much higher than in induction heating. The rotation of the polar groups with the electric field produces internal heating within the adhesive. This internal heating is sufficient to cure the adhesive. As with induction curing, a problem can easily occur where the adhesive is overheated.

RF or microwave cured adhesives must be highly polar. This is generally recognized by resins and curing agents that have a high electrical loss factor. It is this loss that causes internal heating within the adhesive. Conductive fillers are not required, and in fact they are discouraged because of microwave breakdown.

Epoxy adhesives should be 100 percent solids so as to eliminate the possibility of gassing and bubbles occurring in the bond line. However, dielectric heating is often used in the furniture industry to drive off water quickly from water-based adhesive emulsions. Water, being a polar material, heats rapidly in a microwave field, as every cook knows.

Many commercial one- and two-part epoxy adhesives can be used for RF or microwave heating without extensive reformulation. When susceptors are necessary, they are generally polyaniline materials doped with aqueous acid, such as hydrochloric acid. This introduces polar groups and a degree of conductivity into the molecular structure of the adhesive. It is these polar groups that preferentially generate heat when exposed to the high-frequency fields. Such doped materials are also often used to produce thermoplastic gaskets, which can be used as an adhesive or sealant.

Because of the highly polar nature of common epoxy resins and their curatives, most formulations can be cured via a microwave environment. A further advantage is that the uncured material absorbs microwave energy more strongly than that which is cured, so the cure occurs uniformly within the material.

Microwave curing has been found to increase the glass transition temperature of epoxies and in some instances improve the mechanical properties. It is also possible to control the morphology of toughened epoxies more closely with microwaves.[35]

Several studies have shown that a microwave cure cycle can be developed that provides equivalent performance properties to a thermal cure cycle. Table 14.9 shows the processing and performance characteristics of three commercial one-component epoxy adhesives cured via microwave and conventional thermal energy. Certain commercial epoxy adhesives could contain a large number of bubbles due to volatiles present during the cure cycle and the fast rate of cure. Therefore, specifically formulated adhesives for microwave curing may be necessary to optimize performance.

There are several claims in the literature regarding the maximum speed of microwave curing of epoxy adhesives. Claims have been made of ten- to twentyfold cure time reduction when compared to conventional thermal heating. A more practical estimate, however, is that microwave joining using epoxy-based adhesive will significantly reduce the curing time

TABLE 14.9 Curing Parameters and Properties of Microwave Cured Epoxy Adhesives[34]

Adhesive	Microwave ramp rate, °C/min	Microwave time at cure temperature, min	Microwave cure temperature, °C	Modified tensile shear strength, MPa	
				Thermal cure	Microwave cure
EA 9689	13	48	177	43.1	35.2
EA 9391	12	40	149	47.8	25.1
AF-163-2K	10	40	121	57.1	51.7

to one-third to one-quarter of the conventional cure time. This is accomplished while maintaining equal or slightly higher values of the ultimate tensile strength obtained in a single lap shear test.[36] A threshold in the rate of energy deposition or input power level exists. Above a certain energy level, the adhesive will outgas and form voids, which reduce both the adhesive and the cohesive strength of the final product.

The epoxy adhesive formulation may require special additives and curing agents to avoid the effects of overheating and reaction speed. A DGEBA epoxy resin with 4,4′-diaminodiphenylsulfone (DADPS), 4,4′-diaminodiphenylmethane (DDM), and metaphenylene diamine (MPD) can be cured between 200 and 600 W of microwave power. The DADPS curing agent exhibits a slower reaction rate than DDM or MPDA.[37] Coupling to the microwave energy source is made more efficient by the use of various additives in the epoxy adhesive. Carbon black, ZrO_2, and Al_2O_3 have been shown to be particularly effective.[38,39]

14.6 WELDBONDING

Weldbonding is a hybrid method of assembly that utilizes components of both the metallic welding and adhesive bonding processes. Weldbonding is claimed to provide the advantages of both processes while minimizing the disadvantages. The benefits of instant strength and high peel resistance provided by the welds supplement the adhesive bonding advantages of uniform stress distribution, fatigue and vibration resistance, improved strength and durability, and greater design flexibility.

Initially developed in the Soviet Union and used in the fabrication of transport aircraft, weldbonding continues to be used in the assembly of aircraft parts and other large structures. Weldbonding has also found its way into the ground transportation markets. It is in high-volume market segments where the potential of weldbonding seems to be greatest. Weldbonding can be fully automated and utilized with robotic systems. It has been successfully applied on both thin-gauge aluminum and steel substrates and to a lesser extent on titanium.

14.6.1 The Weldbonding Process

Two general processing approaches prevail for weldbonding, and minor variations are possible within each approach. The first and most common method is to apply the adhesive and weld through the joint. This process is sometimes referred to as the *weld-through method*. The second approach, sometimes referred to as the *flow-in method*, is to form the welded joint first and then allow a low-viscosity adhesive to flow into the spaces between the joint

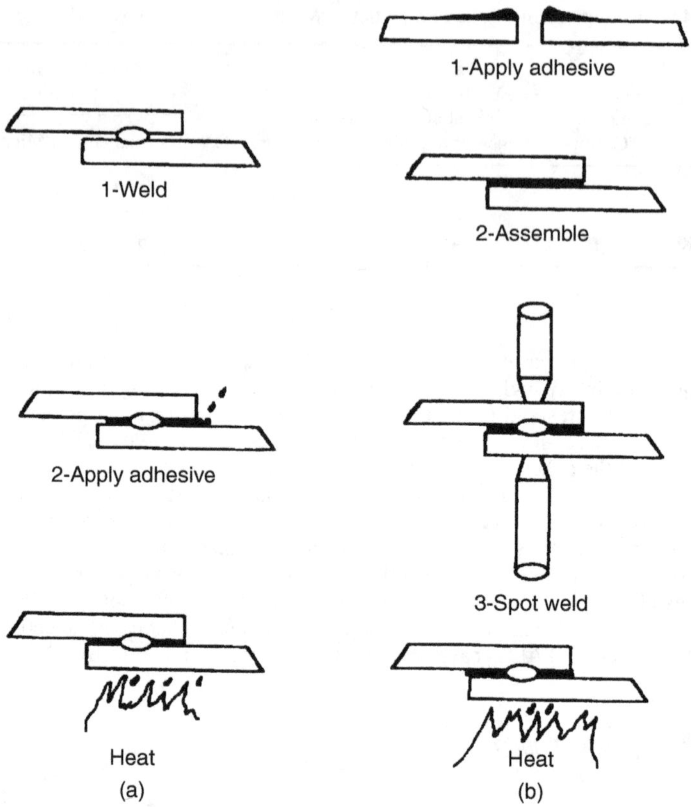

FIGURE 14.8 Schematic illustrations of the two fundamental approaches for producing weldbonded joints: (*a*) filling adhesive into the spaces between previous made spot welds and (*b*) spot welding through preapplied adhesive.[40]

by capillary action. These two fundamental approaches for producing weldbonds are shown in Fig. 14.8.

In these approaches, several different forms of welding can be utilized such as resistance spot welding, gas tungsten spot welding, laser spot welding, and electron beam spot welding. However, resistance spot welding is the most popular and well accepted of these methods.

Weldbonding process specifications have been developed for use by both manufacturers and government agencies. These process specifications give detailed steps to provide optimum weldbonds. The descriptions below provide a generic summary of these processes.

Weld-Through Method. Since this is a hybrid process, techniques common to both welding and adhesive bonding are modified to fit the combined process. The essential steps of the weld-through method of weldbonding are

1. Cutting and fitting of joint members
2. Surface preparation of the substrates for both adhesive bonding and resistance welding
3. Coating of one joint member with an adhesive strip of controlled thickness and width

4. Clamping the joint members into position and spot welding directly through the adhesive layer
5. Spacing of spot welds in a predetermined pattern designed to increase the structural integrity of the joint
6. Final curing of the adhesive either at room temperature or in an oven at elevated temperatures
7. Nondestructive examination of the joint

The joints are generally made by first applying a paste adhesive, much as in conventional adhesive bonding. However, after the adhesive is applied (and generally before it is cured), the joint is assembled and a resistance spot weld is made through the adhesive layer (Fig. 14.9). The welding electrode's force displaces the adhesive to obtain electrical contact between the substrates, and a weld is made in the conventional way. As the heating of the weld is very localized, little damage occurs in the adhesive around the weld.

The adhesive is then cured at either room or elevated temperatures to complete the assembly. Heat curing adhesives are normally used because of production requirements. Typically such adhesives are cured in an oven at up to 180°C for 30 min. Often the cure requirements can be achieved during another processing step farther down the assembly line, such as during a paint baking operation. In this way a separate process for curing the adhesive can be eliminated and costs saved. The spot weld allows the joint to be held together until the assembly reaches the curing station, thereby eliminating the need for fixturing equipment and further saving time and costs.

Flow-in Method. In the flow-in method of weldbonding, spot welding precedes the adhesive application. A normal substrate separation will result from the weld nugget buildup in the joint (Fig. 14.8*a*). A low-viscosity adhesive is then used to infiltrate the clearance between the substrate members, forming a sealed bond line. The adhesive may be drawn into

FIGURE 14.9 Aluminum sheet is joined using weldbonding, a combination of resistance spot welding and weld-through adhesive for reinforcement and sealing of a joint. (*Courtesy: TWI World Center for Materials Joining Technology.*)

the joint by capillary action during cure or by other means such as with vacuum or pressure. Once the joint is infiltrated with adhesive, it is cured at either room or elevated temperature.

Several joint designs have been specifically developed to provide for the flow-in method of hybrid joining. Invented by TWI, AdFAST is a series of specially designed fasteners that allow the adhesive to be introduced into the joint after the structure has been assembled using welds or other mechanical fasteners.[41] These fasteners incorporate a means of controlling the spacing between the top and bottom substrates. This provides easy access for the adhesive, greater bond line control, and improved process reliability and joint quality.

There is a debate among experts as to whether the flow-in method provides better properties and lower cost than the weld-through method. A relatively recent analysis indicates that the flow-in technique is easier to implement and better preserves the microstructure and hardness of the produced weldments.[42] However, the weld-through method is the most commonly used because of its high-speed production advantages.

Important Processing Issues. As with conventional adhesive bonding, there are several important issues that cannot be overlooked with weldbonding. Two of the most important issues are joint design and surface preparation.

It is important in the weld-through method that the adhesive be displaced as much as possible from the local area where the resistance weld is to be made. This provides for optimal welding and reduces the potential of the adhesive contaminating the weld site. The welding electrode force locally brings the adherends together and displaces the adhesive to allow electrical contact between the electrodes. A joint design where the spot weld areas are raised (Fig. 14.10) facilitates flow of adhesive away from the weld location. Such a design also allows adhesive to more easily enter the joint in the case of the flow-in method of weldbonding.

Other techniques can be used to achieve displacement of the adhesive from the weld area. This can be done through (1) use of an adhesive in tape form with a circle cut out from the expected weld areas or (2) masking the weld area before the liquid adhesive is applied to the substrate.

The surface preparation method must be carefully considered, especially if the completed weldbond is to have long-term durability to hostile environments. The surface preparation should provide an optimal surface for *both* adhesion and welding. Thus, the choice of surface treatment is crucial, and there can be a conflict of requirements. The spot welding process requires a low electrical surface resistance, and many adhesive surface preparation processes provide a high surface resistance because of oxide layer buildup. When it is impossible to harmonize on a surface treatment, current practice tends to favor treatments that yield good weld nuggets at the expense of the adhesive bond.

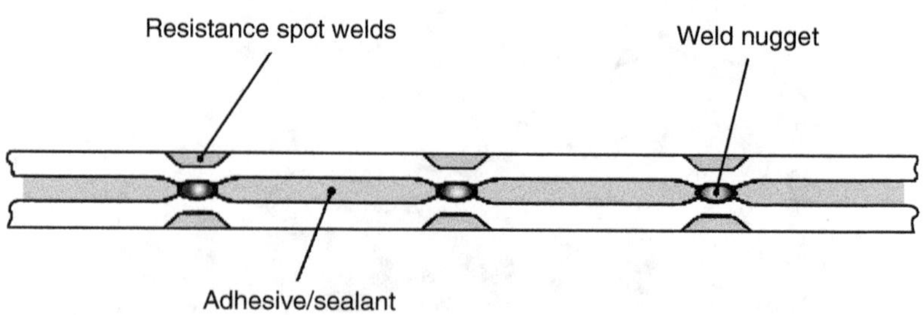

FIGURE 14.10 Joint designs are preferred that allow the substrates to electrically contact and displace the adhesive from the weld locations. (*Courtesy: TWI World Center for Materials Joining Technology.*)

Metal surface resistance should be uniformly low to ensure good, consistent weld nuggets. Because of its surface resistance, the spot welding of aluminum requires 2.5 to 3 times the current required for steel.

Good process control is required with weldbonding to ensure correct joint filling of the adhesive and to avoid weld quality problems. The process needs to be carefully controlled so that health and safety requirements are met. Welding through the adhesive may create hazardous fumes, and little information is available as to the organic compounds that are produced. Suitable ventilation and fume extraction equipment should be provided.

14.6.2 Adhesives for Weldbonding

Epoxy, modified epoxy, and polyurethane adhesives are commonly used for weldbonding applications. These are used as one- or two-part liquids, pastes, or film forms. Generally the adhesive system must provide high bond strength and durability, be easy to process, provide good weldability, and afford corrosion protection to the substrate surfaces. The most important criteria for any weldbonding adhesive are that it (1) have the capability of flowing under pressure of the welding electrodes in order for metal-to-metal contact to occur at the joint interface (weld-through method only) and (2) have sufficient heat resistance to the welding temperatures so as not to detrimentally affect the strength of the final joint.

Most adhesives used for weldbonding are arbitrarily selected from adhesives that were developed for other purposes, resulting in a compromise when used for weldbonding. Adhesives specifically developed for weldbonding should have the following characteristics in addition to those mentioned above:

1. Viscosity should be low enough to be forced out of welding area by the pressure of the electrodes yet sufficiently high that it will not flow out of the joint during the cure cycle. For the flow-in process, the viscosity must be low enough to fill the joint via capillary pressure or with moderate vacuum or positive pressure.
2. Metal fillers are often used to provide low electrical resistance between substrates.
3. An adhesive layer with appropriate thickness and elastic modulus is necessary to obtain reasonable distribution of stresses in the region of a weldbond joint. A thin adhesive layer of high elastic modulus improves the fatigue properties of weldbonded joints.[43]
4. The heat created by the spot welding process should not degrade the adhesive so as to weaken the final bond strength or create thermal decomposition products, which can contaminate and weaken the weld area.
5. The adhesive must remain pliable long enough to allow process completion.
6. The thermal expansion coefficient of the adhesive should closely match that of the substrates.

Film adhesives can be cut out to provide for the absence of adhesive at the weld locations. However, often these adhesives do not flow sufficiently at the cure temperature without pressure (as would be the case in weldbonded joints) to wet the substrates.

Many different adhesives have been tested and used in weldbonding. These include both high-strength structural adhesives, such as modified epoxies, and relatively low-strength adhesives, such as vinyl plastisols or epoxy-polysulfides that are commonly used for sealing and vibration damping. Table 14.10 shows that a variety of adhesives and substrates are compatible with the weldbonding process.

A commonly used adhesive in weldbonding applications is a modified epoxy, one- or two-component paste containing conductive metal filler. Other fillers commonly used in

TABLE 14.10 Strength of Weldbonded Adhesive Joints[44]

Joint alloy	Adhesive	Curing condition	Spot weld, lb	Adhesive bond, lb	Weldbond, lb
2036-T4	None		770		
2036-T4	Polysulfide-epoxy	Ambient		700	1175
2036-T4	High-peel epoxy	Ambient		1385	1090
2036-T4	Polyamide-epoxy	Ambient		735	875
2036-T4	Vinyl plastisol	Not cured	690		
2036-T4	Vinyl plastisol	2 h @175°C		910	1450
2036-T4	One-part epoxy	Not cured	750		
2036-T4	One-part epoxy	2 h @175°C		1610	1270
Steel	None		1440		
Steel	Vinyl plastisol	Not cured	1380		
Steel	Vinyl plastisol	2 h @175°C		900	1800
Steel	One-part epoxy	Not cured	1430		
Steel	One-part epoxy	2 h @175°C		2000	1930

adhesives for weldbonding include fumed silica to provide thixotropy and prevent run-out during cure and corrosion inhibitors for maximum durability in moist environments.

14.6.3 Performance Factors and Opportunities

To assess the advantages of the weldbonded joint, one must look at the properties of the spot weld alone, the adhesive bond alone, and compare these to the properties of the weldbonded joint. One must also be aware of the physical and environmental effects on the joint. Studies show that weldbonded joints can be stronger than joints that are only spot welded or only adhesively bonded. However, metal thickness, surface preparation, adhesive flow and cohesion, and weld quality can influence the results.

Weldbonded aluminum joints (1-mm aluminum alloy 2036-T4) where compared with spot welded aluminum (1-mm aluminum alloy 2036-T4) and spot welded steel (1-mm steel 1010) joints.[45,46] Weldbonded 2036-T4 joints made with a vinyl plastisol adhesive and with a one-part modified epoxy had a fatigue strength about twice that of the spot welded aluminum alloy and approximately the same as that of the spot welded steel. The fatigue strength of the weldbonded aluminum joints with polysufide-epoxy adhesive or with high-peel-strength epoxy adhesive is higher than that of steel spot welds alone.

The degree of acceptance of weldbonding applications has been increasing, as the process has been understood and its mechanical properties developed.[47] The principal advantages that have been claimed for weldbonding include the following.

1. Fatigue endurance is enhanced compared to that of mechanical fasteners or spot welding alone, since the stress concentration factor at the joint is reduced. The adhesive layer results in a more uniform stress field around the weld nugget.
2. There is improved energy absorption compared to either spot welding or adhesives bonding alone.
3. Environmental durability is improved compared to either spot welding or adhesive bonding alone.

4. There is improved tolerance of transient elevated-temperature excursions compared to adhesive bonding alone.
5. Cost savings can be achieved compared to adhesive bonding alone. The spot weld clamps the joint so that expensive fixturing is not required and the joints can be cured in commercial ovens rather than autoclaves.
6. Weldbonding automatically achieves joint sealing, which in mechanically fastened joints requires an additional process.
7. Adherend stresses in weldbonded joints are lower and more uniform than those for comparable spot welded joints. This provides increased in-plane tensile shear and/or compressive buckling load-carrying ability for a given joint design. The presence of the spot weld provides enhanced out-of-plane load-carrying capability compared to adhesive bonding only.

Aircraft applications include fuselage skin panels, wing center sections, leading and trailing edges, wingtips, spoilers, and access or actuated floors. Other indicated aerospace applications are for cryogenic tanks and for rocket shrouds. Weldbonding is important in this industry to achieve the weight savings and structural performance required of aircraft design.

In the automotive industry, spot welding can be used in numerous joints of the structural frame of the vehicle. Currently, weldbonding is not being used to reduce the weld spacing as much as it is being used to gain higher durability and stiffness. Weldbonding is beginning to find use for improved vehicle NVH (noise, vibration, and harshness) and for increased fatigue resistance. The increased stiffness of the vehicle reduces noise and provides a higher driving performance. Weldbonding is also being investigated to improve crash performance. Applications include the bonding of side apertures, the bonding of inner to outer rocker panels, the bonding of cowling, various cross bracing, and the rear shelf panel.

The several disadvantages listed above for weldbonding are being overcome with product and process development. Therefore, a high growth rate of nearly 20 percent annually for weldbonding over the next 5 years is predicted.[48] Future directions for structural weldbonding adhesives in vehicles will be to reduce metal gauge and weight for cost savings and fuel economy.

14.7 OTHER CURING TECHNOLOGIES

The heating technology that can be used to cure epoxy adhesives is very broad. The processes described above are the most widely used of these innovative, indirect heating processes. However, the possible approaches appear to be nearly unlimited. Two less widely used processes are ultrasonic curing and embedded resistance heating processes. Like those described above, the advantage of these processes is that they cure the epoxy adhesive from the inside of the bulk material. Thus, significant advantage can be gained in efficiency and curing time.

14.7.1 Ultrasonic Curing

Ultrasonic welding is a frictional process that has been well established for heat welding thermoplastic parts. Like induction welding, it also has been adapted for curing structural adhesives such as epoxy.

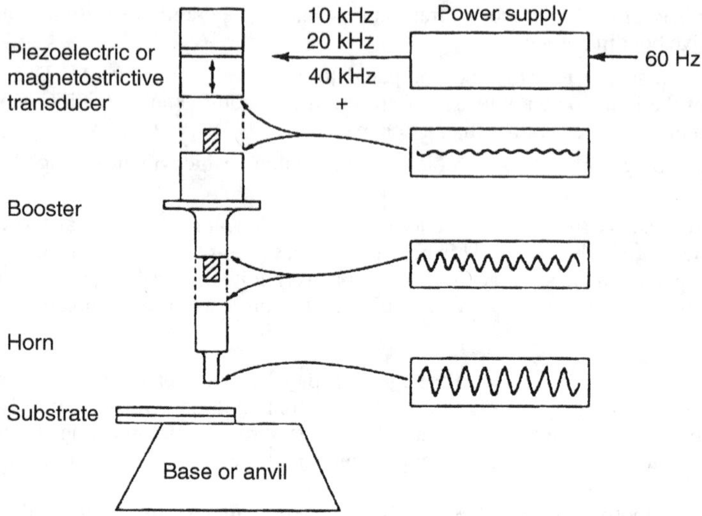

FIGURE 14.11 Equipment used in standard ultrasonic welding process.

The basic parts in a standard ultrasonic welding device are shown in Fig. 14.11. During ultrasonic welding, a high-frequency electrodynamic field is generated which resonates a metal horn that is in contact with one substrate. The horn vibrates the substrate sufficiently fast relative to a fixed substrate that significant heat is generated at the interface. Thus the adhesive is heated at the interface, and a strong and efficient joint can be obtained.

A slow-curing, liquid thermosetting epoxy is not appropriate for ultrasonic bonding. The liquid lubricates the interface and is hammered out of position with little curing action. However, a B-stage epoxy adhesive can be used successfully with ultrasonic activation. Table 14.11 shows that several epoxy adhesives are capable of gaining good strength during a few seconds of ultrasonic activation, and then full-strength bonds can be accomplished with an oven postcure. No clamps were used during the oven cycles with these particular adhesives.

Rapid curing of structural epoxy adhesive by ultrasonic heating has been demonstrated successfully in recent work. The conversion of epoxy groups produced by ultrasonic curing for 50 s was almost 3 times higher than that obtained by thermal heating.[51]

TABLE 14.11 Ultrasonic Curing of Epoxy Adhesives[50]

Adhesive	Form	Adherends	Weld time, s	Tensile shear strength, psi
Shell EPON 927	B-stage	Aluminum-aluminum	8	1460
		Aluminum-glass	8	1070
Shell EPON 9601	B-stage	Aluminum-aluminum	5.5	77
			5.5, plus 2 h at 107°C	4124
Bloomingdale FM47 Type 1	B-stage	Phenolic-phenolic	9	800
			9, plus 1 h at 150°C	1700

14.7.2 Embedded Resistance Curing

In situ heating of epoxy film adhesives may be accomplished by laminating high-resistance filaments into the film and subsequently applying an electric current. This provides the advantage of a one-component adhesive without the requirements of external heat sources. Once the element is heated, the surrounding material melts, flows, and finally crosslinks. After the adhesive cures, the resistance element that is exterior to the joint is cut off.

Heating elements can be anything that conducts current and can be heated through Joule heating. This includes nichrome wire, carbon fiber, woven graphite fabric, and stainless steel foil. Implant materials should be compatible with the epoxy adhesive and the intended application, since they will remain in the bond line for the life of the product.

The resistance heating process can be performed at either constant power or constant temperature. When one is using constant power, a particular voltage and current are applied and held for a specified time. The actual temperatures are not controlled and are difficult to predict. In constant-temperature resistance wire welding, temperature sensors monitor the temperature of the weld and automatically adjust the current and voltage to maintain a predefined temperature. Accurate control of heating and cooling rates is important in bonding some plastics or in welding substrates that have significantly different melt temperatures or thermal expansion coefficients. This heating and cooling control also can be used to minimize internal stresses in the joint due to thermal effects.

Large parts can require considerable power requirements. Resistance welding has been applied to complex joints in automotive applications, including vehicle bumpers and panels, and joints in plastic pipe, and in medical devices. Resistance wire welding is not restricted to flat surfaces. If access to the heating element is possible, repair of badly bonded joints is possible, and joints can be disassembled in a reverse process to which they were made.

REFERENCES

1. "Study Focuses on Radiation Cured Products," *Paint and Coatings Industry*, vol. 12, no. 10, October 1996.
2. Pappas, S. P., *UV Curing, Science and Technology*, vol. 1, Technology Marketing Corporation, Norwalk, CT, 1978.
3. Pappas, S. P., *UV Curing, Science and Technology*, vol. 2, Technology Marketing Corporation, Norwalk, CT, 1985.
4. Holman, R., *UV and EB Curing Formulations for Printing Inks, Coatings, and Paints*, Selective Industrial Training Association, London, 1984.
5. RadTech N.A, Chevy Chase, MD.
6. Carder, C. H., "Radiation Curing Coatings," *Paint and Varnish Products*, vol. 64, no. 8, 1974, p. 19.
7. Zwanenburg, R. C. W., "How to Formulate UV-Curing Coatings," *RadNews Highlights*, Paint Research Association, Middlesex, UK, Spring 1998.
8. Oldring, P. K. T., ed., *Chemistry and Technology of UV and EB Formulations for Coating, Inks, and Paints*, vol. 4: *Formulation*, SITA Technology, London, 1991.
9. Shirane, K., "Radiation-Induced Free Radical Cure of Epoxy Resin," *Journal of Polymer Science*, vol. 17, 1979, p. 139.
10. *H-Nu 470*, Spectra Group, Ltd., Maumee, OH.
11. Valero, G., "Enlightening Developments," *Adhesives Age*, June 2002, pp. 16–17.
12. Janke, C. J., et al., "Electron Beam Curing of Epoxy Resins by Cationic Polymerization," *SAMPE International Symposium*, vol. 41, 1996, p. 196.
13. Buehner, R. W., and Atzinger, G. D., "Waterborne Epoxy Dispersions in Adhesive Applications," Epoxy Resin Formulators Conference, San Francisco, Feb. 20–22, 1991.

14. Amstock, J., *Handbook of Adhesives and Sealants in Construction*, McGraw-Hill, New York, 2002.
15. Waterborne Epoxy Dispersions in Adhesive Applications, Resolution Performance Products LLC, SC2267, Houston, TX, 2000.
16. Waterborne Epoxy Dispersions in Adhesive Applications, Resolution Performance Products LLC.
17. Young, G. C., "Modifying Latex Emulsions with Epoxy Resin Dispersions," *Adhesives Age*, September 1996, pp. 24–27.
18. Waterborne Epoxy Dispersions in Adhesive Applications, Resolution Performance Products LLC.
19. Schields, J., *Adhesives Handbook*, 3d ed., Butterworths, London, 1984.
20. Technical Bulletin, "EMA Bond Process," Ashland Inc., Norwood, NJ, 2004.
21. Turi, D., "Supplier/User Cooperation Yields Beneficial Induction Cure Epoxy," *Adhesives Age*, June 1993, p. 22.
22. Bolger, J. C., and Lysaght, M. J., "New Heating Methods and Cures Expand Uses for Epoxy Bonding," *Assembly Engineering*, March 1971, pp. 46–49.
23. Mittleman, E., "Fast Bonding Cuts Auto Costs," *IEEE Spectrum*, November 1977, pp. 73–75.
24. Stefanides, E. J., "Epoxy Cured by Induction Heating Gives Strong Sheet Metal Joint," *Design News*, June 22, 1987, pp. 102–103.
25. Stefanides, E. J., "Epoxy Cured by Induction Heating Gives Strong Sheet Metal Joint," *Design News*, June 22, 1987.
26. Turi, D., "Supplier/User Cooperation Yields Beneficial Induction Cure Epoxy," pp. 22–25.
27. Bolger, J. C., "New One Component Epoxy Insulation Compounds," *Insulation*, October 1969.
28. Bolger and Lysaght, "New Heating Methods and Cures Expand Uses for Epoxy Bonding."
29. Technical Bulletin, "Ancamine 2441," Air Products and Chemicals Company, Allentown, PA, 2004.
30. Li, C., and Dickie, R. A., "Bonding Adhesive Joints with Radio Frequency Dielectric Heating," *International Journal of Adhesion and Adhesives*, vol. 11, no. 4, October 1991, pp. 241–246.
31. Li and Dickie, "Bonding Adhesive Joints with Radio Frequency Dielectric Heating."
32. Li and Dickie, "Bonding Adhesive Joints with Radio Frequency Dielectric Heating."
33. Technical Bulletin, "MicroCure," Lambda Technologies, Raleigh, NC, 2004.
34. Gaskin, G. B., et al., "Electromagnetic Curing of Epoxy Adhesive Systems," *38th SAMPE International Symposium*, May 10–13, 1993.
35. Howell, B. F., "Modification of Epoxy Resins," in *Polymer Modification: Principles, Techniques, and Applications*, J. J. Meister, ed., Marcel Dekker, New York.
36. Gaskin, "Electromagnetic Curing of Epoxy Adhesive Systems."
37. Boey, F. Y. C., et al., "Microwave Curing of Epoxy Amine System—Effect of Curing Agent on Rate Enhancement," *Polymer Testing*, vol. 18, no. 2, 1999.
38. Paulauskas, F. L., et al., "Adhesive Bonding/Joining Via Exposure to Microwave Radiation," *27th International SAMPE Technical Conference*, September 9–12, 1995.
39. Soesatyo, B., et al., "Effect of Microwave Curing Carbon Doped Epoxy Adhesive–Polycarbonate Joints," *International Journal of Adhesion and Adhesives*, vol. 20, no. 6, 2000, pp. 489–495.
40. Mussler, R. W., "Weld-Bonding: The Best or Worst of Two Processes," *Industrial Robot*, vol. 29, no. 2, 2002, pp. 138–148.
41. *AdFAST Joining*, TWI World Center for Materials Joining Technology, Cambridge, UK, 2004.
42. Darwish, S. M. H., and Ghanya, A., "Critical Assessment of Weld-Bonded Technologies," *Journal of Materials Processing Technology*, vol. 105, 2000, pp. 221–229.
43. Chang, B. H., et al., "A Study on the Role of Adhesive in Weld-Bonded Joints," *Welding Research Supplement*, August 1999, pp. 275s–279s.
44. Anonymous, "Spot Welding Teams Up with Adhesives for Stronger Metal-to-Metal Bonds," *Product Engineering*, May 1975.
45. Anonymous, "Spot Welding Teams Up with Adhesives for Stronger Metal-to-Metal Bonds."
46. Anonymous, "Weldbonding Joins Auto Body Sheet," *Welding Design & Fabrication*, May 1979, pp. 80–81.

47. Schwartz, M. M., *Metals Joining Manual Book*, McGraw-Hill, New York, 1979, pp. 1–32.
48. Lohman, R. J., "Weldbonding Leads Growth of Structural Adhesives in Auto Market," *Adhesives & Sealants Industry*, May 1997, pp. 30–34.
49. Grimm, R. A., "Welding Processes for Plastics," *Advanced Materials and Processes*, March 1995.
50. Hauser, R. L., "Ultra Adhesives for Ultrasonic Bonding," *Adhesives Age*, March 1969.
51. Kwan, K. M., and Benatar, A., "Investigation of Non-Thermal Effects Produced by Ultrasonic Heating on Curing of Two-Part Epoxy Adhesive," Society of Plastic Engineers Annual Technical Conference, Dallas, TX, 2001.

CHAPTER 15
EFFECT OF THE SERVICE ENVIRONMENT

15.1 INTRODUCTION

For an adhesive or sealant bond to be useful, not only must it withstand the mechanical forces that are acting on it, but also it must resist the service environment or the chemical forces that are applied. Thus, one of the most important characteristics of an epoxy adhesive or sealant is its endurance to the operating environment.

Strength and permanence are influenced by many common environmental elements. These include high and low temperatures, moisture or relative humidity, chemical fluids, and outdoor weathering. Table 15.1 summarizes the relative resistance of various types of epoxy adhesives to common operating environments.

The effect of simultaneous exposure to both mechanical stress and a chemical environment is often more severe than the sum of each factor taken separately. Mechanical stress, elevated temperatures, and high relative humidity can be a fatal combination for certain adhesive formulations if all occur at the same time. Add to this the possible cyclic effects of each factor, and one can easily see why it is important to understand the effects of environment on the joint.

In this chapter, first the importance of environmental testing is reviewed. Then the effects of individual and combined environments on epoxy adhesives are considered. The root cause of environmental degradation is studied so that those seeking to design, manufacture, and use epoxy adhesives can make better judgments about the durability of bonded joints. This chapter points to common problem areas, suggests changes in the bonding system to provide greater bond durability, and acts as a foundation for further, more detailed investigation into this important area.

15.2 THE IMPORTANCE OF ENVIRONMENTAL TESTING

15.2.1 Singular and Coupled Stress Effects

Environmental consequences are so severe that it is usually necessary to test preproduction joints under conditions as close to those of the actual service environment as possible. The parameters that will likely affect the durability of a given joint are

- Maximum stress level
- Average constant stress level

TABLE 15.1 Relative Resistance of Epoxy Adhesives to Common Service Environments[1]

Environment	Epoxy + polyamine	Epoxy + anhydride	Epoxy + polyamide	Epoxy-phenolic	Epoxy-polysulfide	Epoxy-nylon	Phenoxy (thermoplastic)
Shear	2	2	2	1	2	1	2
Peel	5	5	2	6	2	1	3
Heat	3	1	6	1	6	6	4
Cold	5	4	2	3	2	2	3
Water	2	3	2	2	1	1	3
Hot water	2	3	6	2	6	6	4
Acid	2	2	3	2	2	2	3
Alkali	2	2	6	2	2	2	2
Oil, grease	2	2	2	3	2	2	3
Fuels	3	2	2	3	2	2	5
Alcohols	1	2	1	2	2	2	5
Ketones	6	6	6	6	6	6	
Esters	6	6	6	6	6	6	
Aromatics	1	2	3	2	2	2	6
Chlorinated solvents					6	6	

Key: 1, excellent; 2, good; 3, fair; 4, poor; 5, very poor; 6, extremely poor.

- Nature and type of environment
- Cyclic effects of stress and environment (rate and period)
- Time of exposure

In applications where possible degrading elements exist, candidate adhesives must be tested under simulated service conditions. Standard lap shear tests, such as ASTM D1002, which use a single rate of loading and a standard laboratory environment, do not yield optimal information on the service life of the joint. Important information such as the maximum load that the adhesive joint will withstand for extended periods and the degrading effects of various chemical environments are addressed by several test methods. Table 15.2 lists common ASTM environmental tests that are often reported in the literature.

Time and economics generally allow only short-term tests to verify the selection of the adhesive system relative to the environment. It is tempting to try to accelerate service life in the laboratory by increasing temperature or humidity, for example, and then to extrapolate the results to actual conditions. However, often too many interdependent variables and modes of potential adhesive degradation are in operation, and a reliable estimate of life using simple extrapolation techniques cannot be achieved.

For example, elevated-temperature exposure could cause oxidation or pyrolysis and change the rheological characteristics of the adhesive. Thus, not only is the cohesive strength of the adhesive weakened, but also its ability to absorb stresses due to thermal expansion or impact is degraded. Chemical environments may affect the physical properties of the adhesive and also cause corrosion at the interface; however, the adhesive may actually become more flexible and be better able to withstand cyclic stress. Exposure to a chemical environment may also result in unexpected elements from the environment replacing the adhesive at the interface and creating a weak boundary layer. These effects are dependent not only on the type and degree of environment but also on the specific epoxy adhesive formulation.

TABLE 15.2 ASTM Test Methods for Determining Environmental Resistance of Adhesives and Sealants

Test title	ASTM test method
Atmospheric Exposure of Adhesive Bonded Joint and Structures, Recommended Practice for	D 1828
Exposure of Adhesive Specimens to High Energy Radiation, Recommended Practice	D 1879
Integrity of Glue Joint in Structural Laminated Wood Products for Exterior Use, Test for	D 1101
Resistance of Adhesives to Cyclic Laboratory Aging Conditions, Tests for	D 1183
Effect of Bacteria Contamination on Permanence of Adhesive Preparations and Adhesive Bonds	D1174
Effect of Moisture and Temperature on Adhesive Bonds, Tests for	D 1151
Effect of Mold Contamination on Permanence of Adhesive Preparation and Adhesive Bonds, Test for	D 1286
Resistance of Adhesive Bonds to Chemical Reagents, Test for	D 896
Strength Properties of Adhesive in Shear by Tension Loading in the Temperature Range of –450 to –57°F, Test for	D 2557
Strength Properties of Adhesives in Shear by Tension Loading at Elevated Temperatures, Test for	D 2295

If there is only one parameter that changes due to environmental exposure, then the application of accelerated test techniques and analysis may yield useful information as to service life. In the electrical insulation industry, for example, Arrhenius plots are often used to predict end of life of insulating materials by simple extrapolation. This can be accomplished because insulation life is primarily dependent on temperature, and other factors are relatively minor. However, there are a multiplicity of aging phenomena that can occur within the adhesive, interface, and adherend and with each having a possible effect on the others. The formidable task of determining the end of life is one of the most difficult challenges in adhesive science and possibly the single item that most greatly inhibits the use of adhesives and sealants in most structural applications.

15.2.2 Accelerated Aging and Life Prediction

Predicting the service life of adhesives is a risky business. The most difficult question ever put to an adhesive consultant is, How long will the adhesive joint last in service? The problem is that an adhesive joint is not made up of just one element. It contains several elements, and some of them interact. In fact, in most adhesive joints at least five elements must be considered: substrate A, interface A, the adhesive, interface B, and substrate B. To understand and predict the rate of degradation of each of these elements is challenging, but it can be done. The most difficult failure situations to predict are those that result from interactive effects.

Thus, it is important to consider and evaluate the adhesive joint as a complete "system," but to do so we must generally separate these elements and look at their individual mechanisms of degradation. The following discussion considers mainly the bulk adhesive component of the joint. The process summarized below is the combination of work that has been described previously in several references.[2,3]

Life prediction methodology embraces all aspects of the numerous processes that could affect the function of the element—in this case the bulk adhesive. The first step is to define the function of the adhesive clearly enough for a failure criterion to be derived. This failure criterion may be an unacceptable reduction in tensile strength, time to creep failure under a given stress, reduction in modulus due to moisture ingression, increase in modulus due to oxidation, unacceptable crack depth, or a variety of other possible criteria. It is also important that the criteria be related to practical adhesive joint performance. This is where it is difficult, and one must presume, at least for this limited analysis, that the adhesive will fail via a bulk (cohesive) property.

After the failure criterion has been defined, the various processes that could cause this failure must be analyzed. For example, an increase in modulus could occur by thermal oxidation, increased postcure crosslinking, or the loss of plasticizer. Whatever the mechanism, each possible process needs to be identified and its rate characterized separately. Only then can interactions between different mechanisms be considered for life prediction.

Finally, the rate of change in the critical property must be measured relative to expected environments of different severity and time intervals. If measurement cannot be made at the service temperature in the time that is available (as is generally the case), then accelerated tests may be used at elevated temperature or increased frequency. However, it is extremely important that care be taken to match the accelerated test conditions to the service conditions in as realistic a way as possible. For example, if accelerated aging by elevated temperatures is being used, the temperature must not be so high as to begin a degradation mechanism that would not normally be seen in service.

After the rate of each process is determined under the accelerated conditions, then the rate at the actual service condition can be determined by extrapolation. The Arrhenius relationship (i.e., plotting the log failure rate against reciprocal temperature) is often used to accomplish this. This can also be expressed by the following equation for the time to failure t_f at absolute temperature T and constant stress σ.

$$\log t_f = C - \log T + a(1/T) - b(\sigma/T)$$

The terms C and b are constants, and $a = E_A/2.303R$, where E_A is the Arrhenius constant. All three terms can be found by graphical procedures.[4]

A comparison of experimentally determined failure times for different stress levels and those predicted by the above equation for epoxy-aluminum lap shear joints aged at 60°C and 95 percent RH is presented in Table 15.3. These results indicate that the reaction rate method is satisfactory for predicting the effects of temperature and stress on the lifetime of adhesive bonds, provided that failure is cohesive within the adhesive. This, of course, should be validated by prototype testing.

After each process is separately understood, interactions can be put together to provide a life prediction of the durability of the component under actual service conditions. One approach is to use multiple regression analysis to develop predictive equations for failure times in which several parameters (e.g., temperature, relative humidity, and stress) are treated as independent variables.[6]

The above analysis applies to degradation processes that relate to the bulk adhesive. Interfacial degradation processes such as corrosion can be similarly determined. Thermal and oxidative stability, as well as corrosion and water resistance, depends on the adherend surface as well as on the adhesive itself. Epoxy-based adhesives degrade less rapidly at elevated temperatures when in contact with glass or aluminum than when in contact with copper, nickel, magnesium, or zinc. The divalent metals have a more basic oxide surface than the higher-valence metal oxides and hence serve to promote dehydrogenation reactions, which lead to anion formation and chain scission.[7]

A diagram that one might use to illustrate a possible set of experimental data to represent all failure modes of an adhesive joint is presented in Fig. 15.1. When the data are closely analyzed and the extent of ultimate service life and proper safety margins are specified, the critical failure mode and time can be defined by identifying the "weakest link"—in this case the corrosion mechanism. If this predicted life is longer than the expected service life of the product, then the material specified for the adhesive joint can be qualified for use.

What is presented above is a very simplistic approach. Joint geometries, for example, may have a significant effect on the rate of degradation, again depending on the environment. As a result, geometric modeling and finite element analyses have been employed with durability studies to assist in life predictions.

TABLE 15.3 Comparison of Experimental and Predicted Failure Times for Epoxy-Aluminum Joints[5]

	Failure time t_f, min	
Applied stress σ, psi	Calculated	Experimental
1540	2455	7079
1760	1230	3548
1980	617	562
2200	316	138
2420	158	182

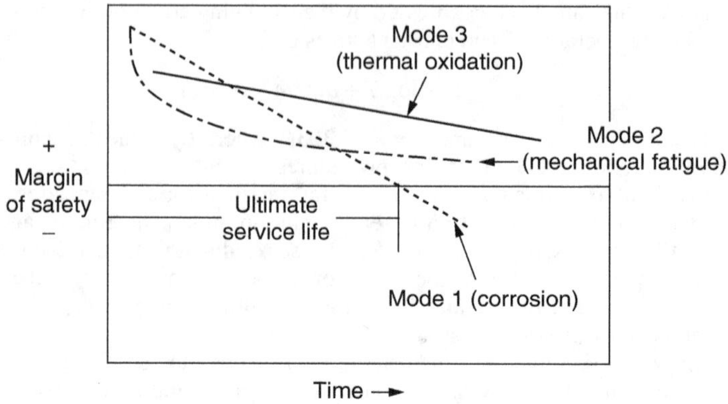

FIGURE 15.1 Determination of critical failure mode.[8]

Life prediction of adhesive joints in service is difficult because several degradation mechanisms may be operating simultaneously. However, a process sequence to use in estimating the service life of adhesive joints is as follows:

1. Define the extent of service condition variables.
2. Identify specific failure mode(s).
3. Determine the rate of change for each failure mode (may require accelerated aging and extrapolation).
4. Define the critical failure mode.
5. Establish the endurance limit of the "system"; establish its reliability.
6. Plan for a margin of safety.

15.3 HIGH-TEMPERATURE ENVIRONMENT

One of the most degrading elements for organic adhesives, including epoxies, is heat. All polymeric materials are degraded to some extent by exposure to elevated temperature. Not only do elevated temperatures lower short-term physical properties, but also properties will likely degrade with prolonged thermal aging at lower temperatures. Thus, several important questions need to be asked of an adhesive if high service temperatures are expected.

- What is the maximum temperature that the bond will be exposed to in service?
- What is the average temperature to which the bond will be exposed?
- How long will the bond be exposed to various temperatures?
- What is the rate of temperature change?
- What is the type of stress and stress levels that the bond will be exposed to during service? How does this relate to temperature?

Ideally, one would like to have a definition of the entire temperature-time relationship representing the adhesive's expected service history. These data would include time at various temperatures, number of temperature cycles, and rates of temperature change.

Certain epoxy adhesive formulations have excellent resistance to high temperatures over short durations (e.g., several minutes or hours). The short-term effect of elevated temperature is primarily one of increasing the molecular mobility of the adhesive. Thus, depending on the adhesive, the bond could actually show increased toughness but lower shear strength. Optimal toughness is often noted at service temperature near the adhesive's glass transition temperature. Certain polymers with lower glass transition temperatures will show softness and a high degree of creep at elevated temperatures.

However, prolonged exposure to elevated temperatures may cause several reactions to occur in the adhesive. These mechanisms can weaken the bond both cohesively and adhesively. The main reactions that affect the bulk adhesive material at elevated temperature are (1) oxidation and (2) pyrolysis. These reactions generally result in brittleness and loss of cohesive strength. Thermal aging can also affect adhesion by causing changes at the interface. These changes include internal stress on the interface due to shrinkage of the polymer, chemical reactions of the substrate, and reduced peel or cleavage strength because of adhesive brittleness.

If heating brings an adhesive above its glass transition temperature, the molecules will become so flexible that their cohesive strength will decrease. In this flexible, mobile condition, the adhesive is susceptible to creep and greater chemical or moisture penetration. Then on prolonged heating at an excessively elevated temperature, the following effects may be noticed:

- Split polymer molecules (chain scission), causing lower molecular weight, degraded cohesive strength, and low-molecular-weight by-products
- Continued crosslinking, resulting in bond embrittlement and shrinkage
- Evaporation of plasticizer, resulting in bond embrittlement
- Oxidation (if oxygen or a metal oxide interface is present), resulting in lower cohesive strength and weak boundary layers

Often thermal stability can be indicated by the weight loss that the bulk adhesive experiences in a high-temperature environment. However, this is not a very good method for predicting joint strength because it neglects the effect of joint geometry, adhesion, interface chemistry, etc. Figure 15.2 is a comparison of cured DGEBA epoxy with various other polymers in terms of weight loss after 15 h at 175°C.

This moderately good thermal resistance is one of the reasons for the wide use of epoxy adhesives in critical high-temperature environments. However, as described below, the elevated-temperature performance of epoxy adhesives is limited by the chemical nature of the epoxy molecule. Additives, modifiers, or improved curing agents only make minor improvements regarding thermal resistance.

Most organic adhesives including epoxies degrade rapidly at service temperatures greater than 150°C. However, several polymeric materials (e.g., polyimides, polybenzimidazoles) have been found to withstand up to 250 to 300°C continuously and even higher temperatures for a short-term basis. To use these materials, one must generally pay a premium in adhesive cost and be able to provide long, high-temperature cures, often with pressure. Long-term temperature resistance, greater than 250 to 300°C, can only be accomplished with inorganic or ceramic-based adhesives.

15.3.1 High-Temperature Requirements of the Base Epoxy Polymer

The base polymer, of course, is a key ingredient in a high-temperature epoxy adhesive system. For an adhesive to withstand elevated temperatures, it must have a high melting or softening point and resistance to oxidation. Materials with a low melting point, such as many of the thermoplastic adhesives, may prove excellent adhesives at room temperature;

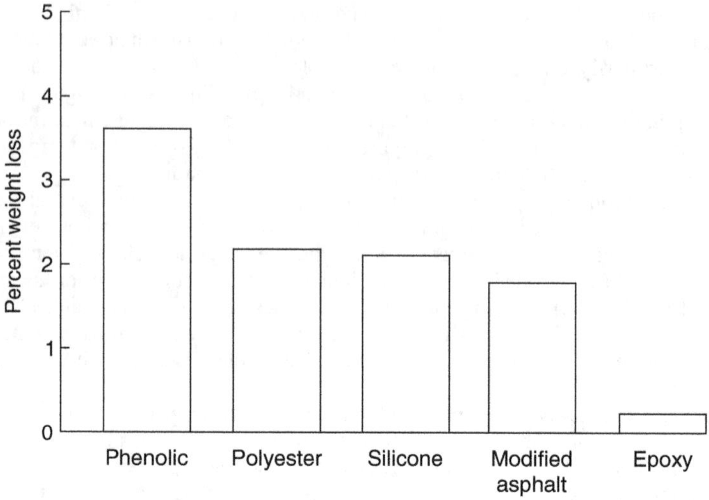

FIGURE 15.2 Comparison of cured DGEBA epoxy with various other polymers in terms of weight loss after 15 h at 175°C.[9]

however, once the service temperature approaches the glass transition temperature plastic flow results in deformation of the bond and degradation of cohesive strength.

Thermosetting adhesives, exhibiting no melting point, consist of highly crosslinked networks of macromolecules. Because of this dense crosslinked structure, they show relatively little creep at elevated temperatures and exhibit relatively little loss of mechanical function when exposed to either elevated temperatures or other degrading environments. Many of these materials are suitable for moderately high-temperature applications.

When one is considering thermosets, the critical factor is the rate of strength reduction due to thermal oxidation or pyrolysis. Thermal oxidation can result in chain scission or crosslinking. Crosslinking causes the polymer to increase in molecular weight, leading to brittleness and decreased elongation. Progressive chain scission of molecules results in losses of weight, strength, elongation, and toughness within the bulk adhesive. Figure 15.3 illustrates the effect of oxidation by comparing epoxy adhesive joints that are aged in both high-temperature air and inert gas (nitrogen) environments. The rate of bond strength degradation in air depends on the temperature, adhesive, rate of airflow, and even the type of adherend.

For reasons as described above, some metal adhesive interfaces are chemically capable of accelerating the rate of oxidation. For example, it has been found that nearly all types of structural adhesives exhibit better thermal stability when bonded to aluminum than when bonded to stainless steel or titanium (see Fig. 15.3).

Pyrolysis is simple thermal destruction of the molecular chain of the base polymer in the adhesive or sealant formulation. Pyrolysis causes chain scission and decreased molecular weight of the bulk polymer. This results in reduced cohesive strength and increased brittleness. Resistance to pyrolysis is predominantly a function of the intrinsic heat resistance of the polymers used in the adhesive formulation. As a result, many of the aromatic and multifunctional epoxy resins that are used as base resins in high-temperature adhesives are rigidly crosslinked or are made of a molecular backbone referred to as a *ladder* structure, as shown in Fig. 15.4.

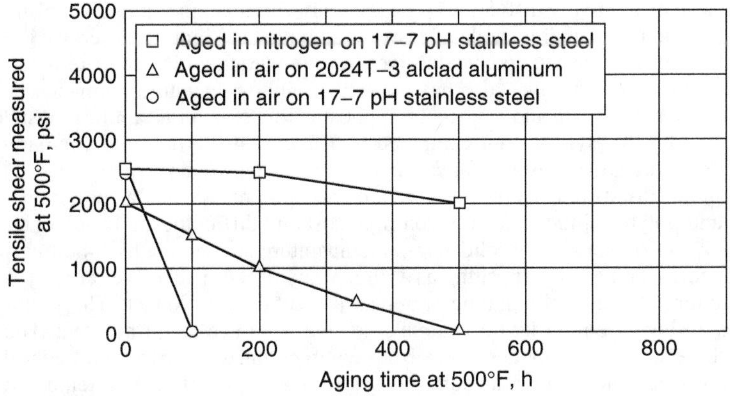

FIGURE 15.3 The effect of 260°C aging in air and nitrogen on an epoxy-phenolic adhesive.[10]

The ladder structure is made from aromatic or heterocyclic rings in the main polymer structure. The rigidity of the molecular chain decreases the possibility of chain scission by preventing thermally agitated vibration of the chemical bonds. The ladder structure provides high bond dissociating energy and acts as an energy sink to its environment. Notice in Fig. 15.4 that to have a complete chain separation (resulting in a decrease in the molecular weight), two bonds must be broken in the ladder polymer; whereas only one needs to be broken on a more conventional linear or branched chain structure.

To be considered a promising candidate for high-temperature applications, an adhesive must provide all the usual functions necessary for good adhesion (wettability, low shrinkage on cure, thermal expansion coefficient similar to that of the substrate, toughness, etc.), and it must also possess

1. High softening point or glass transition temperature
2. Resistance to oxidative degradation
3. Resistance to thermally induced chain scission

High-temperature adhesives are usually characterized by a rigid polymeric structure, high glass transition temperature, and stable chemical groups. The same factors also make

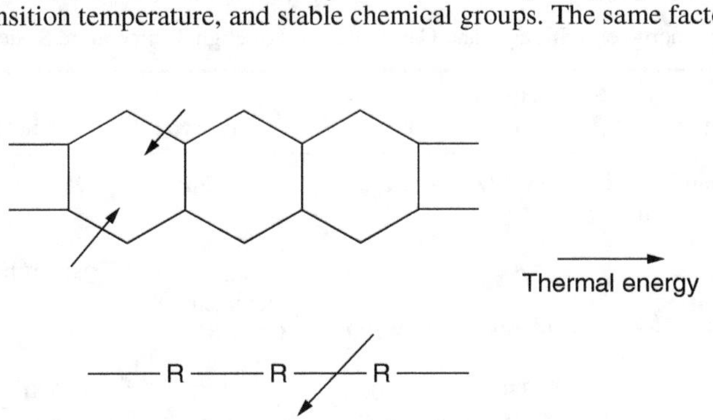

FIGURE 15.4 Degradation of ladder polymer and straight-chain polymer due to thermal aging.

these adhesives relatively difficult to process. Only certain epoxy-phenolic, bismaleimide, polyimide, and polybenzimidazole adhesives can withstand long-term service greater than 177°C. However, modified epoxy adhesives have moderately high short-term temperature resistance. Silicone adhesives also have excellent high-temperature permanence, but they exhibit low shear strength and may not be applicable for "structural" applications. Properties of several adhesive systems are compared in Table 15.4. Figure 15.5 compares various high-temperature adhesives as a function of heat resistance and thermal aging.

The major disadvantages of the more aromatic polymers, such as polyimides, amide-imides, and polybenzimidazole, are their high cost and difficulties in handling and curing. These polymers generally are solid at room temperature and require high-boiling-point solvents for formulating and depositing as a film. Many of these adhesives cure by condensation reactions, thereby eliminating water as a reaction by-product. The elimination of volatiles during cure in order to obtain a void-free bond is a major problem. Thus, only a few of these aromatic polymers have been commercialized as structural adhesives, and work continues at developing an adhesive system having superior high-temperature resistance but with the low cost and convenience of an epoxy.

A second major disadvantage with high-temperature adhesives is that aromatic chemical structures in the molecule result in chain stiffness. As a result, high-temperature adhesives are generally very rigid. They exhibit poor fracture resistance and peel strength at room temperature. Attempts to toughen such resins via the addition of flexible aliphatic chain segments usually involve a serious sacrifice in thermal stability and hot strength.

15.3.2 Additives and Modifiers Commonly Used in High-Temperature Adhesives

The high-temperature resins described above provide the main elements in the adhesive formulator's recipe. However, there are also additives, fillers, etc., that can further enhance the thermal properties of more conventional epoxy adhesives. These additional components improve thermal resistance by providing oxidation resistance, toughening, and control of bond line stress.

Oxidation Resistance. Oxidation in high-temperature adhesive joints involves reaction of the adhesive polymer with oxygen in the air as well as reaction with certain metal surfaces (e.g., ferrous metals). Oxidative degradation is initiated by the action of highly

TABLE 15.4 Short-Term Strength and Cure Properties of High-Temperature Structural Adhesives

Property	Modified epoxy	Epoxy-phenolic	Cyanoacrylate	Polyamide	Silicone rubber
Temperature range, °C	−55–177	−251–260	−40–246	−251–315	−73–232
Optimum cure condition					
Time, min	60	60	Seconds	90	24 h
Temperature, °C	177	177	Room temperature	288–371	Room temperature
Pressure, psi	10–50	10–100	Contact	50	Contact
Tensile shear, psi at					
20°C	4330	3800	3120	3300	275
175°C	2300	2500	970	—	—
260°C	—	2000	430	2300	275

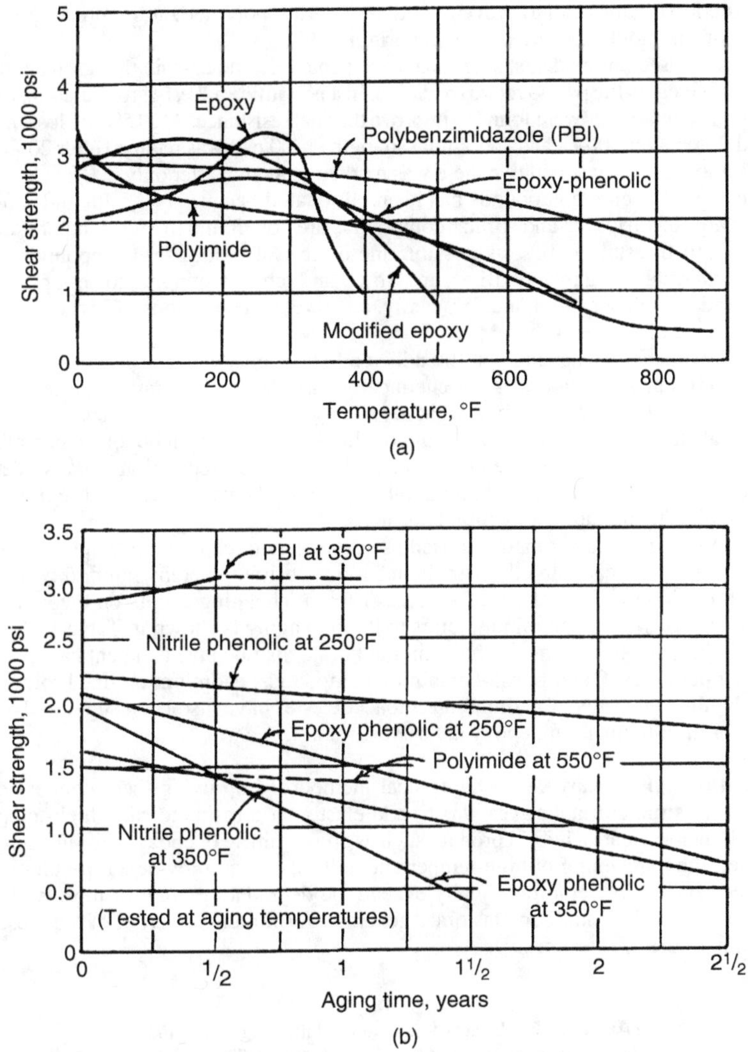

FIGURE 15.5 Comparison of (*a*) heat resistance and (*b*) thermal aging of several high-temperature structural adhesives.[11]

reactive free radicals caused by heat or metallic impurities. The function of an antioxidant is to prevent propagation or the reaction of these free radicals with oxygen to form unstable species.

Antioxidants should be included in high-temperature adhesive formulations in order to achieve optimum thermal aging properties. Antioxidants used in structural adhesives differ from those used to improve thermal stability of thermoplastic materials in that they must be less volatile, resistant to higher temperatures, longer-acting, and of course compatible with the base polymer. Antioxidants used in structural adhesives are generally of inorganic origin, whereas antioxidants used to prevent oxidation during polymerization, processing, or fabrication of thermoplastics are of organic origin.

Arsenic-based antioxidants, such as arsenic pentoxide and arsenic thioarsenate, had been used extensively in the past to retard oxidation. In a polyimide adhesive formulation, for example, arsenic compounds were found to improve thermal resistance. At 315°C no loss in strength was exhibited after 1000 h and substantial strength (1300 psi) was retained after 2000-h exposure. Without the arsenic additive there was marked reduction after only 200 h at 315°C.

The use of arsenic compounds has been eliminated because of health and safety concerns. Antimony trioxide and similar compounds are now found in high-temperature epoxy adhesives to forestall as best as possible the effects of oxidation. Compounds found to improve thermal aging include Bi_2O_3 and Sb_2O_3 and others belonging to Group V and having secondary valances of 3 and 5. Usually, additive concentrations of less than 1 percent by weight are effective.

Oxidative stability depends on the adherend surface as well as on the adhesive itself. Some metal adhesive interfaces are chemically capable of accelerating the rate of oxidation. For example, it has been found that nearly all types of structural adhesives exhibit better thermal stability when bonded to glass or aluminum than when bonded to stainless steel or titanium.[12] For any given metal, the method of surface preparation can also determine oxide characteristics, and hence bond durability. Thus, the use of primers is common practice with high-temperature structural adhesives.

Chelating agents are sometimes used as scavengers to capture undesirable metal ions. These compounds react directly with the metallic substrate, thereby inhibiting its catalytic effects on oxidation. The effect of several different chelating agents on the resistance of epoxy-phenolic bonded aluminum joints to thermal aging is shown in Table 15.5.

Chelating agents such as triethanolamine borate also behave as latent catalysts at elevated temperatures. Other boron-containing compounds, cadmium or zinc bromide diethylenetriamine, and salts of aluminum acetoacetic ester have also been suggested as curing agents for high-temperature epoxy adhesives.

Toughening. For many years, the typical method of improving the toughness of high-temperature structural adhesives was to add elastomeric resins to rigid high-temperature base polymer to create a hybrid product such as epoxy-nitrile or epoxy-polysulfide systems. However, the toughening of high-temperature adhesives can provide a difficult challenge, since the service temperatures usually exceed the degradation point of most rubber additives. Also, the addition of an elastomer generally resulted in lowering of the glass transition temperature of the adhesive.

TABLE 15.5 Effect of Several Chelating Agents on the Resistance of an Epoxy-Phenolic Adhesive to Thermal Aging[13]

Chelating agent at 1% by weight	Shear strength, psi*
None	670
n-Propel gellate	1074
Gallic acid	820
Acetyl acetone	985
Catechu	980
Aluminum triacetonylacetonate	960
Ethylenediamine	835

*Measured on aluminum adherends; tested at 23°C after aging 200 h at 286°C.

However, newer adhesives systems having moderate temperature resistance have been developed with improved toughness without sacrificing other properties. When cured, these structural adhesives have discrete elastomeric particles embedded in the matrix. The most common toughened hybrids using this concept are acrylic and epoxy systems. The elastomer is generally a vinyl- or carboxyl-terminated acrylonitrile butadiene copolymer. These adhesive formulations are discussed in detail in Chaps. 8 and 12.

Within the past several years, improvements in the toughening of high-temperature epoxies and other reactive thermosets, such as cyanate esters and bismaleimides, have been accomplished through the incorporation of engineering thermoplastics. Additions of poly(arylene ether ketone), PEK, and poly(aryl ether sulfone), PES, have been found to improve fracture toughness. Direct addition of these thermoplastics generally improves fracture toughness but results in decreased tensile properties and reduced chemical resistance.

Chemical functionalization of the thermoplastics was found to improve toughness without such detractions. High-molecular-weight resins based on amine-terminated PES oligomers or chain extension of bismaleimide resin with the same amine-terminated PES were found to have improved fracture resistance and reduced thermal shrinkage.[14] Also a mechanism was found to toughen cyanate esters by incorporating epoxy resins, which can react with the ester.[15]

The only high-temperature resin family that retains a moderate amount of flexibility is the polysiloxanes. A significant amount of research has been devoted to trying to marry the properties of siloxanes with epoxy resins to obtain less brittle, high-temperature adhesives. However, these efforts have yet to result in commercial adhesives systems.

Reducing Internal Stress. Internal stresses are common in joints made with high-temperature adhesives. These stresses can be due to

1. The high-temperature curing processes generally used
2. Temperature excursions and cycling between ambient and service temperature
3. Thermal shrinkage that occurs after the adhesive is aged for a period of time at elevated temperatures

Stresses caused by items 1 and 2 above are magnified by the mismatch in thermal expansion coefficients between the adhesive and the substrate. Incorporating fillers into the adhesive formulation can often reduce these stresses. Fillers also reduce the thermal shrinkage during aging by bulk displacement of the polymeric resin.

Flexibilizers generally cannot be used to counteract internal stress in high temperature adhesive because of their relatively low glass transition temperature and thermal endurance properties. However, most high-temperature adhesive systems incorporate metallic fillers (generally aluminum powder) to reduce the coefficient of thermal expansion and degree of shrinkage.

It is usually not possible to match the adhesive's coefficient of thermal expansion to the substrate, because of the high filler loadings that would be required. High loading volumes increase viscosity to the point where the adhesive could not be easily applied or wet a substrate. For some base polymers, filler loading values up to 200 parts per hundred (pph) may be employed, but optimum cohesive strength values are usually obtained with lesser amounts.

Metal fillers for high-temperature adhesives must be carefully selected because of their possible effect on oxidation, as indicated in the previous section. Carrier films, such as glass cloth, are generally used to facilitate the application of the adhesive, but they also provide a degree of reinforcement and lowering of the coefficient of thermal expansion. Thus, they reduce the degree of internal stress experienced at the joint's interface.

15.3.3 High-Temperature Epoxy Adhesive Formulations

Epoxy adhesives are generally limited to applications below 125°C. Figure 15.6 illustrates the aging characteristics of a typical epoxy adhesive at elevated temperatures. The epoxy adhesives using aliphatic polyamine hardening agents are not serviceable above 65°C. The aromatic diamine and monoanhydride cured products are usable at temperatures of 120 to 150°C. Certain epoxy adhesive formulations, however, have been able to withstand short terms at 260°C and long-term service at 150 to 175°C. These systems were formulated specifically for thermal environments by incorporation of stable epoxy coreactants or high-temperature curing agents into the adhesive.

Although conventional bisphenol A epoxy resins are limited to service temperatures under 125°C owing to their molecular structure, a number of approaches have been investigated in the development of high-temperature adhesives with epoxylike processability. Investigations of high-temperature epoxy adhesives have generally taken one of four courses to development:

1. Epoxy coreactants which will increase the temperature resistance of the system (e.g., epoxy-phenolic adhesives)
2. High-temperature curing agents
3. Special epoxy resins
4. Combinations of the above

The number of epoxide groups per molecule and the rigidity of the molecular structure are factors that affect the hot strength of the epoxy adhesive. Thus, epoxy novolac and glycidyl ethers of tetraphenolethane have become important resins for incorporating into high-temperature epoxy adhesives because of their multifunctionality.

Advantages of epoxy-based high-temperature adhesives relative to other adhesive types include relatively low cure temperatures, no volatiles formed during cure, low cost, and a variety of available formulating and application possibilities. The higher-temperature aromatic

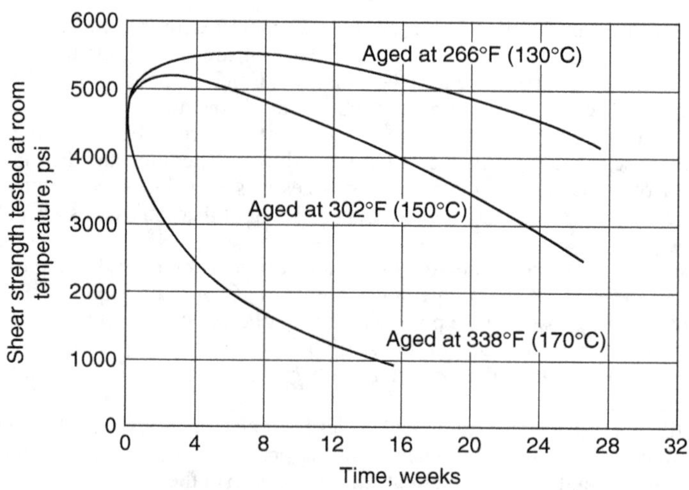

FIGURE 15.6 Effect of temperature aging on typical epoxy adhesive in air. Strength is measured at room temperature after aging.[16]

adhesives, such as polyimides, lose many of these advantages in favor of improved thermal-aging characteristics.

Epoxy Coreactants. One of the most successful epoxy coreactant systems developed thus far is an epoxy-phenolic alloy. The excellent thermal stability of the phenolic resins is coupled with the valuable adhesion properties of epoxies to provide an adhesive capable of 371°C short-term operation and continuous use at 175°C. The heat resistance and thermal-aging properties of an epoxy phenolic adhesive are compared with those of other high-temperature adhesives in Fig. 15.5. Epoxy-phenolic adhesives are generally preferred over other high-temperature adhesives, such as the polyimides and polybenzimidazoles, because of their lower cost and ease of processing.

Two of the first commercial epoxy-phenolic adhesive formulations are described in Table 15.6. These adhesives generally contain more phenolic resin than epoxy resin, and as a result, they have good high-temperature and chemical-resistant properties, but the adhesive suffers from relatively poor peel and impact strengths. The first epoxy-phenolic adhesives were developed at the Forest Products Lab (FPL) as a result of research in the 1950s aimed at high-temperature military aircraft adhesives. The cure cycle to yield optimum properties is about 1 h at 175°C. In addition to good hot strength and moderate thermal stability, epoxy-phenolic adhesives also exhibit good long-term resistance to moisture.

The formulation on the right of Table 15.6 is intended to be applied as a solution. The adhesives solution is brushed on both substrates, precured for 30 min at 93°C, and then pressed at 160°C for 30 min at 50 psi. The 442J formulation was designed to be applied to a glass fabric carrier. When cured, this adhesive provides 2000- to 3000-psi shear strength at room temperature and good strength at 260°C. Typical properties of 442J are compared to those of polyimide and polybenzimidazole adhesives in Table 15.7.

Chelating agents are commonly used in epoxy-phenolic adhesives to stabilize the metal interface. In the epoxy-phenolic formulations given above, quinolinate and gallate are used

TABLE 15.6 First Commercial Epoxy-Phenolic Adhesive Systems[17]

Component	Epoxy-phenolic 422J	Epoxy-phenolic solution coating
	Part by weight	
DGEBA epoxy resin (mol. wt. 1000)	100	
DGEBA epoxy resin (mol. wt. 2000)		100
Phenolic resin	50	500
Powdered aluminum	150	
Dicyandiamide	9	
Copper 8-quinolinolate	1.5	
Methyl ethyl ketone		100
Hexamethylenetetramine		6
Tensile shear strength, psi, on aluminum at		
–58°C	2300	1366
Room temperature	2300	1497
159°C	1900	
232°C		1328
260°C	1600	
316°C		1074
25°C after 100 cycles at 25–250°C	1000	
25°C after 200 h at 250°C	700	

as chelating agents. They provide increased bond durability above 200°C on substrates such as iron and steel.

The epoxy-phenolic adhesives have moderately good peel strength (10 to 15 lb/in), which is acceptable for many structural applications. However, the demands for high-peel-strength, high-temperature adhesives, such as aerospace honeycomb bonding, have resulted in formulations with the addition of thermoplastic modifiers in the epoxy-phenolic formulation. These materials generally sacrifice high-temperature resistance for improved peel strength.

Epoxy Resins. Where long-term service is required, most epoxies are limited to temperatures no higher than 125°C. With the increased demand for high-temperature and chemical resistance, new epoxy resins have been developed which have a higher epoxide and aromatic content. These resins are polyfunctional and are often cured with an aromatic diamine, such as methylene dianiline, metaphenylene diamine, or diamiodiphenylsulfone. Heat distortion temperature up to 165°C can be obtained. Several commercially available high-temperature epoxy resins are shown in Fig. 15.7. The high-performance resins can be crosslinked at elevated temperatures with either an aromatic amine or a catalytic curing agent.

Certain epoxy adhesives based on new multifunctional resins were found to provide strength retention to about 232°C. These epoxy resins have been the subject of research, but have not as yet reached significant commercial status.

Glycidyl polyether of tetraphenylethane is a standard, multifunctional high-temperature epoxy resin, and Fig. 15.8 illustrates the relationship between strength and temperature of this resin cured with two different amines and two different anhydrides. This resin has also been used with diaminodiphenylsulfone (DADPS) curing agents and polyvinyl acetal modifiers (to improve peel strength and toughness). Glycidyl ether of resorcinol has also been shown to have excellent hot strength even when cured at room temperature.

Epoxy novolacs are a class of epoxy adhesives that are often used in high-temperature application because of the resin's polyfunctionality that results in a high degree of crosslinking. They are made by reacting phenolic resins of the novolac type with epichlorohydrin. The response to curing agents is similar to that of the more standard DGEBA epoxy resins.

TABLE 15.7 High-Temperature Adhesives[18]

Tensile shear strength, psi	Polyimide (PI-1101)	Polybenzimidazole	Epoxy-phenolic
Initial at 25°C	2800	4000	4000
At 288°C after			
1 h	1600	3500	2100
10 h	1900	2500	0
1000 h	1400	0	0
At 371°C after			
1 h	1300	2000	900
10 h	1100	2000	0
600 h	1100	0	0
At 25°C after			
30 days, tap water	2100	1900	2600
30 days, 43°C, 100% RH	1900	1900	2900
30 days, salt spray	2000	1900	3200
7 days, isopropyl alcohol	2600	3300	4000

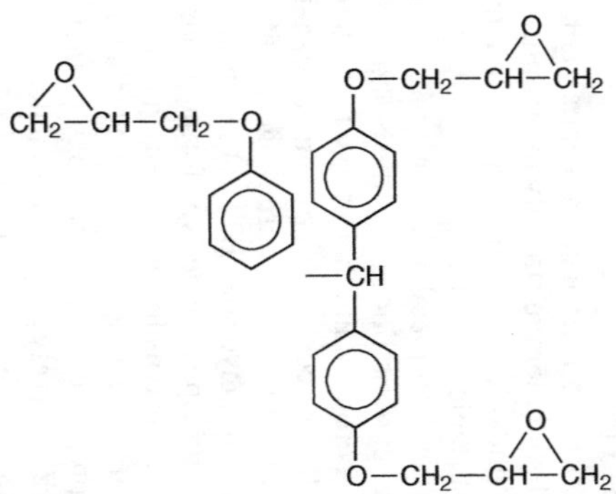

Tactix 742

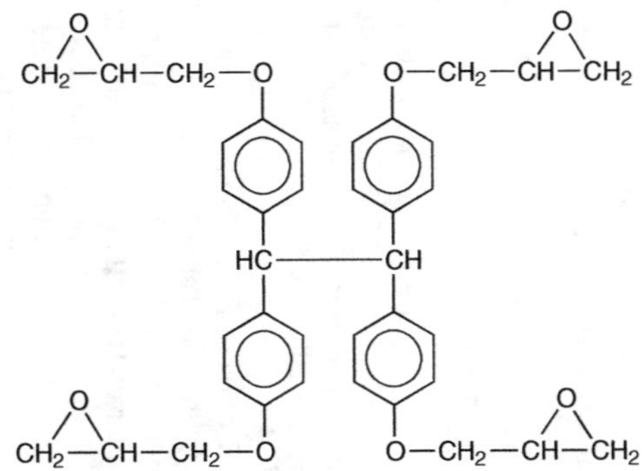

EPON 1031

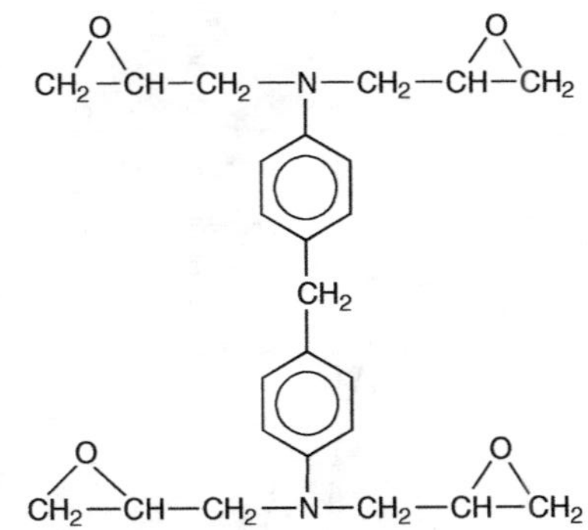

Araldite MY 720

FIGURE 15.7 High-performance epoxy resins: Tacktix 742 (Dow), EPON 1031 (Resolution Performance Products LLC), and Araldite MY 720 (Ciba Geigy).[19]

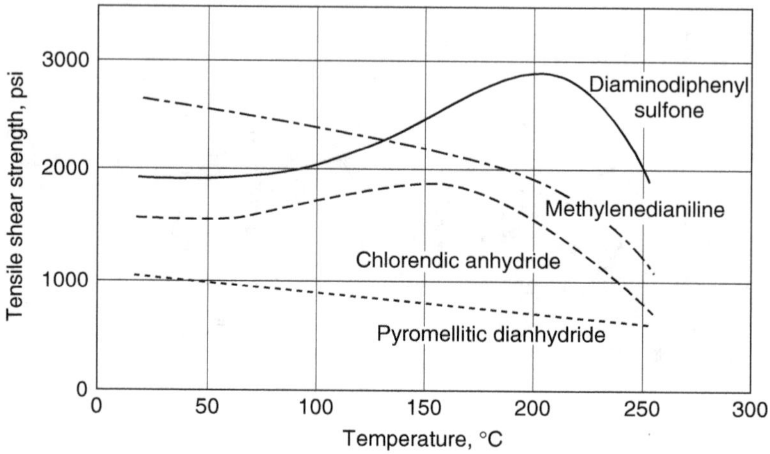

FIGURE 15.8 Strength-temperature relationship of glycidyl ether of tetraphenylethane cured with four different curing agents.[20]

For maximum heat resistance, pyromellitic dianhydride is often used. The composition and properties of metal-to-metal adhesives based on a combination of epoxy novolac and a bisphenol A epoxy resin have been described in Sec. 12.5.1. Depending on the cure temperature and the choice of amine curing agent, strength as high as 3000 psi at room temperature and over 1000 psi in the range of –55 to + 150°C is possible.

Epoxy Curing Agents. The selection of the curing agent can have a significant effect on the heat resistance of an epoxy adhesive. Table 15.8 shows the heat resistance of a conventional liquid DGEBA epoxy resin (EEW: 200) cured with various hardeners. The tensile shear strength is provided after no thermal aging and after 200-h aging at 260°C.

Anhydride curing agents give unmodified epoxy adhesives somewhat greater thermal stability than most other epoxy curing agents. Benzophenonetetracarboxylic dianhydride (BTDA), phthalic anhydride, pyromellitic dianhydride, and chlorendic anhydride allow greater crosslinking and result in short-term heat resistance to 232°C. Long-term thermal endurance, however, is limited to 150°C.

Although they are more difficult to formulate into epoxy adhesive systems, anhydride cured epoxies have somewhat better thermal stability than amine cured systems. Aromatic and cyclic anhydrides, such as phthalic anhydride, pyromellitic dianhydride, and chlorendic anhydride, provide the most stable structures.

A typical formulation for a metal-to-metal adhesive-sealant that is cured with a combination of phthalic anhydride and pyromellitic anhydride is shown in Table 12.6. Table 15.9 shows the high-temperature properties of another epoxy formulation cured with pyromellitic dianhydride. Epoxy formulations cured with pyromellitic dianhydride (PMDA) show good short-term thermal stability in the temperature range of 150 to 230°C.

Benzophenonetetracarboxylic dianhydride (BTDA) is another anhydride curing agent that provides good high-temperature epoxy adhesives formulations. Tables 12.12 and 15.10 show the effect of temperature on the tensile shear strength of a liquid DGEBA cured with BTDA on acid-etched aluminum substrates. Table 15.11 shows the effect of long-term thermal aging on the adhesive properties of several epoxy formulations cured with BTDA.

The BTDA-epoxy blends are relatively easy to formulate, and several high-temperature epoxy adhesive formulations have been developed aimed at specific end properties.

TABLE 15.8 Tensile Shear Strength (Aluminum-Aluminum) of DGEBA Epoxy Resin (EEW: 180–200) Cured with Various Hardeners[21]

		Tensile shear strength, psi							
		25°C		83°C		122°C		260°C	
Curing agent	Cure conditions	Unaged	Aged	Unaged	Aged	Unaged	Aged	Unaged	Aged
DETA	60 min, 95°C	200	195	1625	35	690	0	0	0
DEAPA	90 min, 95°C	1315	895	1450	890	1440	0	0	0
Polyamide (Versamid 115)	90 min, 95°C	2510	340	1285	50	290	20	0	0
BF_3-MEA	2 h, 160°C	1655	320	520	542	165	460	200	90
Phthalic anhydride	3 h, 160°C	1875	355	1500	210	1635	420	130	325
HHPA	3 h, 160°C	1875	845	2925	1435	585	1710	245	130
1,B DAPM	3 h, 150°C	1845	180	2145	170	2235	0	125	235
4,4′ MDA	2 h, 160°C	1775	230	1740	245	1750	900	225	80
Dicyandiamide	2 h, 175°C	2060	145	3025	225	1600	200	175	280

Specimens aged 200 h at 260°C.

TABLE 15.9 Tensile Shear Strength of Pyromellitic Dianhydride Cured Epoxy Adhesives at Elevated Temperatures[22]

Substrate	Treatment	Tensile shear strength, psi, at 25°C	150°C	232°C
Aluminum	Etched	2300	2600	900
Aluminum	Untreated	1800	1400	500
Cold rolled steel	Etched	1700	2000	1100
Cold rolled steel	Untreated	1200	1200	900

TABLE 15.10 Effect of High Temperature on Tensile Shear Strength of BTDA Cured DGEBA Epoxy Adhesive[23]

Test temperature, °C	Tensile shear strength, psi
23	2630
150	1600
260	1200
315	470

Formulation: Anhydride/epoxy = 0.6; 100 pph atomized aluminum, 3 pph Cab-O-Sil. Cure: 2 h at 200°C. Substrates: acid-etched alcad aluminum.

TABLE 15.11 Effect of Heat Aging on the Tensile Shear Strength of BTDA-Epoxy Adhesives[24]

Formulation	Anhydride/epoxy	Tensile shear strength, psi, at 260°C: Initial	After 500 h at 250°C	After 1000 h at 250°C
BTDA/EPON 828	0.6	1220	1090	1040
BTDA/80% EPON 828, 20% DER 438	0.6	700	1020	NA
BTDA/60% EPON 828, 40% DER 438	0.6	675	1080	NA

Formulation: 100 phh atomized aluminum, 3 pph Cab-O-Sil. Cure: 2 h at 200°C. Substrate: acid-etched alcad aluminum.

Fillers were found to significantly increase the room temperature tensile shear strength of BTDA cured epoxy adhesives. However, other modifiers such as phenolic or nylon resin did not significantly improve the peel strength, flexibility, or temperature resistance of the unmodified formulation.

15.4 LOW TEMPERATURES AND THERMAL CYCLING

Many applications for adhesives and sealants require high strength and durability at low temperatures. Many of these same applications also require resistance to thermal cycling between high and low operating temperatures. Unfortunately, the properties of adhesives and sealants at low temperatures are not as well studied or documented as they are at high temperatures.

Typical examples of adhesive and sealant applications requiring low-temperature or thermal cycling performance include

- Cryogenic equipment (advanced superconducting machines and processes exposed to liquid helium at –268°C and liquid nitrogen at –196°C)
- Refrigeration equipment
- Automotive parts (window applications, light fixtures) exposed to outdoor temperature
- Building and construction applications (architectural sealants, road repair compositions, bridge decking)
- Outdoor equipment (light fixtures, electrical enclosures, natural gas, oil transmission line components)
- Space and undersea craft (spacecraft propulsion systems, deep submergence vehicles)
- Electrical equipment (transformers, power electronic chips, conductor coatings)

Because of their ease of use and overall good adhesive properties, epoxy structural adhesives often find themselves in these types of applications.

Many of the applications listed above are affected by more than just diurnal (day/night cycling) or seasonal variations. Outdoor electrical equipment, such as light fixtures, transformers, etc., is subjected to the normal temperature variations that occur during the day or season; however, it is also exposed to the temperature variations that occur when energizing and deenergizing the equipment.

Applications in the electrical and electronics industries are often the most severely stressed due to thermal cycling because of the fast energization and deenergization of the devices employed. A housing for an outdoor lighting fixture in a shopping center parking light is a case in point. The design engineer must design the adhesive joint not only for the maximum temperature difference that occurs during the day and night, but also for the temperature difference that the joint will see when the light is turned on and off. This latter temperature cycling effect could have a degrading influence on the adhesive joint because of the rate of temperature change and the temperatures involved.

These low-temperature environmental effects can be significant factors that contribute to an adhesive system's durability and life. This section discusses the characteristics of epoxy adhesive joints exposed to low temperatures and to thermal cycling and suggests formulations for improving the resistance of adhesives and sealants to these conditions.

15.4.1 The Effect of Low Temperatures on the Joint Strength

Basically, there are two major considerations when one is formulating or selecting adhesives or sealants for low-temperature applications. The first is the effect of the low temperature on the bulk properties of the polymer, and the second is the effect of thermal cycling and resulting internal stresses on the joint interface.

In essence, for optimal properties at low temperatures, the adhesive joint

1. Must be able to absorb stresses and have a high fracture energy at the service temperatures (perhaps as low as −196°C)
2. Equally important, must be able to resist the transition and cycling from high to low temperatures

Bulk Polymer Properties. At its service temperature, the adhesive should be tough and strong. The mechanical energy caused by loading of the joint should be readily distributed throughout the adhesive. Generally, at very low temperatures, bonds are quite brittle and have reduced peel and impact strength, resistance to vibrations, and tensile-compressive fatigue life.

Acceptable bonds at reduced temperatures are partially attributed to molecular transition in the polymer from the plastic to the glassy state. At these transitions the backbone of the polymer chain can vibrate and actually move in restricted motion. These transitions are usually given Greek letters with the higher letters representing transitions at lower and lower temperatures. These are minor transitions compared to the well-known glass transition temperature T_g at which a polymer's physical properties change from that of a glass-like material to that of a tough or leathery material (see Table 3.9). In materials having low-temperature transitions, polymer chain motion can take place at temperatures far below the T_g.

A certain amount of molecular chain flexibility is desirable since it imparts resiliency and toughness to the polymer. This toughness is highest at temperatures around these transition regions. Therefore, polymers that are most resistant to low temperatures are those that have transition temperatures in the low-temperature region.

Unmodified epoxy adhesives have moderately high glass transition temperature depending on the curing agent used and the cure cycle. Therefore, they are generally not considered good candidates for low-temperature applications or applications where there is a great amount of thermal cycling.

It can be expected, then, that one of the major problems in adhesives technology is the development of adhesives that must withstand both elevated temperatures as well as periodic excursions to low temperatures. Several solutions have been developed. Certain adhesive systems, notably blends of epoxy resin with more elastic resins, have been formulated with a very broad glass transition temperature range or with multiple glass transitions at both high and low temperatures. These have found some success in the applications discussed in this chapter.

Flexibilizers and plasticizers can be used to lower the glass transition temperature and improve the low-temperature bond strength of epoxy adhesives. These will also provide a degree of elongation when there are differing coefficients of thermal expansion between the substrates and/or the adhesive. For adjusting the coefficient of thermal expansion, mineral or metallic fillers are normally used. With these modifications, good properties often can be obtained down to about the range of −20 to −40°C. However, optimal properties in the cryogenic temperature range (less than −100°C) dictate the base polymers recommended in Sec. 15.4.2.

Stresses at the Interface. When adhesive systems are used over a wide temperature range, the coefficient of thermal expansion becomes quite important in determining residual stresses in a joint. This is most important in low-temperature applications, since the modulus of elasticity generally increases with decreasing temperature and the adhesive is likely to be less forgiving to stress.

The most significant factors that determine the strength of an adhesive joint when used over a wide temperature range are the following:

- The coefficient of thermal expansion, especially as compared to the coefficient of thermal expansion of the substrates
- The elastic modulus of the adhesive at the service temperature
- The thermal conductivity of the adhesive and the thickness of the bond line

Residual stress resulting from thermal expansion or contraction is due to the differences in the thermal expansion coefficient between the adhesive and adherend and to temperature distribution in the joint due to differences in thermal conductivity.

The adhesive's thermal conductivity is important in minimizing transient stresses during cooling. This is why thinner bond thickness and adhesives or sealants with higher levels of thermal conductivity generally have better cryogenic properties.

Other opportunities for stress concentration in bonded joints that may be aggravated by low-temperature service include trapped gases or volatiles evolved during bonding, residual stresses in adherends as a result of the release of bonding pressure, and elevated-temperature cure (i.e., shrinkage and thermal expansion differences).

These internal stresses are magnified when the adhesive or adherend is not capable of deforming to help relieve the stress. Often these residual stresses are present at room temperature; however, the adhesive strength and resiliency are sufficient to resist them. On excursions to low temperatures, the residual stresses become magnified and could lead to bond rupture.

It is necessary that the adhesive retain some resiliency if the thermal expansion coefficients of the adhesive and adherend cannot be closely matched. At room temperature, a standard low-modulus adhesive may readily relieve stress concentration by deformation. At cryogenic temperatures, however, the modulus of elasticity may increase to a point where the adhesive can no longer effectively release the concentrated stresses. At low service temperatures, the difference in thermal expansion is very important, especially since the elastic modulus of the adhesive generally decreases with falling temperature.

15.4.2 Low-Temperature Epoxy Adhesives and Sealants

Most conventional low-modulus adhesives and sealants, such as polysulfides, flexible epoxies, silicones, polyurethanes, and toughened acrylics, are flexible enough for use at intermediate low temperatures such as −40°C. Low-temperature properties of common structural adhesives used for applications down to −129°C are illustrated in Fig. 15.9, and the characteristics of these adhesives are summarized in Table 15.12.

Flexibilizers are generally employed to improve low-temperature bond strength to −50°C. Good bond strength at cryogenic temperatures has been reported for liquid DGEBA epoxy cured with primary amine curing agents and diethylaminopropylamine, as illustrated in Fig. 15.10. Metaphenylene diamine (MPDA) cured epoxy adhesives have also shown good bond strength (3200 psi on aluminum) at −128°C.[26]

When the epoxy adhesive cannot be made flexible enough, the thermal conductivity and thermal expansion coefficient are controlled by appropriate fillers. General-purpose room temperature cured epoxy-polyamide adhesive systems can be made serviceable at low temperatures by the addition of appropriate fillers to control thermal expansion.

Modified epoxies are generally selected for lower-temperature applications. The unmodified epoxy-based systems are not as attractive for low-temperature applications as some

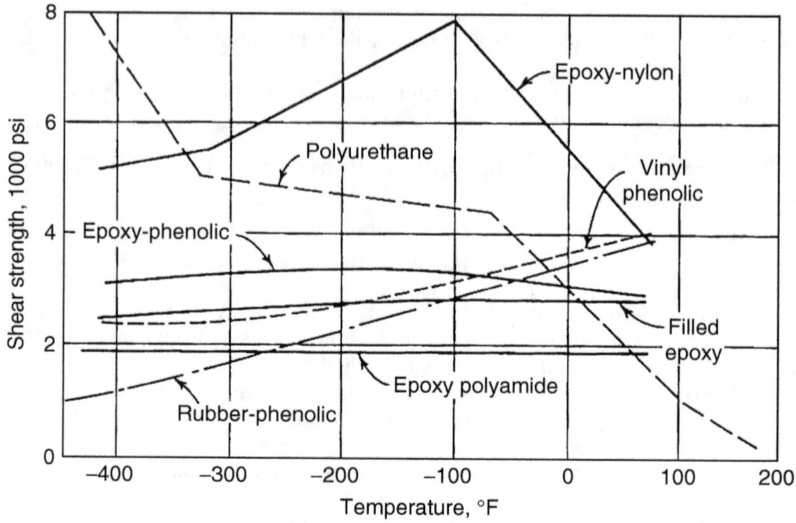

FIGURE 15.9 Properties of cryogenic structural adhesive systems.[25]

others because of their brittleness and corresponding low peel and impact strength at cryogenic temperatures. On a basis of lap shear strength at low temperatures (below −55°C), the epoxy formulations are ranked in decreasing order of shear strength as follows: epoxy-nylon, epoxy-polysulfide, epoxy-phenolic, epoxy-polyamide, and amine cured and anhydride cured epoxy.[28]

Epoxy-nylon adhesives are among the toughest and strongest adhesives and are usually produced as a dry B-staged film. Epoxy-nylon adhesives retain their flexibility and provide 5000 psi shear strength in the cryogenic temperature range. They also have useful impact resistance properties down to −147°C. Peel strengths can be as high as 40 lb/in of width, and resistance to vibration and fatigue is excellent. However, epoxy-urethane hybrid

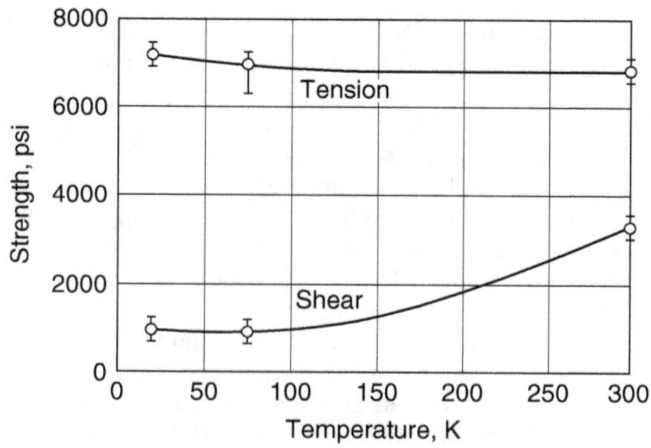

FIGURE 15.10 Tensile shear strength versus temperature for aluminum-filled DGEBA epoxy cured with DEAPA.[21]

TABLE 15.12 Properties of Low-Temperature Structural Adhesives

	Epoxy-nylon	Epoxy-phenolic	Epoxy-polyamide	Filled epoxy
Performance range	–252–80°C	–252–260°C	–253–80°C	–253–175°C
Advantages	Highest shear strength in cryogenic range	Uniform properties; moderate cost	RT cure; easy handling; low cost	Adhesion to many materials; easy handling
Limitations	Moderate peel strength at low temperatures, cannot be used at high temperatures, high cost	Low peel strength and impact resistance	Low peel strength; cannot be used at high temperature	Very low peel strength unless modified
Form available	Supported and unsupported films	Supported films	Two-part liquid and pastes	One- and two-part liquids and pastes
Applications	All types of structural bonding	Large area metal-to-metal bonds, sandwich construction	General-purpose	General-purpose

adhesives are generally noted to have higher peel strength in the cryogenic temperature range. The epoxy-polyurethane hybrid adhesives are especially promising because of the low-temperature properties that are provided by the polyurethane constituent.

Other useful epoxies include epoxy-polysulfide, epoxies modified with nitrile butadiene rubber, and epoxy-phenolic. Epoxy-phenolic adhesives are exceptional in that they have good adhesive properties at both elevated and low temperatures. Sandwich peel for these systems at –55°C is as high as 12 lb/in, and tensile shear strength is retained in the range of 3000 psi.[29] However, as shown in Fig. 15.9, although the tensile shear strength of epoxy-phenolic adhesives remains steady over a very broad temperature range, it is not as great as that for epoxy-nylon or polyurethane adhesives at cryogenic temperatures.

Next to the epoxy-nylons and epoxy-urethanes, the epoxy-polysulfide adhesives show the greatest lap shear tensile strength at temperatures below 0°C. Bonded etched steel substrates show a tensile shear strength of 2900 psi at room temperature, and this increases to 3400 psi when the temperature is reduced to –156°C. Epoxy-nylon and epoxy-polysulfide are the only adhesives that show an increase in strength as the temperature is significantly reduced.

The good low-temperature properties of epoxy-polysulfide adhesives are one reason why they have found significant use in the building and construction industry. Another reason is their excellent flexibility and ability to absorb stress and move with the thermal expansion and contraction of the substrates. However, in these applications the adhesive is generally not formulated for high tensile shear strength but rather for optimum elongation.

15.5 MOISTURE RESISTANCE

Moisture is the substance that often causes the greatest difficulties in terms of environmental stability for bonded or sealed joints. Water can be an exceptional problem because it is very polar and permeates most polymers. Other common fluids, such as lubricants and fuels, are of low or zero polarity and are not as likely to permeate and weaken adhesive or sealant joints.

Moisture can degrade a cured adhesive joint in three distinctive ways.

- Moisture can degrade the properties of the bulk adhesive or sealant itself.
- Moisture can degrade the adhesion properties at the interface.
- Moisture can also degrade physical properties and cause dimension changes of certain adherends.

The hostility of certain moisture environments can be seen in Fig. 15.11 Aluminum joints were bonded with room temperature curing epoxy-polyamide adhesive and aged in a hot, wet (tropical) environment and in a hot, dry (desert) environment. Excellent durability is achieved under dry conditions while significant degradation is caused by the wet conditions.

Ambient moisture can also affect certain types of uncured adhesive, either as it is being mixed and applied to a substrate or as it is stored in a container waiting to be applied. The degradation mechanism *before* cure of the adhesive is discussed in Chap. 3.

15.5.1 Moisture Degradation Mechanism

Even though epoxy adhesives are insoluble in water, they are not immune to water attack. Moisture can affect the strength of the epoxy adhesive joint by

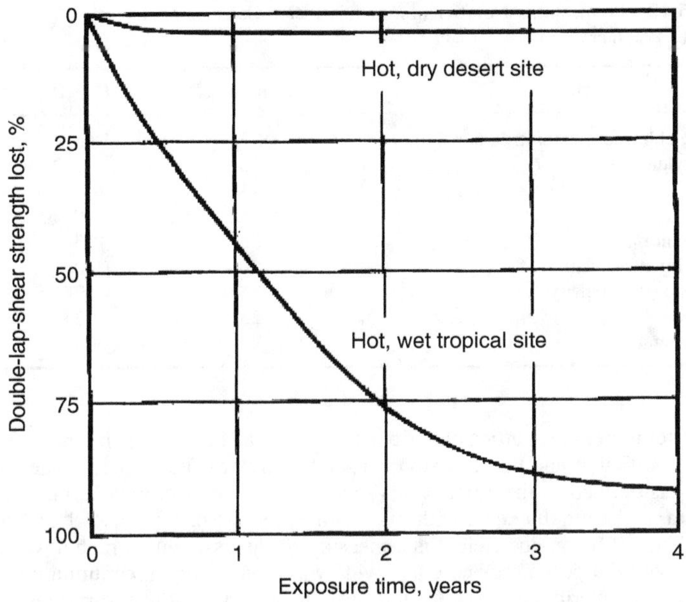

FIGURE 15.11 Effect of outdoor weathering on the strength of aluminum joints bonded with epoxy-polyamide.[30]

- Permeation of the adhesive
- Hydrolysis of some chemical bonds: breaking of bonds within the adhesive molecule and at the substrate-adhesive interface
- Hydration of substrate surfaces, causing bond rupture
- Swelling and increasing internal stresses leading to debonding
- Weakening of the interface between the adhesive resin matrix and internal fillers, if present

Moisture degradation of adhesive bonds occurs within the bulk adhesive material, at the adhesive-adherend interface, and within certain substrates. These degradation mechanisms are discussed below. Particularly insidious is the effect of the combined elements of moisture, stress, and temperature. Unfortunately, this synergistic effect occurs at relatively low temperatures, and such a service environment is common to many adhesive applications. For these reasons, this combined environment is given special focus in Sec. 15.5.2.

Effect on the Bulk Property of the Adhesive. Moisture can alter the properties of the bulk material by changing its glass transition temperature, inducing cracks, or chemically reacting with the polymer—a process called *hydrolysis*. But before these mechanisms occur, the moisture must first find its way into the bulk polymer.

Internal degradation within the bulk adhesive or sealant occurs primarily by absorption of water molecules into the polymer structure. All polymers will absorb water to some extent. Moisture can also enter by wicking along the adhesive-adherend interface or by wicking along the interfaces caused by reinforcing fibers and the resin. Deterioration may occur more quickly in a 100 percent relative humidity (RH) environment than in liquid water because of more rapid permeation of the vapor.

TABLE 15.13 Permeability Coefficient P and Diffusion Constant D of Water into Various Polymers[31]

Polymer	Temperature, °C	$P \times 10^{-9}$	$D \times 10^{-9}$
Vinylidene chloride-acrylonitrile copolymer	25	1.66	0.32
Polyisobutylene	30	7–22	NA
Phenolic	25	166	0.2–10
Epoxy	25	10–40	2–8
Polyvinyl chloride	30	15	16
Polymethyl methacrylate	50	250	130
Polyethylene (low-density)	25	9	230
Polystyrene	25	97	NA
Polyvinyl acetate	40	600	150

The water ingress properties of various polymers can be assessed by values of their permeability coefficient and the diffusion constant of water (Table 15.13). The permeability coefficient is defined as the amount of vapor at standard conditions permeating a sample that is 1 cm^2 and 1-cm thickness within 1 s with a pressure difference of 1 cmHg across the polymer. The diffusion coefficient is a measure of the ease with which a water molecule can travel within a polymer. There is a wide variation in the maximum amount of water absorbed by polymeric materials. Certain systems have very low absorption at lower temperatures, but the rate of absorption increases significantly at higher temperatures.

Epoxies, nitriles, and phenolics show relatively low diffusion rates and are less susceptible to moisture attack than most other polymers. As a result, these materials are often used in adhesive and sealant formulations where resistance to moisture is essential.

Water permeation generally lowers the glass transition temperature of the polymer by reducing the attractive forces between molecules. Data for certain epoxy adhesives cured with different curing agents are given in Table 15.14. The effect of absorbed water on the mechanical properties of cured epoxy adhesives is shown in Table 15.15. Water lowers tensile strength and modulus but increases elongation at break. Since no chemical linkages are broken, these properties generally recover fully when the polymer is dried unless irreversible hydrolysis has taken place.

Some polymeric materials, notably certain anhydride cured epoxies and ester-based polyurethanes, will chemically change or "revert" when exposed to humid conditions for a

TABLE 15.14 Effect of Water Immersion on the Glass Transition Temperature of Epoxy Adhesives Based on DGEBA[32]

	Glass transition temperature, °C		
Hardener	Dry	After initial uptake	After 10 months
DAPEE*	67	37	49
TETA†	99	86	111
DAB‡	161	143	157
DDM§	119	110	130

*Di-(1-aminopropyl-3-ethoxy) ether.
†Triethylenetetramine.
‡1,3-Diaminobenzene.
§4,4′-Diaminodiphenylmethane.

TABLE 15.15 Effect of Water on Mechanical Properties of Epoxy Structural Adhesives[33]

Exposure conditions	Weight gain, %	Tensile strength, MPa	Elongation at break, %	Modulus, MPa	Failure mode
		Epoxy/polyamide			
None	0	73	5	1880	Brittle
3 months at 65% RH	2.9	52	263	623	Ductile
5 days in water at 50°C	9.4	19	260	3.0	Rubbery
5 days in water at 50°C, then dried at 60°C for 2 days	3.3	76	5.7	1980	Brittle
		DGEBA epoxy/DAPEE* hardener			
None		41	7.1	1700	Ductile
24 h in water at 100°C		24	37	1020	Ductile
24 h in water at 100°C, then dried at 65°C for 2 days		53	6.8	1560	Ductile

*Di-(1-aminopropyl-3-ethoxy) ether.

prolonged period. Reversion or hydrolysis results in breaking of the molecular chains within the base polymer. This causes the adhesive or sealant to lose hardness and strength and in the worst cases transform to a fluid during exposure to warm, humid air. Figure 15.12 illustrates the degradation of polymer chains by hydrolytic reaction with water.

The rate of reversion, or hydrolytic instability, depends on the chemical structure of the base polymer, its degree of crosslinking, and the permeability of the adhesive or sealant. Certain chemical linkages such as ester, urethane, amide, and urea can be hydrolyzed. The rate of attack is fastest for ester-based linkages. Ester linkages are present in certain types of polyurethanes and anhydride cured epoxies. Generally, amine cured epoxies offer better hydrolytic stability than anhydride cured types.

Figure 15.13 illustrates the hydrolytic stability of various polymeric materials, determined by a hardness measurement after exposure to high-RH aging. A period of 30 days in the 100°C, 95 percent RH test environment corresponds approximately to a period from 2 to 4 years in a hot, humid climate such as that of southeast Asia. The hydrolytic stability of urethane potting compounds was not believed to be a problem until it resulted in the failure of many potted electronic devices that were noticed first during the military action in Vietnam in the 1960s.

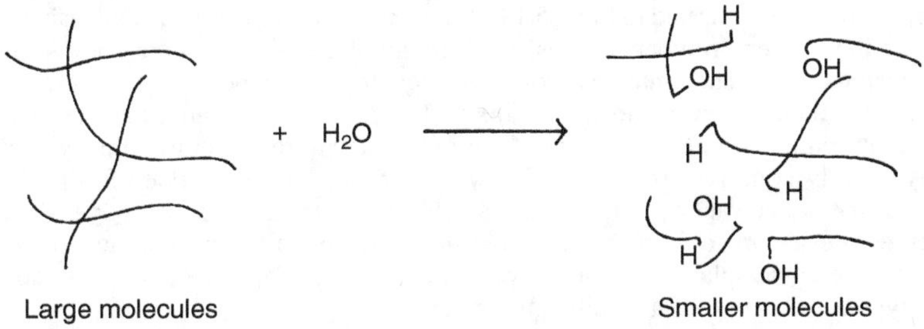

FIGURE 15.12 The degradation of polymer chains by reaction with water is called hydrolysis.[34]

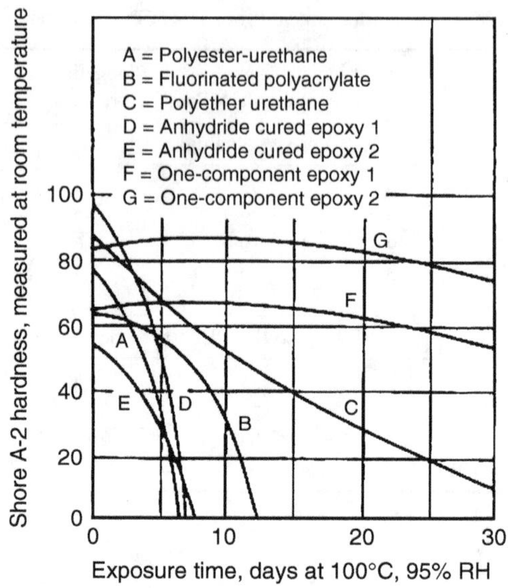

FIGURE 15.13 Hydrolytic stability of potting compounds. Materials showing rapid loss of hardness in this test soften similarly after 2 to 4 years in high-temperature, high-humidity climate zones.[35]

Reversion is usually much faster in flexible materials because water permeates them more easily. Hydrolysis has been seen in certain epoxy, polyurethane, and cyanoacrylate adhesives. The reversion rate also depends on the type and amount of catalyst used in the formulations and the degree of crosslinking. Best hydrolytic properties are obtained when the proper stoichiometric ratio of base material to catalyst is used.

In the case of conventional construction sealants, the polysulfides, polyurethanes, epoxies, and acrylics have all shown various degrees of sensitivity to moisture. Hydrolysis causes the breaking of bonds within the sealant. Thus, the bond strength decreases and cohesive failure results. However, before this occurs, the sealant usually swells and may cause deformation or bond failure before hydrolysis can completely take action.

Effect on the Interface. Water can also permeate the adhesive or sealant and preferentially migrate to the interfacial region, displacing the bulk adhesive material at the interface. This mechanism is illustrated in Fig. 15.14. It is the most common cause of adhesive strength reduction in moist environments. Thus, even structural adhesives that are not susceptible to the reversion phenomenon may lose adhesive strength when exposed to moisture.

The degradation curves shown in Fig. 15.15 are typical for an adhesive exposed to moist, high-temperature environments. The mode of failure in the initial stages of aging usually is truly cohesive. After 5 to 7 days, the failure becomes one of adhesion. It is expected that water vapor permeates the adhesive through its exposed edges. The water molecules are absorbed into the adhesive and preferentially concentrate on the metal adherend, thereby displacing the adhesive at the interface. This effect is greatly dependent on the type of adhesive and the adherend material.

Notice that the moisture resistance of the epoxy-amine adhesive is better than that of the heat-resistant phenolic in Fig. 15.15. The higher initial bond strength of the epoxy is likely due to better adhesion to metal and less internal stress. But the epoxy-amine adhesive

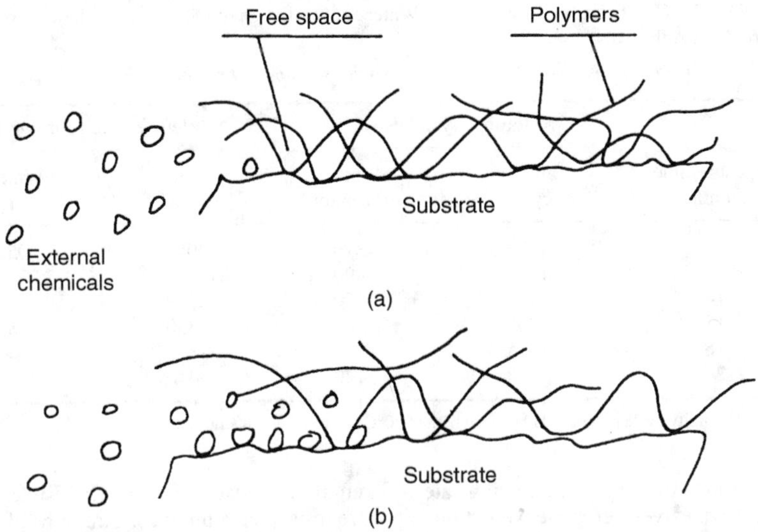

FIGURE 15.14 Competition between an adhesive and other chemicals for surface sites leading to displacement of the adhesive from the surface. (*a*) Adhesive adsorbed at surface sites; (*b*) adhesive displaced from the surface sites.[36]

shows a more rapid rate of decline in bond strength on exposure to the humid environment. This is likely due to the epoxy adhesive's higher degree of water absorption.

Certain adhesive systems are more resistant to interfacial degradation by moist environments than are other adhesives. Table 15.16 illustrates that a nitrile-phenolic adhesive does not succumb to failure through the mechanism of preferential displacement at the interface. Failures occurred cohesively within the adhesive even when tested after 24 months of immersion in water. A nylon-epoxy adhesive bond, however, degraded rapidly under the same conditioning owing to its permeability and preferential displacement by moisture.

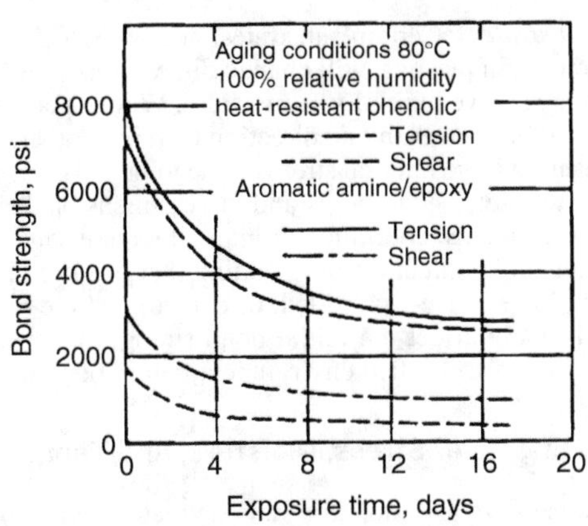

FIGURE 15.15 Effect of humidity on adhesion of two structural adhesives to stainless steel.[37]

TABLE 15.16 Effect of Humidity and Water Immersion on the Shear Strength of Two Structural Adhesives[38]

Substrate is aluminum treated with a sulfuric acid–sodium dichromate etch.

	Nylon-epoxy adhesive, psi		Nitrile/phenolic adhesive, psi	
Exposure time, months	Humidity cycle*	Water immersion	Humidity cycle*	Water immersion
0	4370	4370	3052	3052
2	1170	2890	2180	2740
6	950	1700	2370	2280
12	795	500	2380	2380
18	1025	200	2350	2640
24	850	120	2440	2390

*Humidity cycle of 93% RH between 65 and 30°C with a cycle time of 48 h.

The shape of the curve showing rate of strength degradation in Fig. 15.15 is common for most adhesives being weakened on exposure to wet surroundings. Strength falls most rapidly at the beginning of the aging process and then slows down to a low or zero rate of degradation. The initial rate and overall percent of degradation will vary with the adhesive and surface treatment.

There also appears to exist a critical water concentration within the adhesive below which water-induced damage of the joint will not occur. This also infers that there is a critical humidity for deterioration. For an epoxy system, it is estimated that the critical water concentration is about 1.35 to 1.45 percent and that the critical humidity is 50 to 65 percent.[39,40] Any loss in joint strength by the absorbed water can be restored upon drying if the equilibrium moisture uptake is below the critical water concentration.

Another way moisture can degrade the strength of adhesive joints is through hydration or corrosion of the metal oxide layer at the interface. Common metal oxides, such as aluminum and iron, can undergo hydration. The resulting metal hydrates become gelatinous, and they act as a weak boundary layer because they exhibit very inadequate bonding to their base metals. Thus, the adhesive or sealant used for these materials must be compatible with the firmly bound layer of water attached to the surface of the metal oxide layer.

Effect of the Bulk Substrate. Certain substrates, notably wood but also other materials such as laminates and certain plastics, will change dimensions significantly when exposed to variations in ambient relative humidity or moisture. Wood is an anisotropic material, so dimensional change will be greater in one direction than in another. Such change with relative humidity can result in large internal stress on the joint and even warpage of the assembly. The adhesive must be selected to withstand these dimensional changes.

Maximum bond performance and minimal internal stresses are sometimes achieved if the substrate has a moisture content during bonding that is close to the average moisture content anticipated during service—provided, of course, that the moisture retained in the substrate does not adversely affect the initial bond strength. This may require preconditioning of the substrates in a controlled environment before bonding.

15.5.2 Combined Effects of Stress, Moisture, and Temperature

The interaction of temperature and moisture causes greater degradation than can be attributed to either environment by itself. There is evidence to suggest that exposure to humid environments, at temperatures above about 60°C, can produce permanent damage in epoxy adhesives.[41] This can be explained by the formation of microcavities produced by clusters of water molecules and due to possible hydrolytic reactions.

Mechanical stress accelerates the effect of environment on the adhesive joint. A great amount of data is not available on this phenomenon for specific adhesive systems because of the time and expense associated with stress-aging tests. However, it is known that moisture, as an environmental burden, markedly decreases the ability of an adhesive to bear prolonged stress, especially at slightly elevated temperatures.

This stress-induced effect was first noted in the 1960s, when stressed and nonstressed aluminum lap shear joints were aged in a natural weathering environment in Florida.[42] Stress was applied by flexural bending of lap shear samples and keeping them in that state during the aging period. Depending on the type of adhesive, there was a significant degradation after 1 to 2 years due to stress-weathering, whereas all of the joints that were aged in the nonstressed condition showed little degradation.

Joints made with a flexible adhesive having a low glass transition temperature fail by creep of the adhesive at relatively short service times. Figure 15.16 illustrates the effect of stress-aging on specimens exposed to humidity cycling from 90 to 100 percent and simultaneous temperature cycling from 25 to 50°C. The loss of load-bearing ability of a flexibilized epoxy adhesive is great. The stress on this particular adhesive had to be reduced to 13 percent of its original static strength in order for the joint to last a little more than 44 days in the high-temperature, high-humidity environment.

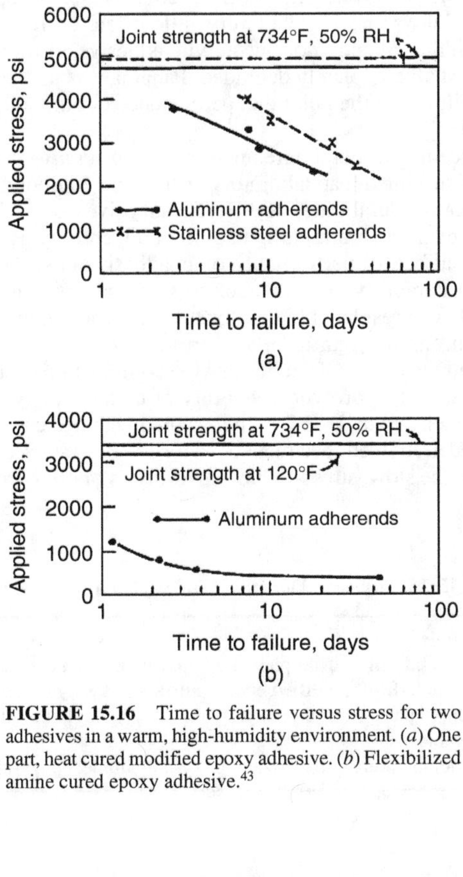

FIGURE 15.16 Time to failure versus stress for two adhesives in a warm, high-humidity environment. (*a*) One part, heat cured modified epoxy adhesive. (*b*) Flexibilized amine cured epoxy adhesive.[43]

TABLE 15.17 Comparison of Long-Term and Accelerated Exposures[44]

	Strength retention, %					
	Florida (3 years)		Panama (3 years)		Saltwater spray (30 days)	
Adhesive type	Stress	No stress	Stress	No stress	Stress	No stress
Vinyl phenolic (1)	60	97	87	83	97	95
Vinyl phenolic (2)	0	78	95	97	100	103
Vinyl phenolic (3)	0	62	75	96	97	100
Acrylic (1)	19	79	72	105	104	103
Acrylic (2)	0	24	15	54	16	94
One-component epoxy	0	57	0	79	106	120

The 8 × 9 in panels stressed by mounting on a steel bending frame to get 0.25-in deflection at the center of a 6-in span; $^1/_2$-in joint overlap is at center of span.

Table 15.17 shows the adverse effect of stress and tropical climates on aluminum joints bonded with various adhesives. If a 30-day saltwater spray were used as an accelerated test to determine the long-term performance of these adhesives in a tropical climate, it would be very misleading. Saltwater spray had very little effect on the strength of stressed or unstressed joints with the exception of one acrylic adhesive. However, the stressed specimens in Florida almost all completely degraded. Panama was not nearly as severe an environment. These data illustrate the point that permanence or durability must be tested in the specific environment.

Several sources seem in general agreement as to the *relative* durabilities of structural adhesives. Results of sustained load laboratory testing and outdoor weathering studies provide the same order for the durabilities of different adhesive classes. These are summarized in Table 15.18. However, such a ranking should be taken only as a general guide, since other factors can also affect the performance of an adhesive or sealant joint. A comparison of the sustained load durability of various classes of one-component epoxy adhesives is given in Table 15.19. The results obtained for two-part room temperature curing epoxies under the same conditions are significantly inferior.

It has also been shown in several studies that the combination of stress, temperature, and moisture can accelerate the hydrolytic instability of certain epoxy adhesives. In an FTIR study of the effect of moisture on DGEBA epoxy cured with nadic methyl anhydride, spectra changes were observed in stressed specimens aged for 155 days at 80°C and 100 percent RH.[47] This was attributed to the slow, stress-induced hydrolysis of ester groups. In another study,

TABLE 15.18 Relative Durabilities of Structural Adhesives[45]

Most durable
175°C cured film (nitrile-phenolic, vinyl-phenolic, novolac-epoxy)
120°C cured film (modified epoxy, nitrile-epoxy, nylon-epoxy)
Heat cured paste (nitrile-epoxy, nylon-epoxy, vinyl-epoxy)
Least durable
Room temperature cured paste (epoxy-polyamide, epoxy-anhydride)

TABLE 15.19 Effect of Adhesives on the Sustained Load Durability of FPL-Etched Aluminum Specimens[46]

Adhesive	Sustained load, psi	Environment	Effect on durability
Epoxy novolac	1550	60°C, 100% RH	All samples failed before 2 years
	900		No failures after 9 years
Epoxy-nitrile	1500	60°C, 100% RH	All samples failed before 2 months
	900		All samples failed before 6 months
Modified epoxy	1500	60°C, 100% RH	All samples failed before 2 years
	600		No failures after 7 years
Epoxy	800	38°C, 100% RH	Partial sample population failed before 8 years
	400		No failures after 8 years

irreversible losses in bond strength of aluminum-epoxy joints were attributed to stress-induced hydrolysis of primary chemical bonds.[48] However, it was observed that such stress-induced reactions do not readily occur below 90°C. Stress-induced hydrolytic effects appear to be greatest under high tensile stress and highly alkaline conditions, less severe under less caustic conditions, and negligible in the absence of stress.

15.5.3 Providing Moisture-Resistant Epoxy Adhesives

Resistance of adhesive joints to moisture can be improved either by preventing water from reaching the interface or by improving the durability of the interface itself. Several methods of minimizing degradation are possible.

1. Selection of a base polymer having a low water permeability and diffusion coefficient
2. Chemically modifying the adhesive or sealant to reduce water permeation
3. Incorporating inert fillers into the adhesive to lower the volume that can be influenced by moisture
4. Coating the exposed edges of the joint with very low-permeability resins
5. Using primers or chemically treating the substrate surface to improve adhesion and thus protect the interface from the intrusion of water
6. Chemically altering the substrate surface prior to bonding to provide for better adhesion and corrosion protection

Both the adhesive formulator and the end user can help to mitigate the degradation effects of moisture on the adhesive bond. The formulator can do this through proper selection of a base polymer along with additives and modifiers that will inhibit moisture ingress, and if moisture penetration does occur, it will not result in an irreversible degradation. The end user can prevent or delay degradation due to moisture by using the proper curing conditions and taking steps to protect the substrate surface (both before and after application of the adhesive) and interface.

Addressing the Bulk Adhesive. All polymers absorb moisture to some extent and in doing so become plasticized by the water molecules. The bulk properties are changed: glass transition temperature, tensile strength, and modulus are lowered, and elongation is increased.

In sealants, swelling and deformation are also noted. These properties generally recover on drying unless hydrolysis takes place. Even if the properties completely recover, the migration of moisture into the adhesive and possible preferential accumulation at the joint interface can result in loss of adhesion.

To improve moisture resistance, the formulator generally must operate on the bulk adhesive. This will occur mainly through modification or change in the bulk polymer and somewhat by modification or change of the fillers and additives in the formulation.

Base Polymer. The water uptake properties of polymers can be assessed by immersing films in water and recording increases in weight. The diffusion coefficient can be obtained from such data.[49] Values of the diffusion coefficient were given in Table 15.13. There is a wide variation in the maximum amount of water absorbed by polymeric resins. Certain systems have a very low absorption at the lower temperatures, but this breaks down at higher temperatures.

It is apparent that the first step toward formulation or selection of a moisture-resistant adhesive is to choose a base polymer that has low diffusion coefficient and permeability to water. This helps in two ways.

- It reduces the rate of diffusion of moisture to the critical interphase between the substrate and the adhesive.
- It reduces the effect on the bulk properties of the adhesive.

There have been many investigations to determine the best chemical structures to provide for resistance to moisture and hydrolysis. Attempts have been made to synthesize epoxy adhesives with improved water resistance by replacing some hydrogen atoms by fluorine.[50] However, the cost of processing of such materials has restricted commercial development. For electronic sealants, it is highly desirable to keep moisture from penetrating into critical areas. Hydrophobic polymers have been developed to accomplish this task. They are siloxyimides, fluorosilicones, fluoroacrylics, phenylated silicone, and silastyrene.

Other things being equal, water permeates fastest through flexible polymers. Hence the moisture pickup is generally much faster for flexible compounds than for more rigid types. Unfortunately those polymers that provide the best resistance to ingress of moisture tend to be rigid, highly crosslinked systems. They form brittle adhesives with relatively low peel and impact strengths. Microvoids can also be formed within the polymer by clusters of water molecules, and a mechanism of damage is evident in thermoset resins in which the rigid crosslinked structure does not allow the matrix to relax after microvoid formation.

Most structural adhesives are, therefore, formulated to provide the best compromise between environmental resistance and the desired mechanical properties. Experience has generally revealed that although the moisture ingress of the adhesive or sealant does affect the durability, it is seldom the dominant factor. Generally, of greater importance is how the moisture influences the adhesive-adherend interface region. Table 15.16 summarizes the moisture resistance and performance properties of some of the more common structural adhesives.

There is great variation within types of adhesives because of differences in chemical linkages, formulation parameters, crosslinking density, etc. For example, the room temperature curing two-part epoxy adhesives are usually considered to have a lower level of performance than the heat cured counterparts. However, investigators have shown that certain two-part, room temperature curing epoxy adhesives can demonstrate excellent durability even after 12 years of tropical exposure.[51,52] The performance of many two-part systems can be improved by a heat treatment following room temperature cure to optimize crosslinking.

Extensive information on durability of adhesive joints is more available on aluminum than on other substrates. Figure 15.17 illustrates typical results showing the effect of adhesive variations on joint durability during marine exposure. Vinyl-phenolics and nitrile-phenolics have

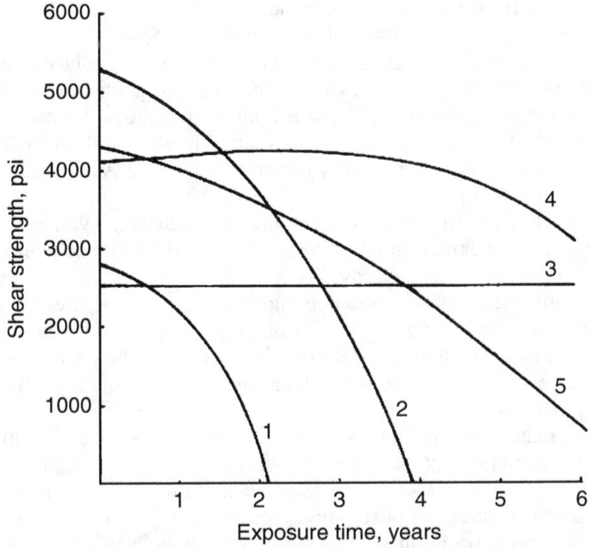

FIGURE 15.17 Effect of adhesive on the durability of etched 6061-T6 aluminum alloy joints exposed to a marine environment. (1) Two-part epoxy, (2) one-part epoxy, (3) nitrile-phenolic, (4) vinyl-phenolic, (5) vinyl-phenolic.[53]

an excellent history of joint durability and rank among the most resistant to environmental deterioration. However, the current trend is to use epoxy-based adhesives, which provide easier processing and higher initial strength.

Certain chemical linkages are susceptible to hydrolytic attack and, if present in an adhesive or sealant, are potential sites for irreversible reaction with water that has diffused into the joint. Such hydrolytic (chemical) degradation causes a permanent reduction in the cured physical properties. The functional groups present in the chains are hydrolyzed, resulting in both chain breaking and loss of crosslinking.

The hydrolytic attack on an ester linkage in the presence of water at high pH is an important example of this mode of degradation. The attack initiates on the electron-deficient atom in a highly polarized bond, as on carbonyl carbon:

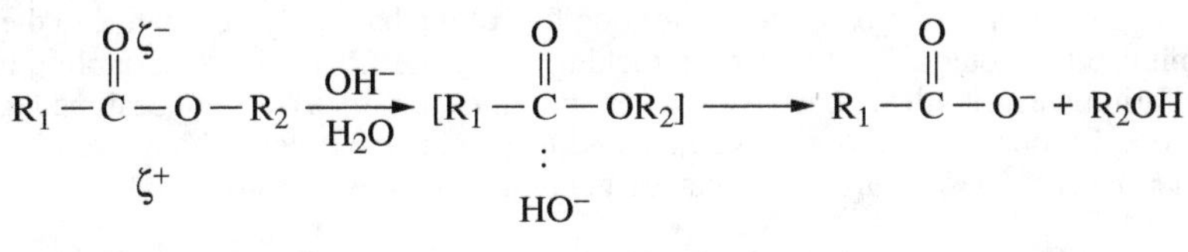

Substitutions of electron withdrawing groups for the aliphatic R_1 and R_2 groups will delocalize the charge on the carbonyl carbon, leading to reduced rates of hydrolysis. Thus, hydrolytic stability increases in the order: aliphatic ester groups < aromatic ester groups < urethane groups < aliphatic amide < urea groups < aromatic amide groups.[54]

Since the reaction between an epoxy resin and an acid anhydride curing agent also produces an ester linkage, anhydride cured epoxies have poorer hydrolytic stability than do amine cured epoxies. It has also been shown that reversion rates of urethanes and anhydride cured epoxies increase as the amount of tertiary amine or other base catalyst increases.[55]

Other things being equal, the rate of hydrolytic attack on any adhesive increases rapidly as the crosslinking density decreases. Epoxy resins cured with flexibilizing anhydrides, derived from long-chain aliphatic acids, will hydrolyze rapidly. Epoxy resins cured with short-chain, highly functional acid anhydrides such as methyl nadic anhydride yield a rigid network having a much lower permeability for water and a greatly reduced rate of hydrolytic attack under alkaline conditions.

Many attempts have been made to develop epoxy adhesive systems with outstanding hydrophobicity to improve their moisture resistance. Barrie, Johncock, and coworkers considered the halogenation of tetraglycidyl 4,4′-diaminodiphenylmethane resin cured with a diaminodiphenylsulfone (DDS).[56] Goobich and Marom[57] investigated the effect of introducing bromine into an epoxy resin formulation employing a brominated coreactant into a system similar to that investigated by Barrie. These modifications were shown to reduce water uptake; however, glass transition temperature and the processability of the adhesive were negatively affected.

Perhaps the greatest success at hydrophobic epoxy adhesive development was the development of fluorinated epoxies together with compatible curing agents. These adhesives resulted in equilibrium water concentrations as low as 0.2%. However, for various reasons these adhesives have not been commercialized.

Additives. Formulation additives do not always have a positive effect on moisture resistance. Therefore, selection of additives must go through a similar assessment process as the base polymer with respect to hydrolytic stability. Any polymeric additive used to modify the properties of the base resin should be considered for its possible effect on moisture resistance and environmental durability.

Sometimes filled adhesives will show better resistance to moisture resistance than unfilled adhesives simply because incorporating inert fillers into the adhesive lowers the organic volume that can be affected by moisture. Aluminum powder seems to be particularly effective, especially on aluminum substrates. The filler can provide a reduction of shrinkage on cure, a reduction of the thermal expansion coefficient, and a reduction of the permeability to water and other penetrants. However, fillers do not always produce more durable bonds.

Many film adhesives have a supporting carrier or reinforcement fabric incorporated into the adhesive to improve handling of the film and provide control of bond line thickness. The carriers are usually glass, polyester, or nylon fabrics of knitted, woven, or nonwoven construction. The difficulty with such carriers is that they can provide an effective way of moisture entering the bulk of the adhesive. Moisture can wick along the fiber-adhesive interface. Nylon carriers should especially be reviewed since they have a strong tendency to absorb moisture.

When fibrous carriers or fillers are needed in structural adhesive formulation, coupling agents are often used to provide for better bonding of the base polymer matrix to the fiber, to minimize the potential for moisture wicking along the fiber surface. Coupling agents, either formulated into the adhesive or coated onto the carrier surface, are designed to bond to hydroxyl groups on the fiber's surface and to be reactive toward the appropriate bulk polymer matrix. These coupling agents are generally organosilanes.

Addressing the Interface. Methods that are available to improve the durability of the interface itself mainly center on surface preparation of the substrate and/or the use of primers or coupling agents. These actions are generally taken by the end user, and they are usually most beneficial on metallic surfaces that are prone to degradation by corrosion. However, simple steps, such as coating the exposed edges of the joint with sealants having very low permeabilities to inhibit wicking of moisture along the interface, have also produced positive results.

Strong chemical bonds between the adhesive and adherend help stabilize the interface and increase joint durability. Aluminum joints formed with phenolic adhesives generally exhibit better durability than those with epoxy adhesives. This is partially attributable to strongly interacting phenolic and aliphatic hydroxyl groups that form stable primary chemical bonds across the interface.

Primers. Primers tend to hinder adhesive strength degradation in moist environments by providing corrosion protection to the adherend surface. A fluid primer that easily wets the interface presumably tends to fill in minor discontinuities on the surface. Organosilane, organotitanate, and phenolic primers have been found to improve the bond strength of many adhesive systems.

Silanes and other coupling agents can be applied to various substrates or incorporated into an adhesive primer to serve as hybrid chemical bridges to increase the bonding between organic adhesive and inorganic adherend surfaces. Such bonding increases the initial bond strength and also stabilizes the interface to increase the durability of the resulting joint.

Some primers will improve the durability of the joint by protecting the substrate surface area from hydration and corrosion. These primers suppress the formation of weak boundary layers that could develop during exposure to wet environments. Primers that contain film-forming resins are sometimes considered interfacial water barriers. They keep water out of the joint interface area and prevent corrosion of the metal surfaces. By establishing a strong, moisture-resistant bond, the primer protects the adhesive-adherend interface and lengthens the service life of the bonded joint.

However, moisture can diffuse through any polymeric primer, and eventually it will reach the interface region of the joint. Therefore, the onset of corrosion and other degradation reactions can only be delayed by the application of a primer unless the primer contains corrosion inhibitors or it chemically reacts with the substrate to provide a completely new surface layer that provides additional protection.

Representative data are shown in Fig. 10.5 for aluminum joints bonded with an epoxy film adhesive and a standard chromate-containing primer. Until recently, standard corrosion-resistant primers contained high levels of solvent, contributing to high levels of VOCs, and chromium compounds, which are considered to be carcinogens. As a result, development programs have been conducted on waterborne adhesive primers that contain low VOC levels and low or zero levels of chromium.

Epoxy-based primers are commonly used in the aerospace and automotive industries. These primers have good chemical resistance and provide corrosion resistance to aluminum and other common metals. Polysulfide-based primers have been developed for applications where a high degree of elongation is necessary. These systems are used where the joint is expected to encounter a high degree of flexing or thermal movement.

Resins, curing agents, and additives used in primer formulation are much like adhesive or sealant formulations except for the addition of solvents or low-viscosity resins to provide a high degree of flow.

Chemical Surface Modification. In considering the interface, one must contemplate not only the possibility of moisture disrupting the bond but also the possibility of corrosion of the substrate. Corrosion can quickly deteriorate the bond by providing a weak boundary layer before the adhesive or sealant is applied. Corrosion can also occur after the joint is made and, thereby, affect its durability. Mechanical abrasion or solvent cleaning can provide adhesive joints that are strong in the dry condition. However, this is not always the case when joints are exposed to water or water vapor. Resistance to water is much improved if metal surfaces can be treated with a protective coating before being bonded.

Approaches for the development of water-resistant surface treatments include application of inhibitors to retard the hydration of oxides or the development of highly crystalline oxides as opposed to more amorphous oxides. Standard chemical etching procedures, which remove surface flaws, also result in improved resistance to high humidity.

A number of techniques have been developed to convert corrosion-prone, clean surfaces to less reactive ones. Three common conversion processes are phosphating, anodizing, and chromating. These processes remove the inconsistent, weak surface on metal substrates and replace it with one that is strong, permanent, and reproducible.[58] Figure 15.18 shows the effect of various pretreatments on the durability of aluminum alloy-epoxy joints subjected to aging in water at 50°C.

One very beneficial chemical pretreatment treatment for aluminum substrates is phosphoric acid anodization (PAA), which provides an oxide coating that is inherently hydration-resistant. Its stability is due to a layer of phosphate incorporated into the outer Al_2O_3 surface during anodization.

The type of conversion process will depend on the substrate, the nature of the oxide layer on its surface, and the type of adhesive or sealant used. The formation of a nonconductive coating on a metal surface will also minimize the effect of galvanic corrosion.

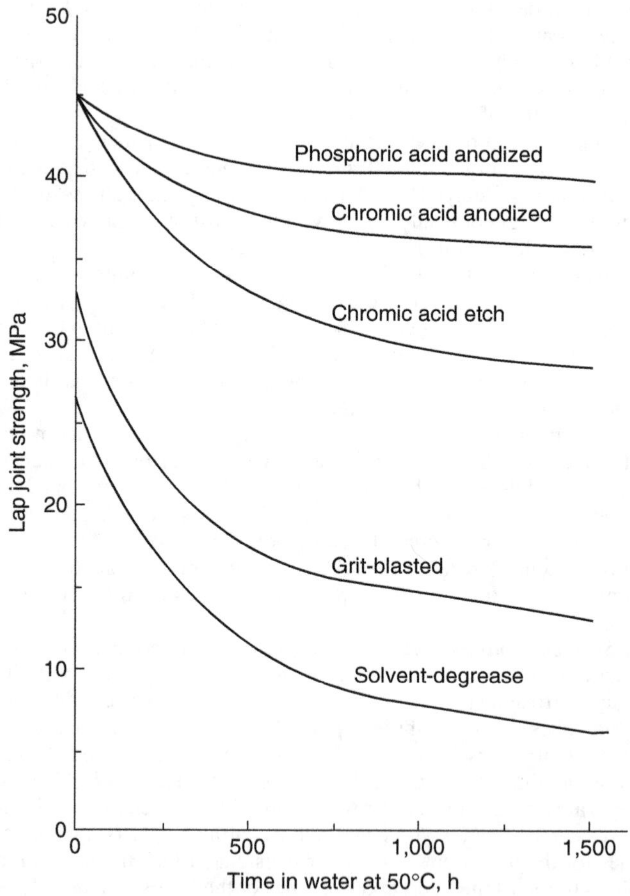

FIGURE 15.18 Effect of surface pretreatment on the durability of aluminum-epoxy joints subjected to accelerated aging in water at 50°C.[59]

15.6 OUTDOOR WEATHERING

Epoxy adhesives and sealants are generally not significantly affected by simple outdoor weathering. However, there are certain circumstances that could affect the permanence of joints exposed to outdoor service. It is important that these be considered early in the design of the adhesive joint and selection of materials.

The outdoor durability of epoxy bonded joints is very dependent on the type of epoxy adhesive, specific formulation, nature of the surface preparation, and specific environmental conditions encountered in service. The data shown in Fig. 15.19, for a two-part room temperature cured polyamide epoxy adhesive with a variety of fillers, illustrates the differences in performance that can occur due to formulation changes. Excellent outdoor durability is provided on aluminum adherends when chromic-sulfuric acid etch or other chemical pretreatments are used.

Differences in performance are also noted depending on the specific nature of the environment. Differences can be expected in joint durability when bonds are exposed to an aggressive wet-freeze-thaw cycle, marine seacoast, or inland environments. Outdoor weathering conditions are often classified by one of the following exposure conditions:

1. Temperate, moderate climate, normally at northern latitudes, often in or close to industrial environment
2. Industrial environment, strong presence of acid, ozone, and other by-products of production
3. Desert, hot, dry climate
4. Tropical, hot, wet climate including jungle exposure
5. Marine, corrosive seacoast environment

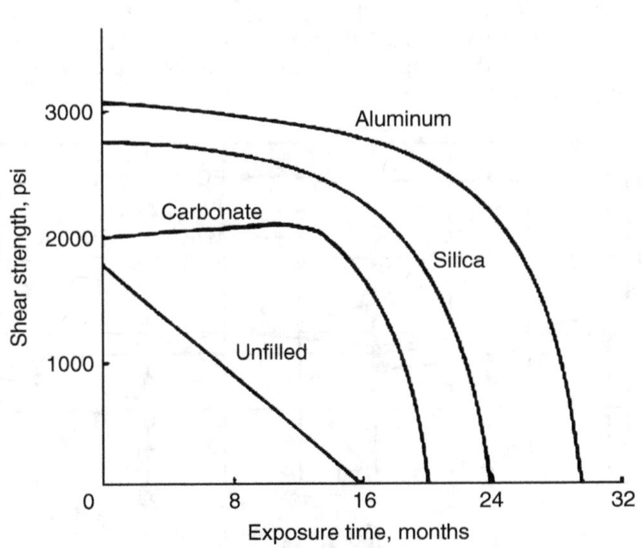

FIGURE 15.19 Durability of polyamide-epoxy two-part adhesives exposed to marine atmosphere (fillers are indicated).[60]

The factors that influence the durability of epoxy adhesives in outdoor weathering conditions are

- Solar radiation (usually uv)
- Moisture (dew, humidity, rain)
- Heat (surface temperature of the material)
- Pollutants (ozone, acid rain)
- Microbiological attack
- Salt water and salt spray

Because these factors vary so widely over the earth's surface, the weathering of adhesives is not an exact science. Furthermore, most materials are weathered by a combination of these factors, and the contribution of each is dependent on the adhesive formulation.

By far, the most detrimental factors influencing adhesives aged in a nonseacoast environment are heat and humidity. The reasons why warm, moist climates degrade many adhesive joints were presented in the last section. Near the seacoast, corrosion due to salt water and salt spray must also be considered when one is designing an adhesive joint. Thermal cycling due to weather, oxygen, ultraviolet radiation, and cold are relatively minor factors with most structural adhesives.

Extensive information on the outdoor durability of bonded aluminum joints is available in the reviews of Minford.[61,62,63] Hartshorn has also provided a catalog of references to outdoor weathering of structural adhesives by adhesive class and exposure conditions.[64]

15.6.1 Nonseacoast Environment

When exposed to weather, structural adhesives may rapidly lose strength during the first 6 months to 1 year. After 2 to 3 years, however, the rate of decline usually levels off at strength that is 25 to 30 percent of the initial joint strength, depending on the climate zone, adherend, adhesive, and stress level. Adhesive systems that are formulated specifically for outdoor applications show little strength degradation over time in a moderate environment. Figure 15.20 shows the weathering characteristics of unstressed epoxy adhesives to the Richmond, Virginia, climate.

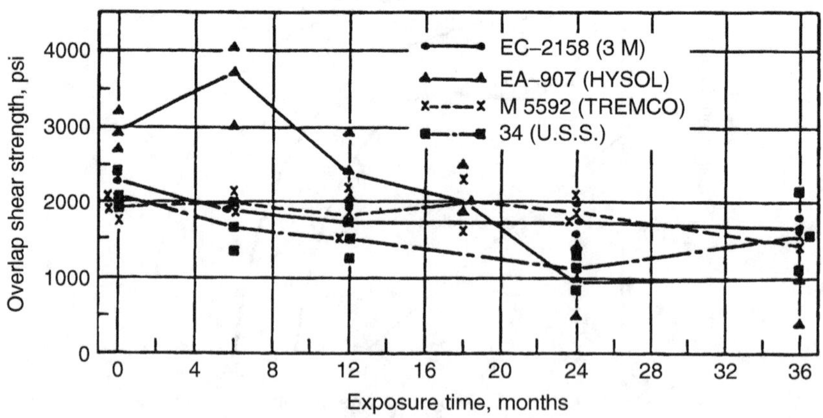

FIGURE 15.20 Effect of outdoor weathering on typical aluminum joints made with four different two-part epoxies cured at room temperature.[65]

The following generalizations are of importance in designing a joint for outdoor service:

1. The most severe locations are those with high humidity and warm temperatures.
2. Stressed panels deteriorate more rapidly than unstressed panels.
3. Stainless steel panels are more resistant than aluminum panels because of corrosion.
4. Heat cured adhesive systems are generally more resistant than room temperature cured systems.
5. For the better adhesives, unstressed bonds are relatively resistant to severe outdoor weathering, although all joints will eventually exhibit some strength loss.

MIL-STD-304 is a commonly used accelerated-exposure technique to determine the effect of weathering and high humidity on adhesive specimens.[66] In this procedure, bonded panels are exposed to alternating cold (–54°C) and heat and humidity (71°C, 95 percent RH) for 30 days. The effect of MIL-STD-304 conditioning on the joint strength of common structural adhesives is presented in Table 15.20. However, only relative comparisons can be made with this type of test; it is not possible to extrapolate the results to actual service life.

15.6.2 Seacoast Environment

For most adhesive bonded metal joints that must see outdoor service, corrosive environments are a more serious problem than the influence of moisture. The degradation mechanism is corrosion of the metal interface, resulting in a weak boundary layer. Surface preparation methods and primers that make the adherend less corrosive are commonly employed to retard the degradation of adhesive joints in these environments.

The quickest method to measure corrosion-dominated degradation in bonded metal joints is to store the bonded parts in a salt spray chamber with a continuous 5% NaCl salt solution spray in 95 percent RH at 35°C. This is a procedure described in ASTM B 117. Usually only several hundred hours of exposure are needed to show significant differences in corrosion resistance of various adhesive joint systems. Figure 15.21 shows the shear

TABLE 15.20 Effect of MIL-STD-304 on Bonded Aluminum Joints[67]

	Tensile shear, psi, at 23°C		Tensile shear, psi, at 73°C	
Adhesive	Control	Aged	Control	Aged
Room temperature cure:				
Epoxy polyamide	1800	2100	2700	1800
Epoxy polysulfide	1900	1640	1700	6070
Epoxy-aromatic amine	2000	0	720	0
Epoxy-nylon	2600	1730	220	80
Resorcinol epoxy-polyamide	3500	3120	3300	2720
Epoxy-anhydride	3000	920	3300	1330
Elevated-temperature cure:				
Epoxy-phenolic	2900	2350	2900	2190
Modified epoxy	4900	3400	4100	3200
Epoxy-nylon	4600	3900	3070	2900

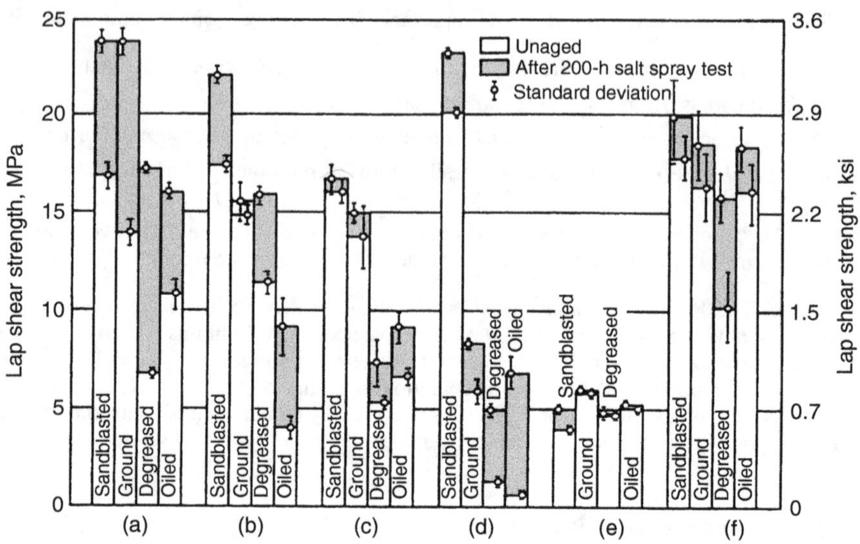

FIGURE 15.21 Shear strength of mild steel joints. (*a*) two-part epoxy, (*b*) two-part acrylic, (*c*) anaerobic acrylic, (*d*) cyanoacrylate, (*e*) PVC plastisol, (*f*) one-part heat cured epoxy.[68]

strength of mild steel joints with various adhesives before and after exposure for 200 h in a salt spray test.

Resistance of the adhesive joint to salt climates depends not only on the type of adhesive but also on the method of surface preparation and on the type of primer used. The good bond durability in saltwater exposure of anodized surface pretreated joints has been shown by several studies.[69]

15.6.3 Epoxy Adhesive Formulations

Room temperature curing, two-part epoxy adhesives are usually considered to have a lower level of outdoor performance than the heat cured adhesives. Their performance, however, can be improved by heat treatment following the room temperature cure.

Epoxy adhesive cured with diethylene triamine, a primary amine, tends to be somewhat water-sensitive. However, under conditions of ambient laboratory storage, there is no significant change in bond strength over a period of 11 years.[70] Unfilled polyamide cured epoxy adhesive survives 11 years of room temperature storage at more than double the shear strength of diethylene triamine cured epoxy with no significant decrease in strength. The initial value of 2408 psi leveled off at about 2900 psi within a few months and for the remainder of the 11 years.

Polyamide cured epoxy adhesives were also found to survive up to 4 years' exposure to industrial atmosphere, high-humidity, and water immersion environments without any appreciable bond degradation.[71] However, the strength of these adhesives were noticed to increase from 1500 to 1590 psi during aging for 3 years in a jungle environment of Panama.[72] Elevated-temperature curing, one- and two-part epoxy adhesives generally have a higher level of durability when exposed to aggressive outdoor environments. These adhesives generally show good performance when sustained loads are less than 50 percent of the initial bond strength.

Diethylaminopropylamine (DEAPA) was studied extensively as a curing agent for epoxy adhesives exposed to outdoor weathering. No changes were noticed during 11 years of ambient laboratory storage, or during 24 weeks of exposure in environments of 50°C, 100 percent RH, and thermal cycling between −18 and 50°C. When exposed to more rigorous environments, the bond strength degraded.

15.7 CHEMICAL RESISTANCE

Many organic adhesives tend to be susceptible to chemicals and solvents, especially at elevated temperatures. Most standard tests to determine chemical resistance of adhesive joints last only 30 days or so. Unfortunately, exposure tests lasting less than 30 days are not applicable to many service life requirements. Practically all adhesives are resistant to these fluids over short time periods and at room temperatures. Some epoxy adhesives even show an increase in strength during aging in fuel or oil over these time periods. This effect is possibly due to either postcuring or plasticizing of the epoxy by the oil.

There are two properties of adhesive joints that protect them from exposure to chemical or solvent environments: high degree of crosslinking and low exposure area.

1. Most crosslinked thermosetting adhesives such as epoxies, phenolics, polyurethanes, and modified acrylics are highly resistant to many chemicals, at least at temperatures below their glass transition temperature.
2. The adhesive bond line is usually very thin and well protected from the chemical itself. This is especially true if the adherends are nonporous and nonpermeable to the chemical environments in question.

Adhesive joint designers will take maximum advantage of this second effect by designing the joint configuration for protection or by specifying a protective coating and/or sealant around the exposed edges of the adhesive.

Figure 15.22 shows the long-term effect of a heat cured one-part epoxy adhesive to various chemical environments. As can be seen, the temperature of the immersion medium is a significant factor in the aging properties of the adhesive. As the temperature increases, the adhesive generally adsorbs more fluid, and the degradation rate increases.

Epoxy adhesives are generally more resistant to a wide variety of liquid environments than other structural adhesives. However, the resistance to a specific environment is greatly dependent on the type of epoxy curing agent used. Aromatic amine (e.g., metaphenylene diamine) cured systems are frequently preferred for long-term chemical resistance.

There is no universally "best adhesive" for all chemical environments. As an example, maximum resistance to bases almost axiomatically means poor resistance to acids. It is relatively easy to find an adhesive that is resistant to one particular chemical environment. It becomes much more difficult to find an adhesive that will not degrade in two widely differing chemical environments. Generally, adhesives that are most resistant to high temperatures have good resistance to chemicals and solvents because of their dense, crosslinked molecular structure.

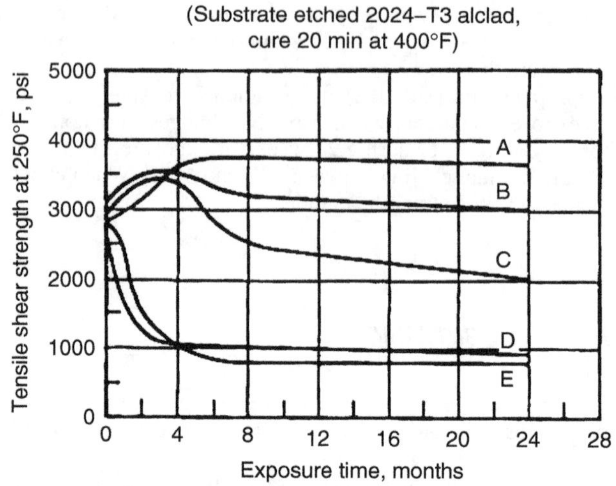

FIGURE 15.22 Effect of immersion in various chemical environments on a one-part heat curing epoxy adhesive (EA 929, Hysol Corp.): (A) gasoline at 23°C, (B) gear oil at 120°C, (C) distilled water at 23°C, (D) tap water at 100°C, (E) 38% Shellzone at 120°C.[73]

From the information reported in the literature with regard to aging of adhesive joints in chemical environments, it can be summarized that

1. Chemical resistance tests are not uniform, in concentrations, temperature, time, or properties measured.
2. Generally, chlorinated solvents and ketones are severe environments.
3. High-boiling-point solvents, such as dimethylforamide and dimethyl sulfoxide, are severe environments.
4. Acetic acid is a severe environment.
5. Amine curing agents for epoxies are poor in oxidizing acids. Anhydride curing agents are poor in caustics.

ASTM D 896-84, "Standard Test Method for Resistance of Adhesive Bonds to Chemical Reagents," specifies the testing of adhesive joints for resistance to solvents and

TABLE 15.21 Standard Test Fluids and Immersion Conditionals for Adhesive Evaluation per MMM-A-132

Test fluid	Exposure conditions
Water	30 days at room temperature
100% RH	30 days in 43°C humidity cabinet
5% Salt spray	30 days in 35°C salt spray cabinet
JP-4 jet fuel	7 days
Anti-icing fluid (isopropyl alcohol)	7 days
Hydraulic oil (MIL-H-5606)	7 days
HC test fluid (70/30 v/v isooctane/toluene	7 days

chemicals. Standard chemical reagents are listed in ASTM D 543, and the standard oils and fuels are given in ASTM D 471. Standard test fluids and immersion conditions used by many adhesive suppliers and specified in MMM-A-132[74] are listed in Table 15.21.

15.8 VACUUM AND OUTGASSING

Adhesive systems may be composed of low-molecular-weight constituents that can be extracted from the bulk adhesive when exposed to a vacuum environment. If these low-molecular-weight constituents also have a low vapor pressure, they may migrate out of the bulk because of exposure to elevated temperatures with or without the presence of a vacuum. This results in an overall weight loss and possible degradation of the adhesive or sealant.

The ability of an adhesive to withstand long periods of exposure to a vacuum is of primary importance for materials used in space travel or in the fabrication of equipment that requires a vacuum for operation. The outgassed constituents can also become a source of contamination and be highly objectionable in certain applications, such as with electronic products, optical equipment, and solar arrays.

The degree of adhesive evaporation is a function of the vapor pressure of its constituents at a given temperature. Loss of low-molecular-weight constituents such as plasticizers or diluents could result in hardening and porosity of adhesives or sealants. Since most structural epoxy adhesives are relatively high-molecular-weight polymers, exposure to pressures as low as 10^{-9} torr is not harmful to the base resin. However, high temperatures, radiation, or other degrading environments may cause the formation of low-molecular-weight fragments that tend to bleed out of the adhesive in a vacuum.

Epoxy and polyurethane adhesives are not appreciably affected by 10^{-9} torr for 7 days at room temperature. However polyurethane adhesives can exhibit significant outgassing when aged under 10^{-9} torr at 107°C.[75] At room temperature a high vacuum does not generally cause significant weight loss in commercial adhesive and sealant materials.

15.9 RADIATION

High-energy particulate and electromagnetic radiation including neutron, electron, and gamma radiation has similar effects on organic adhesives. Radiation initially causes increased hardening due to increased crosslinking. Radiation of sufficient energy causes molecular chain scission of polymers used in structural adhesives, which results in weakening and embrittlement of the bond. This degradation is worsened when the adhesive is simultaneously exposed to both elevated temperatures and radiation. ASTM D 1879, "Standard Practice for Exposure of Adhesive Specimens to High Energy Radiation," specifies methods to evaluate resistance to radiation.

Figure 15.23 illustrates the effect of radiation dosage on the tensile shear strength of several structural adhesives. Generally, heavily crosslinked, heat-resistant adhesives have been found to resist radiation better than less thermally stable systems. Aromatic epoxy resins are more resistant to radiation damage than comparable aliphatic epoxies. Fibrous reinforcement, fillers, curing agents, and reactive diluents will also affect the radiation resistance of adhesive systems. In epoxy-based adhesives, aromatic curing agents offer greater radiation resistance than aliphatic-type curing agents.

Polyester resins and anaerobic adhesives and sealants have also exhibited high radiation resistance. Anaerobic adhesives have several years of long-term exposure in radiation environments due to their use as thread locking sealants in nuclear reactors and accessory equipment.

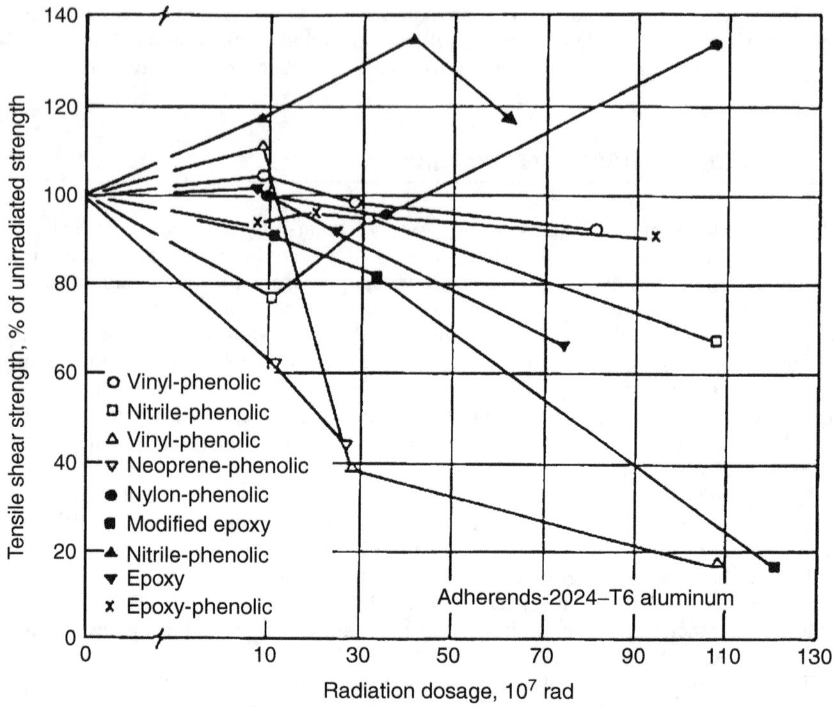

FIGURE 15.23 Percent change in initial tensile shear strength caused by nuclear radiation dosage.[76]

Thread locking grades of anaerobic adhesives have sustained 2×10^7 rads without molecular change or loss of locking torque.[77]

In an early study by McCrudy and Rambosek[78], radiation does not appear to have serious effects on the tensile shear strength of highly crosslinked adhesives. Nitrile-phenolic adhesives are somewhat more resistant to radiation damage than epoxy-based adhesives. The study also showed that thick adhesive layers retain useful strength better than thin glue lines. Ten mils is recommended as the minimum glue line thickness when radiation is an environmental factor.

REFERENCES

1. Waggemans, D. M., "Adhesives Charts," in *Adhesion and Adhesives*, vol. 2, Elsevier, Amsterdam, 1967.
2. Hartshorn, S. R., "The Durability of Structural Adhesive Joints," *Structural Adhesives*, S. R. Hartshorn, ed., Plenum Press, New York, 1986.
3. Lewis, A. F., and Gounder, R. N., "Permanence of Structural Adhesive Joints," *Treatise on Adhesion and Adhesives*, vol. 5, R. L. Patrick, ed., Marcel Dekker, New York, 1988.
4. Levi, D. W., *Journal of Applied Polymer Science: Applied Polymer Symposium*, vol. 32, 1977, p. 189.
5. Levi, D. W., et al., "Effect of Titanium Surface Pretreatment on Adhesive Bonds," *SAMPE Quarterly*, April 1976.

6. Jones, W. C., et al., "Use Multiple Regression Analysis to Develop Preductive Models for Failure Times of Adhesive Bonds at Constant Stress," *Journal of Applied Polymer Science*, vol. 18, 1974, p. 555.
7. Bolger, J. C., and Michaels, A. S., "Molecular Structure and Electrostatic Interactions at Polymer–Solid Interfaces," in *Interface Conversion for Polymer Coatings*, P. Weiss, ed., Elsevier, New York, 1969.
8. Lewis and Gounder, "Permanence of Structural Adhesive Joints."
9. Lee, H., and Neville, K., *Handbook of Epoxy Resins*, McGraw-Hill, New York, 1967.
10. Krieger, R. B., and Politie, R. E., "High Temperature Structural Adhesives," in *Aspects of Adhesion*, vol. 3, D. J. Alner, ed., University of London Press, London, 1967.
11. Kausen, R. C., "Adhesives for High and Low Temperatures," *Materials Engineering*, August–September 1964.
12. Krieger and Politi, "High Temperature Structural Adhesives."
13. Black, J. M., and Bloomquist, R. F., "Metal Bonding Adhesives for High Temperature Service," *Modern Plastics*, June 1956.
14. Wilkinson, S. P., et al., "Reactive Blends of Amorphous Functionalized Engineering Thermoplastics and Bismaleimide/Diallyl Bisphenol A Resins for High Performance Composite Matrices," *Polymer Preprints*, vol. 33, no. 1, 1992, p. 425.
15. Shimp, D. A., et al., "Co-Reaction of Epoxide and Cyanate Resins," 33d SAMPE Symposium and Exhibition, Anaheim, CA, Mar. 7–10, 1988.
16. Burgman, H. A., "The Trend in Structural Adhesives," *Machine Design*, Nov. 21, 1963.
17. Lee and Neville, *Handbook of Epoxy Resins*, pp. 21.43–44.
18. Bolger, J. C., "Structural Adhesives for Metal Bonding," in *Treatise on Adhesion and Adhesives*, vol. 3.
19. Meath, E. R., "Epoxy Resin Adhesives," Chapter 19 in *Handbook of Adhesives*, I. Skeist, ed., van Nostrand Reinhold, New York, 1990, p. 348.
20. Hourwink, R., and Salmon, G., eds., *Adhesion and Adhesives*, 2d ed., Elsevier, Amsterdam, 1986, p. 272.
21. Licari, J. J., "High Temperature Resistant Adhesives," *Product Engineering*, December 1964, p. 104.
22. Licari, "High Temperature Resistant Adhesives," p. 105.
23. Barie, W. P., and Franke, N. W., "High Temperature Epoxy Adhesives Based on BTDA," *Adhesives Age*, October 1971, pp. 36–39.
24. Barie and Franke, "High Temperature Epoxy Adhesives Based on BTDA."
25. Kausen, R. C., "Adhesive for High and Low Temperatures—II," *Materials in Design Engineering*, September 1964, pp. 108–112.
26. McClintock, R. M., and Hiza, M. J., "Epoxy Resins as Cryogenic Structural Adhesives," *Modern Plastics*, June 1958.
27. "Low Temperature Strength of Epoxy Resin Insulation Adhesive," *Insulation*, December 1958.
28. Miska, K. H., "Which Low Temperature Adhesive Is Best for You?" *Materials Engineering*, May 1975.
29. Kausen, "Adhesive for High and Low Temperatures—II."
30. Sung, N. H., "Moisture Effect on Adhesive Joints," *Adhesives and Sealants*, vol. 3, *Engineered Materials Handbook*, ASM International, Materials Park, OH, 1990, p. 622.
31. Sung, "Moisture Effect on Adhesive Joints," p. 622.
32. Comyn, J., *Adhesion Science*, Chapter 10, "Adhesive Joints and the Environment," Royal Society of Chemistry, Cambridge, 1997.
33. Comyn, "Adhesive Joints and the Environment."
34. Schneberger, G. L., "Polymer Structure and Adhesive Behavior," in *Adhesives in Manufacturing*, G. L. Schneberger, ed., Marcel Dekker, New York, 1983.
35. Bolger, J. C., "New One Part Epoxies Are Flexible and Reversion Resistant," *Insulation*, October 1969, pp. 38–44.

36. Schneberger, "Polymer Structure and Adhesive Behavior."
37. Falconer, D. J., et al., "The Effect of High Humidity Environments on the Strength of Adhesive Joints," *Chemical Industry*, July 4, 1964.
38. Falconer et al., "The Effect of High Humidity Environments on the Strength of Adhesive Joints."
39. Brewis, D. M., et al., "The Effect of Humidity on the Durability of Aluminium Epoxide Joints," *International Journal of Adhesion and Adhesives*, vol. 10, 1990, p. 247.
40. Kinloch, A. J., "Interfacial Fracture: Mechanical Aspects of Adhesion Bonded Joints," Review Article, *Journal of Adhesion*, vol. 10, 1979, p. 193.
41. Hartshorn, "The Durability of Structural Adhesive Joints," p. 351.
42. Carter, G. F., "Outdoor Durability of Adhesive Joints under Stress," *Adhesives Age*, October 1967.
43. Sharpe, L. H., "Aspects of the Permanence of Adhesive Joints," in *Structural Adhesive Bonding*, M. J. Bodnar, ed., Interscience, New York, 1966.
44. Olson, W. Z., et al., "Resistance of Adhesive Bonded Metal Lap Joints to Environmental Exposure," Report No. NADC TR59-564, October 1962. Also in DeLollis, N. J., "Durability of Structural Adhesive Bonds: A Review," *Adhesives Age*, September 1977.
45. Hartshorn, "The Durability of Structural Adhesive Joints."
46. Hartshorn, "The Durability of Structural Adhesive Joints."
47. Antoon, M. K., and Koenig, J. L., "*Journal of Polymer Science Polymer Physics Edition*, vol. 19, 1981, p. 197.
48. Orman, S., and Kerr, S., "Effect of Hostile Environments on Adhesive Joints," in *Aspects of Adhesion*, vol. 6, J. Alner, ed., University of London Press, London, 1971, p. 64.
49. Comyn, J., "Adhesive Joints and the Environment," Chapter 10 in *Adhesion Science*, Royal Society of Chemistry, Cambridge, 1997.
50. Shaw, S. J., and Tod, D. A., "Adhesive Bonding in Severe Environments," *Materials World*, vol. 2, 1994, pp. 523–525.
51. Minford, J. D., in "Permanence of Adhesive Bonded Joints," *Durability of Structural Adhesives*, A. J. Kinloch, ed., Applied Science Publishers, London, 1983, p. 135.
52. Minford, J. D., "Durability of Aluminum Bonded Joints in Long-Term Exposure," *International Journal of Adhesion and Adhesives*, vol. 2, no. 1, 1982, p. 25.
53. Minford, *Durability of Structural Adhesives*, p. 135.
54. Bolger, "Structural Adhesives for Metal Bonding."
55. Bolger, "New One Part Epoxies Are Flexible and Reversion Resistant."
56. Barrie, J. A., et al., "Sorption and Diffusion of Water in Halogen Containing Epoxy Resins," *Polymer*, vol. 26, 1985, p. 1167.
57. Goobich, J., and Marom, G., "Moisture Absorption by Tetraglycidyl 4,4′-Diaminodiphenyl Methane/4,4′ Diamino-diphenyl Sulfone Epoxies Containing Brominated Epoxy Copolymers," *Polymer Engineering and Science*, vol. 22, 1982, p. 1052.
58. Petrie, E. M., Chapter 6 "Surfaces and Surface Preparation," in *Handbook of Adhesives and Sealants*, McGraw-Hill, New York, 2000.
59. Shaw, S. J., "Adhesives in Demanding Applications," *Polymer International*, vol. 41, 1996, pp. 193–207.
60. Hartshorn, "The Durability of Structural Adhesive Joints."
61. Minford, J. D., "Effect of Surface Preparation on Adhesive Bonding of Aluminum," *Adhesives Age*, July 1974.
62. Minford, *Durability of Structural Adhesives*, p. 135.
63. Minford, in "Durability of Adhesive Bonded Aluminum Joints," *Treatise on Adhesion and Adhesives*, vol. 3, p. 79.
64. Hartshorn, "The Durability of Structural Adhesive Joints," p. 355.
65. *Adhesive Bonding Aluminum*, Reynolds Metals Company, Richmond, VA.
66. MIL-STD-304, U.S. Department of Defense, Washington, DC.

67. Tanner, W. C., "Adhesives and Adhesion in Structural Bonding for Military Material," in *Structural Adhesive Bonding*, M. J. Bodnar, ed., Interscience, New York, 1966.
68. Brockmann, W., "Durability and Life Assessment and Life Prediction of Adhesive Joints," in *Adhesives and Sealants*, vol. 3.
69. Minford, J. D., "Comparison of Aluminum Adhesive Joint Durability as Influenced by Etching and Anodizing Treatments of Bonded Surfaces," *Journal of Applied Polymer Science*, Applied Polymer Symposia, vol. 32, 1977, pp. 91–103.
70. DeLollis, N. J., "Durability of Structural Adhesive Bonds: A Review," *Adhesives Age*, September 1977.
71. Minford, "Durability of Adhesive Bonded Aluminum Joints," in *Treatise on Adhesion and Adhesives*, vol. 3.
72. Bodnar, M. J., and Wegman, R. F., "Effect of Outdoor Aging on Unstressed, Adhesive Bonded Aluminum to Aluminum Lap Shear Joints," Technical Report No. 3689, Picatinny Arsenal, Dover, NJ, May 1968.
73. "Aerospace Adhesive," EA 929, Hysol Division, Dexter Corp., Technical Bulletin A5-129, Olean, NY, 1970.
74. MIL-STD-132, U.S. Department of Defense, Washington, DC.
75. Landrock, A. H., "Effect of Environment on Durability of Adhesive Joints," Chapter 9 of *Adhesives Technology Handbook*, Noyes Publications, Park Ridge, NJ, 1985.
76. Arlook, R. S., and Harvey, D. G., "Effect of Nuclear Radiation on Structural Adhesive Bonds," Wright Patterson Air Development Center, Report WADC-TR-456-467, August 1956.
77. Pearce, M. B., "How to Use Anaerobics Successfully," Applied Polymer Symposia No. 19, *Symposium on Processing for Adhesive Bonded Structures*, M. J. Bodnar, ed., Interscience, New York, 1972, pp. 207–230.
78. McCrudy, R. M., and Rambosek, G. M., "The Effect of Gamma Radiation on Structural Adhesive Joints," SAMPE National Symposium on the Effects of Space Environment on Materials, St. Louis, MO, May 1962.

CHAPTER 16

EPOXY ADHESIVES ON SELECTED SUBSTRATES

16.1 INTRODUCTION

This chapter identifies and discusses various epoxy adhesives and the processes that have been used to successfully bond or seal specific substrates. There are only a few materials that epoxy adhesives will not bond well. These uncooperative substrates are most notably low-surface-energy plastics, such as the polyolefins, fluorocarbons, and silicones. However, even these materials can be bonded effectively with epoxy adhesives if a prebond surface treating process is used to change the nature of the substrate surface. Of the other substrate materials, there are some that epoxy adhesives will bond more effectively than others. Table 16.1 lists substrates that generally provide excellent epoxy adhesive joints.

The selection of the proper epoxy adhesive formulation depends on

1. The physical nature of the bulk substrate material (porosity, modulus, thermal expansion coefficient, etc.)
2. The physical and chemical nature of the surface (surface energy, weak boundary layer attachment, resistance to corrosion, etc.)

Optimal joints can generally be fabricated by the correct combination of epoxy adhesive formulation (i.e., type of resin, curing agent, fillers, modifiers) and surface treatment process.

It is impossible to avoid a discussion on prebond surface preparation since it is one of the most important factors in the fabrication of a durable and consistent epoxy adhesive joint. Selection of a proper surface preparation is not an easy task, and the actual implementation of the surface treating process in production is equally daunting.

Various substrate surface treatments suggested for use with a common epoxy-substrate joint and service environment combinations are discussed in this chapter. Surface preparation processes for a range of specific substrates and detailed process specifications are provided in App. F. The reader is also directed to several excellent texts that provide prebond surface treatment recipes and discuss the basics of surface preparation, the importance of contamination or weak boundary layers, and specific processes for adhesive systems other than epoxy.[1,2,3]

The discussion in this chapter is organized by the type of substrate material. The substrates are broadly identified as

- Metals
- Plastics (thermosets and thermoplastics)
- Composites

TABLE 16.1 Common Substrates that Provide for Good Epoxy Adhesion

Metal	Plastic	Elastomeric	Other
Aluminum	ABS	Nitrile	Glass and ceramics
Copper	Epoxy	Neoprene	Wood
Stainless steel	Polycarbonate	Urethane	Composites
Nickel	Phenolic		Honeycomb
	Polyester		Concrete

- Foams
- Elastomers
- Wood and wood products
- Glass and ceramics
- Sandwich and honeycomb structures

Specific substrates are described under each classification. Of course, adhesive applications are not limited to joints having only one type of substrate. Metal-to-plastic, aluminum-to-steel, metal-to-wood, glass-to-metal, and an infinite variety of other joint configurations are all possible. In these applications the nature of each substrate needs to be understood and considered in the overall selection of an adhesive formulation and bonding process.

This chapter certainly does not consider all possible substrates. However, the guidance that is offered should be sufficient for the user to select "candidate" joining processes and epoxy adhesive materials, no matter what substrate or combination of substrates is involved.

16.2 METAL BONDING

Metals, having a relatively high surface energy, are generally considered easy to bond. However, several problems could occur when one is working with metallic substrates. These are related to the following characteristics:

1. Durability related to environmental effects on the substrate surface and the interface of the adhesive joint
2. Variation in the surface chemistry depending on alloy, processing, preconditioning, etc.
3. Relatively low thermal expansion coefficient and high thermal conductivity compared to most epoxy adhesives

One difficulty in bonding metals is the durability of the joint. It is not so much a problem of making a strong joint as one of keeping it that way throughout its expected service life. A weld may have strength of only 600 lb, but it is likely to remain that strong for 5 to 10 years afterward. An epoxy adhesive, on the other hand, may have 3 to 4 times the initial strength of a weld, but it could weaken when exposed to high humidities, cycled between hot and cold temperatures, or immersed in salt water and then dried. By definition, a structural adhesive must be able to withstand these conditions without significant deterioration.

A second difficulty in bonding metals lies in understanding the surface. One of the important points to consider in bonding metals is that only the surfaces are involved. Adhesives and sealants are active only on the molecular surface layer that forms the joint interface and on any surfaces contained in the porosity of the metal itself. Thus, if unprepared steel is being bonded, it is not the bulk iron-carbon alloy that is being bonded, but the iron oxide layer on the surface (presuming that the metal surface was cleaned of contaminants). Similarly with aluminum, the actual bond is to aluminum oxide rather than the pure metal.

The physical and chemical nature of the metal substrate surface depends on the bulk alloy composition, processing conditions used, and any preconditioning environments that the substrate may be exposed to during fabrication and storage. Once the joint is assembled, the physical and chemical nature of the adhesive-metal interface may change due to the environment and to the chemical nature of the adhesive.

The base metal is highly reactive in most cases and forms various oxides, sulfides, and hydrates when exposed to the atmosphere. As a result, it is necessary to consider not only the bulk metal but also the ability to bond to its hydrated oxide. One must consider the inherent nature of the adhesive force existing between the base metal and its oxide. The final joint will be no stronger than its weakest link. The weakest link will likely be determined by:

1. The cohesive strength of all the materials involved
2. The forces between the adhesive and the metal oxide
3. The strength of the metal-oxide bond to the base metal

The actual metal surface that takes part in the bonding is illustrated in Fig. 16.1. Adhesives recommended for metal bonding are in reality used for metal-oxide bonding. They must be compatible with the firmly bound layer of water attached to surface metal-oxide crystals. Even materials such as stainless steel and nickel or chromium are coated with transparent metal oxides that tenaciously bind at least one layer of water.

The nature and characteristic of these oxide layers depend on the base metal and the conditions that were present during its formation. With steel, for example, the oxide adhesion to the base metal is very weak. In the case of aluminum, however, the oxide is extremely stable and clings tightly to the base metal. In fact, it adheres so well that it serves as a protective coating for the aluminum, which is one reason why aluminum is a corrosion-resistant metal. Certain metals possess surfaces that interact more effectively with one type of adhesive than with another. This is the reason why adhesive formulators need to know as much as possible about the surfaces being assembled.

One of the benefits of surface treatment prior to adhesive bonding is that it not only removes weak boundary layers such as contamination, but also provides a more consistent surface to which the adhesive can bond. Common surface treatments used for metal substrates are characterized generally in Table 16.2.

A third difficulty in bonding metal surfaces is that they have a higher thermal coefficient of expansion and thermal conductivity than most epoxy adhesive systems. As explained in other chapters of this book, the difference in rates of thermal expansion results in internal stresses in the adhesive joint, especially when the adhesive bond is cured at elevated temperatures or when it is exposed to low temperatures or repeated thermal cycling.

The sections below describe the characteristics of various metal substrates with regard to epoxy adhesives. Only the more common substrates and those that are most reactive (to either the adhesive or the environment) are discussed. Metal substrates not covered are generally easy to bond with epoxy adhesives, and there is little change in properties with environmental aging.

16.2.1 Aluminum

Aluminum is an almost ideal substrate for adhesives. It has high surface energy and is very resistant to most environments. It is also a material with good formability and high

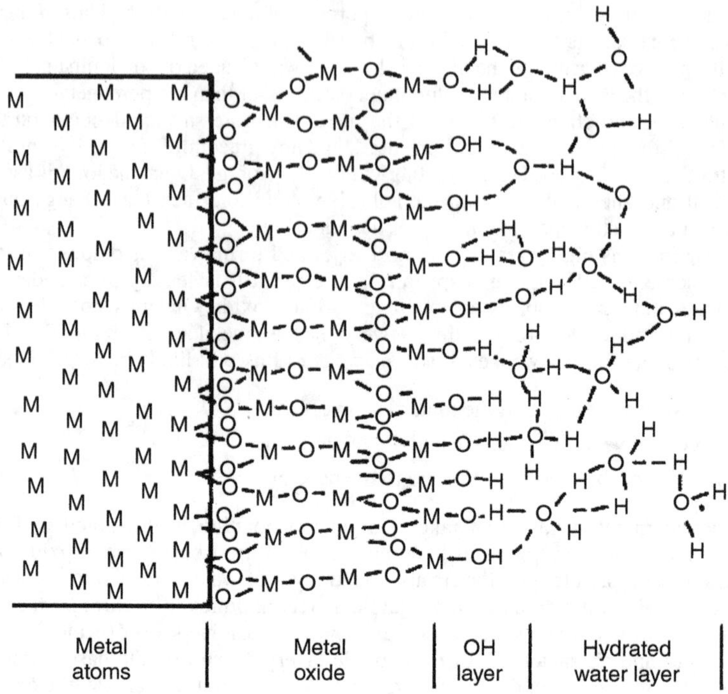

Note:

1. The oxide layer is typically 40–80 Å thick.
2. The hydrated layer is tightly bound.
3. A pure metal surface is rarely available for bonding with adhesives.

FIGURE 16.1 Metal surfaces are actually hydrated metal oxides.[4]

TABLE 16.2 Characterization of Common Surface Treatments of Metals[5]

Treatment type	Possible effects of treatment
Solvent	Removal of most organic contamination
Mechanical	Removal of most organic contamination. Removal of weak or loosely adhering inorganic layers, e.g., mill scale. Change in topography (increase in surface roughness). Change in surface chemistry
Conversion coating	Change in topography (increase in surface roughness). Change in the surface chemistry (e.g., the incorporation of a phosphate into the surface layers)
Chemical (acid etching, anodizing)	Removal of organic contamination. Change in topography (increase in surface roughness). Change in the surface chemistry. Change in the thickness and morphology of metal oxide

strength-to-weight ratio that can benefit greatly from properties offered by adhesive joints. As a result, adhesive bonded aluminum joints are commonly used in the aircraft and automotive industries. Bonding of aluminum most likely comprises the majority of the applications for epoxy adhesives. Aluminum joints are also commonly used in adhesive studies and for comparison of different adhesive materials and processes. Adhesive manufacturers' literature generally describes the properties of bonded aluminum joints.

However, the corrosion resistance of aluminum as well as the durability of joints made with epoxy adhesives is very dependent on the type of aluminum alloy used. Bonds made with relatively corrosion-resistant 6061-T6 aluminum alloy will last about 4 times as long as equivalent joints made with 2024-T3 alloy when exposed to marine environments.

However, with the proper combination of surface treatment and adhesive, these differences in durability to aggressive environments can be minimized, as shown in Table 16.3. The initial shear strength and permanence depend on the type of alloy and the pretreatment used. Note that the data presented here show only the relative differences in joint strength for one specific epoxy adhesive and are not representative of other adhesive formulations. There was no attempt to maximize any of these values through choice of the adhesive.

The oxide layer that forms on aluminum is more complex than with other metal substrates. Aluminum is a very reactive surface, and oxide forms almost instantaneously when a freshly machined aluminum surface is exposed to the atmosphere. Fortunately, the oxide is extremely stable, and it adheres to the base metal with strength higher than could be provided by most adhesives. The oxide is also cohesively strong and electrically nonconductive. These surface characteristics make aluminum a desirable metal for adhesive bonding, and they are the reasons why many adhesive comparisons and studies are done with aluminum substrates.

The strength of an epoxy adhesive aluminum joint can be improved by cleaning the surface to remove contaminants or by converting the existing surface to a new surface that may be more consistent. In the case of the aluminum surface, chemical conversion can also protect the base metal from corrosion and enhance the durability of the bonded joint to various service environments. The most common surface preparations that have been used for bonding aluminum can be generally segregated into three groups:

- Simple cleaning and abrading
- Chemical etch cleaning
- Primers and conversion coatings

TABLE 16.3 Effect of Surface Treatment on Aluminum Alloy Joints Bonded with a Heat Cured Epoxy Adhesive (EC-3443, 3M Company)[6]

	Tensile shear strength, psi								
	Alloy 2036			Alloy 6151			Alloy X5085		
	A	B	C	A	B	C	A	B	C
Initial, no aging	1930	1850	2200	2550	2690	2530	2250	2270	2110
After 3 months' aging at 23°C, 85% RH	1390	1050	2400	1350	1550	2410	1860	1890	1910
After 3 months' aging at 52°C, 100% RH	420	350	1110	920	1050	1100	1260	1520	1230
After 3 weeks' aging in 5% salt spray solution	0	0	1410	80	150	2090	690	530	1210

A, Mill finish; B, vapor-degreased; C, Alodine 401–45.

Included in the simple cleaning and abrading category are (1) solvent wiping, (2) vapor degreasing, or (3) either of these methods combined with mechanical abrading. In each instance, care must be taken to ensure that the cleaning materials themselves do not become unknowingly contaminated, thus providing ineffectual cleaning or cross-contamination, resulting in poor bond performance.

Mechanical abrasion or sandblasting is commonly used for treating aluminum surfaces prior to adhesive bonding because of its simplicity and economics. However, chemical treatments such as etchants produce higher reliability and longer service life in a bonded assembly. If aluminum adherends are first cleaned, then sandblasted, and finally chemically treated, the surface area is increased, the contaminants are removed, and the initial and long-term strengths are generally excellent. However, this three-step process is often not necessary when only moderate strengths (500 to 2000 psi) are required or if the finished adhesive joint will not be exposed to aggressive service environments (especially high-moisture or high-humidity conditions). Useful bonds in these low- to medium-strength applications can be achieved simply with cleaning and/or abrasion.

When one is bonding aluminum to itself or to other materials, the optimal surface preparation should be determined for the application based on the initial strength and durability required, and then the process must be rigidly followed. Overspecifying the strength requirements should be avoided since it could result in the selection of a surface preparation process that is time-consuming, difficult to control, and expensive. Table 16.4 serves as a guideline for selecting the pretreatment to try first.

The parameters of the pretreating process and their control can be very important in ensuring consistency from joint to joint. It has been reported that the temperatures used for rinsing and drying aluminum should not exceed 70°C since different oxides are formed above and below this temperature. The oxide (bayerite, β-$Al_2O_3 \cdot H_2O$) formed at lower temperatures supposedly provides better adhesion.[8] It has also been recognized that the use of distilled or deionized water rather than tap water for cleaning and etching baths is preferred for greater bond strength and consistency.

In selecting a pretreatment process for aluminum or any other substrate, both the initial strength and the permanence in a specific operating environment must be considered. Mechanical abrasion is a useful pretreatment in that it removes the oxide and exposes bare aluminum. When this is done, however, many of the benefits of the protective oxide layer are lost. For example, if bare abraded aluminum is bonded, the reactive metal at the joint interface can potentially become hydrolyzed and oxidized, which will displace the adhesive. Hence, this bonded joint may initially be much stronger than one made with unabraded metal, but it will deteriorate rapidly when exposed to a harsh environment such

TABLE 16.4 Effect of Substrate Treatment on Strength of Aluminum Joints Bonded with Epoxy Adhesives[7]

Surface treatment	Type of bond
Solvent wipe (MEK, MIBK, trichloroethylene)	Low- to medium-strength
Abrasion of surface (sandblasting, coarse sandpaper, etc.) plus solvent wipe	Medium- to high-strength
Hot vapor degrease (trichloroethylene)	Medium-strength
Abrasion of surface plus vapor degrease	Medium- to high-strength
Alodine treatment	Low-strength
Anodize	Medium-strength
Caustic etch	High-strength
Chromic acid etch (sodium dichromate–sulfuric acid)	Maximum-strength

as heat and humidity. This is why pretreatments that modify the oxide layer or create a new, stable oxide layer are especially desirable when permanence is a primary consideration. They improve bondability *and* maintain protection.

To protect the aluminum joint from the effects of the environment, especially water and corrosion, an artificially thickened oxide layer is generally formed on the surface. Historically, chemical etching as a surface preparation has provided the surest way of obtaining durable adhesive bonds with aluminum.

While various acidic or caustic procedures can be employed with or without vapor degreasing, the most recognized etching pretreatment for bonding aluminum has been the sulfuric acid–dichromate solution used by the aircraft industry and described in ASTM D 2651. This process is sometimes known as the *FPL etch*, named after its developers, Forest Products Laboratories.[9] The first step in this process is vapor degreasing followed by alkaline cleaning and then chemical immersion. The substrates are finally forced air dried. There are several modifications of this treatment including a pastelike etching solution to allow for parts that cannot be immersed in the acid solution and a chromate-free etching process (designated PT2) for improved environmental and occupational health and safety perspectives.[10]

Other important methods of pretreating aluminum for adhesive bonding include anodizing and chromate conversion coating.[11] In anodizing, the aluminum is immersed in various concentrations of acids (usually phosphoric or chromic) while an electrostatic charge is applied. The oxide reacts with the etchant to form a compound that protects the surface and is compatible with the adhesive. In this way the aluminum oxide is retained, but it is rendered more receptive to bonding. It has been shown that an anodized surface on aluminum alloy can constitute a very durable surface for epoxy adhesive bonding with excellent resistance to seacoast or other saltwater types of exposure. Examples of widely used anodizing processes are the Boeing phosphoric acid anodize (PAA) process[12] and the chromic acid anodize (CAA) process.[13]

The landmark U.S. Air Force Primary Adhesively Bonded Structure Technology (PABST) program in the late 1970s demonstrated that properly designed and manufactured bonded fuselage panels made from the correct aluminum materials can actually operate safely at higher stress levels than comparable rivet-joined aluminum structures.[14,15] The results of this program show phosphoric acid anodizing as an optimal way to achieve durable aluminum bonds.

- PAA is the most durable pretreatment for aluminum that is processable within reasonable production tolerances.
- PAA with a corrosion-resistant primer provides the best corrosion resistance.
- The adhesive should be selected on the basis of durability as defined by slow cyclic testing in a hot and humid environment.

The chemical conversion coating method is also commonly used to treat aluminum substrates prior to bonding. Chemical conversion coating is an amorphous phosphatization process wherein the aluminum is treated with a solution containing phosphoric acid, chromic acid, and fluorides. Chromate conversion coatings on aluminum constitute an effective way to enhance the surface bondability and also improve the corrosion resistance of the bond line.[16] The resulting durability observed with mechanical abrasion and chemical conversion coatings is variable and depends on the particular processing conditions, whereas anodizing and etching processes produce consistent and generally durable aluminum joints.

The durability of epoxy bonded aluminum joints that were immersed in water is shown in Fig. 16.2. The anodized and grit-blasted surface treatments, although giving different initial joint strengths, showed no deterioration after 2 years' exposure. Both the vapor-degreased and conversion coating treatments were significantly degraded by the moist environment.

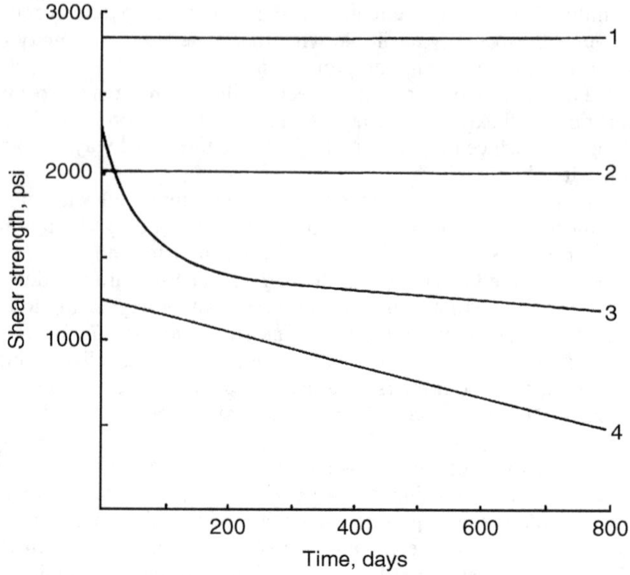

FIGURE 16.2 Effect of surface treatment on the durability of epoxy-aluminum joints exposed to room temperature water immersion. (1) Anodized, (2) grit-blasted plus vapor degrease, (3) vapor degrease, (4) chromate conversion coating.[18]

Exposure of similarly prepared specimens to a more aggressive soak-freeze-thaw cycle gave rise to even greater differences in performance with only the anodized treated aluminum joint showing a high percentage of joint strength after a 2-year period.[17]

However, the recognition that chrome is carcinogenic has forced alterations in surface treatment processes and formulation changes in primers. Chromic acid anodizing and FPL etching are being phased out in many locations. Work on sulfuric acid anodizing and sulfuric boric acid anodizing is now in progress.[19] In the automotive industry, a pretreatment has been developed for aluminum coil that is nontoxic and compatible with weldbonding. This proprietary treatment is claimed to be as effective as chromium-based pretreatment processes on exposure to salt spray.[20]

The usual approach to good bonding practice is to prepare the aluminum surface as thoroughly as possible, then wet it with the adhesive as soon afterward as practical. In any event, aluminum parts should ordinarily be bonded within 48 h after surface preparation. However, in certain applications this may not be practical, and primers are used to protect the surface between the time of treatment and the time of bonding. Primers are also applied as a low-viscosity solution which wets a metal surface more effectively than more viscous, higher-solids-content adhesives. Corrosion-resistant epoxy primers are often used to protect the etched surface during assembly operations. Primers for epoxy adhesive systems are described in Chap. 10.

The chemical cleaning methods used for aluminum will have slightly different effectiveness with different aluminum alloys. The permanence of these bonds will also depend on the type of alloys used because of their different corrosion rates under extreme environmental conditions. The yield strength of the alloy also has an influence on bond strength when stressed in shear. The peel test is usually considered a more meaningful test method for measuring the surface treatment effectiveness. The yield factor of the adherend is never approached because of relatively poor peel strength of the adhesive.

Once the surface considerations are taken care of, there are many types of epoxy adhesives that can bond well to aluminum. The selection will depend on the strength needed, the type of stress involved (e.g., peel or shear; static or dynamic), and the operating environment. Reynolds Metals Company[21] offers some general observations in selecting an adhesive for aluminum bonding.

- Bonds to aluminum are generally stronger than bonds to steel.
- A chromic-sulfuric acid etch gives the best resistance to weathering and saltwater environments.
- Room temperature curing epoxies offer the best saltwater resistance.
- Higher strengths are usually obtained with heat curing epoxies than with room temperature curing epoxies.
- Modified phenolic films give the highest peel and shear strength combinations.
- The most severe adhesive environment is a hot, humid climate (temperature 30 to 50°C; humidity +90 percent).
- Structural adhesives are strong in shear and weakest in peel and cleavage.
- Heat curing adhesives are less sensitive to surface preparation than room temperature curing adhesives.

All of the commercial epoxy adhesives presented in App. B bond well to aluminum and to a wide variety of other materials. Sell[22] has ranked a number of aluminum adhesives in order of decreasing durability as follows: nitrile-phenolics, high-temperature epoxies, elevated-temperature curing epoxies, elevated-temperature curing rubber-modified epoxies, vinyl epoxies, two-part room temperature curing epoxy paste with amine cure, and two-part urethanes.

The starting formulations presented in Table 16.5 are designed for general-purpose bonding of aluminum where other substrates may also be involved. Note that aluminum powder is a key ingredient in these formulations to provide for a closer match in coefficient of thermal expansion between the adhesive and the substrate.

16.2.2 Beryllium

Beryllium and its alloys (e.g., beryllium copper) have gained interest in the aerospace industry and specialty sports equipment industry in recent years. Brazing or riveting can be used for joining, but these methods are expensive, and distortion or highly stressed areas may be encountered. The metal must be handled with care when the processing produces dust, chips, scale, slivers, mists, or fumes, since airborne particles of beryllium and beryllium oxide are toxic with latent health effects. Abrasives and chemicals used with beryllium must be disposed of properly.

The gain in the popularity of beryllium has been due in large part to adhesive bonding. Bonding permits one to take advantage of the excellent combination of physical and mechanical properties of beryllium and minimizes the inherent problems of high notch sensitivity and low ductility. Adhesive bonding is generally applied where high strength-to-weight ratios are important. Since beryllium retains significant strength at temperatures up to 540°C, many of the adhesives and surface preparations developed for beryllium have been used to exploit the metal's high-temperature properties in aerospace applications.

Beryllium is a very reactive metal, and it reacts quickly with methyl alcohol, fluorosolvents, perchloroethylene, and blends of methyl ethyl ketone and fluorosolvents. Beryllium

TABLE 16.5 Starting Formulations for Epoxy Adhesive for Bonding Aluminum[23]

	Parts by weight		
Component	A	B	C
Part A			
DEGEB epoxy resin (EPON 828, Resolution Performance Products)	100	100	50
Ground tubular alumina (Alumina T-60/T-64, Alcoa)	230		
Aluminum powder		100	50
Part B			
Aromatic amine eutectic (60/40 blend of MDA and MPDA)	12		
Aliphatic amine (EPI-CURE 3234, Resolution Performance Products)	5		
Aliphatic amine (EPI-CURE 3245, Resolution Performance Products)			5
Polyamide (EPI-CURE 3125)		54.5	
Property			
Mix ratios, part A : part B			
• By weight	19.4 : 1.0	100 : 27	100 : 5
• By volume	9.6 : 1.0	2 : 2.1	—
Viscosity, cP, at 25°C	Paste	Paste	Paste
Working life at 25°C	1–2 h	60 min	60 min
Cure schedule	Elevated temperature	Room temperature	Room temperature
Tensile shear strength, psi, measured at 25°C on aluminum when cured			
• 7 days at 25°C	—	2050	1050
• 10 min at 204°C	2250		
• 100 min at 121°C	2300		

can be pitted by long exposure to tap water containing chlorides or sulfates.[24] Proprietary coatings are used to provide a corrosion-resistant barrier.[25]

A number of prebonding surface preparations for bonding beryllium and its alloys with epoxy adhesives have been suggested in the literature. One procedure is to degrease the substrate with trichloroethylene, followed by immersion in the solution listed below for 5 to 10 min at 23°C.

Sodium hydroxide	20 to 30 pbw
Distilled water	170 to 180 pbw

The substrate should then be washed in tap water and rinsed with distilled water. The final step in the process is an oven-dry of 10 min at 121 to 177°C.[26] Several other surface preparation procedures for beryllium have been reported to have merit. Epoxy and epoxy hybrid adhesives have been found to provide high strengths on sulfuric acid–sodium dichromate etched beryllium.[27]

16.2.3 Copper

Copper substrates are commonly bonded with epoxy adhesives in the microelectronics and marine industries. Compared to aluminum substrates, copper when bonded with epoxy adhesives provides lower initial strength. Depending on the adhesive and the type of test used, this can be as much as 50 percent lower. Similar to aluminum joints, copper joints bonded with epoxy adhesives can show poor durability in moist environments unless the interface is protected.

Copper is used in three basic forms: pure, alloyed with zinc (brass), and alloyed with tin (bronze). Copper and copper alloys are difficult to bond satisfactorily, especially if high shear and peel strengths are desired. The primary reason for this difficulty is that the oxide that forms on copper develops rapidly (although not as fast as the rate of oxide development on aluminum). The copper oxide layer is weakly attached to the base metal under usual conditions. Thus, if clean, bare copper substrates were bonded, the initial strength of the joint would be relatively high, but on environmental exposure an oxide layer could develop which will reduce the durability of the joint.

Cleaning and mechanical abrasion are often used as pretreatments for copper and its alloys where low to medium bond strengths are acceptable. For optimum bond strength and permanence, the oxide layer must be specifically "engineered" for adhesive bonding. This is often done through what has been called the "black oxide" coating process or through chromate conversion coatings.

The black oxide processes form microrough morphologies on the copper surface. One such process uses a commercially available product[28] to form a strong black oxide coating on the copper surface. This process requires alkaline cleaning and acid immersion before the copper surface is immersed at elevated temperatures in a solution containing the proprietary product. The process steps are critical, and time and temperature must be carefully controlled. Parts should be joined within 4 h after chemical treating. Another black oxide coating process, Method E in ASTM D 2651, is intended for relatively pure copper alloys containing 95 percent copper.

The sodium dichromate–sulfuric acid process has been found by some to be superior to other ferric chloride methods for the prebond treatment of copper. This dichromate–sulfuric acid method is also defined by ASTM D 2651. Nitric acid and ferric chloride etching processes have also been found to be useful for copper, brass, and bronze substrates in certain applications.

Copper may also react with certain chemicals in the adhesive formulations at elevated cure temperatures. Copper has a tendency to form brittle amine compounds with some epoxy curing agents including certain amines and anhydrides. However, room temperature curing epoxy adhesives will generally give very good adhesion to copper and copper alloy substrates. Table 16.6 shows the effect of heat aging on tensile shear strength of DGEBA epoxy cured with two curing agents.

For elevated-temperature cures, however, dicyandiamide and melamine curing reactions have been shown to be beneficial in epoxy-based adhesives when used on copper substrates. These epoxy adhesives perform about as well in copper joints as in aluminum or steel joints. The formulations show significantly increased time to adhesive failure on either bare or alkaline permanganate-treated copper.[30]

Brass is an alloy of copper and zinc, and bronze is an alloy of copper and tin. Sandblasting or other mechanical means of surface preparation may be used for both of these copper alloys. Surface treatment combining mechanical and chemical treatment with a solution of zinc oxide, sulfuric acid, and nitric acid is recommended for maximum adhesion properties. Adhesives similar to those recommended for copper may be used on brass and bronze substrates.

TABLE 16.6 Effect of Heat Aging on Tensile Shear Strength of Copper Joints Bonded with a DGEBA Epoxy Cured with Tri-2-ethylhexoate Salt of DMP-30 and DEAPA[29]

	Tensile shear strength, psi, at 100°C			
	DMP-30 salt cured, aged at		DEAPA cured, aged at	
Aging time, weeks	120°C	160°C	120°C	160°C
0	1300	1300	900	900
2		900		920
4		800		900
5	1150		1250	
6		700		850
8		600		750
10	900	530	1175	
15	875		1150	
20	850		1100	
25	850		1100	

16.2.4 Magnesium

The surface preparation and bonding methods developed for magnesium and magnesium-based alloys are closely associated with corrosion prevention because magnesium is one of the most reactive of all metals. Numerous processes have been developed for both painting and bonding of magnesium. The choice of a bonding process will be determined by criteria such as high bond strength, high corrosion resistance, or both.

The position of magnesium in the electrochemical or galvanic series indicates that it has a potential of reacting chemically with a great number of other metals when there is electrical connection. The conductive path could be caused by direct metal-to-metal contact, an aqueous solution in which there is an electrolyte (e.g., chloride ions in solution), or by other ways. The key in any assembly involving magnesium is to design and assemble the parts in such a manner that the conductive path is eliminated.

Magnesium will react with moisture to form a magnesium hydroxide that, in itself, provides some corrosion protection. However, this coating is not quite uniform and may contain carbonates and sulfates or other ionic materials that are present in the surrounding environment. Magnesium is quite resistant to alkalies, and under controlled conditions this factor is used to advantage in surface preparation. Organic compounds, in general, do not react with magnesium as do the aqueous electrolytes. Anhydrous methyl alcohol is an exception—it can react readily with magnesium.

Minor constituents found in magnesium alloys can play a significant part in adhesive bonding. Not all magnesium alloys react to protective surface treatments in the same way. Knowledge of the alloy composition must be coupled with a careful surface preparation selection process. Grain structure of the metal can also influence the nature of the surface preparation. The temper of the alloys, such as strain-hardened (W), annealed (O), or fabricated (F), and the various ways of treating (T1 through T10) indicate the manner of treatment an alloy has received. These different surfaces have a bearing on the effectiveness of the surface preparation process.

Removal of light oil or light chromate coatings (used to protect the magnesium during shipment and storage), mill scale, lubricants, welding fluxes, etc., must occur before the desired surface preparation is applied. Three cleaning methods, used either separately or in combination, constitute the necessary cleaning before the final surface preparation: solvent degreasing, mechanical cleaning (galvanic reactions from the grit must be avoided), and chemical treatment (acid pickling using chromic, nitric, or phosphoric acid solutions). It is not uncommon for the acid pickling or cleaning procedure to suffice as the final surface preparation process. An alkaline detergent cleaning process is described in ASTM D 2651 Method A. A hot chromic acid cleaning process that can be combined with the alkaline cleaning process is given by ASTM D 2651 Method B.

Light anodic treatment and various corrosion preventive treatments produce good surfaces on magnesium for adhesive bonding (ASTM D 2651 Method D). These treatments were developed by magnesium alloy producers, such as Dow Chemical Company (Dow 17 and Dow 7) and others. Details are available from the ASM *Metals Handbook*[31] and in MIL-M-45202, Type I, Class 1, 2, and 3.[32] Some surface dichromate conversion coatings and wash primers designed for corrosion protection can also be used for adhesive bonding of magnesium (ASTM D 2651 Method E). Details are found in ASM *Metals Handbook* and MIL-M-3171.[33]

The thickness of the surface coating influences the relative degree of protection for the magnesium against corrosive elements. Heavy anodic coatings of about 0.001 in are used for maximum corrosion resistance. However, such thick coatings are generally not optimal for adhesive bonding because cohesive failure in the coating results. Thinner coatings of 0.0001 to 0.0003 in. obtained from the Dow 17 anodic surface treatment are better for adhesive bonding. Thin films also provide greater coating flexibility with less likelihood of cracking upon flexure.

A problem unique to magnesium substrates is the formation of loose particles (termed *smut*) on the surface of the metal. The source of the smut can be varied but is usually related to the environment during the cleaning process. Such loose, gray, black, or brown smut particles are believed responsible, in some cases, for erratic adhesive strength results as well as lower bond strengths. These loose particles generally can be rubbed off with a solvent-dampened pad.

High-humidity and salt spray environments have been found to cause the greatest decrease in bond strength of magnesium adhesive joints. Of the several surface treatments that have been evaluated, the Dow 17 surface preparation (ASTM D 2651 Method C) provides the best overall performance with all adhesives under these types of environmental conditions.[34]

Almost any epoxy adhesive can be used on magnesium provided that proper surface protection is maintained. In view of the corrosion potential, water-based adhesives or adhesives that allow water permeation may be expected to cause problems with magnesium substrates.

16.2.5 Nickel

Nickel and its common alloys such as Monel (nickel-copper), Inconel (nickel-iron-chromium), and Duranickel (primarily nickel) can be bonded with procedures that are recommended for stainless steels.[35] A simple nitric acid process has also been used consisting of solvent cleaning, immersion for 4 to 6 s at room temperature in concentrated nitric acid, rinsing with cold deionized water, and finally drying. Also, a chromium trioxide–hydrochloric acid process consisting of a 60- to 80-s immersion in acid solution has been suggested. If immersion is impossible, this latter solution may be applied with a cheesecloth after solvent cleaning. The solution is applied to approximately 1 ft^2 of the substrate surface at a time, and it remains on the part for approximately 1 min.

Epoxies are commonly used to bond nickel substrates. However, the nickel alloys often find applications at temperatures higher than most organic adhesives are capable of resisting.

Thus, there is relatively little development work done on optimizing pretreatments for nickel alloy substrates.

16.2.6 Plated Parts (Zinc, Chrome, and Galvanized)

One of the major problems associated with bonding plated parts is the different surface conditions that can be caused by variations in plating equipment, process methods, and solution concentrations. These variables result in plated surfaces with broad conditions of surface finish and inconsistent metallurgical and adhesion properties.

Nickel-plated parts should not be heavily etched or sanded. Roughening the surface to obtain good bonds is unnecessary, as excellent bonds can be made to the smoothest of surfaces as long as it is properly cleaned. The most likely surface treatments are solvent cleaning, vapor degreasing, or soap cleaning. A recommended practice for surface-treating plated parts is light scouring with a nonchlorinated commercial cleaner, rinsing with distilled water, and drying at temperatures under 50°C. The substrate is then primed or bonded as soon as possible after surface preparation.

Chromium and chrome-plated alloys can be etched in a 50% solution of concentrated hydrochloric acid for 2 to 5 min at 90°C. Zinc and galvanized metal parts can be similarly immersed for 2 to 4 min at room temperature in such a solution at 15% concentration. In both cases, the part should be primed or the adhesive applied as soon as possible after surface treatment.

Another metal for which adhesive bonding is widely utilized is galvanized steel. Adhesive bonding is a preferred method of joining this material because it is difficult to weld. Galvanized steel is nominally zinc-coated steel, but in reality it is much more complex. Aluminum, lead, tin, and magnesium are also present, and these contribute to the difficulty in bonding this material. Magnesium oxide, in particular, has virtually no bond strength to magnesium or zinc. The most common way to prepare galvanized steel for bonding is through repeated detergent washings. This removes the magnesium oxide progressively with the aluminum oxide, so that eventually only zinc oxide remains, which is well attached and relatively easily bonded.

16.2.7 Steel and Iron

Because of their widespread use in industry, steel and iron are frequently bonded. Like the surfaces of most metals, their surfaces actually exist as a complex mixture of hydrated oxides and absorbed water. Unfortunately, iron oxides are often not the best surface for adhesives because the oxides may continue to react with the atmosphere after an adhesive has been applied, thus forming weak crystal layers. Iron oxides are more difficult to "engineer" than aluminum, copper, or titanium oxides. As a result, grit blasting is used in most applications. Although it provides adequate adhesion and durability for many applications, grit blasting does not provide great durability in severe environments. Conversion coatings are often used in these cases.

Bonding operations frequently require the mechanical or chemical removal of loose oxide layers from iron and steel surfaces before adhesives are applied. To guard against slow reaction with environmental moisture after the bond has formed, iron and steel surfaces are often phosphated prior to bonding. This process converts the relatively reactive iron atoms to a more passive, "chemically stable" form that is coated with zinc or iron phosphate crystals. Such coatings are applied in an effort to convert a reactive and largely unknown surface to a relatively inert one whose structure and properties are reasonably well understood.

Corrosion protection is critical in bonding steel—even more critical than for many other metallic adherends. The initial adhesion to steel is usually good but deteriorates rapidly during environmental conditioning. Thus, corrosion-preventing primers are usually recommended because they protect the surface against changes after bonding, and the time lapse from cleaning to bonding is not as critical. Steel alloys will form surface oxides in a very short time. Drying cycles after cleaning can be critical. During these processes an alcohol rinse after a water rinse tends to accelerate drying and reduce surface layers that are undesirable.

Mild steel (carbon steel) may require no more extensive treatment than degreasing and abrasion to give moderate-strength adhesive bonds. Tests should be carried out with the actual adhesive to be used to determine whether a chemical etch or other treatment is essential.[36,37,38] Figure 15.21 illustrates the initial and residual strengths obtained with joints of mild steel employing different adhesives and different surface treatments. From these results it may be concluded that the durability of steel bonds produced with a elevated-temperature curing epoxy adhesive is better than that of specimens produced with room temperature curing adhesive systems.

The general sequence of surface preparation for ferrous surfaces such as iron, steel, and stainless steel consists of the following methods: degreasing, acid etch or alkaline clean, rinse, dry, chemical surface treatment, and priming. The chemical surface treatment step is not considered a standard procedure, but it is sometimes used when optimum quality joints are required. It consists of the formation of a corrosion-preventing film of controlled chemical composition and thickness. These films are a complex mixture of phosphates, fluorides, chromates, sulfates, nitrates, etc. The composition of the film may be the important factor that controls the strength of the bonded joint.

Stainless steel, or corrosion-resistant steel, has a high chromium content (11% or higher) as the primary alloying element. The stable oxide film that exists on stainless steel will tightly bind ions from prior manufacturing steps. Rinsing after treatment seems to be especially critical in ensuring the successful bonding of stainless steel surfaces; but even bond quality on well-rinsed surfaces may not be comparable with that on other, more easily bonded materials.

There have been a large number of surface preparation methods reported to give excellent bonds with stainless steels. In addition to mechanical methods, strong acids and strong alkalies are used. A wet abrasive blast with a 200-grit abrasive followed by thorough rinsing to remove the residue is an acceptable procedure for some uses but does not produce high bond strengths. Acid treatments are usually used to produce strong bonds with most adhesives. Passivation in nitric acid solution and concentrated sulfuric acid saturated sodium dichromate solution both produce high bond strength but with low or marginal peel strength. Such joints may fail under vibration stress, particularly when a thin stainless steel sheet is bonded with a brittle adhesive.

Acid etching can be used to treat types 301 and 302 of stainless. These processes result in a heavy black smut formation on the surface. This material must be removed if maximum adhesion is to be obtained. The acid etch process produces bonds with high peel and shear strengths. The 400 series of straight chromium stainless steels should be handled in the same manner as the plain carbon steels. The various types of precipitation hardening (PH) stainless steels each present an individual problem. Processes must be adopted or developed for each type.

Once properly treated, there are practically thousands of organic adhesive compounds that are available for bonding steel alloys. Epoxies are the most common of structural adhesive for bonding steel. Figure 15.15 shows the effect of humidity on the adhesion of two structural epoxy adhesives used to bond stainless steel.

In high-volume industries such as automotive and appliances, there is a desire to minimize or eliminate any surface preparation process for steel. Special adhesive systems have been developed that bond well to steel coming direct from the mill. This substrate is usually coated with light oil for protection against corrosion. Toughened epoxy, epoxy-urethane hybrid, and thermosetting acrylic adhesives have been found suitable for this application. These adhesives usually work with oily surfaces through either (1) thermodynamic displacement of the oil at the surface by the adhesive or (2) absorption of the oil into the body of the adhesive as a plasticizer. Bond strengths of an epoxy-urethane adhesive to oily steel surfaces as well as other substrates commonly found in the auto industry are shown in Table 7.9.

16.2.8 Titanium

Titanium is widely used in aerospace applications that require high strength-to-weight ratios at elevated temperatures. As a result, a number of different prebonding surface preparation processes have been developed for titanium. These generally follow the same sequence as for steel and other major industrial metal substrates: degrease, acid-etch or alkaline-clean, rinse and dry, chemical surface treatment, rinse and dry, and finally prime or bond. Mechanical abrasion is generally not recommended for titanium surfaces.

If chlorinated solvents are used with titanium surfaces, they must be completely removed prior to bonding. Chlorinated solvents give rise to stress corrosion cracking in the vicinity of welds. Welding of titanium often occurs in the same plant as adhesive bonding, and it is sometimes done on the same parts. So the best practice is to avoid the use of chlorinated solvents completely. Several airframe manufacturers that fabricate titanium alloys no longer permit the use of chlorinated solvents.

Different titanium alloys are attacked by acid etching solutions at different rates. Titanium containing lower percentages of alloy elements is generally more resistant; so treating times need to increase. Extreme caution must be used when one is treating titanium with acid etchants that evolve hydrogen. In strongly acidic etching solutions, and particularly in sulfuric acid pickling solutions, there can be appreciable hydrogen pickup during treatment. Hydrogen pickup on surfaces of titanium can cause embrittlement. Immersion times must be closely controlled and minimized.

Shaffer et al. provides an excellent review of adhesive bonding titanium including several surface preparation processes.[39] Of 31 surface preparation procedures for adhesive bonding of titanium alloy, including several anodizing processes, studied by General Dynamics,[40] a phosphate-fluoride (PF) process was selected as optimal. The process was modified by Picatinny Arsenal and is described in MIL-A-9067. This method gives excellent bond durability for both 6,4 titanium and chemically pure (CP) titanium. The former, however, shows a loss of lap shear strength after 5 years' outdoor weathering. The CP titanium does not show this effect.

Alkaline peroxide (AP) surface treatments of titanium have led to bonded structures that possess high adhesive strength and improved resistance at elevated temperatures.[41,42] The use of this process, however, has declined in recent years due to the instability of its components and long treatment times at room temperature (up to 36 h).[43] Titanium surfaces produced by electrochemical anodization typically provide the best initial strengths and long-term durability. This treatment provides a porous cellular oxide morphology that is very compatible with epoxy adhesives. The most commonly used anodic process is chromic acid anodization (CAA).[44,45,46] Recently a sodium hydroxide anodization (SHA) treatment has gained favor because of bond durabilities equal to or greater than those provided by the CAA process,[47] and it is a more environmentally friendly process.

A proprietary alkaline cleaner, Prebond 700, appears to be satisfactory for a number of metal adherends including titanium and is recommended as a versatile one-step surface preparation process.[48] A proprietary alkaline etch solution, Turco 5578, is available from

Turco Product Division, Purex Corporation. When combined with vapor degreasing and alkaline cleaning, this process offers very high-strength bonds on titanium. Plasa Jell is also a proprietary chemical marketed by Semco Division, Products Research and Chemical Corp. This formulation is available either as a thixotropic paste suitable for brush application or as an immersion solution for tank treatment.

The VAST process was developed by Vought Systems of LTV Aerospace Corporation; VAST is the acronym for Vought Abrasive Surface Treatment. In this process, the titanium is blasted in a specially designed chamber with a slurry of fine abrasive containing fluorosilicic acid under high pressure. The aluminum oxide particles are about 280-mesh, and the acid concentration is maintained at 2 percent. The process produces a gray smut on the surface of 6,4 titanium alloy that must be removed by a rinse of 5 percent nitric acid. The resulting joint strength is claimed to be superior to that provided by the unmodified phosphate fluoride process, but is slightly lower than that provided after the Turco 5578 alkaline etch.

Adhesives recommended for bonding titanium substrates include epoxies, nitrile-epoxy, nitrile-phenolic, polyimide, and epoxy-phenolic. Epoxy adhesives have generally been selected when a combination of high initial strength and durability is required. Representative data are shown in Fig. 16.3 for an epoxy adhesive and several proprietary and nonproprietary treating processes. Keith[49] has covered all aspects of titanium adhesive bonding, including adhesive selection.

16.3 PLASTIC BONDING

Plastics are usually more difficult to bond with adhesives than are metal substrates. Plastic surfaces can be unstable and thermodynamically incompatible with the adhesive. The actual bonding surface may be far different from the expected substrate surface. The plastic part can possess physical properties that will cause excessive stress in the joint. The operating environment can change the adhesive-plastic interface, the base plastic, the adhesive, or all three.

However, even with these potential difficulties, adhesive bonding can be an easy and reliable method of fastening one type of plastic to itself, to another plastic, or to a nonplastic substrate. Pocius et al. provides an excellent treatise on the use of adhesives in joining plastics.[51]

The physical and chemical properties of both the solidified adhesive and the plastic substrate affect the quality of the bonded joint. Major elements of concern are the thermal expansion coefficient, modulus, and glass transition temperature of the substrate relative to the adhesive. Special consideration is also required of polymeric surfaces that can change during normal aging or on exposure to operating environments.

Significant differences in the thermal expansion coefficient between the substrate and the adhesive can cause severe stress at the interface. This is common when plastics are bonded to metals because of the difference in thermal expansion coefficients between the substrates. Residual stresses are compounded by thermal cycling and low-temperature service. Selection of a resilient adhesive or adjustments in the adhesive's thermal expansion coefficient via fillers or additives can reduce such stress.

Bonded plastic substrates are commonly exposed to peel because the part thickness is usually small and the modulus of the plastic is low. As a result, tough adhesives with high peel and cleavage strengths are usually recommended for bonding plastics. The requirement of flexibility is especially important for thermoplastics because of their lower modulus and greater thermal expansion coefficient than thermosetting plastics. Table 16.7 shows starting formulations and properties for several epoxy adhesives that are recommended for bonding plastic substrates.

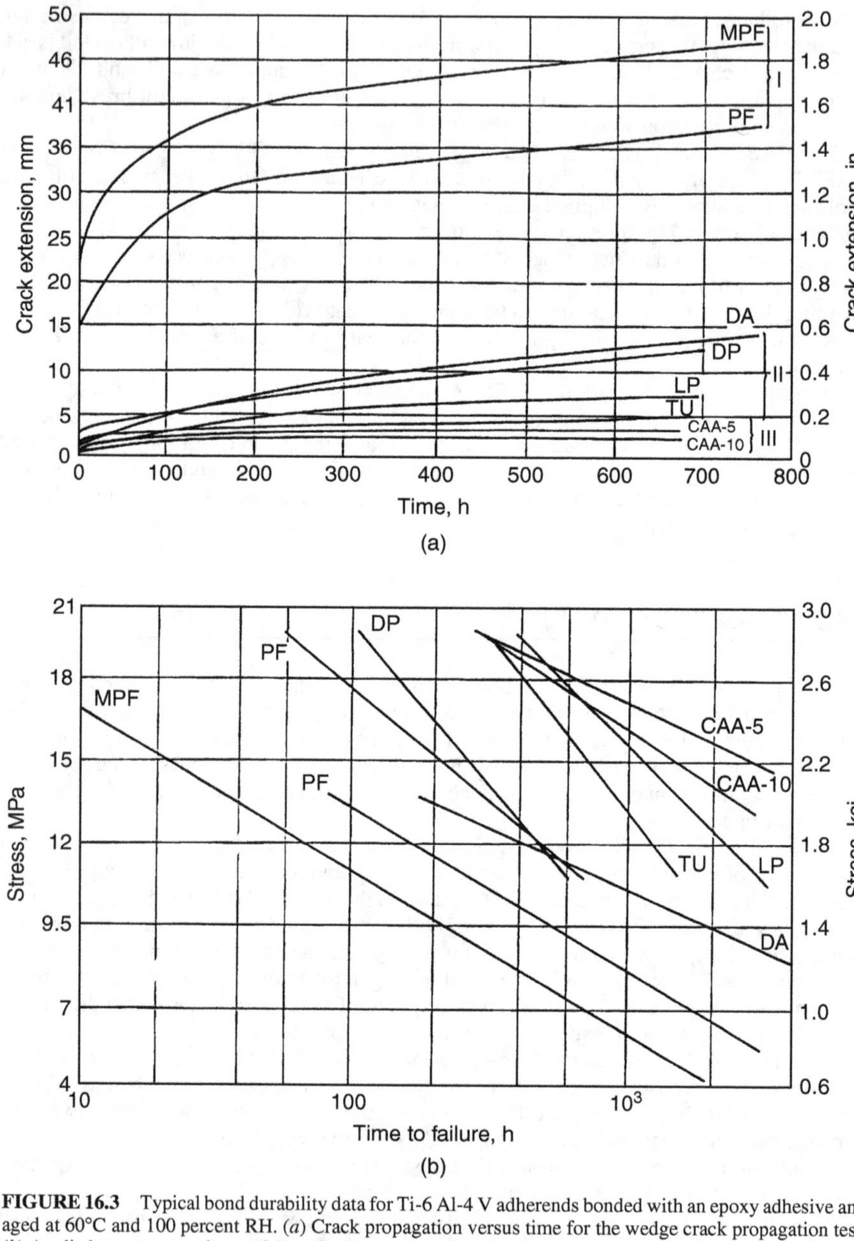

FIGURE 16.3 Typical bond durability data for Ti-6 Al-4 V adherends bonded with an epoxy adhesive and aged at 60°C and 100 percent RH. (*a*) Crack propagation versus time for the wedge crack propagation test. (*b*) Applied stress versus time to failure for the lap shear geometry. PF—phosphate fluoride; MPF—modified phosphate fluoride; DP—PasaJell 109 dry hone; LP—PasaJell 107 liquid hone; CAA-5—5% solution; CAA-10—10% solution; TU—Turco 5578 etch; DA—Dapcotreat.[50]

TABLE 16.7 Starting Formulations for Several Epoxy Adhesives Recommended for Bonding Plastic Substrates[52]

	Parts by weight						
Component	A	B	C	D	E	F	G
Part A							
Modified epoxy resin (EPON 8132, Resolution Performance Products)	75				100	100	100
DGEBA resin (EPON 828)	25	75					
CTBN modified bisphenol A resin (EPON 58005)		25					
Modified epoxy resin (EPON 862)			75				
CTBN modified bisphenol F resin (EPON 58003)			25				
Elastomer modified bisphenol A resin				100			
Fumed silica (Cab-O-Sil M5)	2	2	2	2	2	2	2
Part B							
Amidoamine (EPI-CURE 3055, Resolution Performance Products)	42	42	46	42			
Polyamide (EPI-CURE 3125)					50		
Polyamide (EPI-CURE 3163)						138	
Polyamide (EPI-CURE 3164)							86
Property							
Gel time, min, at 25°C, 100 g	345	165	178	142	193	113	109
Thin film setting time, h	16	11	13	13	10	20.5	34.5
Tensile shear strength, psi, after 7 days at 25°C on							
• PVC	691	677	754	776	308	549	529
• Polyurethane	479	435	759	573	718	906	730
• Polyethylene terephthalate	255	254	487	506	525	800	632
Tensile shear strength, psi, after 30-min cure at 140°C on							
• PVC	1216	1281	939	951	661	969	674
• Polyurethane	319	550	385	407	307	583	429
• Polyethylene terephthalate	768	653	890	796	964	1244	848
• Nylon	434	—	414	444	650	725	466
Glass transition temperature, °C	60	86	75	70	50	45	40

Structural adhesives must have a glass transition temperature higher than the operating temperature or preferably higher than that of the part that is being bonded, to avoid a cohesively weak bond and possible creep problems at elevated temperatures. Modern engineering plastics, such as polyimide or polyphenylene sulfide, have very high glass transition temperatures. Common adhesives have a relatively low glass transition temperature, so that the weakest thermal link in the joint may often be the adhesive. The use of an adhesive too far *below* the glass transition temperature could result in low peel or cleavage strength. Brittleness of the adhesive at very low temperatures could also manifest itself in poor impact strength.

Plastic substrates could be chemically active, even when isolated from the operating environment. Many polymeric surfaces slowly undergo chemical and physical change.

The plastic surface, at the time of bonding, may be well suited to the adhesive process. However, after aging, undesirable surface conditions may present themselves at the interface, displace the adhesive, and result in bond failure. These weak boundary layers could come from the environment or from within the plastic substrate itself. Plasticizer migration and degradation of the interface through uv radiation are common examples of weak boundary layers that can develop with time at the interface.

Moisture, solvent, plasticizers, and various gases and ions can compete with the cured adhesive for bonding sites. The process by which a weak boundary layer preferentially displaces the adhesive at the interface is called *desorption*. Moisture is the most common desorbing substance, being present both in the environment and within many polymeric substrates. Solutions to the desorption problem consist of eliminating the source of the weak boundary layer or selecting an adhesive that is compatible with the desorbing material. Excessive moisture can be eliminated from a plastic part by drying the part before bonding. Additives that can migrate to the surface can possibly be eliminated by reformulating the plastic resin. Also, certain adhesives are more compatible with oils and plasticizers than others. For example, the migration of plasticizer from flexible polyvinyl chloride can be counteracted by using a nitrile-based adhesive. Epoxy-nitrile rubber and to a lesser extent epoxy-urethane hybrid adhesives are capable of absorbing the plasticizer without degrading.

16.3.1 Thermosetting Plastic Substrates

Thermosetting plastics cannot be dissolved in solution and do not have a melting temperature since these materials are crosslinked. Therefore, they cannot be heat- or solvent-welded. In some cases, solvent solutions or heat-welding techniques can be used to join thermoplastics to thermoset materials. However, most thermosetting plastics are not particularly difficult to bond with adhesive systems. They are generally bonded with many different types of adhesives such as epoxies, acrylics, and urethanes. Since thermosetting parts are often highly filled and rigid, a flexible adhesive is not so important as one that can resist the service environment and provide practical joining processes.

In general, unfilled thermosetting plastics tend to be harder, more brittle, and not as tough as thermoplastics. Thus, it is common practice to add filler to thermosetting resins. These fillers can affect the nature of the adhesive bond (either positively or negatively) and are a possible source of lot-to-lot and supplier-to-supplier variability.

Thermosetting plastics, being chemically crosslinked, shrink during cure. Sometimes the cure is not entirely complete when the part is bonded. In these cases, cure of the part can continue during the bonding operation or even on aging in service, resulting in shrinkage and residual stresses in the joint. Depending on the nature of the crosslinking reaction, volatile by-products could also generate due to postcuring of the part and could provide materials for a weak boundary layer.

The surface of thermoset materials may be of slightly different chemical character than the material beneath the surface because of surface inhibition during cure or reaction of the surface with oxygen and/or humidity in the surroundings. By abrading the surface, a more consistent material is available for the adhesive to bond.

Abrasion and solvent cleaning are generally recommended as a surface treatment for thermosetting plastics. Frequently a mold release agent is present on thermoset materials and must be removed before adhesive bonding. Mold release agents are removed by a detergent wash, solvent wash, or solvent wipe. Clean, lint-free cloth or paper tissue is commonly used, and steps must be taken to ensure that the cleaning materials themselves do not become contaminated. Cleaning solvents used for thermosetting materials are acetone, toluene, trichloroethylene, methyl ethyl ketone (MEK), low-boiling petroleum ether, and isopropanol.

Similar surface abrasion processes can be applied on all thermosetting plastics. Mechanical abrasion methods consist of abrasion by fine sandpaper, carborundum or alumina abrasives, metal wools, or steel shot. The following surface treatment procedure is usually recommended for most thermosetting plastics:

1. Solvent-degrease (with MEK or acetone).
2. Grit- or vapor-blast, or abrade with 100- to 300-grit emery cloth.
3. Wash with solvent.

The roughness of the abrasion media can vary with the hardness of the plastic. Usually, this is not a critical parameter except where decorative surfaces are important.

Adhesives commonly used on thermosetting materials include epoxies, urethanes, cyanoacrylates, thermosetting acrylics, and a variety of nonstructural adhesive systems. The following discussion includes a very brief description of various thermosetting substrate materials, the properties that are critical relative to epoxy adhesion, and any special processes that should be noted for the particular substrate.

Alkyds. Alkyd resins consist of a combination of unsaturated polyester resins, a monomer, and fillers. Alkyd compounds generally contain glass fiber filler, but they may also include clay, calcium carbonate, alumina, and other fillers. Alkyds have good heat, chemical, and water resistance, and they have good arc resistance and electrical properties. Alkyds are easy to mold and economical to use. Postmolding shrinkage is small. Their greatest limitation is extremes of temperature (above 175°C) and humidity.

Alkyd parts are generally very rigid, and the surfaces are hard and stiff. Surface preparation for alkyd parts consists of simple solvent cleaning and mechanical abrasion. Epoxies, urethanes, cyanoacrylates, and thermosetting acrylics are commonly used as structural adhesives.

Diallyl Phthalate. Diallyl phthalates are among the best of the thermosetting plastics with respect to high insulation resistance and low electrical losses. These properties are maintained up to 200°C or higher and in the presence of high-humidity environments. Also, diallyl phthalate resins are easily molded and fabricated. There are several chemical variations of diallyl phthalate resins, but the two most commonly used are diallyl phthalate (DAP) and diallyl isophthalate (DAIP). The primary difference is that DAIP will withstand somewhat higher temperatures than will DAP. Both DAP and DAIP have excellent dimensional stability and low shrinkage after molding. Surfaces are hard and tough, and they pick up very little moisture. DAP parts are ordinarily molded or laminated with glass fibers. Only filled molding resins are commercially available.

Typical surface preparation calls for cleaning with acetone, MEK, or other common solvent. Once clean, the substrate is then mechanically abraded with sand, grit or vapor blast, or steel wool. The surface is again wiped clean with fresh solvent. Typical adhesives that are employed include epoxies, urethanes, and cyanoacrylates. Polysulfides, furanes, and polyester adhesives have also been suggested.

Epoxy. Epoxy resins are one of the most commonly used thermosetting materials. They offer a wide variety of substrate properties depending on base resin, curing agent, modifiers, fillers, and additives. Epoxies show good dimensional stability, electrical properties, and mechanical strength. They have good creep resistance and will operate over a wide temperature range. However, high temperatures tend to oxidize epoxies after long periods. Loose surface particles caused by UV exposure (chalk) could provide a weak boundary layer. Both filled and unfilled grades are available. Common fillers include minerals, glass, silica, and glass or plastic microballoons. Epoxy composites are available with continuous or discontinuous reinforcing materials of several types.

The common surface preparation treatment for epoxy resins is to wipe with solvent, mechanical abrasion, and final solvent cleaning. Epoxy parts can be most easily bonded with an epoxy adhesive similar to the material being bonded. Urethanes, cyanoacrylates, and thermosetting acrylics have also been used when certain properties or processing parameters are required.

Phenolic, Melamine, and Urea. The phenolics are heavily commercialized thermosetting materials that find their way into many applications. They have an excellent combination of physical strength and high-temperature resistance. They have good electrical properties and dimensional stability. Like epoxies and diallyl phthalate, phenolic resins are often found to contain fillers and reinforcement.

Phenolics are formed by a condensation reaction. This results in the formation of water during cure. If a phenolic substrate is not completely cured, it may continue to cure during processing or when exposed to elevated temperatures in service and liberate additional water vapor. This water vapor could form a weak boundary layer and drastically reduce the strength of a bonded joint.

The phenolics can be surface-treated by any of the standard processes for thermosetting materials. Solvent cleaning and mechanical abrasion are commonly employed for high joint strengths. The parts must be completely dry and cleaned of any mold release.

Phenolic parts can be bonded with a wide variety of adhesives. With many adhesives it is possible that the bond strength of the joint will be greater than the strength of the adherend. Phenolics are used in many high-temperature applications. Adhesives with good high-temperature properties are required when these parts are bonded. For this reason, high-temperature epoxy, nitrile-phenolic, and urea-formaldehyde adhesives are commonly used. Neoprenes and elastomeric contact adhesives are used for bonding decorative laminate.

Melamine (melamine formaldehyde) resins and urea (urea formaldehyde) resins are similar to the phenolics. They are hard, rigid materials that have excellent electrical and abrasion-resistant characteristics. Melamine parts are also noted for high impact resistance and resistance to water and solvents. Only filled melamine resins are available.

As a result of their properties, melamines are often used as decorative laminates. The melamine resins cure via an addition reaction mechanism so no reaction by-products can be produced on postcure as with the phenolic resins. The specific surface preparation for adhesive bonding and the preferred adhesives for bonding melamine and urea parts are similar to those suggested for phenolic resins.

Polyimides. Polyimide plastics have exceptional thermal stability. Some types of polyimides are able to withstand temperatures up to 480°C for short periods. Polyimides are available as both thermosetting and thermoplastic materials. Certain types of thermosetting polyimides can be cured either by a condensation reaction or by an addition reaction mechanism. Polyimides are available in many forms such as fabricated parts, molding resins, films, and coatings. Polyimides are often filled for greater physical properties and dimensional stability. In certain bearing products, polyimides are filled with low-friction materials such as graphite, polytetrafluoroethylene (PTFE), and molybdenum disulfide. These low-surface-energy fillers could interfere with adhesion when they are present at the interface.

Polyimide parts can be bonded using either a standard thermosetting substrate surface treatment listed above or one of a number of specialty processes developed for higher adhesive strength and permanence. DuPont recommends that Vespel polyimide parts be bonded by a process consisting of solvent cleaning in trichloroethylene, perchloroethylene, or trichloroethylene. Because of the high-temperature and solvent resistance of this plastic, parts have also been cleaned by refluxing in the solvent and by ultrasonic agitation. Once clean, the parts are abraded with a wet or dry abrasive blast, solvent-cleaned again, and dried before bonding.

A sodium hydroxide etch process has been developed for polyimide parts that require maximum adhesive strength.[53] The parts are first degreased and then etched for 1 min at 60 to 90°C in 5 percent solution of sodium hydroxide in water. After etching, the parts are rinsed in cold water and air-dried.

Thermosetting polyimide materials can be bonded with any structural adhesive. Unfortunately, the high-temperature strength of the adhesives is generally not as good as the high-temperature characteristics of the polyimide. Where the maximum strength is needed at elevated temperatures, high-temperature curing epoxy, phenolics, or polyimide adhesive formulations are generally applied. Where high strengths are not crucial, silicone adhesives are commonly employed.

Polyester. Polyesters are also common thermosetting resins that are used in many applications such as automobile and boat parts, fiberglass structures, and molding compounds. Polyesters are generally heavily filled and/or contain reinforcing fibers such as glass. Polyester resins are somewhat lower in cost than epoxy resins and can be formulated to have very short molding times. They have good resistance to water, good electrical properties, good resistance to oils and solvents, and high strength-to-weight ratios. Polyester parts exhibit a high shrinkage rate.

Adhesive bonding of polyester parts could be complicated by the addition of a variety of additives in the resin to enhance curing, mold release, and surface gloss. In certain polyester resin formulations that are cured in contact with the atmosphere, waxes are included to shield the free radical polymerization process from inhibition by oxygen. Certain polyester mold release agents are formulated directly into the resin (internal mold release). For a glossy finished part appearance, thermoplastic polyolefins are sometimes incorporated into the polyester formulation. Surface preparation must take into consideration these possible weak boundary layers. The recommended surface preparation is simple solvent cleaning and mechanical abrasion. Generally, the same surface preparations as recommended for epoxies and phenolics are recommended for polyesters.

Polyester parts are frequently bonded with epoxy, polyester, or polyurethane adhesives. Polyester adhesives, however, form a rigid bond and exhibit a high shrinkage on curing, resulting in internally stressed joints. A primer is usually not necessary for polyester parts. Good results have been obtained in the automotive industry with two-part epoxies and one- and two-part urethanes.[54] Good outdoor weather resistance of polyester fiberglass-aluminum bonds has been reported for epoxy, acrylic, and silicone adhesives.[55]

Silicone. Silicones are a family of unique polymers that are partly organic and partly inorganic. Silicones have outstanding thermal stability. Silicone polymers may be filled or unfilled. They can be cured by several different mechanisms and are available as both flexible and rigid resins. They are low-surface-energy materials and are generally difficult to bond with adhesives. In fact, the poor bonding characteristics of silicone surfaces are often used profitably for baked-on release agents and graffiti-resistant paints. Silicone resins are also used for molding compounds, laminates, impregnating varnishes, high-temperature paints, and encapsulating materials.

Rigid silicone resins are used as organic coatings, electrical varnishes, laminates, and circuit-board coatings. The noncoating products are generally filled or reinforced with mineral fillers or glass fibers. Thermosetting molding compounds made with silicone resins are finding wide application in the electronic industry as encapsulants for semiconductor devices.

Rigid silicone resins also have a low surface energy and are difficult materials to bond with adhesives or sealants. Parts should be clean and dry, and the best bonding candidate is another silicone resin (either rigid or flexible) plus a compatible primer. Flexible silicones elastomers are discussed in Sec. 16.6.

Thermosetting Polyurethane. Polyurethanes can be furnished as either thermosetting or thermoplastic material. The thermosetting variety can be obtained in a number of different densities, rigidities, and forms, depending on the curing agents and reaction mechanism. They are excellent for low temperatures including cryogenic applications and have good chemical resistance, skid resistance, and electrical properties. Polyurethanes are limited to 125°C maximum temperature applications. Polyurethanes have good electrical properties and are often used in electrical applications. Polyurethanes are elastomeric materials that deform under pressure. This could cause internal strains to be frozen into the joint if the part is bonded under pressure.

Substrate cleaning usually involves the light sanding of a clean, dry bonding surface. A primer (urethane or silane) is sometimes used to improve adhesion. Urethanes are generally bonded with a flexible epoxy or a urethane adhesive system.

16.3.2 Thermoplastic Substrates

Unlike the thermosetting resins, the thermoplastic resins will soften on heating or on contact with solvents. They will then harden on cooling or on evaporation of the solvent from the material. This is a result of the noncrosslinked chemical structure of thermoplastic molecules. The following are important characteristics of thermoplastic resins that can affect their joining capability.

- Many thermoplastic compounds are alloyed and really consist of two or more resins.
- Additives and mold release agents are commonly employed in the formulation.
- Thermoplastics exhibit a relatively high degree of water absorption and mold shrinkage.
- Dimensional changes due to moisture migration, thermal expansion, etc., are generally greater than with other materials.
- The properties of the surface, such as surface energy and crystallinity, may be different from those of the bulk (this is especially true for thermoplastics that are molded at very high temperatures).
- Many grades of the same material are available (high-flow, high-density, etc.).

Thermoplastic materials often have a lower surface energy than do thermosetting materials. Thus, physical or chemical modification of the surface is necessary to achieve acceptable bonding. This is especially true of the crystalline thermoplastics such as polyolefins, linear polyesters, and fluoropolymers. Methods used to increase the surface energy and improve wettability and adhesion include

- Oxidation by chemical means or flame treatment
- Roughening of the surface by electrical discharge
- Gas plasma treatment

The reasons that these surface treatments improve the adhesion of epoxy adhesive to plastic substrates are summarized in Table 16.8.

As with metal substrates, the effects of plastic surface treatments decrease with time, so it is important to carry out the priming or bonding as soon as possible after surface preparation. The surface preparation methods suggested in App. F are recommended for conventional adhesive bonding. Greater care must be taken in cleaning thermoplastics than

TABLE 16.8 Characterization of Common Surface Treatments for Polymers[56]

Pretreatment type	Possible effects of pretreatment
Solvent	Removal of contaminants and additives. Roughening (e.g., trichloroethylene vapor with polypropylene). Weakening of surface regions if excessively attacked by solvent
Mechanical	Removal of contaminants and additives. Roughening of surface
Oxidative (flame, corona, acid etching)	Removal of contaminant and additives. Introduction of functional groups. Change in topography (e.g., roughening with chromic acid treatment of polyolefins)
Plasma	Removal of contaminants and crosslinking (if inert gas used). Introduction of functional groups if active gases such as oxygen are used. Grafting of monomer to polymer surface after activation, e.g., by argon plasma

with thermosets. Thermoplastic parts can be attacked or can swell on contact with certain solvents. Therefore, cleaning solvent selection must be made depending on the materials being joined.

Unlike thermosets, many thermoplastics can be joined by solvent cementing or thermal welding methods. Solvent cementing and thermal welding do not require abrasion or chemical treatment of the plastic surfaces. The surfaces must be clean, however, and free of impurities that could cause a weak boundary layer. Bond strengths achieved by solvent or thermal welding are generally as high as or higher than those of adhesive bonding. Bond strengths are often greater than 80 percent of the strength of the substrate material.

Acrylonitrile-Butadiene-Styrene (ABS). ABS plastics are derived from acrylonitrile, butadiene, and styrene. ABS materials have a good balance of physical properties. There are many ABS modifications and many blends of ABS with other thermoplastics that can affect adhesion properties. ABS resin can be bonded to itself and to other materials with adhesives, by solvent cementing, or by thermal welding.

When solvent welding or thermal welding is not practical or desired, adhesive systems can be used. Adhesive types such as epoxies, urethanes, thermosetting acrylics, nitrile phenolics, and cyanoacrylates permit ABS to be bonded to itself and to other substrates. The best adhesives have shown strength greater than that of ABS; however, these adhesives provide very rigid bonds.

For low- to medium-strength bonds, simple mechanical abrasion is a suitable surface preparation if the substrates are cleaned first. This surface preparation has found success in most ABS applications. A silane primer such as Dow Corning A-4094 or General Electric SS-4101 may be used for higher strength.[57] For maximum joint strength, a warm chromic acid etch of the ABS substrate is suggested.[58]

Electroplated ABS is used extensively for many electrical and mechanical products in many forms and shapes. Adhesion of the electroplated metal to the ABS resin is excellent; therefore, the electroplating does not generally have to be removed to have a good-quality joint.

Acetal. Acetals are among the group of high-performance engineering thermoplastics that resemble nylon somewhat in appearance but not in properties. Acetals are strong and tough and have good moisture, heat, and chemical resistance. There are two basic types of acetals: the homopolymers by DuPont and the copolymers by Celanese. The copolymers are more stable in long-term, suitable for high-temperature service, and are more resistant to hot water.

The homopolymers are more rigid, stronger, and have greater resistance to fatigue. Both types of acetal resins are degraded by uv light.

Because acetals absorb a small amount of water, the dimensions of molded parts are affected by their water content, which varies with the relative humidity of the environment. The absorbed water should be considered when bonding because it could migrate to the interface during exposure to elevated temperatures from the joining process or from the service environment. This released moisture could create a weak boundary layer. Dimensional changes must also be considered when acetal parts are bonded to other substrates and then exposed to changing temperature and humidity conditions.

The acetal surface is generally hard, smooth, glossy, and not very easy to bond with adhesives. The nonstick and solvent-resistant nature of acetal requires that the surfaces be specially prepared before adhesive bonding can occur. Once prepared, the surface can then adhere to like substrates or to others.

Because of the solvent and chemical resistance of acetal copolymer, special etching treatments have been developed for surface preparation prior to adhesive bonding. A chromic acid etch and a hydrochloric acid etch have been suggested. Acetal parts that have been formed by heat treatment or machining should be stress-relieved before etching.

Acetal homopolymer can be effectively bonded with various adhesives after the surface is sanded, etched with a chromic acid solution at elevated temperature, or "satinized." *Satinizing* is a patented process developed by DuPont for preparing Delrin acetal homopolymer for painting, metallizing, and adhesive bonding. In this process, a mild acidic solution produces uniformly distributed anchor points on the adherend surface. Adhesives bond mechanically to these anchor points, resulting in strong adhesion to the surface of the homopolymer. Performance of various epoxy adhesives on acetal homopolymer is shown in Table 16.9. Oxygen plasma and corona discharge treatment have also shown to be effective on acetal substrates.[59] Epoxies, nitrile, and nitrile phenolics can be used as adhesives with acetal homopolymer substrates that are treated first.

Epoxies, isocyanate cured polyester, and cyanoacrylates are used to bond acetal copolymer. Generally, the surface is treated with sulfuric-chromic acid. Epoxies have shown 150- to 500-psi shear strength on sanded surfaces and 500- to 1000-psi on chemically treated surfaces. Two-component epoxies give slightly lower bond strengths. However, they bond acetal to itself and to many other materials.

TABLE 16.9 Performance of Adhesives with Acetal Homopolymer[60]

		Tensile shear strength, psi				
		Acetal to acetal		Acetal to aluminum	Acetal to steel	
Adhesives	Curing conditions	Sanded	Satinized	Satinized	Sanded	Satinized
Modified epoxy	24 h or 1 h at 95°C	150	450	400	250	830
Epoxy-polyamide	24–48 h at 25°C	150	850	600	200	600
Epoxy A	8 h or 30 min at 65°C	500	600	60	300	50
Epoxy B	1 h at 65°C		400	150		200
Epoxy C	48 h or 2 h at 75°C	500	500	600	500	600

Acrylic. Acrylic resins (polymethyl methacrylate) have exceptional optical clarity and good weather resistance, strength, electrical properties, and chemical resistance. They have low water absorption characteristics. However, acrylics are attacked by strong solvents, gasoline, acetone, and similar organic fluids.

Adhesive bonded acrylic joints usually give lower strength than solvent- or heat-welded joints. Cyanoacrylate, epoxy, and thermosetting acrylic adhesives offer good adhesion but poor resistance to thermal aging. Epoxy adhesive will generally provide tensile shear strength on acrylic substrates of greater than 3 MPa.[61]

Surface preparation for bonding can be accomplished by wiping with methanol, acetone, MEK, trichloroethylene, isopropanol, or detergent; abrading with fine abrasive media; wiping with a clean, dry cloth and repeating a clean solvent wipe. Care must be exercised, however, when one is contacting solvent with molded acrylic parts. Stress cracking or crazing can occur due to a solvent attack at highly stressed areas within the molded part. The tendency for stress cracking can be significantly reduced by annealing the part at a temperature slightly below the heat distortion temperature.

Cellulosics (Cellulose Acetate, Cellulose Acetate Butyrate, Cellulose Propionate, Ethyl Cellulose, Cellulose Nitrate). Cellulosics are among the toughest of plastics. However, they are temperature-limited and are not as resistant to extreme environments as other thermoplastics. The four most prominent industrial cellulosics are cellulose acetate, cellulose acetate butyrate, cellulose propionate, and ethyl cellulose. A fifth member of this group is cellulose nitrate.

Cellulosic resins are formulated with a wide range of plasticizers for specific properties. The extent of plasticizer migration should be determined before cementing cellulose acetate (and, to a much lesser extent, butyrate and propionate) to cellulose nitrate, polystyrene, acrylic, or polyvinyl chloride. Plasticizer migration in some cases will cause crazing or softening of the mating material or degradation of the adhesive.

Cellulosics are normally solvent-cemented unless they are to be joined to another substrate. In these cases, conventional adhesive bonding is employed. Polyurethane, epoxy, and cyanoacrylate adhesives are commonly used to bond cellulosics. Surface treatment generally consists of solvent cleaning and abrasion. Cellulosics can be stress-cracked by uncured cyanoacrylate adhesives and some components of acrylic adhesives. A recommended surface cleaner is isopropyl alcohol.

Chlorinated Polyether. This thermoplastic resists most solvents and is attacked only by nitric acid and fuming sulfuric acids. Thus, it is not capable of being solvent-cemented. Chlorinated polyether parts can be bonded with epoxy, polyurethane, and polysulfide-epoxy adhesives after treatment with a hot chromic acid solution. Tensile shear strength of 1270 psi has been achieved with an epoxy-polysulfide adhesive.

Fluorocarbons. There are eight types of common fluorocarbons. They differ primarily by the concentration and arrangement of fluorine atoms along their molecular chain. Chemical types, suppliers, and trade names are given in Table 16.10. Like other plastics, each type of fluorocarbon is available in several different grades. They differ principally in the way they are processed and formed, and their properties vary over the useful temperature range.

The original basic fluorocarbon, and perhaps the most widely known one, is tetrafluoroethylene (TFE). It has the optimum electrical and thermal properties and almost complete moisture resistance and chemical inertness. However, TFE does cold-flow or creep at moderate loading and temperatures. Filled modifications of TFE resins are available; these are generally stronger than unfilled resins. Fluorinated ethylenepropylene (FEP) is similar to TFE except that its operating temperature is limited to 200°C. FEP is more easily processed,

TABLE 16.10 Chemical Types of Fluorocarbons

Fluorocarbon	Common designation	Trade names and suppliers
Tetrafluoroethylene	TFE	Teflon TFE (DuPont) Halon TFE (Allied Chemical)
Fluorinated ethylenepropylene	FEP	Teflon FEP (DuPont)
Ethylene-tetrafluoroethylene copolymer	Copolymer of ethylene and TFE	Tefzel (DuPont)
Perfluoroalkoxy	PFA	Teflon PFA
Chlorotrifluoroethylene	CTFE	Kel-F (3M)
Ethylene-chlorotrifluoroethylene copolymer	E-CTFE	Halar E-CTFE (Allied Chemical)
Vinylidene fluoride	PVDF	Kynar (Pennsalt Chemicals)
Polyvinyl fluoride	PVF	Tedlar (DuPont)

and it can be molded, which is not possible with TFE. Ethylene-tetrafluoroethylene copolymer (ETFE) is readily processed by conventional methods including extrusion and injection molding. Perfluoroalkoxy (PFA) is a class of melt-processable fluoroplastics that perform successfully in the 260°C area. Chlorotrifluoroethylene (CTFE) resins are also melt-processable. They have greater tensile and compressive strengths than TFE within their service temperature range. Ethylene-chlorotrifluoroethylene copolymer (E-CTFE) is a strong, highly impact-resistant material. E-CTFE retains its strength and impact resistance down to cryogenic temperatures. Vinylidene fluoride (PVF_2) is another melt-processable fluorocarbon with 20 percent lower specific gravity compared to that of TFE and CTFE. Thus, it is economical to use PVF_2 in parts requiring a useful temperature range from –62 to 150°C. Polyvinyl fluoride (PVF) is manufactured as film and has excellent weathering and fabrication properties. It is widely used for surfacing industrial, architectural, and decorative building materials.

Because of their high thermal stability and excellent resistance to solvent, fluorocarbons cannot be joined by solvent cementing, and they are very difficult to join by thermal welding methods. Because of their inertness and low surface energy, they also tend to be difficult materials to join by adhesive bonding. Surface treatment is necessary for any practical bond strength to the fluorocarbon parts.

Surface preparation consists of wiping with acetone; treating with commercially available sodium naphthalene solutions; washing again with acetone and then with distilled or deionized water; and drying in a forced-air oven at 40°C. Because of the hazardous nature of these surface treatment chemicals, the user must exercise extreme caution and follow all the manufacturers' recommendations. The sodium naphthalene etching process removes fluorine atoms from the surface of the substrate, leaving it with a higher surface energy and a surface that is more wettable by conventional adhesives. The sodium naphthalene treated fluorocarbon surface is degraded by uv light and should be protected from direct exposure.[62]

Sodium napthalene solutions in tetrahydrofuran solvent have been used to provide surface treatment of fluorocarbon surfaces for several decades. A commercial surface treatment based on this process has been Tetra-Etch.[63] Recently the manufacturers of Tetra-Etch introduced Fluoro-Etch, which has several advantages. Fluoro-Etch requires no refrigeration, is easier to handle, and offers significant cost, efficiency, and safety advantages. It offers a 1-year shelf life and is manufactured in both the United States and Europe.

Flexible epoxy adhesives generally give good bond strengths to treated fluorocarbon surfaces. Table 16.11 shows the effect of surface treatments on the bondability of epoxy

TABLE 16.11 Effect of Surface Treatment on Bondability of Epoxy Adhesive to Teflon (Tetrafluoroethylene) and Kel-F (Chlorotrifluoroethylene)[65]

	Tensile shear strength, psi			
Substrate	No treatment	Abraded	Treated*	Abraded and treated*
Kel-F 270	380	1120	1570	3010
Kel-F 300	550	1250	2820	2910
Kel-F 500	780	1580	2030	2840
Teflon TFE			1150	

*Treated with sodium naphthalene etch solution.

adhesives to polytetrafluoroethylene (Teflon) and polychlorotrifluoroethylene (Kel-F). Plasma treatment can also be used to increase the wettability of fluorocarbon surfaces,[64] but the joint strength is not as high as with the sodium naphthalene etching treatment.

Nylon (Polyamide). Nylons, also known as polyamides, are strong, tough thermoplastics with good impact, tensile, and flexural strengths from freezing temperatures up to 150°C. They provide excellent low-friction properties and good electrical resistivities. Four common varieties of nylon are identified by the number of carbon atoms in the diamine and dibasic acid that are used to produce that particular grade. They are referred to as nylon 6, nylon 6/6, nylon 6/10, and nylon 11. These materials generally vary by their processing characteristics and their dimensional stability.

All nylons absorb some moisture from environmental humidity. Moisture absorption characteristics must be considered in designing and joining these materials. They absorb from 0.5 to 2 percent by weight of moisture after 24-h water immersion. Freshly molded objects contain less than 0.3 percent moisture since only dry molding powder can be successfully molded. Once molded, these objects absorb moisture when they are exposed to humid air or water. The amount of absorbed moisture increases until an equilibrium condition is reached based on the relative humidity of the environment. Equilibrium moisture contents of two commercial nylon resins for two humidity levels are as follows:

	Zytel 101	Zytel 31
50 percent RH air	2.5%	1.4%
100 percent RH air (or water)	8.5%	3.5%

The absorbed water can create a weak boundary layer under certain conditions. Generally, parts are dried to less than 0.5 percent moisture before bonding. Caution must be observed when one is mating the nylon to another substrate. The nylon part grows and shrinks due to ingress and egress of moisture from within the substrate. This leads to stresses at the interface that may cause warpage and degradation of the bond strength.

Various commercial adhesives have been used to provide bond strength with nylon on the order of 250 to 1000 psi. Priming of nylon adherends with a composition based on resorcinol formaldehyde, isocyanate modified rubber, and cationic surfactants has been reported to provide improved joint strength. Some epoxy, resorcinol formaldehyde, phenol-resorcinol, and rubber-based adhesives have been found to produce satisfactory joints between nylon and metal, wood, glass, and leather. Exposure of nylon 6 to oxygen and helium plasmas for 30 s to 1 min improved the adhesion of two-part epoxy adhesives.[66]

Polycarbonate. Polycarbonates are among those plastic materials that are grouped as engineering thermoplastics because of their high-performance characteristics in engineering designs. Polycarbonates are especially outstanding in impact strength, having strengths several times higher than those of other engineering thermoplastics. Polycarbonates are tough, rigid, and dimensionally stable. Polycarbonate resins are available in either transparent or colored grades. An important molding characteristic is the low and predictable mold shrinkage, which sometimes gives polycarbonates an advantage over nylons and acetals for close-tolerance parts. Polycarbonates are also alloyed with other plastics to increase strength and rigidity. Polycarbonates are somewhat hygroscopic, like nylon, so it is important to keep the humidity low before bonding.

Polycarbonate plastics are generally joined by solvent cementing or thermal welding methods. However, caution needs to be used when one is solvent-cementing because these plastics can stress-crack in the presence of certain types of solvents. When one is joining polycarbonate parts to metal parts, a room temperature curing adhesive is suggested to avoid stress in the interface caused by differences in thermal expansion. When adhesives are used to join polycarbonate, the epoxies, urethanes, and cyanoacrylates are generally chosen. Adhesive bond strengths with polycarbonate are generally 1000 to 2000 psi.

Recommended solvents for cleaning are methanol, isopropanol, petroleum ether, heptane, and white kerosene. Ketones, toluene, trichloroethylene, benzol, and a number of other solvents including paint thinners can cause crazing or cracking of the polycarbonate surface. After cleaning, possible surface preparations include flame treatment or abrasion. In flame treatment, the part is passed through the oxidizing portion of a propane flame. Treatment is complete when all surfaces have a high gloss, and then the part must be cooled before bonding.

Polyolefins (Polyethylene, Polypropylene, Polymethylpentene). This large group of polymers, classified as polyolefins, is basically waxlike in appearance and extremely inert chemically. They exhibit a greater decrease in strength at lower temperatures than the higher-performance engineering thermoplastics. All these materials come in various processing grades. Polyethylenes are relatively soft with thermal stability in the 90 to 125°C range depending on grade. Polypropylenes are chemically similar to polyethylene, but they have somewhat better physical strength at lower density. Mold shrinkage is less of a problem and more predictable than with the other polyolefins. Polymethylpentene is a rigid, chemically resistant polyolefin that has greater thermal stability than the other polyolefins. It is used in microwave-compatible containers, autoclavable medical products, and the like.

Because of their excellent chemical resistance, polyolefins are impossible to join by solvent cementing. Because of their very low surface energy, polyolefins can only be adhesively bonded after surface treatment processes. The most common way of joining polyolefins is by thermal welding techniques.

To obtain a usable adhesive bond with polyolefins, the surface must be treated. A number of surface preparation methods, including flame, chemical, plasma, and primer treatments, are in use. Figure 16.4 illustrates the epoxy adhesive strength improvements that can be made by using various prebond surface treatments to change the critical surface tension of polyethylene.

The chromic acid etch method, similar to the FPL etch developed for treating aluminum, had been recognized as one of the more effective ways of surface-treating polyolefin parts. More recently plasma treatment has been recognized as the optimum surface treatment for polyolefins when high bond strength is the only criterion. Plasma surface treatment provides the strongest adhesive joints with conventional adhesives. The difficulties with plasma treatment are that it is a batch process and requires investment in equipment. Gas plasma treatment is effective on geometrically complex parts.

Epoxy and nitrile-phenolic adhesives have been used to bond polyolefin plastics after plasma surface preparation. Shear strengths in excess of 3000 psi have been reported on

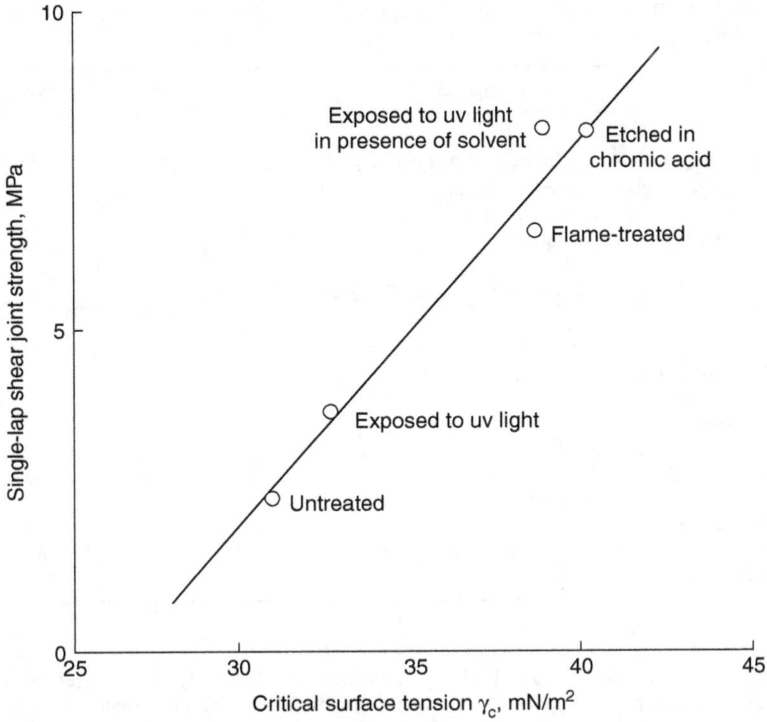

FIGURE 16.4 Tensile shear strength of polyethylene pretreated by various methods before bonding with an epoxy adhesive.[67]

polyethylene treated for 10 min in an oxygen plasma and bonded with an epoxy adhesive.[68] Table 16.12 presents the tensile strength of epoxy-bonded polyethylene and polypropylene joints after plasma surface treatments.

Flexibilized epoxy adhesives have moderate strength on flame and corona treated polyolefin substrates. Elevated cure temperature results in better adhesion because of more efficient wetting of the substrate surface. Table 16.13 shows a starting formulation for an epoxy adhesive that develops high peel strength to many difficult-to-bond substrates such as polyethylene, thermoplastic rubber, and polyester film.

TABLE 16.12 Tensile Shear Strength of Plasma Treated Polyolefin Substrates Bonded with Epoxy-Polyamide Adhesive[69]

Adherend	Tensile shear strength, psi	
	Control	After plasma treatment
High-density polyethylene-aluminum	315	3500
Low-density polyethylene-aluminum	372	1466
Polypropylene-aluminum	370	3080

TABLE 16.13 Starting Formulation for an Epoxy Adhesive for Polyolefin, Polyester, and Thermoplastic Substrates[70]

Component	Part by weight
Part A	
DGEBA epoxy resin (EPON 828, Resolution Performance Products)	80
Butyl glycidyl ether reactive diluent (HELOXY 61 Resolution Performance Products)	20
Adhesion promoter (Silquest A-172)	3
Part B	
Polyamide (EPI-CURE 3125, Resolution Performance Products)	180
Property	
Mix ratio, part A : part B	
• By weight	4 : 7
• By volume	1 : 2
Working life at 25°C, 1 qt	3 h
Viscosity, cP, at 25°C	6900
T-peel strength, lb/in of width, cured 20 min at 93°C	
• Polyethylene to aluminum	23–24
• Polyethylene to thermoplastic rubber	8.8
• Polypropylene to aluminum	12

Polyphenylene Oxide. The polyphenylene oxide (PPO) family of engineering thermoplastics is characterized by outstanding dimensional stability at elevated temperatures, broad service temperature range, outstanding hydrolytic stability, and excellent dielectric properties over a wide range of frequencies and temperatures. Several grades are available for a wide range of engineering applications. The PPO materials that are commercially available today are polystyrene-modified. PPOs are usually joined by solvent cementing or thermal welding.

Polyphenylene oxide joints must mate almost perfectly; otherwise solvent welding provides a weak bond. Thermal welding techniques found suitable for modified polyphenylene oxide parts include heat sealing, spin welding, vibration welding, resistance wire welding, electromagnetic bonding, and ultrasonics.

Various adhesives can be used to bond polyphenylene oxide to itself or to other substrates. Parts must be prepared by sanding or by chromic acid etching at elevated temperature. Methyl alcohol is a suitable solvent for surface cleaning. The prime adhesive candidates are epoxies, modified epoxies, nitrile phenolics, and polyurethanes. Epoxy adhesive will provide tensile shear strength on abraded polyphenylene oxide substrates of 600 to 1300 psi and 1300 to 2200 psi on etched (chromic acid) substrates.[71]

Thermoplastic Polyesters (Polyethylene Terephthalate, Polybutylene Terephthalate, Polytetramethylene Terephthalate). Thermoplastic polyesters have achieved significant application in film and fiber forms. For years polyethylene terephthalate (PET) was the primary thermoplastic polyester available. This material is best known in its film form as Mylar. A new class of high-performance molding and extrusion grades of thermoplastic polyester has been made available and is becoming increasingly competitive among engineering plastics. These polymers are denoted chemically as polybutylene terephthalate (PBT) and polytetramethylene terephthalate. These newer thermoplastic polyesters are highly crystalline with melting points above 220°C.

These plastics are quite inert; thus, compatibility with other substrates does not pose major problems. The terephthalates have high tensile and tear strengths, excellent chemical resistance, good electrical properties, and an operating temperature range from –55 to 200°C. These materials are generally joined with adhesives, and surface treatments are used to enhance adhesion, if required.

Commonly used adhesives for both PET and PBT substrates are isocyanate cured polyesters, epoxies, and urethanes. Surface treatments recommended specifically for PBT include mechanical abrasion and solvent cleaning with toluene. Gas plasma surface treatments and chemical etch have been used where maximum strength is necessary.

Polyethylene terephthalate cannot be solvent-cemented or heat-welded. Adhesives are the prime way of joining PET to itself and to other substrates. Only solvent cleaning of PET surfaces is recommended as a surface treatment. The linear film of polyethylene terephthalate (Mylar) provides a surface that can be pretreated by alkaline etching or plasma for maximum adhesion, but often a special treatment such as this is not necessary. An adhesive for linear polyester has been developed from a partially amidized acid from a secondary amine, reacted at less than stoichiometric with a DGEBA epoxy resin, and cured with a dihydrazide.[72]

Polyimide (PI), Polyetherimide (PEI), Polyamide-imide (PAI). These aromatic resins are a group of high-temperature engineering thermoplastics. They are available in a variety of forms including molded parts, coatings, and film. These are commercially available plastics that provide the highest service temperatures. They are also one of the strongest and most rigid plastics available. Parts made from these resins have thermal expansion coefficients very similar to those of metals, and generally there is no problem in bonding them to metals.

Polyimide parts can be either thermosetting or thermoplastic. Even the thermoplastic variety must be considered like a thermoset because of its high-temperature resistance and resistance to solvents. Polyimide parts are generally joined with adhesives and are not commonly joined by solvent cementing or thermal welding. Bonding is generally accomplished with moderate surface preparation processes and high-temperature adhesives. There are certain grades of polyamide-imide that are used as a bearing material and have inherent lubricity. These are more difficult to bond.

Polyimide parts can be bonded with epoxy adhesives. Only solvent cleaning and abrasion are necessary to treat the substrate prior to bonding. Selection of an adhesive for high-temperature service could be critical since the plastic substrate will generally have a higher thermal rating than the adhesive.

Parts molded from polyetherimide can be assembled with all common thermoplastic assembly methods. Adhesives that are recommended include epoxy, urethane, and cyanoacrylate. However, service temperature must be taken into consideration in choosing an adhesive because PEI parts are generally used for high-temperature applications. Good adhesion can be effected by simple solvent wipe, but surface treatment by corona discharge, flame treatment, or chromic acid etch will provide the highest bond strengths.

Polyamide-imide parts can be joined mechanically or with adhesives. They have too great a resistance to solvents and too high a thermal stability to be solvent-cemented or thermally welded. A variety of adhesives including amide-imide, epoxy, and cyanoacrylate can be used to bond polyamide-imide parts. Polyamide-imide parts are relatively easy to bond, and only solvent cleaning and mechanical abrasion are necessary as a surface preparation for good bonds. Plasma surface preparation has also been shown to provide excellent bonds. Polyamide-imide parts should be dried for at least 24 h at 150°C in a desiccant oven. Thick parts (over $^1/_4$ in) may require longer drying times to dispel casual moisture prior to bonding.

Polyetheretherketone (PEEK), Polyaryletherketone (PAEK), and Polyetherketone (PEK). Polyetheretherketone is a high-performance material developed primarily as a coating, but it is also available as film and molded parts. PEEK can be either adhesive-bonded or thermally welded using ultrasonic, friction, or hot plate welding techniques. When one is welding PEEK, it must be remembered that the melting point is very high, and considerable amounts of energy must be put into the polymer during welding to achieve a good bond.

PEEK parts can be bonded with epoxy, cyanoacrylate, silicone, and urethane adhesives. The epoxies give the strongest bond. High-temperature epoxy adhesives may be necessary in high-temperature applications. Surfaces must be clean, dry, and free of grease. Isopropanol, toluene, and trichloroethylene can be used to clean PEEK surfaces prior to mechanical abrasion. Surface roughening and flame treatment or etching may improve bond strengths. PEEK composites that have been surface-treated by plasma activation show excellent bond strength with epoxy adhesives.[73] Polyaryletherketone and polyetherketone parts can also be joined with adhesives by the methods described above.

Polystyrene. Polystyrene homopolymer is characterized by its rigidity, sparkling clarity, and ease of processability; however, it tends to be brittle. Polystyrenes have good dimensional stability and low mold shrinkage, and they are easily processed at low costs. They have poor weatherability, but they are chemically attacked by oils and organic solvents. Resistance is good, however, to water, inorganic chemicals, and alcohol. Impact properties are improved by copolymerization or by grafting polystyrene chains to unsaturated rubber such as polybutadiene (SBR) or acrylonitrile (SAN). Rubber levels typically range from 3 to 12 percent by weight. Commercially available, impact-modified polystyrene is not as transparent as the homopolymers, but it has a marked increase in toughness. Polystyrene properties also can be varied extensively through its polymerization process. They can even be crosslinked to produce higher-temperature material.

Polystyrene resins are subject to stresses in the fabrication and forming operations, and often they require annealing to minimize such stresses for optimized final product properties. Parts can usually be annealed by exposing them to an elevated temperature that is approximately 5 to 10°C lower than the temperature at which the greatest tolerable distortion occurs.

Polystyrene is ordinarily bonded to itself by solvent cementing, although conventional adhesive bonding, thermal welding, and electromagnetic bonding have been used. When polystyrene is bonded to other surfaces, conventional adhesive bonding is usually employed.

Generally only solvent cleaning and abrasion are necessary for surface preparation of polystyrene parts. Methanol and isopropanol are acceptable solvents for solvent cleaning of polystyrene. For maximum bond strength the substrates can be etched with sodium dichromate–sulfuric acid solution at elevated temperature. Table 16.14 shows the results of a study on the durability of joints formed between polystyrene and aluminum with different types of adhesives exposed to different environments.

TABLE 16.14 Joint Strengths for Polystyrene-Aluminum Tensile Shear Joints Exposed to Various Environments[74]

		Percent strength retention after exposure		
Adhesive	Initial tensile shear strength, MPa	52°C, 100 percent RH	Salt fog	Seacoast
Epoxy	2.3	100	18	30
Acrylic	2.3	100	57	71
Urethane	1.7	23	14	22
Silicone	0.14	25	—	—

Polysulfone. Polysulfone is a rigid, strong engineering thermoplastic that can be molded, extruded, or thermoformed into a wide variety of shapes. Characteristics of special significance are its high heat deflection temperature, 170°C at 264 psi, and long-term resistance to temperatures in the 150 to 170°C range. This material can be joined by adhesive bonding, solvent cementing, or thermal welding.

A number of adhesives have been found suitable for joining polysulfone to itself or to other materials. No special surface treatments are generally required other than simple solvent cleaning, although an elevated-temperature sodium dichromate–sulfuric acid etch has been used at times for maximum joint strength. A general-purpose room temperature curing epoxy adhesive, EC-2216 from 3M Company, offers good bond strength at temperatures to 85°C. For higher-temperature applications, heat cured epoxy adhesives are recommended.

Polyethersulfone (PES). Polyethersulfone is a high-temperature engineering thermoplastic with outstanding long-term resistance to creep. It is capable of being used continuously under load at temperatures up to about 180°C and in some low-stress applications up to 200°C. Certain grades are capable of operating at temperatures above 200°C. Some polyethersulfones have been used as high-temperature adhesives. PES is especially resistant to acids, alkalies, oils, greases, aliphatic hydrocarbons, and alcohol. It is attacked by ketones, esters, and some halogenated and aromatic hydrocarbons.

For adhesive bonding of PES to itself or to other materials, epoxy adhesives are generally used. Cyanoacrylates provide good bond strength if environmental resistance is not a factor. Parts made from PES can be cleaned using ethanol, methanol, isopropanol, or low-boiling petroleum ether. Solvents that should not be used are acetone, MEK, perchloroethylene, tetrahydrofuran, toluene, and methylene chloride.

Polyphenylene Sulfide (PPS). Polyphenylene sulfide is a semicrystalline polymer with a high melting point of 288°C, outstanding chemical resistance, thermal stability, and nonflammability. There are no known solvents below 190 to 200°C. This engineering plastic is characterized by high stiffness and good retention of mechanical properties at elevated temperatures. PPS resins are available as filled and unfilled compounds and as coatings. Polyphenylene sulfide resin itself offers good adhesion to aluminum, steel, titanium, and bronze and is used in nonstick coatings that require a baking operation of near 288°C.

Polyphenylene sulfide parts are commonly bonded together with adhesives. A suggested surface preparation method is to solvent-degrease the substrate in acetone, sandblast, and then repeat the degreasing step with fresh solvent. The polyphenylene sulfide surface that forms next to a mold surface is more difficult to bond than a freshly abraided surface. This is possibly due to a different chemical surface structure that forms at high temperature when the resin is in contact with the metal mold surface.

Adhesives recommended for polyphenylene sulfide include epoxies, and urethanes. Joint strengths in excess of 1000 psi have been reported for abraded and solvent-cleaned surfaces. Somewhat better adhesion has been reported for machined surfaces. The high heat and chemical resistance of polyphenylene sulfide plastics makes them inappropriate for either solvent cementing or heat welding.

Polyvinyl Chloride. Polyvinyl chloride (PVC) is perhaps the most widely used type of plastic in the vinyl family. PVC is a material with a wide range of flexibility. One of its basic advantages is the way it accepts compounding ingredients. For instance, PVC can be elasticized with a variety of plasticizers to produce soft, yielding materials to almost any desired degree of flexibility. Without plasticizers, PVC is a strong, rigid material that can be machined, heat-formed, or welded by solvents or heat. It is a tough material, with

TABLE 16.15 Tensile Shear Strength, MPa, for PVC Bonded to Various Substrates[75]

	Adhesive			
Substrate	Epoxy	Acrylic	Urethane	Elastomeric
PVC	6.3	11.0	5.6	0.28
Fiberglass	9.3	9.9	5.4	0.41
Steel	9.1	9.9	5.1	0.28
Aluminum	8.5	9.5	4.6	0.35
ABS	3.4	6.9	1.3	0.62
PPO	4.3	6.4	2.0	0.52
Acrylic	3.5	9.5	2.1	0.17

high resistance to acids, alcohol, alkalies, oils, and many other hydrocarbons. It is available in a wide range of forms and colors. Typical uses include profile extrusions, wire and cable insulation, and various foam applications. It is also made into both rigid and flexible film and sheets.

Plasticizer migration from the vinyl part into the adhesive bond line can degrade the strength of the joint. Adhesives must be tested for their ability to resist the plasticizer. PVC can be made with a variety of plasticizers. An adhesive suitable for a certain flexible PVC formulation may not be compatible with a PVC from another supplier. Nitrile rubber adhesives have been found to be very resistant to plasticizers and are often the preferred adhesive for flexible PVC films. However, certain epoxy adhesive formulations have also been found to provide excellent adhesion to flexible PVC substrates. Several such starting formulations are presented in Table 16.7. A comparison of the performance of several classes of adhesive when bonding PVC to itself and to various other materials is given in Table 16.15.

16.4 COMPOSITES

Modern structural composites are a blend of two or more components. One component is generally made of reinforcing fibers, either polymeric, carbon, or ceramic. The other component is generally made up of a resinous binder or matrix that is polymeric in nature. The fibers are strong and stiff relative to the matrix. Composites are generally orthotropic materials (having different properties in two different directions). When the fiber and matrix are joined to form a composite, they both retain their individual identities and both directly influence the composite's final properties. The resulting composite is composed of layers (laminates) of the fibers and matrix stacked to achieve the desired properties in one or more directions. The reinforcing fiber can be either continuous or discontinuous in length. The fiber's strength and stiffness are usually much greater than those of the matrix material. The commonly commercially available fibers are

- Glass
- Polyester
- Graphite

- Aramide
- Polyethylene
- Boron
- Silicon carbide
- Silicon nitride, silica, alumina, alumina silica

Glass fiber composites are the most common type of composite. However, graphite, aramide, and other reinforcements are finding applications in demanding aerospace functions and in premium sporting equipment such as fishing rods, tennis rackets, and golf clubs.

The resin matrix can be either thermosetting or thermoplastic. Thermosetting resins such as epoxy, polyimide, polyester, and phenolic are used in applications where physical properties are important. Polyester and epoxy composites make up the bulk of the thermoset composite market. Of these two, polyesters dominate by far. Reinforced with glass fiber, these are known as *fiberglass-reinforced plastics* (FRPs). FRPs are molded by layup and spray-up methods or by compression molding either a preform or sheet molding compound (SMC).

Thermoplastic matrix composites are generally employed where high-volume and economic considerations exist such as in the automotive and decorative paneling industries. Thermoplastic resin-based composites range from high-priced polyimide, polyethersulfone, and polyetheretherketone to the more affordable nylon, acetal, and polycarbonate resins. Practically all thermoplastics are available in glass-reinforced grades.

Resin-based composites are usually defined as either *conventional* or *advanced*. Conventional composites usually contain glass or mineral fiber reinforcement, and sometimes carbon fiber, either alone or in combination with others. Conventional composites are usually produced in stock shapes such as sheet, rod, and tube. There are many methods of processing composite materials. These include filament winding, layup, cut fiber spraying, resin transfer molding, and pultrusion.

Advanced composites is a term that has come to describe materials that are used for the most demanding applications, such as aircraft, having properties considerably superior to those of conventional composites and much like metals. These materials are "engineered" from high-performance resins and fibers. The construction and orientation of the fibers are predetermined to meet specific design requirements. Advanced composite structures are usually manufactured in specific shapes. An advanced composite can be tailored so that the directional dependence of strength and stiffness matches that of the loading environment.

Thermoset composites are joined by either adhesive bonding or mechanical fasteners. Thermoplastic composites offer the possibility of thermal welding techniques, adhesives, or mechanical fasteners for joining. Composites are also often joined with a combination of mechanical fasteners and adhesives. Many manufacturers distrust adhesive bonds in applications where joints undergo large amounts of stress (e.g., aircraft structures). Mechanical fasteners must be sized to avoid fiber crushing and delamination; adhesives must balance strength and flexibility. The joining of composite materials involves some special problems not faced with other materials.

A significant advantage is that adhesive bonding does not require the composite to be drilled or machined. These processes cut through the reinforcing fibers and drastically weaken the composite. A cut edge will also allow moisture or other chemicals to wick deep into the composite along the fiber-matrix interface, thereby further weakening the structure.

Much of what we know about bonding to composite materials has come through the aerospace industries. The early studies on adhesives, surface preparation, test specimen preparation, and design of bonded composite joints reported for the PABST Program[76] gave credibility to the concept of a bonded aircraft and provide reliable methods of transferring loads between composites and metals or other composites.

Adhesives that give satisfactory results on the resin matrix alone may also be used to bond composites. The three adhesives most often used to bond composites are epoxies, acrylics, and urethanes.

Epoxies are especially reliable when used with epoxy-based composites because they have similar chemical characteristics and physical properties. Room temperature curing adhesives are often used to bond large composite structures to eliminate expensive fixturing tools and curing equipment required of higher-temperature cure adhesives. However, room temperature epoxies require long cure times, so they are not suitable for large, high-speed production runs. Some of the lower-temperature composite materials are sensitive to the heat required to cure many epoxies. Epoxies are too stiff and brittle to use with flexible composites.

Structural adhesives that are commonly used for composites are supplied in two basic forms: semisolid B-stage film and thixotropic pastes. The film adhesives are cast or extruded onto carrier fabrics or films and partially cured to a semisolid. They can easily be handled, cut, and applied to the joint area. There is no need for mixing, metering, or dispensing of liquid components. In use, these adhesive systems are activated by heat and pressure. The semisolid B-stage film liquefies briefly on application of heat and then cures to an insoluble state. Epoxy, polyimides, epoxy-nylons, epoxy-phenolic, and nitrile-phenolic adhesives are available as B-stage film.

Paste adhesives are supplied as either one- or two-component adhesive systems. They can be used in applications where pressure cannot be applied. Some two-part pastes cure at room temperature after the appropriate proportions are mixed. Epoxy, urethane, and acrylic adhesives are all available as paste adhesives.

Surface preparation of composite parts for adhesive bonding will depend on the specific adherend and adhesive. Recommended surface preparations of many composites simply consist of a solvent wipe to remove loose dirt, oil, and mold release, followed by a mechanical abrading operation. Abrasion should be done carefully to avoid damaging the composite's surface fibers. A degree of abrasion is desired so that the glaze on the resin surface is removed but the reinforcing fibers are not exposed. Many surface abrasion methods have been applied to composite parts, and all these have some merit. These abrasion processes include light sanding, grit honing, vapor honing, Scotch-Briting (3M Company), and other methods. In some cases, a primer may be used to coat the composite before the adhesive is applied.

One surface preparation method that is unique for composites employs a "peel" or "tear" ply.[77] Utilization of the peel ply is illustrated in Fig. 16.5. With this technique, a closely woven nylon or polyester cloth is incorporated as the outer layer of the composite during its production layup. This outer ply is then torn or peeled away just before bonding. The tearing or peeling process fractures the resin matrix coating and exposes a clean, fresh, roughened surface for the adhesive. This method is fast and eliminates the need for solvent cleaning and mechanical abrasion.

Although the fabrication and joining technology that has been developed for thermoset composites is well advanced and established, the enabling technology for bonding thermoplastic composites is still in the relatively early stages of development. However, since thermoplastic composites can be softened by the action of heat and solvent, welding techniques that are unsuitable for use with thermosets can be used for joining thermoplastic composites.

Reinforced thermoplastic parts are generally abraded and cleaned prior to adhesive bonding. However, special surface treatment such as used on the thermoplastic resin matrix may be necessary for optimum strength. Care must be taken so that the treatment chemicals do not wick into the composite material and cause degradation. It may not be a good idea to use chemical surface treatment without first verifying that the treatment does not degrade the substrate.

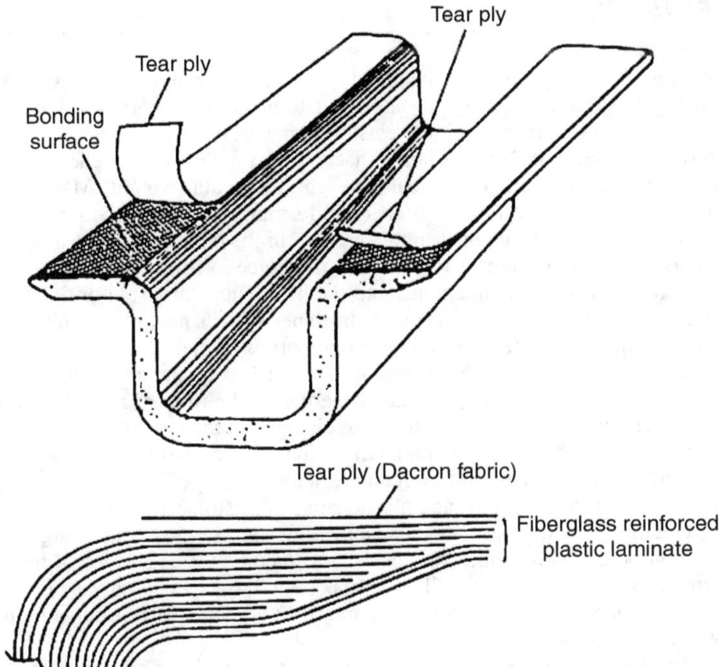

FIGURE 16.5 Structural reinforced composite with tear ply to produce fresh bonding surface.[78]

16.5 PLASTIC FOAMS

Plastic foams are manufactured from thermoplastic and thermoset resins in various forms. The main pitfall in joining plastic foam is that of (1) causing the foam to swell or collapse by contact with a solvent or monomer and (2) having the adhesive alter the properties of the foam through its absorption into the foam. Adhesion and joining are usually not a serious problem because of the porous nature of the foam.

Note that there are closed-cell and open-cell foams. Adhesives may spread or wick deeply into the open-cell variety, thereby affecting the resulting mechanical properties of the foam and perhaps even weakening the foam. When foam is bonded to another less porous substrate, the adhesive could be applied to the nonfoam substrate to minimize the wicking and ingress of the adhesive into the body of the foam. With the closed-cell variety, the adhesive cannot wick deeply into the foam, but usually the foam's skin must be machined or abraded to allow for some surface roughness for the adhesive to mechanically attach.

There are also low-surface-energy foams, such as polyethylene, that require either surface treatment or special adhesives for bonding. Fortunately, extremely strong bonds are generally not required because the foam has a relatively low cohesive strength. Therefore, simple cleaning is generally the only surface preparation required. The surface treatments that are recommended are those that are described in the previous section for the parent plastics. However, the possibility of wicking of the chemical compounds into the foam and degradation of the foam must be considered.

16.6 ELASTOMERS

There are over 30 broad groups of chemical types of elastic polymers. These are arranged by ASTM D 1418 into categories of materials having similar chemical chain structures. There are several problems with joining elastomer materials.

One problem is the significant variation that can exist within a given chemical type. This is due to differences in average molecular weight, molecular weight (MW) distribution, polymerization processes, variation of structural arrangement, copolymer or terpolymer ratios, etc. There also can be significant differences in the compounding recipe for a given product. This is due to differences and latitude of choice in fillers and reinforcing agents, liquid plasticizers, compounding agents, and the like. Thus, almost an infinite number of compounds are possible for a given generic elastomer such as neoprene or nitrile, and each variation can significantly affect the adhesion properties of the material.

Elastomer materials specifications usually do not focus on the adhesive properties, but mainly address the chemical and physical properties of the rubber. Thus, the supplier has wide latitude within the specification to make changes in the compound formulation that could be disastrous to the adhesive bond. One solution is to qualify every new lot of elastomer material for adhesion as well as the more standard properties.

Besides curing systems, fillers, and plasticizers, an elastomeric compound may contain protective chemicals such as antioxidants, antiozone agents, waxes, and fungicides. Some of these are purposefully designed to "bloom" or to come to the surface of the elastomer either during processing or on aging. These weak boundary layers often cannot be removed prior to bonding because their supply to the interface is relatively unlimited by the capacity of the bulk elastomer.

Another problem in joining elastomers with adhesives is that since they are deformable materials, it is easy to develop internal stresses at the bond interface. These stresses could adversely affect the bond strength and permanence of the joint. Minimal pressure to achieve close substrate contact with the adhesive is all that is necessary when bonding with elastomers.

Bonding of already vulcanized elastomers to themselves and to other materials is generally completed by using a flexible thermosetting adhesive such as epoxy-polyamide or

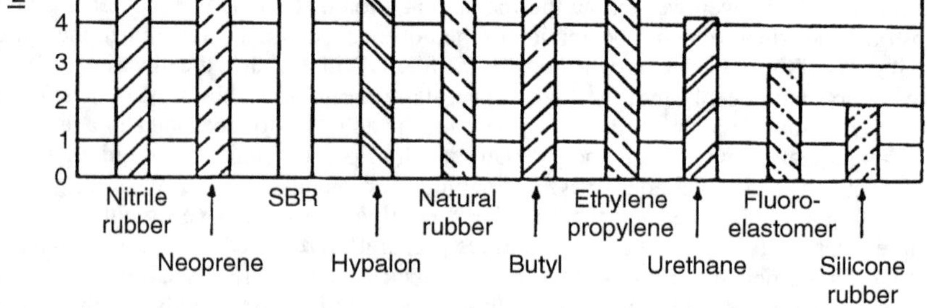

FIGURE 16.6 Bondability index of common and specialty elastomers.[79]

polyurethane. These adhesives offer excellent bond strength to most elastomers. Figure 16.6 illustrates a bondability index of common and specialty elastomers.

Surface treatments consist of washing with solvent, abrading, or, in the most demanding applications, cyclizing with acid. The most common elastomers to be bonded in this way include nitrile, neoprene, urethane, natural rubber, SBR, and butyl rubber. It is more difficult to achieve good bonds with silicones, fluorocarbons, chlorosulfonated polyethylene, and polyacrylate.

16.7 WOOD AND WOOD PRODUCTS

Wood is an important structural material consisting of a cellulosic composition with a highly porous nature. Adhesives are commonly employed to bond wood in the furniture industry. They are increasingly being used in laminating and veneering of wood-based products onto composite panels. Sealants are commonly applied to wood framing members in the construction industry. Several properties are unique to these materials that will affect their ability to be joined.

Wood is an anisotropic material. That is, wood and other cellulosic materials have different properties when measured along different axes. Tensile and shear strengths of wood are greater along the longitudinal direction, parallel with the wood fibers.

Wood is also a hygroscopic material because of its cellulosic nature. Generally, dimensional stability as a result of changing moisture content is highest in the longitudinal direction, somewhat less in the tangential direction, and least in the radial direction. Maximum performance is achieved if the wood has a moisture content during bonding that is close to the average moisture content anticipated during service. In this way, the internal stresses induced in the joint as a result of moisture change in service will be at a minimum. Therefore, wood substrates are generally preconditioned to a known moisture content before bonding. Wood changes little in dimension as a result of temperature changes, unless such changes also affect the internal moisture content.

Wood of different species will vary considerably in density or specific gravity. Higher-density wood of any species is usually stronger than wood of a lower density. Higher-density wood swells and shrinks more for a given moisture content than wood of lower density. This means that internal stresses on high-density wood joints are greater as a result of swelling and shrinkage effects than between two pieces of wood of lower density with the same joint design. Such internal stresses may be sufficiently large to pull the joint apart. Therefore, higher-quality joints must be developed when one is bonding higher-density wood. Generally, it is desirable to have the adhesive joint as strong as or stronger than the wood itself, particularly for structural applications.

Several important rules need to be remembered when one is joining wood. Good-fitting joints are critical. A thin, even layer of adhesive will form a strong bond between two pieces of wood, but a thick cushion of adhesive generally weakens the joint. To achieve a successful edge joint, the long mating surfaces must be perfectly tight all along their length. One should not rely on clamps and pressure to pull warped or deformed parts together. The fit of a mortise-and-tenon joint should also be precise. If the parts must be forced together, there will be no room for the glue between the pieces, and the joint will be starved. If the gap is too large, the adhesive layer will be too thick and stresses will develop. The adhesive layer should be approximately the thickness of a sheet of notebook paper. To ensure that there is sufficient glue in a joint, spread a thin layer on both mating surfaces.

The principal objective in surface preparation of wood for bonding is to provide a clean, undamaged surface that is flat and smooth and will permit the two pieces to mate properly when the liquid adhesive is applied between them. It is important not to damage the wood in cutting or sawing prior to bonding. Damaged wood fibers provide a cohesively weak area near to the adhesive, and they will become a failure mode under stress just as the weakest link in a chain.

Thin glue lines are preferred in bonded wood joints. The prevailing theory is that penetration of the adhesives into the wood is not necessary for good adhesion. Good bonding need occur only to the outside cellulose fibers. Thick bond lines tend to give erratic joint quality because of uneven stresses introduced in the joint. Thus, the best surface finish is one that does not tear wood fibers and provides sufficiently close tolerances that the parts fit together uniformly and closely. Usually, such surface preparation can be accomplished by standard wood machining operations.

The major synthetic adhesives used for bonding wood include urea, phenol, and melamine formaldehyde: resorcinol formaldehyde, phenol resorcinol, and polyvinyl acetate emulsions. More recently one-component, moisture cured polyurethane adhesives have become popular for bonding wood. Natural adhesives such as casein and animal glues are also often used for general-purpose wood bonding. Epoxies have been used for certain specialized wood joining applications such as when wood is bonded to metal substrates.

16.8 GLASS AND CERAMICS

Glass and ceramic substrates are generally high-surface-energy materials, and most adhesives wet them readily. One problem in bonding optically clear glass is to select an adhesive that is optically clear and does not change the optical characteristics of the glass. Another problem is that shrinkage of the adhesive or differences in thermal expansion coefficients could provide high internal stresses that cause catastrophic failure of the brittle glass substrate.

Adhesives used for glass substrates are generally transparent, heat-setting resins that are water- and uv-resistant to meet the requirements of outdoor applications. They are usually flexibilized systems so as not to place stress on the glass substrate either after cure or during thermal cycling. These adhesives include polyvinyl butyl, phenolic butyral, phenolic nitrile, neoprene, polysulfide, silicone, vinyl acetate, and flexible epoxy adhesives. Optical adhesives used for bonding glass lenses are usually styrene modified polyesters and styrene monomer-based adhesives. Epoxies, especially UV cured epoxy adhesives, are also beginning to be used in this application.

Adhesives used for bonding glass and glazed ceramics should have minimum VOC content, since these substrates are nonporous. Epoxy adhesives, being 100 percent solids materials, are probably best for these applications. Thermosetting acrylic is good for bonding glass and ceramics to thermoplastic polymers and to metal surfaces. Where large areas and materials with greatly differing coefficients of thermal expansion are being bonded, polysulfides or urethanes should be considered, since these adhesives have very high elongation.

Usually solvent cleaning is the only surface preparation needed for bonding glass. However, (1) solvent cleaning coupled with light sandblasting or (2) a chromium trioxide etch followed by ultrasonic detergent bath has been suggested for highest bond strength. Mechanical or chemical surface treatment can degrade the optical transmission quality of glass.

Most ceramic surfaces, like metal surfaces, have an adsorbed layer of water that may cause difficulty in bonding. If this water is partially removed by heating prior to bond formation, atmospheric moisture may diffuse into the bond during service and displace adhesive molecules from the interface since they will be preferentially adsorbed on the glass interface. To overcome this problem, coupling agents, typically silanols, have been developed to enhance the attraction between the adhesive and the glass or ceramic surface. Coupling agents can also provide a degree of stress release at the interface and reduce the thermal expansion-related stresses between the substrate and the adhesive.

Adhesives for glass- or ceramic-to-metal seals should never become fully rigid because thermal expansion rates for adhesives are much higher than those for the bonded materials. Temperature changes can lead to high stresses, which can cause cracking and joint failure. Nitrile rubber epoxies, phenolic resin blends, silicones, cyanoacrylates, and anaerobics are some of the important organic adhesives that can be used to make good glass- or ceramic-to-metal seals. The silicones can be used from –73 to 232°C while maintaining good flexibility, but they are relatively costly. The cyanoacrylates are especially useful in bonding glass to metals, but they are less useful for joining porous ceramics because the porosity will absorb most of the adhesive and inhibit its cure.

16.9 HONEYCOMB AND OTHER STRUCTURAL SANDWICH PANELS

Structural sandwich panels represent what is probably one of the most common applications for structural adhesives. Facing materials can be made to adhere to core materials, such as aluminum or paper honeycomb, to give a high composite strength-to-weight ratio.

Adhesives with filleting properties are required for honeycomb cores. A modified phenolic is often used with aluminum honeycomb for high strength, while a neoprene- or nitrile-based organic solvent type of adhesive is often used with impregnated paper honeycomb. Epoxy adhesives are also commonly used in the fabrication of honeycomb sandwich panels.

Facings of metal, paper, or plastic may be combined with a variety of solid-core materials. The adhesive must transmit shear stress from the facing to the core. It must also be able to resist the tendency of the thinner facing to buckle under design loads or under thermal stresses caused by variations in the panel components' coefficients of thermal expansion.

Adhesives for sandwich structures such as honeycomb and faced cores must have different properties than other adhesives. To achieve a good attachment to an open-cell core, such as honeycomb, the adhesive must have a unique combination of surface wetting and controlled flow during its early stages of cure. This controlled flow prevents the adhesive from flowing down the cell wall and leaving a low-strength top-skin attachment and an overweight bottom-skin attachment. The adhesive must resist being squeezed out from between faying (close-fitting) surfaces when excessive pressure is applied during cure to a local area of the part. Many adhesives are formulated to achieve good core filleting and are subsequently given controlled flow by adding an open-weave cloth or fibrous web, cast within a thicker film of adhesive. This cloth, called a *scrim cloth*, then prevents the faying surfaces from squeezing out all the adhesive, which would result in an area of low bond strength.

The adhesive for sandwich construction must also have good toughness. It must be resistant to loads that act to separate the facings from the core under either static or dynamic conditions. Experience has shown that greater toughness in the bond line usually equates to greater durability and longer service life. If the facing is flexible enough that it can be rolled on a drum, the climbing drum peel test (ASTM D 1781) is the most common test to measure toughness in sandwich construction. The resulting value of peel strength will vary considerably depending on the properties of the facing, toughness of the adhesive, amount of adhesive used, density of the core, cell size of the core, direction of peel, adequacy of surface preparation, and degradation of the joint.

Common adhesives for sandwich construction include nitrile phenolic films, modified epoxy pastes and films, epoxy-nylon films, polyimide films, and modified urethane liquid and pastes. Accessory materials that are commonly used in the joining process are core-splicing adhesives, syntactic foams, tapes, and corrosion-inhibiting primers.[80]

16.10 CONCRETE

Concrete provides a substrate surface that changes during its cure and also during environmental exposure. The main difficulty with bonding concrete is due to the presence of moisture. This moisture can be retained in fresh concrete and may be present in old concrete due to the environment.

Concrete is a substrate whose surface characteristics are also likely to be affected by the environment in which it cures. The surface can be different depending on the temperature and humidity conditions during the cure of the concrete.

Once it is cured, concrete has several surface characteristics that are problems for bonding or sealing. The concrete surface is extremely alkaline and will destroy any hydrolysis-sensitive materials that are present at the interface. It often has a weak, porous surface layer that must be penetrated or removed before being bonded. Thus, sealers and primers are commonly used to moisture-proof and strengthen the concrete surface prior to bonding.

The substrate should be free from contaminants such as cement laitance, oils, waxes, greases, and curing compounds. While the old concrete does not need to be thoroughly dry, any standing water must be removed before the adhesive is applied.

Adhesives commonly used on concrete must be formulated so that they cure well in the presence of moisture. The adhesive should have a viscosity that allows it to penetrate the concrete for mechanical bonding. Epoxy and epoxy-polysulfide adhesives are especially effective adhesives for concrete. The starting formulations given in Table 16.16 are for an ambient cure 100 percent solids epoxy and a waterborne epoxy adhesive.

The waterborne emulsion is formulated so that the mixed adhesive can be easily cleaned with water, yet it provides good water resistance properties of a conventional epoxy adhesive. Bond strength of these adhesives on new concrete is greater than the strength of the concrete itself.

TABLE 16.16 Starting Formulations for Epoxy Adhesives To Bond Concrete[81]

Component	Parts by weight	
	A	B
Part A		
DGEBA epoxy resin (EPON 828, Resolution Performance Products)	100	
Epoxy resin emulsion (EPI-REZ WD-510, Resolution Performance Products)		100
Super white silica (C.K. Williams Co.)	Optional	
Fumed silica (Cab-O-Sil TS 720, Cabot Corp.)		1
Part B		
Amidoamine (EPI-CURE 3072, Resolution Performance Products)	35	35
Tap water (added after mixing part A and part B)		53
Property		
Viscosity, cP, at 25°C	4000 (unfilled)	40,000
Gel time at 25°C	2 h	5 h

REFERENCES

1. Petrie, E. M., Chapter 16 "Bonding and Sealing Specific Substrates," in *Handbook of Adhesives and Sealants*, McGraw-Hill, New York, 2000.
2. Snogren, R. C., *Handbook of Surface Preparation*, Palmerton Publishing Co., New York, 1974.
3. Wegman, R. F., *Surface Preparation Techniques for Adhesive Bonding*, Noyes Publications, New York, 1989.
4. Schneberger, G. L., "Adhesives for Specific Substrates," Chapter 21 in *Adhesives in Manufacturing*, G. L. Schneberger, ed., Marcel Dekker, New York, 1983.
5. Brewis, D. M., "Pretreatment of Metals Prior to Bonding," in *Handbook of Adhesion*, D. E. Backham, ed., Longman Scientific and Technical, Essex, England, 1992.
6. *Adhesive Bonding of Aluminum Automotive Body Sheet Alloys*, The Aluminum Association, Inc., New York, NY, 1975.
7. *Adhesive Bonding of Aluminum*, Reynolds Metals Company, Richmond, VA, 1966.
8. Murphy, J. F., and Page, H. A., *American Chemical Society Division Paint Plastics Printing Ink Chemical Papers*, vol. 15, no. 1, 1955, p. 27.
9. Eichner, H. W., and Schowalter, W. E., Forest Products Laboratory Report 1813, Madison, WI, 1950.
10. Wilson, I., et al., "Pretreatment for Bonded Aluminum Structures," *Advanced Materials and Processes*, August 1997.
11. Minford, J. D., "Surface Preparations and Their Effect of Adhesive Bonding," *Adhesives Age*, July 1974.
12. Kabayashi, G. S., and Donnelly, J. P., Boeing Company Report DG-41517, February 1974.
13. Bijlmer, P. F. A., "Influence of Chemical Pretreatments on Surface Morphology and Bondability of Aluminum," *Journal of Adhesion*, vol. 5, 1973, p. 319.
14. Thrall, E., "Bonded Joints and Preparation for Bonding," AGARD Lecture Series, no. 102, October 1979.
15. Shannon, R. W., et al, "Primary Adhesively Bonded Structure Technology (PABST) General Materials Property Data," Douglas Aircraft Co., McDonnell Douglas Corp., Air Force Flight Dynamics Laboratory, Technical Report AFFDL-TR-77-107, September 1977.
16. Kim, G., and Ajersch, F., "Surface Energy and Chemical Characteristics of Interfaces of Adhesively Bonded Aluminum Joints," *Journal of Materials Science*, vol. 24, 1994, pp. 676–681.
17. Hartshorn, S. J., "Durability of Adhesive Joints," in *Structural Adhesives*, S. R. Hartshorn, ed., Plenum Press, New York, 1986. Also in Minford, J. D., *Treatise on Adhesion and Adhesives*, vol. 5, R. L. Patrick, ed., Marcel Dekker, New York, 1981, p. 45.
18. Hartshorn, "Durability of Adhesive Joints."
19. Browne, J., "Aerospace Adhesives in the 90s," *38th International SAMPE Symposium*, May 10–13, 1993.
20. Wilson, I., et al., "Pretreatment for Bonded Aluminum Structures," *Advanced Materials and Processes*, August 1997.
21. *Adhesive Bonding Aluminum*, Reynolds Metals Company, Richmond, VA, 1966.
22. Sell, W. D., "Some Analytical Techniques for Durability Testing of Structural Adhesives," *Proceedings 19th National SAMPE Symposium and Exhibition*, vol. 19, *New Industries and Applications for Advanced Materials Technology*, April 1974.
23. Resolution Performance Products, Starting Formulations 4011 and 4012, Houston, TX, 2004.
24. Magariello, E., and Hannon, M., "Bonding Beryllium Copper Alloys with High Temperature Adhesives," *Adhesives Age*, March 1971.
25. Berylcoat D available from Brush Wellman, Incorporated, Cleveland, OH.
26. Schields, J., *Adhesives Handbook*, 3d ed., Butterworths, London, 1984.
27. Cagle, C. V., "Surface Preparation for Bonding Beryllium and Other Adherends," in *Handbook of Adhesive Bonding*, C. V. Cagle, ed., McGraw-Hill, New York, 1973.

28. Ebanol C Special, Enthane Company, New Haven, CT.
29. McGuiness, E. W., "Properties of Epoxy Adhesive at Elevated Temperatures," *Epoxy Resin Symposium of SPE*, Minneapolis, MN, October 1958.
30. Bolger, J. C., et al., "A New Theory of Improving the Adhesion of Polymers to Copper," Final Report, INCRA Project No. 172, International Copper Research Association, New York, August 1971.
31. American Society for Metals (ASM), *Metals Handbook*, 9th ed., vol. 2, *Properties and Selections: Nonferrous Alloys and Pure Metals*, Materials Park, OH, 1979.
32. Military Specification, MIL-M-45202C, *Magnesium Alloy, Anodic Treatment of*, April 1981.
33. Military Specification, MIL-M-3171C, *Magnesium Alloy, Processes for Pretreatment and Prevention of Corrosion on*, March 1974.
34. Eickner, H. W., "Adhesive Bonding Properties of Various Metals as Affected by Chemical and Anodizing Treatments of the Surfaces," Forest Products Laboratory Reports 1842 and 1842-A, February 1955 and August 1960, Forest Products Laboratory, Madison, WI.
35. Keith, R. E., et al., "Adhesive Bonding of Nickel and Nickel Base Alloys," NASA TMX-63428, October 1965.
36. Cagle, "Surface Preparation for Bonding Beryllium and Other Adherends."
37. Landrock, A. H., "Processing Handbook on Surface Preparation for Adhesive Bonding," Picatinny Arsenal Technical Report 4883, Picatinny Arsenal, Dover, NJ, December 1975.
38. Devine, A. T., "Adhesive Bonded Steel: Bond Durability as Related to Selected Surface Treatments," U.S. Army Armament Research and Development Command, Large Caliber Weapon Systems Laboratory, Technical Report ARLCD-TR-77027, December 1977.
39. Shaffer, D. K., et al., "Titanium as an Adherend," Chapter 4 in *Treatise on Adhesion and Adhesives*, vol. 7, J. D. Minford, ed., Marcel Dekker, New York, 1991.
40. Johnson, W. E., et al., "Titanium Sandwich Panel Research and Development," General Dynamics, Fort Worth, TX, Final Report AMC-59-7-618, vol. 3, Contract AF 33(600)-34392, November 1959.
41. Allen, K. W., et al., *Journal of Adhesion*, vol. 6, 1974, p. 153.
42. Cotter, J. L., and Mahoon, A., "Development of New Surface Pretreatments Based on Alkaline Hydrogen Peroxide Solutions for Adhesive Bonding of Titanium," *International Journal of Adhesion Adhesives*, vol. 2, 1982, p. 47.
43. Shaffer, et al., "Titanium as an Adherend."
44. Ditchek, B. M., et al., "Bondability of Ti Adherends," *Proceedings 12th SAMPE Technical Conference*, 1980, p. 882.
45. Brown, S. R., *Proceedings 27th National SAMPE Symposium*, vol. 363, 1982.
46. Clearfield, H. M., et al., "Surface Analysis of Ti-6Al-4V Adhesive Bonding Systems," *Journal of Adhesion*, vol. 23, 1987, p. 83.
47. Shaffer, et al., "Titanium as an Adherend."
48. Morita, W. H., "Titanium Tankage Program—Titanium Tankage Development," North American Aviation, Inc., Space and Information Systems Division, Technical Documentary Report AFRPL-TR-64-154M, November 1964.
49. Keith, R. E., "Adhesive Bonding of Titanium and Its Alloys," *Handbook of Adhesives Bonding*, C. V. Cagel, ed., McGraw-Hill, New York, 1973.
50. Clearfield, H. M., et al., "Surface Preparation of Metals," in *Adhesives and Sealants,* vol. 3, *Engineered Materials Handbook*, ASM International, Materials Park, OH, 1990.
51. Pocius, A. V., et al., "The Use of Adhesives in the Joining of Plastics," in *Treatise on Adhesion and Adhesives*, J. D. Minford, ed., Marcel Dekker, New York, 1991.
52. Resolution Performance Products, Starting Formulations 4009 and 4014, Houston, TX, 2004.
53. Schields, J., *Adhesives Handbook*, 3d ed., Butterworths, London, 1984.
54. Lupton, D. C., "Selection of an Adhesive for Bonding FRP Automotive Parts," *International Journal of Adhesion Adhesives,* vol. 3, no. 2, 1983, pp. 155–158.

55. Minford, J. D., *Proceedings of International Symposium Physiochemical Aspects of Polymer Surfactants*, vol. 2, 1983, pp. 1139–1160.
56. Brewis, D. M., "Pretreatment of Polymers," in *Handbook of Adhesion.*
57. Landrock, A., "Surface Preparation of Adherends," in *Adhesives Technology Handbook*, Noyes Publishing Co., Park Ridge, NJ, 1985.
58. Snogren, *Handbook of Surface Preparation.*
59. Snogren, *Handbook of Surface Preparation*, Chapter 12.
60. *Delrin Acetal Resins Design Handbook*, DuPont, Technical Bulletin, Wilmington, DL.
61. Toy, L. E., "Plastics/Metals Can They Be United?" *Adhesives Age*, vol. 17, no. 10, 1974.
62. Meier, J. F., and Petrie, E. M., "The Effect of Ultraviolet Radiation on Sodium Etched Polytetrafluoroethylene Bonded to Polyurethane Elastomer," *Journal of Applied Polymer Science*, vol. 17, 1973, pp. 1007–1017.
63. Tetra-Etch and Fluoro-Etch, Action Technologies, Inc., Pittston, PA.
64. Rose, P. W., and Liston, E., "Gas Plasma Treatment of Polymers Prior to Adhesive Bonding," *SPE Annual Technical Conference*, 1985, p. 685.
65. DeLollis, N. J., and Montoya, O., "Surface Treatment for Difficult to Bond Plastics," *Adhesives Age*, January 1963.
66. Bresin, R. L., "How to Obtain Strong Adhesive Bonds via Plasma Treatment," *Adhesives Age*, March 1972.
67. Morris, C. E. M., "Strong, Durable Adhesion Bonding: Some Aspects of Surface Preparation, Joint Design, and Adhesive Selection," *Materials Forum*, vol. 17, 1993, pp. 211–218.
68. Bresin, "How to Obtain Strong Adhesive Bonds via Plasma Treatment."
69. Hall, J. R., et al., "Activated Gas: Plasma Surface Treatment of Polymers for Adhesive Bonding," *Journal of Applied Polymer Science*, vol. 13, 1969, pp. 2085–2096.
70. Resolution Performance Products, Starting Formulation 4014, Houston, TX, 2004.
71. Abolins, V., and Eckert, J., "Adhesive Bonding and Solvent Cementing of Polyphenylene Oxide," *Adhesives Age*, July 1967.
72. Wear, R. L., et al, "Resinous Compositions," U.S. Patent 2,970,972.
73. Ballmann, A., et al., "Surface Treatment of Polyetheretherketone Composite by Plasma Activation," *Journal of Adhesion*, vol. 46, 1994.
74. Minford, *Proceedings of International Symposium Physiochemical Aspects of Polymer Surfactants*, pp. 1161–1180.
75. Toy, "Plastics/Metals Can They Be United."
76. Potter, D. L., *Primary Adhesively Bonded Structure Technology (PABST) Design Handbook for Adhesive Bonding*, Douglas Aircraft Co., McDonnell Douglas Corp., Long Beach, CA, January 1979.
77. Schields, *Adhesives Handbook*, 3d ed.
78. Snogren, *Handbook of Surface Preparation.*
79. Cox, D. R., "Some Aspects of Rubber to Metal Bonding," *Rubber Journal*, April/May 1969.
80. TSB 124, Bonded Honeycomb Sandwich Construction, Hexcel Corp., Dublin CA. (Also in *Advanced Composites* March/April 1993.)
81. Resolution Performance Products, Starting Formulations 4006 and 4028, Houston, TX, 2004.

CHAPTER 17
PROCESSING OF EPOXY ADHESIVES

17.1 INTRODUCTION

Epoxy adhesives have achieved their commercial success due in no small way to their processing capabilities. Epoxy chemistry is compatible with (1) a variety of formulating techniques and processes accessible to the adhesive manufacturer and (2) a variety of application and curing methods accessible to the end user.

The role of the adhesive formulator has become so important that it has been associated with the creation of "new" or "engineered" adhesives. New products actually come both from the development of new polymeric materials such as resins and curing agents and from the creation of more versatile and usable adhesives through selecting, blending, filling, and modifying existing raw materials. With adhesive systems the easiest and most economical method of producing new products is arguably through formulating.

Epoxy resins are not ideal adhesives in their natural form; so they are mixed with materials to improve and enhance their properties and, thereby, make them more useful in a variety of applications. This process is called *compounding* or *formulating*. Compounding is the combining of a base epoxy resin with curing agents, modifiers, additives, reinforcement, fillers, and other polymers to make the base polymer perform better, cost less, and process more easily.

Under this broad definition there are two groups that perform compounding: (1) the suppliers of the base resins or adhesives and (2) the end users of the adhesive system.

Most often the compounding is done by the formulators who are specialists in compounding methods and equipment. They can marry a set of specifications to a formula and provide the product at lowest cost. These independent compounders can be subdivided into two main groups: the proprietary compounder and the custom compounder. Proprietary compounders compound and sell their own special formulations, and custom compounders do compounding for someone else.

Adhesive compounding also occurs at the end-user level. This is generally done when the end user cannot find a product to meet his or her individual needs, to reduce cost associated with a middleman, or to make use of other materials (e.g., scrap resin, fillers, etc.) that are generated during the course of business. Epoxy adhesives are more conducive to the end user or in-house processor than other polymeric adhesives because of the ease with which the liquid resins can be mixed with other ingredients. Many of these operators buy raw materials from various suppliers and do their own blending and formulating.

The role of the epoxy resin manufacturer depends on the nature of the product and the markets. It is often more convenient for the resin manufacturer to "modify" a base resin, such as providing an adduct, a graft polymer, or a blend. In this case the modified resin would likely be shipped to a formulator for incorporation of other fillers and modifiers and for packaging with a curing agent. The resin supplier may also be interested in supplying a completely formulated adhesive product, but generally the market is not large enough.

Because of the nature of the epoxy resin and the other raw materials used in formulating an epoxy adhesive, compounding requires a wide range of mixtures from powders, to liquids, to pastes. The final product may also be supplied in several forms, such as pellets, film, and tape, and in several packaging systems, such as two-component kits, cartridges, and bubble packs. This requires a corresponding broad range of processing knowledge and equipment. The types of compounding machinery and equipment have increased in number and variety at a rate that is almost equivalent to the addition of new raw materials. This machinery has become more sophisticated and versatile and in some instances more specialized.

Compounding operations also require extensive testing from quality control checks on incoming resins and additives through reports and testing on outgoing products. The variety of test equipment to characterize physical, chemical, environmental, thermal, and rheological properties is likewise large.

The following section looks at the processes and equipment commonly used in the compounding of epoxy adhesive systems. Since they can be utilized by either the formulator or the end user, there will be no strict division of the discussion by user. Hazards and safety issues related to these materials and processes are discussed somewhat in this chapter, but a more thorough discussion is found in Chap. 18. Test methods, quality control processes, and standards that are commonly used at the formulation level are addressed in Chaps. 19 and 20.

17.2 COMPOUNDING PROCESSES

Just as there are many formulations, there are many ways to compound and process the requisite raw materials into an adhesive formulation. The methods chosen and how they are used also influence the final physical properties of the adhesive. The important processes to be considered in compounding and use of epoxy adhesives can be listed as follows:

Processes primarily related to the formulator

1. Storage of raw materials
2. Incorporation of fillers, additives, and modifiers via proper mixing equipment
3. Packaging

Processes primarily related to the end user

4. Storage of formulated materials
5. Transferring the formulated product
6. Proper metering and mixing of ingredient concentrations
7. Application of adhesive to the substrate
8. Curing

These processes and associated equipment are discussed below.

17.2.1 Storage of Raw Materials

Liquid epoxy resins are stable for long periods at room temperature. Resins that are cut with diluent or solvent should be stored in tight containers to prevent loss of volatiles. Also, certain resins and curing agents (e.g., anhydride curing agents) can be degraded by exposure to ambient moisture and should be stored in tightly closed containers. Moisture absorption by these materials may affect their rate of cure and other properties.

Storing of materials below room temperature can significantly extend the shelf life of epoxy products. One-component adhesives are generally stored in a freezer at –40°C and shipped in the frozen condition. Prolonged exposure to cold storage conditions, particularly with unformulated epoxy resins, can cause crystallization of the individual components. Formulated adhesive can also, on occasion, form crystals that look like small lumps or agglomerates. If this is the case, the epoxy components should be heated in their containers to 40 to 55°C until the crystals are dissolved. The cover should be loose-fitting but kept on the product. Once they are at temperature, stir the components with a spatula until homogeneous.

Moderate elevated-temperature storage of epoxy resins to reduce viscosity or prevent resin crystallization has little effect on most resins, even after several months of storage time. Storage at temperatures above 55°C is satisfactory for periods of 2 or 3 days for processing purposes.

The storage and transfer of raw chemicals requires special considerations relative to the equipment that come into contact with epoxy resins. Carbon steel is generally satisfactory for pipe and tank construction and will not rust beneath the epoxy resin. Rust occuring above the level of the epoxy and accidentally introduced into the resin may be removed by filters. Epoxy coated steel pipe, aluminum or stainless steel pipe, or epoxy fiberglass composite tubing is used to prevent rusting downstream of the filter.

17.2.2 Incorporation of Fillers, Modifiers, etc.

When one is incorporating ingredients into the epoxy resin or curing agent, the order of addition is most important. The objective is to maintain as liquid a mix as possible at the beginning, and then progress stepwise to a stiffer plastic mixture. The completeness of the mixing may be limited by the most viscous or difficult to incorporate ingredient.

Fillers are normally incorporated during the final stages of the adhesive compounding process. Fillers should be dried to remove any residual moisture before being added to either the epoxy resins or the curing agent component. Air circulating ovens are most commonly used for the drying processes, but dielectric heating can also be used in certain cases.

It is usually most convenient to add the fillers to a warm resin. However, diluents and solvents are sometimes employed to lower the viscosity of the system to allow a higher concentration of filler in the mixture.

After the filler is mixed in, the formulation should be allowed to return to room temperature and stand for a suitable time until all bubbles have escaped the system. This can be facilitated by a vacuum degassing process.

For some applications, filler may be incorporated as a third component at the time the epoxy component is mixed with the curing agent at the bonding site. This is usually the case with highly filled systems such as are commonly used in the construction industry. It is generally done to incorporate high levels of fillers or to negate the settling properties experienced with some fillers.

Fillers may be added to low-viscosity resins by conventional sigma-blade mixer or the same type of equipment used for mixing paint systems. Three roll mills are generally employed for more viscous systems to obtain thorough wetting of the filler by the resin matrix.

In the epoxy adhesive industry, fillers are generally added as a batch process rather than a continuous process. Batch systems are especially suited to short-run operation in which additives and materials are often changed. They allow for greater control of residence time, shear, and temperature, and they can accommodate a feed rate that is not free-flowing.

Mixing equipment is generally classified in terms of the shear or mixing intensity. High-shear mixers are used for producing filled compounds. But most often, low- or medium-shear mixers suffice for compounding epoxy adhesives.

Medium-Shear Mixers. High-speed dispersers are designed to use the high peripheral speed of a mixing element to produce the blend. These machines depend on a high-speed impeller turning in a mixing vessel. The differences among the various models on the market relate to two components: the design of the mixing elements and the design of the mixing chamber.

Some mixing chambers have curved bottoms that enhance movement of materials. Others have internal baffles or some similar design to increase turbulence and eliminate dead spots in the chamber. Shafts can enter the chamber from the top (allowing easier cleanup) or from the bottom (allowing easier vacuum degassing).

The basic high-speed dispersing impeller blade is a flat disk with teeth or similar serration around the diameter, as shown in Fig. 17.1. On some models, disks may be without teeth, or several may be stacked up with small gaps between disks so that material between the disks is sheared as the impeller turns. Sometimes two shafts are used in the same mixing chamber; their disks may interlock and counterrotate, thereby enhancing shear.

Low-Intensity Mixers. Low-intensity liquid mixers are used for simple jobs involving low-viscosity resins. Impellers are of various designs resembling marine or aircraft propellers, long slender screws, shafts with paddles, and shafts with two or more impellers. (See Fig. 17.2.) These units are intended to generate greater motion within the mixture and greater turbulence at the impeller than the flat blades of the high-speed disperser, which are intended to go fast enough to shear material, break down agglomerates, and disperse solids in liquids.

These low-intensity units are less energy-intensive, requiring lower horsepower, but they do not achieve mixing of the same quality in comparable cycle times. They are limited in use to such jobs as blending additives and fillers into low- to medium-viscosity resins.

17.2.3 Packaging

Often the adhesive formulator or supplier can provide added value to the end user and participate in the sharing of this benefit. Perhaps one of the more innovative business solutions has been in the ingenious packaging of adhesive systems, especially reactive epoxy adhesives.

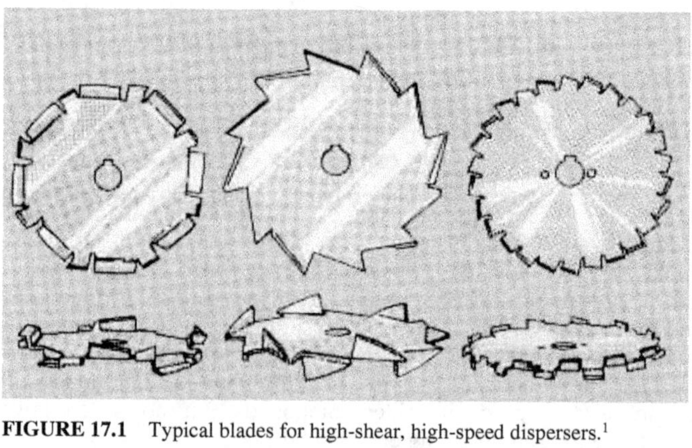

FIGURE 17.1 Typical blades for high-shear, high-speed dispersers.[1]

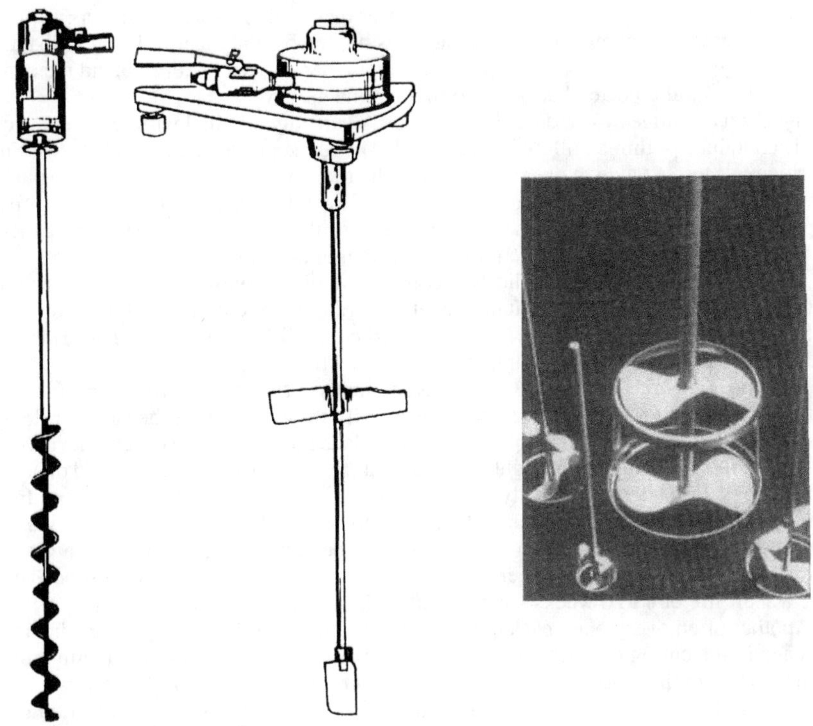

FIGURE 17.2 Mixing elements for low-intensity mixer designs.[2]

The adhesive user cares only about achieving a reliable assembly with minimal cost. The end user would rather not be bothered with handling messy and sometimes hazardous components, carefully metering them out, mixing the components together adequately without air entrapment, and finally applying the product to the substrate. These are all necessary steps in the production of an adhesive bond; however, they have been almost made to vanish by innovative packaging. Although the cost of the packaging is certainly more than that of conventional containers, the savings that result for the end user can be shared between the adhesive supplier and end user by premium pricing.

The two-component epoxy adhesive, of course, can be packaged as a supply kit with a matched set of components. In this method metering is eliminated, and the container of one of the components can be used as a mixing vessel. There are also several well-recognized preproportioned adhesive containers that incorporate means for mixing and dispensing—all within the same package. These include side-by-side syringes with attachable plastic static mixers, coaxial cartridges, and bubble or divider packs with a removable hinge that can be used for spreading and mixing the products.

Innovative "packaging" systems that might be overlooked at first include

- Barrier and injection cartridges
- Frozen adhesives
- Solid and preformed adhesives

The last two methods are especially ingenious in their simplicity.

Barrier cartridges generally have a foil barrier that is held in place inside a cartridge to separate the reactive components. The barrier is broken by a plunger rod similar to the one in a cartridge gun. The head on the plunger rod also serves as a mixer to blend the components with several repeated back-and-forth movements.

Injection cartridges work on a similar basis. A cartridge is filled with resin, and a curative is contained within a hollow injection rod. The rod is pushed against a base within the cartridge to cause a valve to open at the end of the rod, and then a piston is used to push the curative out of the rod. A mixing head at the end of the injection rod mixes the components. A plunger is attached to the other end of the cartridge to extrude the mixed adhesive. Examples of the barrier- and injection-type packaging are shown in Fig. 17.3.

Another package is a semisolid two-component ribbon with hardener and base resin identified by different colors. Suitable lengths of ribbon are cut off and kneaded together by hand until the colors combine to form a third color. Once this is completed, the paste-like material is applied to the joint in a conventional manner.

In some industries, notably the electronics industry, an often overlooked packaging method is sometimes used—frozen adhesives. With these adhesives the supplier does the metering and mixing. The mixed product is then placed into a syringe or other form of container, and it is immediately brought to a temperature of −67 to −73°C. The cold delays any chemical reaction and significantly increases the shelf life of the adhesive system. Frozen containers are packed in dry ice and shipped overnight to the end user. Generally a warm water bath, microwave, or hot air is used to thaw the adhesive and bring it to a workable consistency. In this process, premetered and mixed room temperature curing adhesives can have a shelf life of 4 to 6 weeks when stored in the frozen condition.

Another often overlooked packaging method is a *B stage*. The B stage is the intermediate stage in the curing of certain thermosetting resins such as epoxies cured with aromatic amines. The components are metered, mixed, and allowed to solidify. However, at this stage the solid is not fully cured to a thermosetting structure. This B-staged solid can be heated and caused to flow, thereby wetting the substrate and crosslinking. The B-staged material can be cast to form a certain shape, such as gasket, rod, or film. These preformed shapes facilitate storage and application to the joint. The B-staged solid may also be ground into a powder for press-forming or application via electrostatic coating devices. With a B-staged solid adhesive, the end user avoids the need to meter and mix, safety and health issues are minimized, waste is greatly reduced or eliminated, and with a preformed adhesive even the application is almost complete.

17.2.4 Storage of Formulated Adhesives by the End User

The method by which the adhesive or sealant is stored can affect the quality of the bond even before the containers are open. The manufacturer's storage directions are usually found in the technical bulletins describing the particular adhesive or sealant. As a general rule, all adhesives and sealants should be stored in a cool, dry place until it is time for their use. Certain adhesives are very tolerant of storage conditions, but some types may have to be stored at low temperature or under special conditions. Some adhesive systems are affected by light or moisture, and others will require periodic agitation to make sure that the components do not settle irreversibly.

As organic resins age in storage due to exposure to temperature, moisture, light, or other environments, the viscosity of the system will usually undergo noticeable change. Once the storage life is exceeded, the system will generally be too viscous to use, or else it will be unusable from separation of its components. Figure 17.4 shows the increase in viscosity of a paste adhesive as a function of storage time at room temperature. Ordinarily, the shelf life

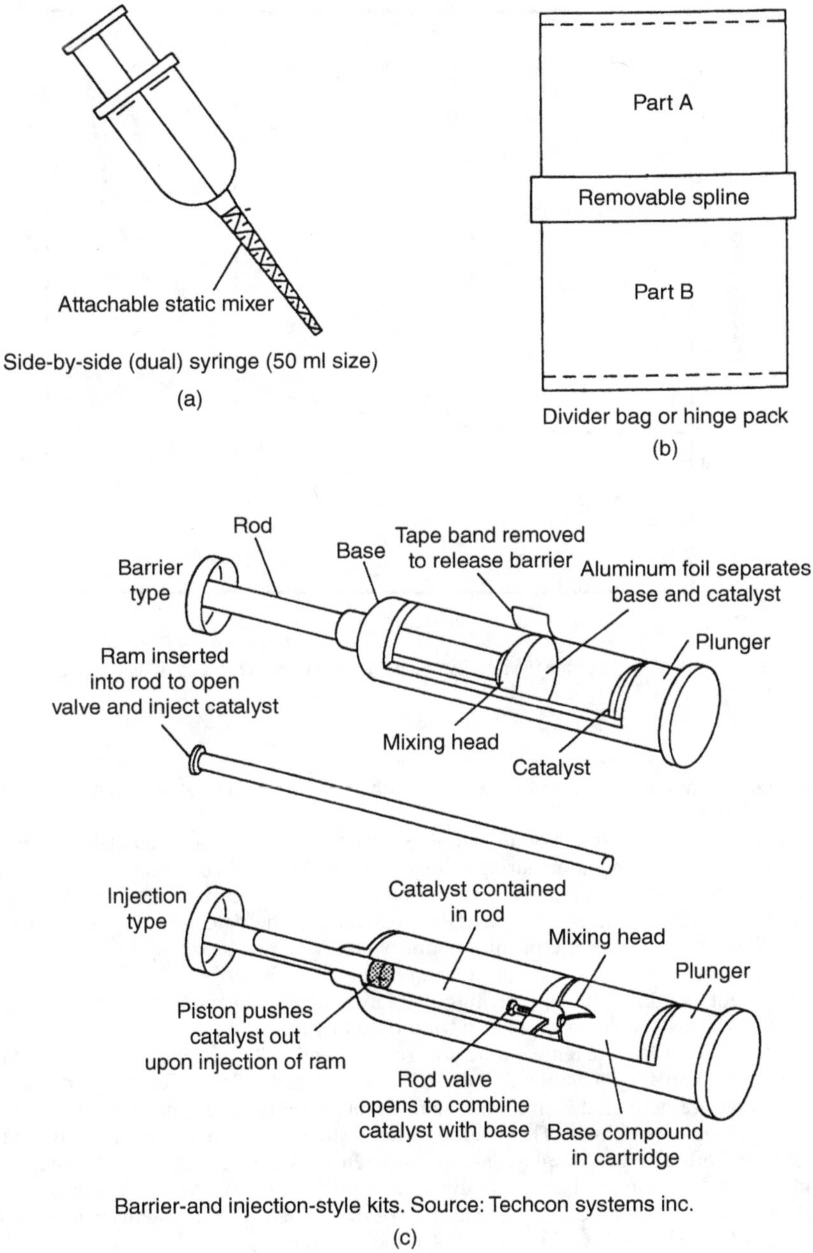

FIGURE 17.3 Several varieties of preproportioned adhesive containers: (*a*) side-by-side syringe with a static mixer, (*b*) bubble pack with hinge, and (*c*) barrier style cartridge.[3]

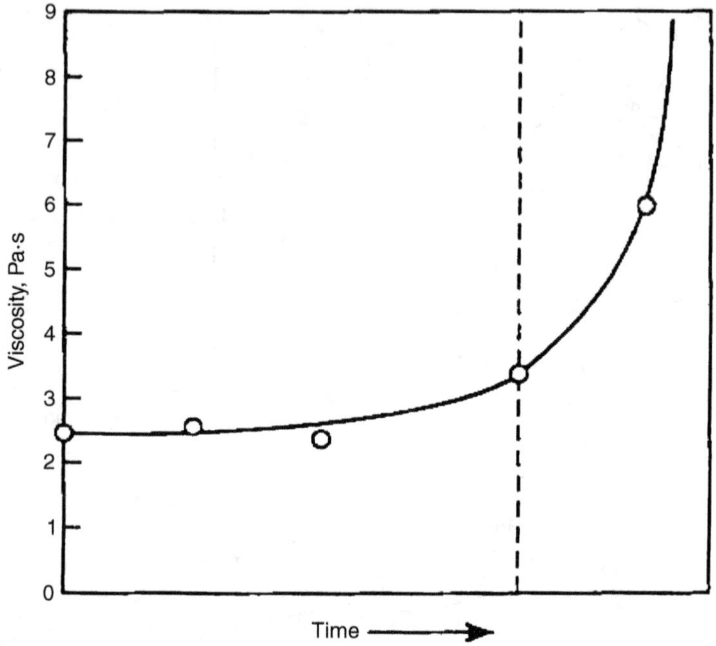

FIGURE 17.4 Increase in viscosity of a paste adhesive as a function of time at room temperature.[4]

is chosen so that the majority of the viscosity change will occur just after the expiration of the adhesive's shelf life.

Quite often the containers of an adhesive or sealant will bear a special label stating the safe upper limit of temperature during storage. Containers should be checked when received to see if such a label is present. As soon as the container is received, the user should place an additional label on it, stating the date received and the date when the shelf life will expire. This will provide useful information when it comes to scrapping old product or maintaining the freshest product in inventory. A record should also be kept of the manufacturer's lot number for possible future reference. Containers should be kept tightly closed, and labels should be kept clean for proper identification.

Three levels of storage temperature are common with adhesive and sealants: room temperature (15 to 30°C), refrigerated (0 to 5°C), and frozen (–20°C). Generally adhesives or sealants that are delivered as multiple-component systems have long shelf lives and are storable for very long times. They can usually be stored for 6 to 12 months at room temperature and often longer. Single-component systems, where the catalyst and base resin are combined in the same product, generally require refrigerated or frozen storage depending on the type of catalyst used and the reactivity of the system. These systems may have storage lives of only several weeks or months at the conditions specified.

Extended exposure of the uncured material to temperatures or conditions outside those recommended by the manufacturer will cause change in physical properties of the uncured material and will likely reduce its resulting cohesive and adhesive strength. The reactions that occur due to ambient storage conditions are described in Chap. 3. Figure 17.5 shows the effect of aging conditions on the tensile shear strength of an epoxy film adhesive.

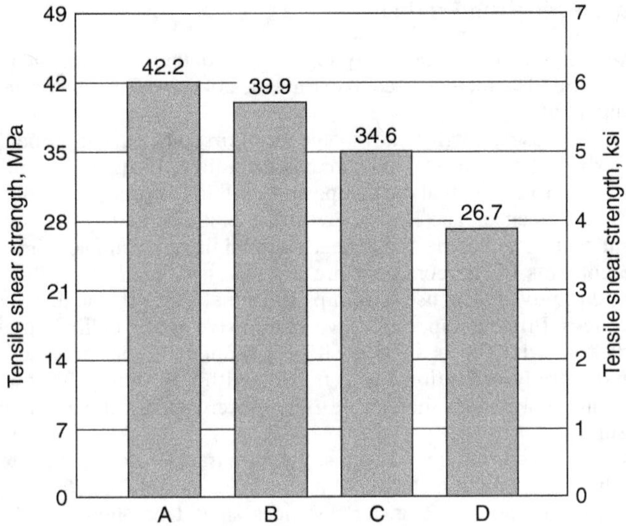

FIGURE 17.5 Effect of aging conditions on epoxy film adhesive before cure: (A) fresh adhesive, (B) after aging 90 days at 24°C, (C) after 90 days at 32°C, (D) after 1-h exposure to condensing humidity.[5]

ASTM D 1337 describes a standard test method for measuring storage life of adhesives by consistency and bond strength.

The safe storage of adhesives and sealants that contain flammable, corrosive, or hazardous substances must also be considered. Many organic solvents are flammable, and certain adhesives will generate copious amounts of heat when reacted. For this reason, containers used for two- or three-part materials should be stored separately to prevent accidental mixing of the components in the case of spillage.

The more hazardous adhesives and sealants should be stored in a small building or area separate from the main building or storage area in the plant. An outside storage building is recommended if several drums of flammable materials are to be stored. The storage facility should be built of fire-resistant materials and a fire extinguishing system installed. For storage of just a few drums of flammable materials, insulated metal storage cabinets are available that will protect the drums by closing their doors automatically in the event of a fire. An electrical ground should be provided to all drums so that a potentially explosive spark cannot occur. The most appropriate internal location for a flammable storage area is a room with brick or masonry walls and fire doors that open outward. Some adhesive products contain corrosive compounds that require special containers and need to be separated from other materials to avoid interaction.

When removed from refrigerated storage, the adhesive must be brought to its application temperature, usually room temperature. The containers should remain closed until they are at temperature. A material is most vulnerable to moisture collection when it is removed from cold storage and the package or container is not properly sealed. Moisture condenses on the adhesive and quickly reduces its capabilities. Containers that are not completely used when originally opened, should be sealed immediately after use to prevent loss of volatile ingredients or contamination from moisture in the air. Equipment used to remove the contents from the containers, such as spatulas, flasks, and cups, should be cleaned between each use.

17.2.5 Transferring the Product

Bulk adhesives and sealants are normally supplied in containers that range in volume from 1 qt to 55-gal drums. The method used to extract the adhesive from the container should be carefully thought out.

If a filled resin is used, it will be advantageous to first place the unopened can on a paint shaker or roller and vibrate or roll the contents for a time. If the resin is in a drum, drum rollers may be used to ensure that the components within the drum are uniformly dispersed before the drum is opened. The least aggressive method should always be used so as to obtain uniform distribution of the components with minimal addition of air into the product.

Once the contents of the container are evenly distributed, the adhesive or sealant material can be removed. The use of pumps for transferring the adhesive or sealant has several drawbacks. Filled compounds have an abrasive action on the moving parts. This is particularly true with fillers such as silica. Compounds that polymerize easily will tend to begin their polymerization due to friction within the pump. This will gum up the moving parts and make them inoperable. Therefore, transfer is best done under some type of pressure.

Although it may require special designs, pressure transfer does eliminate the problems that occur with gear pumping methods. Many resinous systems may be moved from their original drums by air pumps. Transfer distances should be short, and lines should be smooth and noncorrosive. In some instances, barrel warmers and heated lines may be required to reduce pressures required to transport materials. Under no circumstances should the same transfer equipment be used on different resins without complete and thorough cleaning. This will avoid the contamination of one resin system with the other.

17.2.6 Metering and Mixing of Components

Metering and mixing are important processes for the epoxy adhesive end user. Accurate metering of components and thorough mixing of all components are required for the adhesive to attain its ultimate cured properties. Metering and mixing are often done by hand, although automatic systems are available and preferred in many circumstances. Automated equipment is especially useful in continuous or high-volume bonding applications.

Metering. The components of multiple-part adhesive or sealant systems must be measured out carefully. The concentration ratios have a significant effect on the quality of the joint. Strength differences caused by varying the curing agent concentration are most noticeable when the joints are tested at elevated temperatures or after exposure to water or solvents.

Exact proportions of resin and hardener must be weighed out on an accurate balance or in a measuring container for best adhesive quality and reproducibility. Possible problems that can occur by not adhering to the proper mixing proportions can include incomplete polymerization (too little catalyst), brittleness (excessive catalyst), and corrosion of metallic adherends (excessive catalyst).

Some two-component adhesives (e.g., fatty polyamide cured epoxies) have less critical mixing ratios, and component volumes may often be measured by eye without too adverse an effect on the ultimate bond strength. In fact, with certain types of adhesives the mixing ratio is used as a parameter to effect the final properties of the cured adhesive. For epoxy systems cured at room temperature with a polyamide hardener, flexibility and toughness can be improved within limits by increasing the polyamide resin concentration. The temperature and chemical resistance of the adhesive is generally improved by decreasing the amount of polyamide hardener used, but at the cost of decreasing toughness.

If the components require weighing, the weighing equipment should be clean and calibrated. Containers used for weighing should have their tare weights plainly marked on the outside. Containers for weighing adhesive must not be interchanged with those for weighing water or any other ingredients.

Mixing. The measured components must be mixed thoroughly. Mixing should be continued until no color streaks or density stratifications are noticeable. Caution should be taken to prevent air from being mixed into the adhesive through overagitation. This can cause foaming of the adhesive during cure, resulting in porous bonds. If air does become mixed into the adhesive, vacuum degassing may be necessary before application.

In mixing, cleanliness is almost as important as accuracy. Scoops, weighing pans, mixing spatulas, and other accessories should be kept clean. Organic solvents are recommended for cleaning. Adhesives that appear to be similar may be ruined if one is contaminated with the other. Water will spoil many adhesives and sealants.

Only enough adhesive should be mixed to work with before the adhesive begins to cure. The working life of an adhesive is defined as the period of time during which a mixed adhesive remains suitable for use. Working life is decreased as the ambient temperature increases and the batch size becomes larger. One-part and some heat curing, two-part adhesives have very long working lives at room temperature, and application and assembly speed or batch size is not critical.

The optimum batch size depends on the curing agent used in the epoxy adhesive system and the temperature at which the system is worked. Epoxy adhesive systems generally maintain a workable viscosity for the majority of their pot life. However, once the exotherm begins, the viscosity of the epoxy system increases rapidly to a point where it gels. For the more highly exothermic curing agents, the size of the batch will establish working life.

For a given type of mixing equipment, the larger the mass, the longer will be the mixing time. This reduces the time available for application of the adhesive. Alternatively, a system of batch replenishment may be employed, but this is sometimes difficult because of the exotherm of the resident batch causing acceleration of the additive batch.

For mechanical mixing often a variable-speed mixer, preferably air-driven, and a container are all that is required. An air-driven mixer is safer, less costly, and more easily regulated than an electrically driven mixer. For the most reliable mixing operation, the temperature of the compound in the mixing vessel should be uniform. For adhesives or sealants that are stored at cold temperature, adequate mixing viscosity can be reached quickly by the use of a hot liquid jacket around the vessel.

Special mixing blades have been designed to minimize air entrapment and to scour the sides and bottoms of containers. These blades are made for standard container sizes. In this case often a fixed blade is used along with a rotating platform. Mixing equipment is described in Sec. 17.2.2.

For a large-scale bonding operation, hand mixing is costly, messy, and slow; repeatability is entirely dependent on the operator. Equipment is available that can meter, mix, and dispense multicomponent adhesives on a continuous or shot basis. Figure 17.6 illustrates a standard adhesive metering mixing and dispensing machine for multicomponent systems. An important advantage for using such equipment is that it allows the use of short-working-life materials by mixing only the smallest practical quantity continuously. This offers shorter curing cycles, resulting in faster assembly times. A disadvantage of this type of equipment is that it should be used continuously for best results. When the equipment must be shut down, it requires thorough cleaning. Often the equipment is supplied with an automatic solvent flush to make sure that all resin is purged from the lines and from internal mixing devices.

For maximum strength glue lines, it is necessary to remove all air from the adhesive before using it. Entrapped air represents a source of voids in the cured bond line and, hence, the possibility of a failure. Trapped air is especially noticeable when the adhesive requires

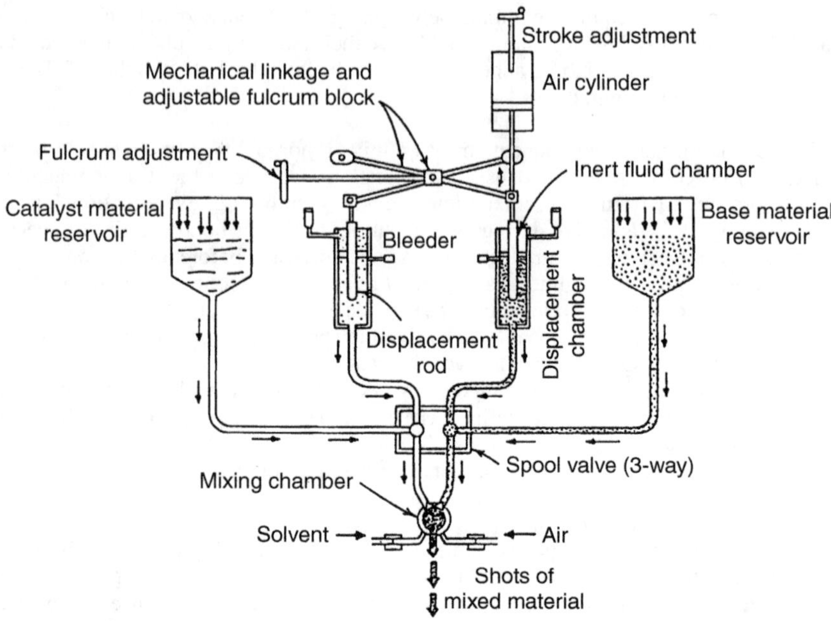

FIGURE 17.6 Operating schematic of an automatic metering, mixing, and dispensing machine for adhesive systems.[6]

an elevated-temperature cure; significant amounts of trapped air can cause the adhesive to foam when exposed to elevated temperatures.

When one is hand-mixing liquid adhesives or sealants, two methods are commonly employed to degas the mixed material: (1) centrifuging and (2) subjecting the mixed system to a vacuum. The latter is the more popular method and requires only a vacuum pump and vacuum chamber. The vacuum pump should be capable of pulling a vacuum of 29 in Hg at a high rate of speed.

Vacuum degassing consists of placing the mixed material (in the container in which it was mixed) into a vacuum chamber. The level of the mixture in the container should be low enough to allow for a foam head to rise and break during application of a vacuum. Usually the volume of the chamber is 5 to 10 times the volume of the adhesive to be mixed. If the adhesive level is too high, the foam head will not break.

It is desirable to know when the foam head has broken, so the vacuum tank should be provided with two sight windows: one for illumination and one for observation. It is best to repeatedly break the vacuum during the degassing operation so that multiple foam heads rise and break. Care should be taken to avoid a "rolling boil" effect. Once all the air is removed from the adhesives, it will become very difficult to cause a foam head to rise. It is best to try to minimize the vacuum time, to reduce the loss of any volatile materials from the mixture. Applied vacuum along with agitation and vibration will effectively remove all entrapped air.

Air can also be eliminated from an adhesive mixture by using a centrifuge. This process is generally employed only with low-viscosity, unfilled adhesives. The centrifuging process consists of rotating the materials at 1000 to 3000 rpm for 1 to 2 min. This is generally sufficient to remove any entrapped air from liquid adhesive systems.

17.2.7 Applying the Adhesive

The adhesive or sealant should be mixed and furnished for application in clean containers. The containers should be small enough that no adhesives or sealant will be left beyond the time when the resin becomes unusable due to its working life. Disposable waterproof cardboard cups or aluminum containers work very well. Containers may be marked to show the hour at which the contents were mixed. Any unused adhesive should then be collected at the end of the working life period recommended by the manufacturer.

Header systems with manual or automatic dispensing valves have been designed to carry adhesives and sealants over long distances within the plant. Distances of as much as 300 ft are not uncommon with drops at points of application. Figure 17.7 shows a modular header system. Adhesives and sealants are nearly ideal materials for application by robots. Robotic application is commonly used in the automobile industry to increase quality and to reduce labor and material cost.[7,8]

The epoxy adhesives are usually applied by simple extrusion from dispensing nozzles, brushing, or troweling. However, spray equipment may be used in certain applications. Conventional spray equipment is used with solvent-borne epoxy adhesives, but frequent cleaning is generally required. For fast-reacting systems, dual-nozzle spray equipment is available.

The type of pumps used to dispense epoxy adhesives depends on whether it is a one-part or two-part adhesive. One-part epoxy adhesives are dispensed using direct metering extrusion pumps. An electric motor pushes a follower plate into a drum of adhesive, which is then extruded through a hose to the dispensing valve. This technique is used for medium- and high-volume production rates.

Two-part adhesives are generally dispensed by volumetric pumps for semiautomated and automated medium- to high-volume assemblies. Metered dispensing systems are easily capable of processing adhesives with mix ratios of 1 : 1 up to 15 : 1.

The method used to apply the adhesive or sealant can have as much to do with the success of the joining operation as the kind of material applied. The selection of an application or dispensing method depends primarily on the form of the adhesive: liquid, paste, powder,

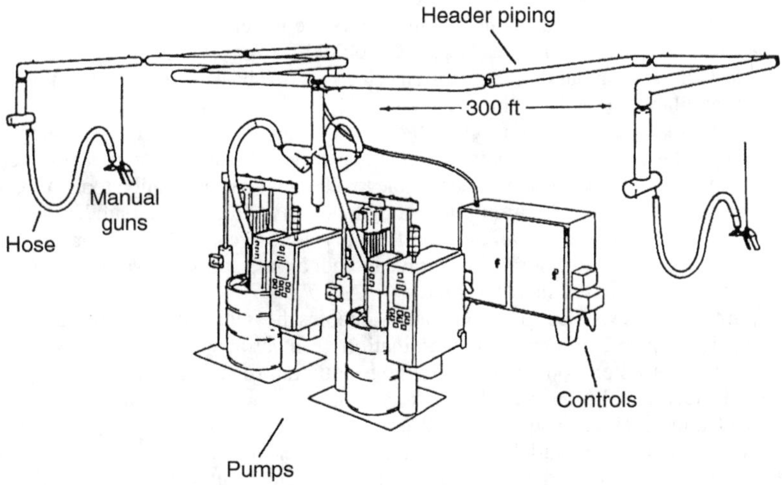

FIGURE 17.7 Modular header system.[9]

or film. Other factors influencing the application method are the size and shape of parts to be bonded, the areas where the adhesive is to applied, and the production volume and rate. One of the prime functions of production equipment is to keep the production employee removed from contact with the adhesive materials.

Liquids. Liquids are the most common form of adhesive, and they can be applied by a variety of methods. Liquids have an advantage in that they are relatively easy to transfer, meter, and mix. They also tend to wet the substrate easily and provide uniform bond line thickness. However, they have the disadvantages of sometimes being messy, requiring cleanup, and having a relatively high degree of waste.

Brushes, simple rollers, syringes, squeeze bottles, and pressurized glue guns are manual methods that provide simplicity, low cost, and versatility. These methods are probably the most widely used because of their simplicity. Manual dispensing methods allow application of adhesive to only a small segment of a surface and are particularly effective for small or irregular parts or low-volume production.

It is possible to mechanize brush methods by using hollow brushes through which the liquid adhesive is fed. This improves production rates and coating uniformity and cuts waste. Uniform films are generally difficult to achieve, but operators can become proficient with time.

Squeeze bottles, oil cans, and pressurized glue guns permit precise and speedy adhesive application. By adjusting the pressure, the rate of adhesive flow can be matched to the production rate. These devices can apply the adhesive or sealant inside a blind hole or limited-access area. The tip of the applicator can be used to deliver multiple spots of adhesive.

Silk screen application is often used when the adhesive has to be applied to specific controlled areas. The liquid adhesive is forced through pores in a cloth or screen. It is possible to coat only selected areas by masking parts of the screen so that adhesive does not pass through in the unwanted areas. Adhesives generally must be specifically formulated for silk screen processing. Very low-viscosity adhesives, with flow characteristics similar to those of coatings, are best for silk screening operations.

One of the most attractive precision dispensing methods for low-viscosity adhesives is the use of microjet printing technology.[10] This technology is based on piezoelectric demand mode ink-jet printing, which can produce droplets of polymeric resins 25 to 125 μm in diameter, at rates up to 1000 drops per second.

Spraying, dipping, and mechanical-roll coaters are generally used on large production runs. Dipping is also used to cover large areas quickly. Areas that do not require adhesive can be masked off. Manual dipping is often effective; and baskets, screens, and perforated drums are available to facilitate immersion of the part into the adhesive bath.

Spraying usually is evaluated against other methods on the basis of cost. Spraying applies the liquid resin fast, reduces the required drying time through better evaporation of solvent, and is capable of reaching areas that are inaccessible to other manual application tools. When automated, spraying is particularly useful for coating long runs of identical or similar products.

Several spraying methods can be used to apply adhesives, including conventional air spray, hydraulic cold airless spray, hot spray, and hot airless spray. The main types of spray equipment are summarized in Table 17.1.

Heating the adhesive before atomization enables heavier adhesive buildup, reduces overspray losses, and minimizes contamination from atmospheric water vapor. Spray methods can be used on both small and large production runs. The liquid to be sprayed is generally in solvent solution. Sizable amounts of product may be lost from overspray. Two-component adhesives are usually mixed prior to placement in the spray gun reservoir. Application systems are available, however, that meter and mix the adhesive within the spray gun barrel. This is ideal for fast-reacting systems, but guns must be thoroughly cleaned to avoid buildup of polymerized product.

TABLE 17.1 Main Types of Spray Equipment

Type	Characteristics
Air spray	Utilizes air pressure to spray a fine mist of atomized adhesive on the substrate. Generally unsuitable for small parts. Maintenance is high due to nozzle clogging and overspray. Several passes necessary for thick coating buildup
Hot spray	Adhesive is first heated, permitting the use of high-viscosity materials. Heavier coatings are possible than with air spray. Drying is accelerated due to evaporation of solvent in the spray
Airless spray	Uses hydraulic pressure rather than air pressure. Saves energy and overspray. Can be applied both hot and cold

When large, flat surfaces and webs of materials are to be coated and when production reaches rates of 200 to 300 pieces per day, machine methods of application should be considered over the manual methods described above. Mechanical-roller methods are commonly used to apply a uniform layer of adhesive via a continuous roll. Such automated systems are used with adhesives that have a long working life and low viscosity. Various machine methods of adhesive application are illustrated in Fig. 17.8, and several are described below.

Bench coaters are roughly 20 to 35 percent faster than hand-brushing and cut waste by up to 20 to 35 percent. They start to be useful at rates of 200 pieces per day and can reach 12,000 pieces per day. Roller sizes range from 4 to 26 in, and they come with various surfaces from smooth for thin applications to increasingly coarse for heavier adhesive layers.

Floor-mounted roll coaters are noted for their extreme efficiency with large webs and flat sheets. Their waste is also very low, less than 2 percent in some cases. Several types are available because no single model or type of roll coater is best for all the variables of the material surface, coating formulation, and end use. Table 17.2 summarizes the roll coaters that are commonly used, and these are illustrated in Fig. 17.8.

Pastes and Mastics. Bulk adhesives such as pastes or mastics are the simplest and most reproducible adhesive to apply. They produce heavy coatings that fill voids, bridge gaps, or seal joints. They can be in the form of high-viscosity extrudable liquid or a trowelable mastic.

These systems can be troweled on or extruded through a caulking gun. Little operator skill is required. Since the thixotropic nature of the paste prevents it from flowing excessively, application is usually clean, and not much waste is generated. With trowel application, the depth of notch and the spaces between them help regulate the amount of adhesive applied. Shallow, rounded, and closely spaced notches are often used with lower-viscosity adhesives to allow the lines of adhesive to flow together and form a continuous, unbroken film.

High-viscosity liquids, pastes, and mastics are ideal adhesives for application by robots. Robotic application is commonly used in industries such as automotive and consumer products to increase quality and to reduce labor and material costs.[12,13]

Robots are fast gaining prominence in the field of material dispensing, especially in the application of adhesives and sealants. The use of robots provides the end user with the following benefits:

- More uniform application of the adhesive bead, resulting in higher quality
- Decreased adhesive or sealant materials costs
- Better working environment because the operator avoids constant direct contact with volatile material
- Better utilization of machinery investments and increased process flexibility

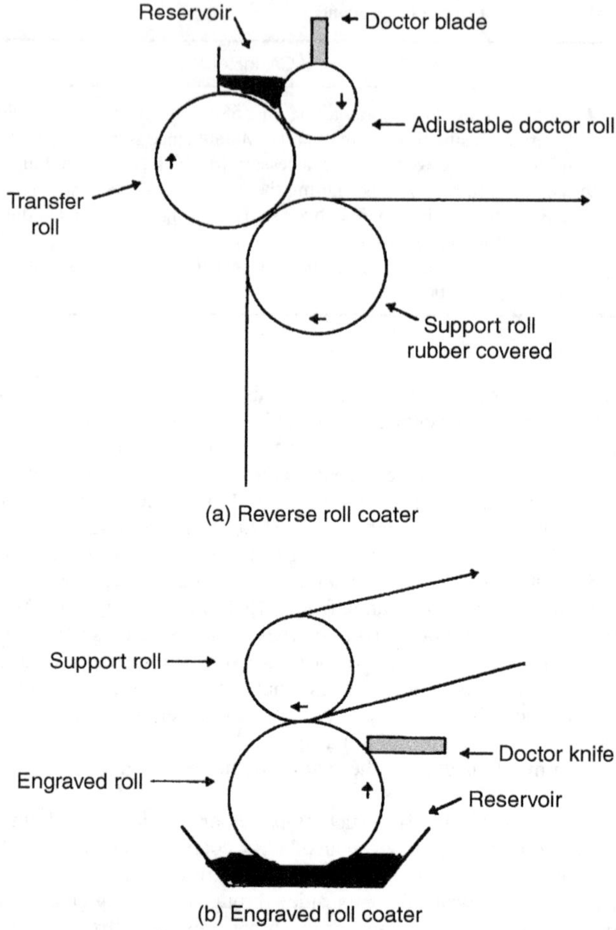

FIGURE 17.8 Methods of machine application of adhesive coatings.[11]

Several areas are important when one is considering a robotic system for dispensing adhesive or sealants. These are summarized in Table 17.3. Monetary justification of robot installation has been and will continue to remain an issue for the manufacturing industries. Several installations have seen payback periods of less than 2 years.

Powders. Powdered adhesive systems are not common, although powder resin coatings are ordinarily used for protective and decorative coating. B-staged powder adhesives may be applied dry to a natural or primed surface. They can be applied electrostatically as powder coatings are, or they can be applied gravimetrically by being dispersed onto a hot substrate surface. Heat and pressure must be applied for full curing, and considerable care taken to obtain uniform thickness over large areas. If coating uniformity is poor, large variations of joint strength may result with powder adhesives. The main advantages of using powder adhesives are their cleanliness and minimal waste. Spills and overspray may be

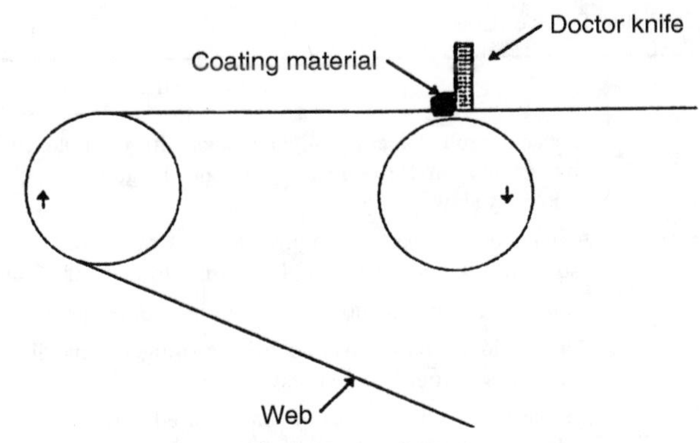

(c) Knife-over-roll coater

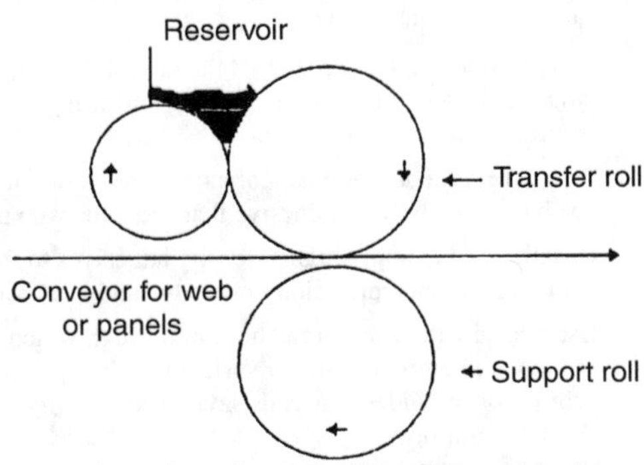

(d) Squeeze roll coater

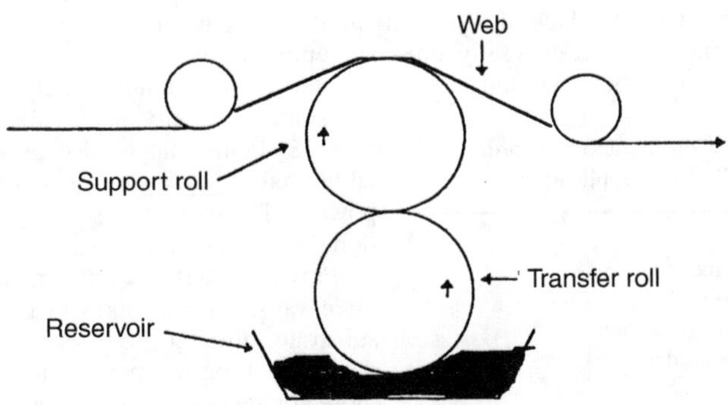

(e) Kiss roll coater

FIGURE 17.8 (*Continued*).

TABLE 17.2 Common Types of Coaters

Type	Characteristics
Kiss	No backup roll; coater is in light contact with substrate; contact is controlled by web tension. Best with adhesives having good flow characteristics and relatively slow drying rates
Pressure or squeeze	Also called nip coaters; can provide greater penetration into open-web structures. Often used to apply coatings to both sides simultaneously
Reverse roll	Popular for web precision coating because of better control over thickness
Two-roll reverse	For additional control over adhesive coating; can handle wet film thicknesses from 1 to 20 mils
Dip roll	Simplest of the basic types of coaters; used mostly with low-viscosity adhesives; works well where impregnation or saturation is required
Engraved roll	Provides accurate control over coating uniformity; best for extremely low adhesive weights
Air blade	Similar to the kiss coater in that the adhesive is applied where the web touches it; possible to apply coating uniformly at high speeds; limited use with solvent-based adhesives
Bead	Two rollers set at specific clearance; a rod between the bottom roll and the web makes a bead of adhesive that coats the web passing above it
Knife spreader	Usually used to apply high-viscosity adhesives to flexible webs; web is drawn between a precision knife edge and a rubber roller or flat-bed plate
Flow or curtain	Use a head with a slot in the bottom through which the adhesive is poured or lowered onto the parts or surfaces as they pass beneath the head by a conveyor; provides materials savings and ability to coat irregular surfaces uniformly

collected and reused, if not contaminated. The main disadvantage is the cost of the application equipment and the lack of manual methods of application.

Powder adhesives are generally one-part, epoxy-based systems that do not require metering and mixing but often must be refrigerated for an extended shelf life. Elastomeric types are generally not available in powder form. Systems that need a primer require an additional coating step prior to applying the powder. Equipment cost for powder application tends to be high.

TABLE 17.3 Considerations regarding Robotics in Adhesives Application

Type of robot
Pumping equipment
Interface controls
Type of adhesive or sealant
Gun and nozzle holders
Safety fencing
Bead detection device
Arrangement and flexibility of arrangement
Regulating equipment
Area of dispensing
Size and shape of bead
Weight of robot arm tooling
Flow control

Powder adhesives are generally applied in three ways: sifting or spraying the powder on the substrate, dipping the substrate into the powder, or melting the powder to liquid form for more conventional application. When sifted or sprayed onto a substrate, the powder falls onto the preheated substrate, melts, and adheres. Electrostatic spraying of adhesive powder has also been used to apply a B-staged powder to various substrates. A preheated substrate could also be dipped into the powder and then extracted with an attached coating of adhesive.

This method helps to ensure even powder distribution. Lastly, the powder can be melted into a paste or liquid and applied by conventional means. Once the powder is adhered to the substrate, the assembly is then mated and cured under heat and pressure according to recommended processes.

Films. Both structural and nonstructural adhesives are commonly available in film form. Adhesives applied in the form of dry films offer a clean, hazard-free operation with minimum waste and excellent control of film thickness. However, the method is generally limited to parts with flat surfaces or simple curves. Optimum bond strength requires curing under heat and pressure, which may involve considerable equipment and floor space, particularly for large parts. Film material cost is high in comparison to liquids, but waste or material loss is the lowest of any application method.

Practically any solid or B-stageable epoxy adhesive that has adequate stability at room temperature can be supplied in the form of a film. This film may be pressure-sensitive, thermoplastic, or, in applications where the greatest structural strength is required, a thermosetting type. Epoxy, epoxy hybrid, phenolic, and other thermosetting film adhesives are commonly used in many industrial applications. Elastomeric adhesives are available in dry films in both solvent and heat reacting types. Both elastomer- and resin-based adhesives are also supplied as pressure-sensitive adhesives, but they have lower strength than the solvent or heat reacting types. Epoxy adhesives generally do not have pressure-sensitive characteristics.

The use of dry adhesive films is expanding more rapidly than other forms because of their following advantages:

1. High repeatability—no mixing or metering, constant thickness
2. Easy to handle—low equipment cost, relatively hardware-free, clean operation
3. Very little waste—preforms that can be cut to size
4. Excellent physical properties—wide variety of adhesive types available

Application requires a relatively high degree of care to ensure nonwrinkling and removal of separator sheets. Films are often supported on scrim that distributes stress in the cured joint and ensures a uniform bond line thickness throughout the bonded area.

Characteristics of typical film adhesives vary widely depending on the type of adhesive used. Epoxy film adhesives are made in both unsupported and supported types. The carrier for supported films is generally fibrous fabric or mat.

17.3 BONDING EQUIPMENT

After the adhesive is applied, the assembly must be mated as quickly as possible to prevent contamination of the adhesive surface. The substrates are held together under pressure and heated if necessary until cure is achieved. The equipment required to perform these functions must provide adequate heat and pressure, maintain constant pressure during the entire cure cycle, and distribute pressure uniformly over the bond area. Of course, many adhesives cure with simple contact pressure at room temperature, and extensive bonding equipment is not necessary.

It is important when one is selecting equipment to give consideration to life-cycle costs which include original investment cost and operation and maintenance cost, such as energy cost, repair, and troubleshooting. A checklist of minimal equipment considerations should include

- Simplicity of operation
- Compatibility with adhesive or sealant
- Ease of disassembly and cleanup
- Trouble detection devices and self-diagnostics
- Service backup relative to spare parts, local resources, etc.
- Knowledge of life expectancy of system and its components
- Clearly defined warranty coverage
- Knowledge of the equipment production rates
- Ease of pressure regulation
- A means of checking the proportions of different adhesive components
- Temperature controls and indicators.[15]

17.3.1 Pressure Equipment

Pressure devices should be designed to maintain constant pressure on the bond during the entire cure cycle. They must compensate for thickness reduction from adhesive flow-out or thermal expansion of assembly parts. Thus, screw-actuated devices such as C-clamps and bolted fixtures are not acceptable when constant pressure is important. However, spring clamp fixtures perform well on assemblies machined to distribute mechanical loads. Spring pressure can supplement the clamps during heat curing to compensate for variation in pressure when the adhesive begins to flow and adherends expand. Deadweight loading may also be applied in many instances. However, this method is sometimes impractical, especially when a heat cure within an oven is necessary.

Pneumatic and hydraulic presses are excellent tools for applying constant pressure. Steam or electrically heated platen presses with hydraulic rams are often used for adhesive bonding. Some units have multiple platens, thereby permitting the bonding of several assemblies at one time.

Large bonded areas, such as on aircraft parts, are usually cured in an autoclave. The parts are mated first and covered with a rubber or film blanket to provide uniform pressure distribution. The assembly is then placed in an autoclave, which can be pressurized and heated. This method requires heavy capital equipment investment.

Vacuum bagging techniques can be an inexpensive method of applying pressure to large parts. A film or plastic bag is used to enclose the assembly, and the edges of the film are sealed airtight. A vacuum is drawn on the bag, enabling atmospheric pressure to force the adherends together. Vacuum bags are especially effective on large areas because the size is not limited by pressure equipment. Pressures, of course, are limited to atmospheric pressure.

17.3.2 Heating Equipment

Many structural adhesives require heat as well as pressure to cure. Even with conventional room temperature curing systems, most often the strongest bonds are achieved by an elevated-temperature cure. With many adhesives, tradeoffs between cure times and temperature are permissible. Generally, the manufacturer will recommend a certain curing schedule for optimum properties.

If a cure of 60 min at 150°C is recommended, this does not mean that the assembly should be simply placed in a 150°C over for 60 min. The temperature is to be measured at the adhesive bond line. A large part will act as a heat sink and may require substantial time for the adhesive in the bond line to reach the necessary temperature. In this example, total oven time would be 60 min in addition to whatever time is required to bring the adhesive up to 150°C. Bond line temperatures are best measured by thermocouples placed very close to the adhesive. In some cases, it may be desirable to place the thermocouple in the adhesive joint for the first few assemblies being cured.

Oven heating is the most common source of heat for bonded parts, even though it involves long curing cycles because of the heat sink action of large assemblies. Ovens may be heated with gas, oil, electricity, or infrared units. Good air circulation within the oven is mandatory for uniform heating. Temperature distribution within an oven should always be checked before items are placed in the oven. Many ovens will have significant temperature distributions and dead spaces in corners where air circulation is not uniform. The number of items placed in the oven will also affect the time for the bond to get to temperature. The geometry and size of the part may affect air circulation and cause variations in temperature distribution.

Heated-platen presses are good for bonding flat or moderately contoured panels when faster cure cycles are desired. Platens are heated with steam, hot oil, or electricity and are easily adapted with cooling-water connections to further speed the processing cycle. Strip or cartridge heaters are useful for localized heating in small areas. They adapt easily to flat dies, pressure bars, specially shaped dies, and other pressure fixtures. They are often used for tack-bonding small areas to facilitate handling prior to complete cure.

Induction and dielectric heating (Chap. 14) are the fastest heating methods because they focus heat at or near the adhesive bond line. Workpiece heating rates greater than 55°C/s are possible with induction heating. For induction heating to work, the adhesive must be filled with metal particles, or the adherend must be capable of conducting electricity or being magnetized, as shown in Fig. 14.5.

Dielectric heating (radio frequency and microwave) is also an effective way of curing adhesives if at least one substrate is a nonconductor. Metal-to-metal joints tend to break down the microwave field necessary for dielectric heating. This heating method makes use of the polar characteristics of the adhesive materials. Dielectric heating is used in the furniture industry to drive off the water and harden water-based adhesives. Dielectric heating is also being explored as a rapid method of curing adhesives on substrates such as glass, plastics, and composites.[15]

Both induction and dielectric heating involve relatively expensive capital equipment outlays, and the bond area is limited. Their most important advantages are assembly speed and the fact that an entire assembly does not have to be heated to cure only a few grams of adhesive.

An interesting development that eliminates the need for fixturing equipment and heating equipment for specifically curing the adhesive joint is *weldbonding*. Weldbonding is adhesive bonding combined with spot welding to make fast, strong, reliable joints. The process produces joints that are stronger, more durable, and more fatigue-resistant than when either method is used by itself. This process has found its way into the automotive industry for use on aluminum and steel substrates. A variety of adhesives and substrates are compatible with this method. Epoxy adhesives capable of weldbonding are described in Chap. 14.

REFERENCES

1. Hockmeyer Equipment Corporation, Harrison, NJ.
2. *Plastics Compounding*, 1991–1992 Redbook, Edgell Communications, Inc., Cleveland, OH.

3. Devlin, W., "Metering and Mixing Equipment," in *Adhesives and Sealants*, vol. 3, *Engineered Materials Handbook*, ASM International, Materials Park, OH, 1990.
4. Van Twisk, J., and Aker, S. C., "Storing Adhesive and Sealant Materials," in *Adhesives and Sealants*.
5. Van Twisk, and Aker, "Storing Adhesive and Sealant Materials."
6. Graco Pyles Corporation, Minneapolis, MN.
7. "Robotic Dispensing of Adhesives and Sealants," *Handbook of Adhesives*, 3d ed., I. Skeist, ed., Van Nostrand Reinhold, New York, 1990.
8. Dueweke, N., "Robotics and Adhesives—An Overview," *Adhesives Age*, April 1983, pp. 11–16.
9. Killick, B. R., "Dispensing High Viscosity Adhesives and Sealants," in *Adhesives in Manufacturing*, G. L. Schneberger, ed., Marcel Dekker, New York, 1983.
10. Hayes, D. J., et al., "Development and Application by Ink Jet Printing of Advanced Packaging Materials," in *Proceedings of the International Symposium on Advanced Materials Process, Properties and Interfaces*, 1999, pp. 88–92.
11. Petrie, E. M., *Handbook of Adhesives and Sealants*, McGraw-Hill, New York, 2000, pp. 744–745.
12. "Robotic Dispensing of Adhesives and Sealants," in *Handbook of Adhesives*.
13. Trees, B., "Auto Adhesive Extrusion Systems," *Adhesives and Sealants Industry*, May 2002.
14. Dueweke, N., "Robotics and Adhesives—An Overview," *Adhesives Age*, April 1983, pp. 11–16.
15. Raulauskas, F. L., "Adhesive Bonding/Joining via Exposure to Microwave Radiation," *27th International SAMPE Technical Conference*, October 9–12, 1995.

CHAPTER 18
HEALTH AND SAFETY ISSUES

18.1 INTRODUCTION

Completely cured epoxy adhesives are almost nontoxic, although caution must be exercised when machining, grinding, etc., joints bonded with epoxies due to the proliferation of airborne particles. Also, decomposition products may exhibit substantial vapor pressures and pose health and safety issues. However, it is the exposure to the uncured components that can be the most harmful and is the subject of this chapter.

Four primary factors must be considered in all adhesive bonding or sealing operations: toxicity, flammability, hazardous incompatibility, and equipment. The adhesive or sealant must be carefully considered because not only can it create health and safety issues within the factory but also it can provide issues within the community relative to release of volatiles and waste disposal.

The epoxy raw materials can be harmful and hazardous. These materials could be toxic, flammable, and irritating to skin. It is generally the duty of the formulator to make sure that these materials are worked within a safe environment. Once they are formulated, the end user also has responsibilities with regard to safe handling of epoxy adhesive products.

Some materials involved in the epoxy adhesive formulations are more dangerous than others, and some affect certain parts of the body whereas others do not. Amine curing agents, for example, are generally the most irritating to the skin and eyes. They may also be considered to be strong caustics, and they produce serious local injury on short exposure. If they are worked in a hot condition, their fumes pose additional hazards.

In a two-component epoxy product, the epoxy resin and the curing agent are packaged separately and must be mixed together just before use. Each component can be hazardous. In a single-component product, the resin and curing agent are supplied in a premixed form. Single-component systems are generally safer because the hazardous chemicals are already combined and because their time of exposure is less.

Solid epoxy adhesives, such as powder, preforms, or film, are the least toxic because they have very low vapor pressures. However, they still can present a hazard when they are heated to a liquid and more volatile state to cure.

Accessory materials and equipment used to process and apply epoxy adhesives can also be hazardous. Solvents that are used either to dilute the epoxy adhesive or to clean equipment after use can be both dangerous and toxic. Mixing equipment can run at very fast speeds, and caution needs to be exercised when one is operating such equipment. Curing equipment often runs at high temperatures and can create a fire or explosion hazard. Electrical hazards are also associated with most equipment.

Because of the wide variety of materials, equipment, and processes used with epoxy adhesive systems, the following discussion provides only a general guide. For more detailed information, the reader should contact the manufacturer of each material or equipment item used. The discussion that follows focuses primarily on

- The general effects of exposure to epoxy adhesive materials
- The specific materials used and their health and safety characteristics
- Common processes used with epoxy adhesive systems
- Processes to control and minimize health and safety issues
- First aid
- Emergency procedures

18.2 EFFECTS OF EXPOSURE TO EPOXY ADHESIVE MATERIALS

All adhesives, solvents, chemical treatments, and like materials must be handled in a manner preventing toxic exposure to the workforce. Methods and facilities must be provided to ensure that the maximum acceptable concentrations of hazardous materials are never exceeded. These values are prominently displayed on the material's Material Safety Data Sheet (MSDS), which must be maintained and available for the workforce.

There is no medical or laboratory test that can accurately measure the amount of epoxy components in the human body, and most of the chemicals found in epoxy systems are not stored in the body. Regional environmental regulators set and enforce workplace exposure limits. These exposure limits are generally set for some of but not all the epoxy components that one may contact.

Table 18.1 shows legally permissible exposure limits (PELs) for some chemicals commonly found in epoxy adhesive systems. These specific levels have been set by the California Division of Occupational Safety and Health (Cal/OSHA). The maximum permissible exposure limits may change from region to region. Legally, an individual's exposure may be above the PEL at times, but only if it is below the PEL value at other times so that the average exposure for any 8-h work shift is not greater than the PEL for the specific chemical. Methods of reducing exposure are described in Sec. 18.5.

TABLE 18.1 Permissible Exposure Limits (PELs) for Some Common Epoxy Resin System Chemicals (California/OSHA)[1]

Chemical name	Common abbreviation	Cal/OSHA PEL,* ppm
n-Butyl glycidyl ether	BGE	25
Isopropyl glycidyl ether	IGE	50
Phenyl glycidyl ether	PGE	1
Diethylenetriamine	DETA	1
Toluene		100
Xylene		100
2-Ethoxyethanol		5
2-Methoxyethanol		5
Methyl ethyl ketone	MEK	200
Phthalic anhydride	PA	1

*PELs are measured as parts of chemical per million parts of air (ppm).

Older epoxy resins were noticed to cause skin cancer in laboratory animals. This was most likely due to the epichlorohydrin. Most newer epoxy resins, which contain less epichlorohydrin, do not seem to cause cancer in animals. Certain curing agents, such as metaphenylene diamine (MPDA) and diaminodiphenyl sulfone (DADPS), and certain glycidyl ethers are carcinogenic in laboratory animals. It is not known if these materials cause cancer in humans.

The chemicals in epoxy adhesive systems can affect human health when they come into contact with the skin or if they evaporate and form a mist or vapor in the air. The main effects of overexposure are irritation of the eyes, nose, throat, and skin; skin allergies; and asthma. The solvent additives and other high-vapor-pressure materials can cause other effects such has headaches, dizziness, and confusion. The effects of these chemicals on the various parts of the human body are summarized below.[2]

Lungs. Vapor inhalation with most liquid resins is not considered a problem unless the resins are heated or the epoxy component has a very high vapor pressure, such as with solvents or certain diluents. Vapor and spray mists of most epoxy chemicals can irritate the lungs. Some people develop asthma from the curing agents. Symptoms of asthma include chest tightness, shortness of breath, wheezing, and coughing. These symptoms may occur at times other than when in contact with the epoxy component. Once a person becomes allergic to curing agents, even the dusts from sanding or grinding, future contact will generally always cause a reaction.

Skin. Most epoxy resins are not acutely irritating to the skin. However, they are capable of causing skin sensitization. Susceptibility to skin irritation and sensitization varies from person to person. The epoxy resins are considered to be milder skin sensitizers than amine-type curing agents or epoxy functional reactive diluents.

Skin irritation symptoms can include redness, swelling, flaking, and itching on the hands, face, or other areas of contact. Some people develop a skin allergy. Skin allergies may develop after just a few days of contact or after many years of exposure, depending on the individual. In the event of skin contact, thoroughly wash the affected area with copious amounts of soap and water. Remove contaminated clothing and launder before reuse.

Eyes, Nose, and Throat. All the epoxy resins are low in acute oral toxicity. However, most epoxy chemicals and their vapors can irritate the eyes, nose, and throat. Some people develop headaches as a result of this irritation. Eye contact with epoxy resins could result in slight, transient irritation. However, curing agents, solvents, and diluents can cause severe irritation. In event of eye contact, the eyes should immediately be flushed with plenty of water and then medical attention sought.

Nervous System. Solvents inhaled or absorbed through the skin can affect the central nervous system in the same way as drinking alcohol does. Symptoms of solvent overexposure include headache, nausea, dizziness, slurred speech, confusion, and loss of consciousness.

Reproductive System. Epoxy resins and curing agents themselves do not seem to affect pregnancy and reproduction in humans. However, some of the diluents and solvents that are used may affect reproduction. Two solvents sometimes found in epoxy resin systems (ethoxyethanol and methoxyethanol) have been determined to cause birth defects in laboratory animals and reduced sperm counts in men. Some diglycidyl ethers also damage the testes and cause birth defects in test animals. It is not known whether these materials cause the same effect in humans.

Most other solvent additives have not been adequately tested to determine if they affect reproduction. However, it is known that solvents inhaled by a woman can reach a developing fetus and may contaminate the woman's breast milk. It is generally recommended that pregnant and nursing women minimize their exposure to solvents.

18.3 MATERIALS USED AND THEIR EFFECT ON HEALTH AND SAFETY

The main components of epoxy adhesive systems may contain materials that are hazardous and/or affect the health of those who come into contact with them. In general, epoxy adhesives have hazardous properties, but can be handled safely. The hazards associated with the specific adhesive being handled depend on the hazardous nature of the components.

Because different material components in epoxy adhesives have different properties and characteristics when it comes to health and safety, one should always try to find out what chemicals are in the products being used. This can be done by consulting the specific manufacturer's Material Safety Data Sheet (MSDS).

An MSDS lists the hazardous chemical contents of a product, describes its health and safety hazards, and gives methods for its safe use, storage, and disposal. The MSDS also includes information on fire and explosion hazards, reactivity, first aid, and procedures for handling leaks and spills.

Generally, environmental regulations require that an employer have all MSDSs for any workplace product that contains hazardous substances, and these must be made available to employees on request. MSDSs can also be obtained from the manufacturer of the material, and many are available over the Internet.

Potential hazards associated with the various components of epoxy resins are reviewed below and in Table 18.2. This information was summarized from professional societies such as the Society of Plastics Industry (SPI) and manufacturers' brochures.[3,4] The reader is directed to similar sources or to the manufacturers themselves for additional information.

Epoxy Resins. Liquid epoxy resins are mild to moderate irritants to the skin, eyes, and mucous membranes. The irritant potential is increased by their sticky nature, which tends to lead to prolonged skin contact. These resins are generally mild to moderate dermal (skin) sensitizers in susceptible individuals.

Solid epoxy resins are not readily absorbed through the skin and present a low risk of skin irritation. Direct contact with solutions of these resins can cause mild to moderate irritation of the skin and the eyes, principally because the solvents remove the protective layer of fat that is on the surface of the skin. When crushed into a fine powder, epoxies should be considered an irritant dust, and inhalation and skin contact should be avoided. Solid resins are generally considered to be low to mild sensitizers.

Resins that are modified by the addition of reactive diluents or solvents can be more serious irritants. These resins should be handled with the same precautions as taken when one is using chemical solvents. Their sensitizing potential tends to increase with decreasing molecular weight. Epoxy components with significant volatility could cause irritation to skin, eyes, and respiratory tract, but inhalation is normally not a hazard except under certain conditions of use, such as heating, spraying, or applications with large surface areas.

Curing Agents and Catalysts. Curing agents are generally more hazardous to health than the unmodified epoxy resins. Polyamide-type curing agents are considered to present a low degree of health hazard compared to other curing agents. Catalytic-type curing agents are difficult to generalize because of the differences in chemical makeup. Specific information on these, as well as all materials used in epoxy adhesives, should be requested from the manufacturer prior to use.

TABLE 18.2 Typical Exposure Effects Associated with Epoxy Resin Systems[5]

Epoxy resin system components	Examples and types	Dermal exposure	Inhalation exposure	Ingestion exposure
Liquid epoxy resins	Based on the reaction product of epichlorohydrin and bisphenol A or bisphenol F	Mild to moderate irritants Mild to moderate sensitizers	Low volatility Exposure unlikely unless heated, sprayed, or spread over large unventilated areas	Low toxicity
Solid epoxy resins	Based on the reaction product of epichlorohydrin and bisphenol A or bisphenol F	Mild to moderate irritants and mild sensitizers Not readily absorbed through skin	Low volatility Exposure unlikely unless crushed or ground	Low toxicity
Modified liquid epoxy resins	Liquid epoxy resins with added reactive diluents or solvents	Mild to moderate irritants Moderate to strong sensitizers	Low volatility Exposure unlikely unless heated, sprayed, or spread over large unventilated areas	Low toxicity
Aliphatic and cycloaliphatic amine curing agents		Irritants, sensitizers, corrosive, absorbed through skin	Respiratory irritants	High toxicity
Aromatic amine curing agents		Sensitizers, long-term health effects, absorbed through skin	Respiratory irritants	Moderate to high toxicity
Anhydride curing agents		Corrosive, severe sensitizers	Dusts may be sensitizers	High toxicity
Reactive diluents	Glycidyl ethers	Moderate to strong sensitizers	Moderate volatility, exposure possible	Low toxicity
Solvents	Acetone, methyl ethyl ketone, toluene, xylene, glycol, ethers, alcohol	Defats and dries skin Some may be absorbed May carry other components through skin	High volatility, exposure possible Irritation Central nervous system depression, e.g., dizziness, loss of coordination	Low to high toxicity, long-term effects
Fillers	Fiberglass, pigments, calcium carbonate, powdered metals	Some may be absorbed	Dust inhalation	Low toxicity

The aliphatic amines, cycloaliphatic amines, and anhydride curing agents may cause irritation or damage to the skin, eyes, and lungs. Certain aliphatic and cycloaliphatic amines are skin sensitizers. Solid anhydride curing agents may cause sensitization in workers exposed to the curing agent dust. The aromatic amines are not strong irritants, but several are skin sensitizers. Certain aromatic amines may absorb through the skin and cause damage to organs such as the liver and interfere with the blood's ability to carry oxygen. Aromatic amines seriously stain skin and surrounding materials.

Solvents. Solvents commonly used in epoxy resin applications present a flammability hazard. These solvents present other special health hazards. Contact with solvents will cause defatting and drying of the skin, which enhances the chance for skin irritation. Some solvents are absorbed directly though the skin, and absorption may be enhanced if the skin is abraded or irritated. They also have the ability to dissolve other epoxy resin system chemicals and carry them through the skin. The inhalation of solvent vapors or mists may cause respirator irritation and depression of the central nervous system.

Reactive Diluents. Reactive diluents are generally considered to present a high degree of hazard. They are capable of causing skin and eye irritation and present a significant hazard from inhalation due to their high vapor pressure. Most reactive diluents fall into a chemical family known as the glycidyl ethers. These materials vaporize more readily than the epoxy resins and, therefore, have an increased potential for inhalation exposure. These diluents are also likely to be much stronger sensitizers than the epoxy resins.

Fillers. Fillers vary in their degree of hazard from handling. Some are considered to be essentially nonhazardous. However, dusts of glass, silica-bearing powders, and powdered metals may present a serious hazard from inhalation and/or explosion. The explosion hazard is present due to the high surface areas of finely divided fillers.

Fillers present a potential inhalation and dermal contact hazard. They can cause mechanical damage to the skin, which may aggravate the irritant effect of other chemicals and additives. When fillers are handled in a liquid epoxy matrix or in a cured epoxy, their inhalation hazard is low. However, inhalation exposure to fillers can occur when they are handled in the dry state or when one is machining or grinding cured epoxy products. Inhalation exposure to fillers such as crystalline silica or fiberglass may result in delayed lung injury. Asbestos fillers have long been abandoned from use for these reasons.

18.4 PROCESSES

Potential exposure to the chemicals used in epoxy adhesive formulations varies with the type of process or task. Closed systems with engineering controls are used to prevent workers from overexposure. However, occasionally open areas with limited controls are encountered, and the potential for exposure increases.

Dermal exposure is the most likely route of exposure, and this generally occurs from tasks involving hand contact with the adhesive. However, if certain curing agents or solvents are being used, inhalation exposure may also be a problem. Inhalation and dermal risks are most significant with any task involving the use of solvents or curing agents. The risk of exposure from the ingestion of epoxy materials is generally minimal.

Several of the more common processes that occur with the formulation and end use of epoxy adhesives are described in Table 18.3. Potential exposure risks for these processes are characterized as dermal, inhalation, or ingestion. Comments regarding the effect of the process on potential exposure are also included.

TABLE 18.3 Exposure Potential of Epoxy Adhesive Production and Application Processes

Process	Exposure potential			Comments
	Dermal	Inhalation	Ingestion	
Production of prepreg or film	High	Medium	Low	
Flooring, grouting, and hand application	High	High	Low	Large surface areas and high temperatures may increase possibility for inhalation exposure.
Coating	High	High	Low	Spraying causes generation of aerosols which increase the potential for inhalation exposures.
Handling: unloading, mixing, and pouring	High	Medium	Low	
Cutting, machining, and finishing	High	High	Low	These tasks generate dust which increases the potential for inhalation exposure.
Cleanup	High	High	Low	Potential for hazardous effects increases due to solvent use.
Spraying	High	High	Low	Spraying causes generation of aerosols which increase the potential for inhalation exposure.
Brushing	High	Medium	Low	

18.5 WORKPLACE PROCESSES TO LIMIT EXPOSURE

There are a number of processes that can be used to limit the exposure to chemicals that are commonly used in formulating epoxy adhesives. In general, the employer is required to protect his or her workers from being exposed to any hazardous chemical over the permissible exposure level.

Chemical manufacturers, formulators, and distributors must make health, safety, and environmental protection an integral part of the product. Several guides have been created to help develop and implement policies and practices that ensure protection through the product life cycle. One of these, developed by the Epoxy Resin Formulators Group of the Society of Plastics Industry, has been found to be most useful for the epoxy formulator.[6]

In general the workplace processes that can be used to limit or prevent potential exposure to hazardous chemicals are:

- Training
- Substitution
- Engineering controls
- Protective equipment and clothing
- Good housekeeping

18.5.1 Training

Those engaged in formulating or using epoxy adhesive should conduct continuing training programs for all personnel involved in handling of epoxy materials or those who can possibly come into contact with epoxy materials. Planning for employee and plant safety has value only when it is interpreted and practiced by the people involved.

Continued instruction of all employees must be given concerning the consequences of contact, as well as the precautions necessary for some operations (storage, handling, compounding, packaging, and disposal). The training program should address at a minimum the following items.[7]

- Labels, material safety data sheets, and product information bulletins
- Health and safety hazards
- Emergency procedures
- First aid
- Workplace controls
- Personal protective equipment
- Safe handling procedures

The training of individuals should emphasize several areas that are unique to epoxy resins. Some of these are described below.

Where flammable solvents and adhesives are used, they must be stored, handled, and used in a manner preventing any possibility of ignition. Proper safety containers, storage areas, and well-ventilated workplaces are required.

Certain adhesive materials are hazardous when mixed together. Epoxy and polyester catalysts, especially, must be well understood, and the user should not depart from the manufacturers' recommended procedure for handling and mixing. Certain unstabilized solvents, such as trichloroethylene and perchloroethylene, are subject to chemical reaction on contact with oxygen or moisture. Only stabilized grades of solvents should be used.

Certain adhesive systems, such as heat curing epoxy and room temperature curing polyester, can develop very large exothermic reactions on mixing. The temperature generated during this exotherm is dependent on the mass of the materials being mixed. Exotherm temperatures can get so high that the adhesive will catch fire and burn. Adhesive products should always be applied in thin bond lines to minimize the exotherm until the chemistry of the product is well understood. Never use elevated temperature curing sealants or adhesives for casting or for application in excessively thick cross-sections without first consulting the manufacturer. These materials are formulated to be applied and cured in thin cross-sections so that the heat generated by exotherm can easily be dissipated.

Safe equipment and proper operation are, of course, crucial to a workplace. Sufficient training and safety precautions must be installed in the factory before any bonding process is established.

18.5.2 Substitution

The most effective way to reduce hazardous chemical exposure is to use a safer chemical if one is available. Unfortunately this is also one of the most difficult methods of control.

Substitution is a difficult task because the specific ingredients will control the application and end-use properties of the epoxy adhesive. Any substitution of material generally requires that the complete product be reverified with production and prototype testing. Of course, the health and safety issues of any alternative material must also be carefully considered to ensure that it is actually safer.

There are several elements to consider when looking for a substitute material or when selecting materials for original formulation. For example, one may be able to choose an epoxy resin system that:

- Contains little or no free epichlorohydrin (a known cancer causing agent in laboratory animals). Generally, the MSDA will indicate how much epichlorohydrin is in the resin.
- Uses curing agents that are less irritating than the simpler aliphatic amines. (e.g., polyamides or amidoamines).
- Contains higher molecular weight resins and has a lower vapor pressure. Resins with high molecular weight are less likely to be vaporized, transported through the air streams, and cause inhalation problems.
- Has a reduced solvent content or is solvent free to minimize health effect due to solvents. There have been significant efforts in developing water borne and UV/EB cured epoxy resins to eliminate solvents.
- Does not contain loose fillers in a dry form.

18.5.3 Engineering Controls

In addition to training, engineering controls are the most effective way of minimizing exposure to harmful chemicals. Engineering controls include process or equipment modifications that reduce the amount of potentially hazardous materials to which an employee may be exposed. Isolation and ventilation are the primary methods of control. Engineering controls also include the maintenance, policing, and changing of work practices when necessary.

Isolation, or enclosure of a process or work operation to reduce exposure is a standard industrial hygiene method. The use of epoxy resins, for example, could be isolated to designated areas that are separate from the remainder of the plant. Examples of isolation are spray booths, enclosed curing and mixing rooms, and glove bag systems. Isolation is an ideal method for use with adhesives systems since the isolation also can prevent unwanted materials, generally contamination (e.g., moisture, mold release, dust) from getting to the work piece and degrading the strength of the finished joint.

Ventilation is the standard method of controlling exposure to airborne vapors of epoxy resins and solvents. Ventilation involves controlling air flow to reduce exposure. Local exhaust ventilation systems capture the vapor at the source and either filter or remove it from the work area. The ventilation system needs to be designed so that vapors, aerosol, and dusts are pulled away from, and not into, the breathing zone of the workers. A constant supply of fresh, noncontaminated air should be available at all times.

Examples of local exhaust ventilation system are drawdown exhaust tables, slot hoods, dust extraction systems, and portable vapor and dust collectors. Care must be taken that the ventilation systems are utilized and maintained as designed. The condition of the filters and air flow rate should be checked periodically as well as the condition of the duct work, motors, belts, etc.

Certain work processes, such as the heating or curing of epoxy resin systems, can be isolated, enclosed, or automated to reduce exposure. Heating of epoxies during compounding or cure can cause components to evaporate more quickly. The higher the temperature, the greater the amount of contaminant released into the air. Therefore, the lowest possible temperatures to perform the functions should be used, and adequate ventilation must be maintained around these areas. Electrostatic spray systems can reduce the amount of work place contamination and waste resulting from sprayed epoxy resin systems.

18.5.4 Protective Equipment and Clothing

When engineering controls cannot sufficiently reduce exposures, protective personal equipment and clothing must be used. An industrial hygienist or other knowledgeable person should be consulted to ensure that the equipment and clothing are appropriate and used correctly. The most common forms of personal protective equipment include eye protection, gloves, aprons, and respirators.

Eye protection can be provided by safety glasses with side shields, chemical goggles, full face respirators, and face shields with glasses or goggles. The type of eye protection required will depend on the hazard assessment of the specific applications. When there is danger of vapor, aerosol, or dust exposure, the eyes must be protected by chemical goggles at a minimum.

Cotton, leather, or rubber gloves should be worn to protect the hands from repeated contact with the materials. There is not much acute danger with many of these systems, but repeated contact over long periods of time can sensitize the skin and produce unpleasant reactions such as itchiness, redness, swelling and blisters.

Selecting the appropriate glove materials with epoxy resins systems involves matching the characteristics of the glove with the requirements of the production task. Glove resistance characteristics can be classified as either physical or chemical. Physical characteristics of production tasks are dexterity; wet grip; and cut, tear, puncture, and abrasion resistance. Chemical characteristics are dependent on the aggressiveness of the resins, solvents, and other materials.

The most common chemical characteristics are permeation and degradation. Glove materials that provide excellent resistance to epoxy resins, curing agents, and solvents are ethyl vinyl alcohol laminates and butyl rubber. Nitrile and neoprene gloves have less resistance to solvents and certain epoxy curing agents. Gloves should b replaced whenever signs of degradation are noticed. Typical signs of chemical degradation include swelling, softening, cracking, or discoloration of the glove material.

Similarly protective clothing should be made of materials that will provide protection from the chemicals in the epoxy adhesive systems or chemicals used in associated processing. The same chemical resistance characteristics as applied to gloves apply to chemically resistant clothing. A wide assortment of disposable aprons, coveralls, lab coats, and sleeves is also available.

Respirators come in two primary types: air supplied or air purifying. Air supplied respirators provide the user with an external supply of clean breathing air, while air purifying respirators make use of adsorbents and fillers to remove chemical vapors and particulate from the air. Respirators are generally required where solvent or dust levels are high, where irritating odors are present, and where materials that are respiratory sensitizers are common.

18.5.5 Good Housekeeping

Personal hygiene is also important in minimizing exposure levels to harmful chemicals. Contaminated clothing should be immediately removed and laundered or disposed of. Absorbent articles of clothing such as belts and shoes are particularly troublesome. Provisions should be made for storage, laundering, and disposal of contaminated clothing.

When one comes into contact with epoxy resins or any of the chemicals used with epoxy resins, he or she should immediately wash off any components that get on the skin. Do not use solvents for personal clean-up, but use soap and water or a commercial cleaner. Particular attention needs to paid to fingernails and the area around the nail bed. After washing, a skin conditioner or lotion should be used to help keep the skin in good condition and protected.

Good housekeeping also maintains that separate eating, drinking, and smoking facilities should be maintained that are isolated from the workplace. Refrigerators and freezers used for the storage of epoxy adhesive components must not be used to store edible items or drinks.

18.6 FIRST AID

First aid should be an important component of any training program. First aid materials and facilities should be maintained in a near-by but isolated area, away from hazardous operations and materials. The specific type of first aid for various problems is generally indicated on the MSDS.

If the skin or clothing becomes contaminated with epoxy components, remove the contaminated clothing and wash the exposed area with soap and water for at least 15 mins. Seek medical attention immediately if irritation or other complications develop. If the eyes become contaminated, they should be washed out with copious quantities of clean water and medical advice should be sought. Eye wash facilities should be provided in areas of the work place where such accidents have a potential to happen.

If respiratory distress is experienced, immediately remove the individual from the contaminated area to fresh air. If the person is not breathing, artificial respiration should be provided. Seek immediate medical attention. If breathing is difficult, transport the individual to a medical care facility for treatment, and if available, give the individual supplemental oxygen.

If a chemical is accidentally ingested, seek immediate medical attention. If the victim is conscious, give water. Do not induce vomiting unless directed to do so by a physician or as directed by the MSDS.

18.7 EMERGENCY PROCEDURES

Emergency procedures are most generally required with epoxy adhesives with respect to flammability and spill contamination and clean-up.

Most resinous materials used in epoxy adhesive are organic and will burn when sufficient heat and oxygen are supplied. A common measure of the flammability is the flash point temperature. This value indicates the minimum temperature at which flammable conditions are produced in controlled laboratory experiment at atmospheric pressure. Solvents, diluents, and other materials used with epoxy resins commonly increase the hazard of flammability and/or explosion.

Fires involving epoxy resins can be extinguished with foam, dry powder, or carbon dioxide. Water is not normally an effective extinguishing agent for these resins. When burning, these resins give off toxic by-products, such as carbon monoxide gas. Therefore, avoid breathing fumes or gases resulting from a fire. Fire fighters should use an organic vapor respirator or self-contained breathing apparatus.

Any spills of epoxy resin or epoxy components should be cleaned up immediately. The immediate concern is to protect personnel, prevent a possible fire hazard, and contain the spill until it is cleaned up. Persons engaged in spill clean-up should be protected from vapors and from skin contact by wearing appropriate protective clothing and equipment. Persons engaged in spill clean-up should also be aware of proper disposal techniques for the materials used in the process.

For small spills of epoxy resin liquid or solution, apply an absorbent material or a high surface area material such as sand to the spill. Shovel the mass with absorbed epoxy into a suitable container. The residue that is left should then be cleaned with hot soap water or steam. For spills of material containing solvent, keep spark producing equipment away form the spill site and shut off or remove all potential sources of ignition.

For larger spills, employees should stay upwind of the spill to avoid inhalation of epoxy components. Evacuate and rope off the spill area, and shut off all potential sources of ignition. The spill should be contained with a dike, and excess resin should be collected in suitable containers for final disposal. Hot soapy water or steam may be used for cleaning up the residue from floors or equipment. The use of solvents during clean-up should be avoided because of the hazardous nature of the solvent.

Liquid and solutions of epoxy resins must be prevented from entering sewers or drains, or any body of water, including rivers, streams, or lakes. Flexible emergency dikes, sometimes known as "pigs," can be used to block entry of the spill into sewers and drains. If spilled materials does enter drains or water ways, local authorities should be notified at once.

REFERENCES

1. *Epoxy Resin Systems*, California Department of Health Services, Oakland, CA, June 1989.
2. *Epoxy Resin Systems*.
3. *Epoxy Resin Systems Safe Handling Guide*, The Society of the Plastics Industry, New York, September 1997.
4. *Hazards and Handling Precautions*, Dow Plastics, Dow Chemical Company, Midland, MI.
5. *Epoxy Resin Systems Safe Handling Guide*.
6. *Product Stewardship Guide*, Epoxy Resin Formulators Group, The Society of the Plastics Industry, New York, 2001.
7. *Epoxy Resin Systems Safe Handling Guide*.

CHAPTER 19
QUALITY CONTROL AND SPECIFICATIONS

19.1 INTRODUCTION

This chapter discusses quality control and the writing and use of specifications that apply to epoxy adhesive systems. These subjects are of interest to both the formulator and the end user.

They are of interest to the formulator as they apply to raw materials and to the final adhesive product. The formulator must understand not only how to control the quality of the adhesive product but also how to use specifications as they relate to raw materials or to the end user or customer.

Similarly, the end user must develop a plan to ensure the consistency and quality of the entire bonding processes. Specifications that are employed by the end user will apply both to the formulated adhesive and to the final bonded assembly that is produced.

Quality control testing processes that are commonly used with epoxy adhesives are discussed in Chap. 20. Tests included are those used by the formulator as well as those employed by the end user of the adhesive system.

19.2 QUALITY CONTROL

Both the formulating of adhesives and bonding with adhesives are complex, multiple-part processes, complete with interacting and sometimes unexpected parameters that may contribute to success or failure of the final product. Thus, it is important that the quality control process consider the *entire* operation from receipt of materials to final product testing.

A generalized flowchart for the quality control process in formulating epoxy adhesives is shown in Fig. 19.1. A flowchart for controlling the quality of the adhesive bonding process is shown in Fig. 19.2. It must be realized that in both cases, the decisions made in one phase of the process may affect the subsequent phases. Therefore, all the individual phases must be carefully coordinated and controlled.

Although sophisticated testing and analysis equipment is available, many of these processes do not require advanced equipment. Simple equipment, visual examination, and common sense are the main tools in most quality control departments. They are supported by strong specifications and proper training.

Quality control encompasses all the processes and activities that ensure adequate quality in the final product from receipt of materials, through manufacture, to final product test. Quality control is very important when one is using adhesives because once a product is fully

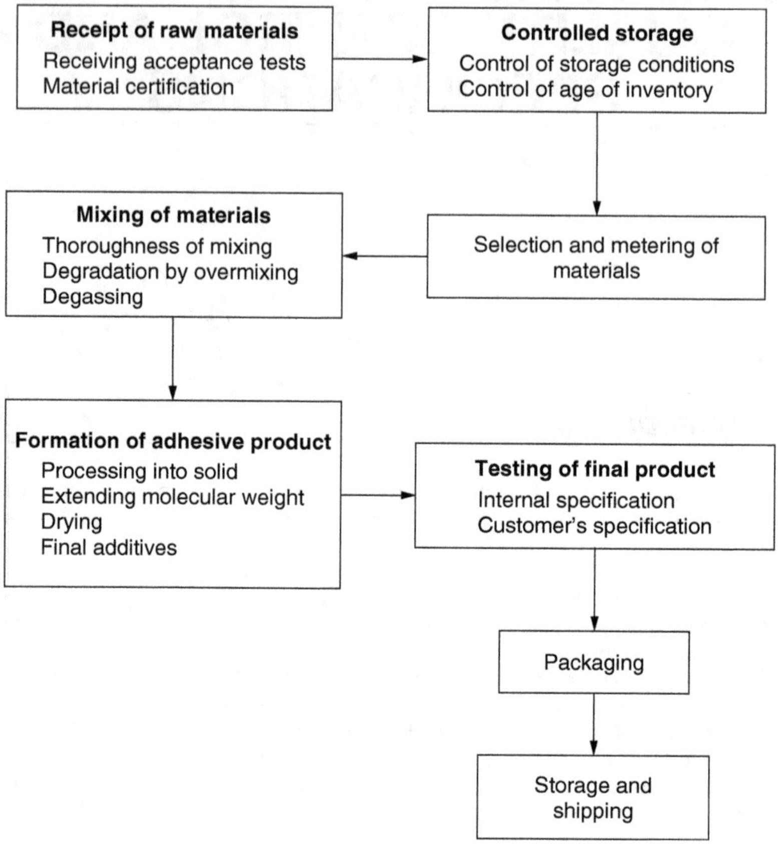

FIGURE 19.1 Generalized flowchart of quality control process in manufacturing epoxy adhesives.

bonded, joints are difficult to take apart or correct. By the time it takes to notice that one step in the formulation or bonding process is out of control, significant costs could be incurred.

For adhesives, quality control must be defined in the broadest sense. This consists of defining the means to prevent problems as well as to detect problems. These preventive measures may include training, controlling manufacturing procedures, incoming inspection of materials, and visual or physical examination of the finished product. Good quality control begins even before the receipt of materials. This usually begins with proper training of personnel and conditioning of the manufacturing area.

The total control program monitors and provides appropriate feedback and corrective actions. Generally, activity within each business process has its own defined quality control plan. These activities include but are not limited to

- Receiving inspection
- Manufacturing processes
- End product inspection

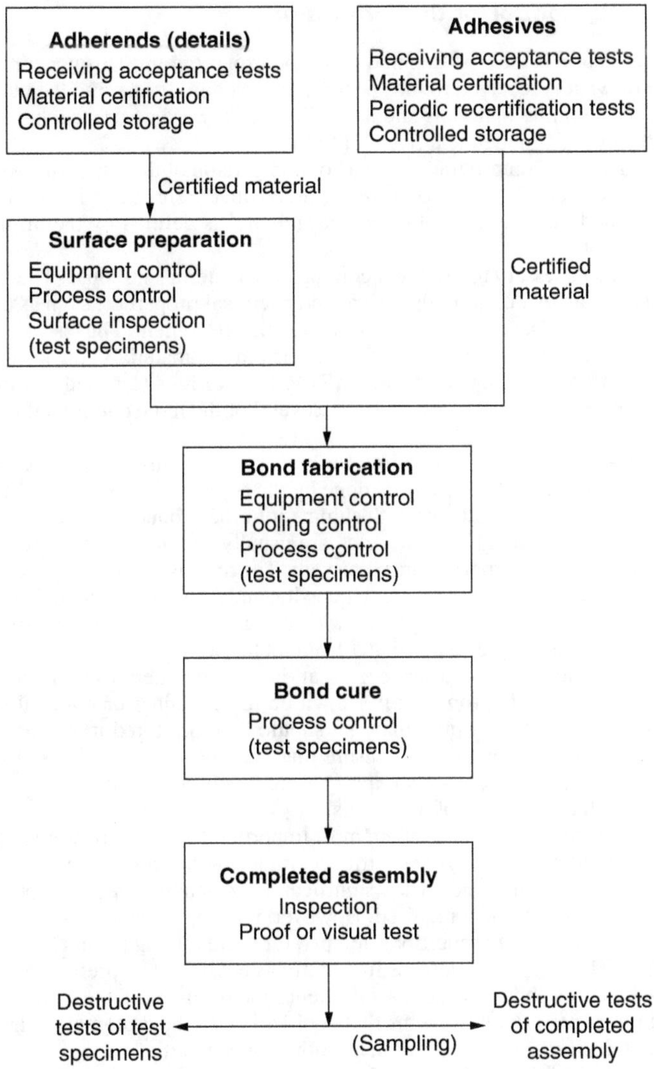

FIGURE 19.2 Flowchart of a quality control system for adhesive bonding.

Other nonmanufacturing-related processes, such as order development, field service, and marketing, can and should be subject to quality control procedures as well.

The manufacturing-related activities are discussed below with regard to the epoxy adhesive formulator and end user. It is sometimes difficult to divide these processes between the formulator and the end user, since many will apply to both.

19.2.1 Quality Control for the Formulator

Receiving inspection is the first step in the total control program. This generally consists of comparison of what is received to the specification used in developing the purchase order. Acceptance tests on raw materials should be directed toward assurance that the incoming materials are correct and identical from lot to lot.

A receiving inspection program consists of comparison of the purchase order with what is received. If the vendor's test reports are required with the shipment, it is verified that they were received and that the test values are acceptable, as defined by the original material specification.

It may also be necessary to test the incoming bulk materials in-house. These inspections usually consist of an evaluation of physical and chemical properties, such as color, viscosity, percentage of solids, weight per gallon, and moisture content. The tests should be those that can quickly and accurately detect deficiencies in the materials' physical or chemical properties or differences from lot to lot. ASTM lists various test methods that are commonly used for acceptance of raw materials. Several of the more common test methods are described in Chap. 20.

Because of the cost and time required for extensive in-house testing, the trend today is to have most of the quality control tests done by the supplier. The supplier then provides a certified test report to the customer with shipment of the product. The test program that is used by the supplier, internal controls, etc., are usually verified and approved by the customer on a periodic basis, such as once per year. The approval tests are used to determine the adequacy of the particular materials in question and the production methods that will be employed. The tests that are done with every lot and submitted on delivery usually are made to ascertain the consistency of the product from lot to lot.

Once identified and approved for receipt, all incoming material should be labeled with a receipt date. This date is prominently shown on the container or material while it is in inventory. The approved incoming material should then be stored in a controlled storage facility. Precautions must be taken to ensure that the oldest material is used first and that the shelf life of the product does not expire before it is used. The date stamp on the product is the most reliable method of doing this.

The manufacturing processes that are most important in formulating are dispensing and mixing. It is difficult to provide for specific recommendations regarding the quality control of these processes because they vary significantly depending on the type of materials employed and the final product required. However, it is important that all parameters in the process, such as component proportions, temperature, and mixing parameters, be monitored and controlled. This should be done as frequently as possible, and preferably on a continuous basis. All records should be kept a sufficient time so that possible comparisons can be made in the future regarding problems that could arise for the customer of the adhesive or the customer's customer (i.e., the recipient of the bonded part).

Final inspection of the adhesive products is used to satisfy the customer's specification and to ensure the formulator of the quality and consistency of the product. The test methods employed will vary considerably depending on the industry for which the product is intended and the specific customer.

19.2.2 Quality Control for the End User

A flowchart for the adhesive bonding quality control system within a large aircraft company is illustrated in Fig. 19.2. This system is designed to ensure reproducible bonds and, if a substandard bond is detected, to make suitable corrections. However, good quality control begins with proper training and conditioning of the manufacturing operation for the use of adhesives and sealants.

Premanufacturing Processes. The human element enters the adhesive bonding process probably more than in any other fabrication technique. An extremely high percentage of defects can be traced to poor workmanship or lack of understanding regarding adhesion. This generally shows up first in the surface preparation steps but may also arise in any of the other steps necessary to achieve a bonded assembly. This problem can be largely overcome by proper motivation and education. All employees from design engineer to laborer to quality control inspector should be somewhat familiar with adhesive bonding technology and aware of the circumstances that can lead to poor joints.

In addition to the staff, the operational facilities must be well prepared prior to the use of the adhesive. The plant's bonding area should be as clean as possible prior to receipt of materials. The basic approach to keeping the assembly area clean is to segregate it from the other manufacturing operations either in a corner of the plant or in isolated rooms. The air should be dry and filtered to prevent moisture or other contaminants from gathering at a possible joint interface. The cleaning and bonding operations should be separated from each other. If mold release is used to prevent adhesive flash from sticking to bonding equipment, great care must be taken to ensure that the release does not contaminate the adhesive or the adherends. Spray mold releases, especially silicone release agents, have a tendency to migrate to undesirable areas.

As with the quality control plan for the formulator described above, acceptance tests on adhesives should be directed toward assurance that incoming materials are identical from lot to lot. However, this plan should also extend to the receipt and control of incoming adherends and other materials used in the bonding process.

Important factors regarding incoming adherends are the chemical and physical properties of the material. This can be especially important with adhesive bonding because different metal alloys have different surface oxidation, and different elastomers and plastics can have different additives and modifiers. With elastomeric and plastic substrates, lot-to-lot differences should be tested. Often a supplier will change formulations but still be within the requirements of the specification. The difference in formulation may have a profound effect on the quality of the ultimate adhesive bond.

Another important factor with regard to adherends is the uniformity and tolerances allowed in the joint fit-up. Improper tolerances on the geometry of the joint can lead to adhesive-starved areas or areas where the bond is so thick that unwanted peel and cleavage forces can easily develop.

The control over storage conditions should be similar to that described above for the formulator. However, the end user must also store the substrate materials. This should be done in an isolated area so that the substrates cannot become contaminated with the shop environment. If stored in a cold area, the substrates must be brought up to room temperature before the surface treatment or adhesive is applied. Moisture can condense on cold substrates, leading to contamination.

In addition to tests performed on the incoming materials, test specimens may be made to verify the strength of the adhesive joint. The quality control tests should be those that can quickly and accurately detect deficiencies in the adhesive's physical or chemical properties. ASTM lists various test methods that are commonly used for adhesive acceptance.

The test specimens should be stressed in directions that are representative of the forces that the bond will see in service, i.e., shear, peel, tension, or cleavage. If possible, the specimens should be prepared and cured in the same manner as actual production assemblies. If time permits, specimens should also be tested in simulated service environments, e.g., high temperature, humidity.

Surface Treatment. Generally, some sort of surface preparation is required for reliable adhesion. The extent of the actual surface preparation process varies according to the performance characteristics desired, the nature of the adherend, and the time and cost considerations.

Surface preparations must be carefully controlled for reliable production of adhesive-bonded parts. If a chemical surface treatment is required, the process must be monitored for proper sequence, bath temperature, solution concentration, and contaminants. If sand or grit blasting is employed, the abrasive must be changed regularly. An adequate supply of clean wiping cloths for solvent cleaning is also mandatory. Checks should be made to determine if cleaning cloths or solvent containers have become contaminated.

The specific surface preparation can be checked for effectiveness by the water-break free test. After the final treating step, the substrate surface is checked for a continuous film of water that should form when deionized water droplets are placed on the surface. A surface that is uniformly wet by distilled water will likely also be wet by the adhesive since the specific surface energy of water is 72 dyn/cm and of most organic adhesives is 30 to 50 dyn/cm. However, this test tells little about weak boundary layers or other contaminants that may be present on the substrate's surface but still be capable of wetting with water.

After the adequacy of the surface treatment has been determined, precautions must be taken to ensure that the substrates are kept clean and dry until bonding. The adhesive or primer should be applied to the treated surface as quickly as possible.

The Bonding Process. Before the actual assembly operation, the cleanliness of the shop and tools should be verified. The shop atmosphere should be controlled as closely as possible. Temperature in the range of 18 to 32°C and relative humidity from 20 to 65 percent are best for almost all bonding operations. All parts should be fitted together without adhesive or sealant to indicate possible production problems due to fit. The suitability of fit is established by either visual inspection or direct measurement with gauge or shim. It is desirable that the extremes in mechanical tolerances also be noted and that test specimens be made with the worst possible fit to ensure that the bonding process will always provide reliable joints.

The adhesive metering and mixing operation should be monitored by periodically sampling the mixed adhesive and testing it for adhesive properties. Simple viscosity measurements, flow tests, and visual inspection of consistency are the best methods of monitoring conformance. A visual inspection can also be made for air entrapment and degree of mixing. The quality control engineer should be sure that the oldest adhesive is used first and that the specified shelf life has not been exceeded.

Innovative formulations have been developed to help the end user monitor the efficiency of the mixing operation. If the epoxy resin component and the curing agent component are different colors, the end user can easily visually determine if the products are mixed thoroughly. Fluorescent dyes that change colors as the adhesive cures have been added to some adhesives. Also pigments that irreversibly change color when encountering the heat of exotherm are used in some adhesives to indicate that cure has taken place. However, these methods of determining the state of mixing and cure lack accuracy and should be used as guidance only.

The amount of the applied adhesive and the final bond line thickness must be monitored because they can have a significant effect on joint strength. Curing conditions should be monitored for pressure, heat-up rate, maximum and minimum temperatures during cure, time at the required temperature, and cool-down rate. The primary concerns are to ensure the following:

- The adhesive is metered and mixed as required.
- The adhesive is applied uniformly and with the correct coating thickness.
- The pot life and open time (time between adhesive application and joint assembly) of the adhesive are not exceeded.
- The joints are assembled correctly and in the proper sequence.

- The mixing and application of the adhesive are done under the specified environmental conditions (temperature and relative humidity).
- The parts are held together under the specified pressure and time until cured.
- Cure time and temperature are as required and measured at the adhesive bond.

Bond Inspection. After the adhesive or sealant is cured, the joint area can be inspected to detect gross flaws or defects. This inspection procedure can be either destructive or nondestructive. The nondestructive type of tests can be visual or use advanced analytical tests. These types of bond inspections are described below.

Destructive Testing. Destructive testing generally involves placing a sample of the production run in simulated or accelerated service and determining if it has similar strength properties to a specimen that is known to have a good bond and adequate service performance. It is desirable to fabricate a standard test specimen in the same production cycle as the part actually being bonded. The test specimen could be test coupons (standard ASTM test specimens), extensions of actual parts (i.e., tabs that can be removed from the part and tested), or special test specimens that are close to the actual part design but amenable to mechanical testing. If a special test specimen is used, the specimen should be designed for a test method that is easy to perform and indicative of the way the part will be loaded in service; but it should also be designed so that the geometry and mass do not depart too severely from those of the actual production part. The test specimen is then tested either immediately after bonding or after a simulated environmental cycle.

The joint area and mode of failure should be examined closely, for this generally leads to clues that are indicative of problems. The causes and remedies for faults revealed by such mechanical tests and visual inspection after testing are described in Table 19.1.

Test specimens such as those mentioned above are often used to verify the quality of the first article through the production line and then to periodically test articles for conformance. This type of testing procedure detects discrepancies affecting the entire lot of parts, but it cannot evaluate factors that affect individual joints or specific areas within a particular joint.

Visual, Nondestructive QC Testing of Joints. There are several forms of nondestructive testing (NDT) that can be used to determine the quality of the final joint. Those that are most appropriate for on-line quality control are described immediately below. More sophisticated nondestructive testing processes are described in the Chap. 20.

VISUAL EXAMINATION: A trained eye can detect a surprising number of faulty joints by close inspection of the adhesive around the bonded area, even if the substrate is not transparent or translucent. Close examination of the adhesive or sealant that is visually apparent (generally around the edges of the joint) can lead to useful conclusions. Unfilled areas and voids can sometimes be detected by noting lack of adhesive or sealant material. Misalignment of parts is readily visible.

The texture of the adhesive around the edges of the joint can be a clue to the effectiveness of the curing process and whether air was entrapped in the adhesive. The adhesion of the adhesive flash to the substrate can also be qualitatively measured by attempting to pry the flash away from the substrate. Table 19.2 lists the characteristics of faulty joints that can be detected visually, their causes, and possible remedies.

TAP TEST: One of the first nondestructive methods used to evaluate the quality of an adhesive joint was by tapping on the bonded joint and assessing the resulting tone. Tone differences indicate inconsistencies in the bonded joint. This could be due to insufficient cure, voids, or other problems. Simple tapping of a bonded joint with a coin or light hammer can indicate an unbonded area. Sharp, clear tones indicate that adhesive is present and adhering to the substrate to some degree; dull, hollow tones and inconsistencies indicate a void or unattached area.

TABLE 19.1 Faults Revealed by Mechanical Tests of Tested Specimens

Fault	Cause	Remedy
Thick, uneven glue line	Clamping pressure too low	Increase pressure. Check that clamps are seated properly
	No follow-up pressure	Modify clamps or check for freedom of moving parts
	Curing temperature too low	Use higher curing temperature. Check that temperature at the bond line is above the minimum specified throughout the curing cycle
	Adhesive exceeded its shelf life, resulting in increased viscosity	Use fresh adhesive
Adhesive residue has spongy appearance or contains bubbles	Excess air stirred into adhesive	Vacuum-degas adhesive before application
	Solvent not completely dried out before bonding	Increase drying time or temperature. Make sure drying area is properly ventilated
	Adhesive material contains volatile components	Seek advice from manufacturer
	A low-boiling constituent boiled away	Curing temperature is too high
Voids in bond (i.e., areas that are not bonded), clean bare metal exposed, adhesive failure at interface	Joint surfaces not properly treated	Check pretreatment process; use clean solvent and wiping rags. Wiping rags must not be made from synthetic fiber. Make sure cleaned parts are not touched before bonding. Cover stored parts to prevent dust from settling on them
	Resin may be contaminated	Replace resin. Check solids content. Clean resin tank
	Substrates distorted	Check for distortion; correct or discard distorted components. If distorted components must be used, try adhesive with better gap-filling properties
Adhesive can be softened by heating or wiping with solvent; adhesive can easily be indented with fingernail pressure	Adhesive not cured properly	Use higher curing temperature or extend curing time. Temperature and time must be above minimum specified throughout the curing cycle. Check mixing ratios and thoroughness of mixing. Large parts act as a heat sink, necessitating longer cure times

TABLE 19.2 Faults Revealed by Visual Tests of Area around the Joint

Fault	Cause	Remedy
No appearance of adhesive around edges of joint or adhesive bond line is too thick	Clamping pressure too low	Increase pressure. Check that clamps are seating properly
	Starved joint	Apply more adhesive
	Curing temperature too low	Use higher curing temperature. Check that temperature is above minimum specified
Adhesive bond line too thin	Clamping pressure too high	Lessen pressure
	Curing temperature too high	Use lower curing temperature
	Starved joint	Apply more adhesive
Adhesive flash breaks away easily from substrate	Improper surface treatment	Check treating process; use clean solvent and wiping rags. Make sure cleaned parts are not touched before bonding
Adhesive flash is excessively porous	Excess air stirred into adhesive	Vacuum-degas adhesive before application
	Solvent not completely dried out before bonding	Increase drying time or temperature
	Adhesive material contains volatile components	Seek advice from manufacturer
Adhesive flash can be softened by heating or wiping with solvent	Adhesive not properly cured	Use higher curing temperature or extend curing time. Temperature and time must be above minimum specified. Check mixing

The success of the tap test depends on the skill and experience of the operator, the background noise level, and the type of structure. Some improvement in the tap test can be achieved by using a solenoid-operated hammer and a microphone pickup. The resulting electric signals can be analyzed on the basis of amplitude and frequency. However, the tap test, in its most successful mode, measures only the qualitative characteristics of the joint. It tells whether adhesive is in the joint or not, providing an acoustical path from substrate to substrate; or it tells if the adhesive is undercured or filled with air, thereby causing a mechanically damped path for the acoustical signal. The tap test provides no quantitative information and no information about the presence and/or nature of a weak boundary layer.

The success of the tap test, although limited, has led to the use of ultrasonics to determine bond quality. Ultrasonic methods are at present the most popular NDT technique for use on adhesive joints. Ultrasonic testing measures the response of the bonded joint to loading by low-power ultrasonic energy. Short pulses of ultrasonic energy can be introduced on one side of the bonded structure and detected on the other side. This is called *through-transmission testing*. An unbonded area, void, or highly damped adhesive (undercured or filled with air) prevents the ultrasonic energy from passing efficiently through the structure. A number of different types of ultrasonic inspection techniques have been adopted to test adhesive bonds.[1]

PROOF TESTS: If a high degree of reliability is required, it is necessary to proof-test the production unit. The proof test should simulate actual service conditions in the manner in which the joint or structure is loaded, and the stress level should be higher than that expected in service. The duration of the proof test should reflect the expected life of the joint, but usually this is not possible. The proof test should be designed so that it is normally a nondestructive test, unless the bond is unexpectedly weak. Care must be taken to design the proof test so that it does not overstress the part and cause damage that will result in a reduced service life.

A common example of a proof test is to apply a cleavage load to a bonded honeycomb sandwich by placing an instrument between the face and core and applying a predetermined force perpendicular to the core. If there is no bond disruption due to this test, it is supposed that the product will meet all its service requirements. Other common proof tests used with sealants are leak-testing with a mobile and easily detected gas such as helium or by the application of hydrostatic pressures.

19.3 SPECIFICATIONS

Specifications are, perhaps, the cornerstone of any good quality control system. Specifications are intended to define what is needed for all parties involved. These parties include outside suppliers as well as internal departments within the factory. Table 19.3 is a short list of dos and don'ts for specification writing. The language of the specification should be clear, exact, brief, technical, and to the point.

Specifications are a necessary part of a quality control program. A specification simply is a statement of the requirements that the adhesives or sealant must meet to be accepted for use. A specification is an agreement between supplier and user. Conformance to a specification does not mean that the adhesive or sealant will perform perfectly in service. It only means that the product conforms to the specification. Specifications not only account for the adhesive or sealant, but also define the adherends and the accessory processes for preparing the adherends and the joint assembly.

The specification writer must try to put into the specification the requirements that, if met, will provide the greatest likelihood of success. These requirements should be standard tests and acceptable test limits that are agreed upon by both the supplier and the user. The tests should be indicative of how the adhesive is used in production and how the finished joint is to be used in service. Tests that are not directly applicable to the specific application should not be included. Tests should not be used simply because they are standard test methods or have been used in the past.

The language of the specification is extremely important. The specification must be unambiguous and use direct and simple words and phases. For example, the words *must*

TABLE 19.3 Dos and Don'ts of Specification Writing[2]

Do:	Don't:
• Use simple words.	• Use words having a double meaning.
• Use the exact meaning of technical terms.	• Make the specification too voluminous.
• Use short sentences.	• Put in requirements that cannot be justified.
• Use the same word throughout, never synonyms.	• Use trade names.
• Use the same mathematical system throughout.	• Make limits too tight or too loose.
	• Define requirements loosely.

and *should* have very different meanings and connotations in a specification. Proprietary or trade names should be avoided, if possible, since they may be changed with a company's merger or acquisition. The typical specification has the following format.[3]

- Title
- Scope
- General requirements
- Performance requirements
- Test methods
- Controls
- Reference documents
- Approved source list

Specifications from other sources (e.g., ASTM, military) may be used if they are applicable, and often test methods from ASTM, etc., are used within a customer specification. The trend is to use well-established ASTM or ISO test methods where possible rather than nonstandard or proprietary tests.

Specifications may require different categories of testing. For example, there may be extensive series of tests required for initial verification or qualification of the adhesive and supplier. These tests would be used to approve a certain product at the onset. Other receiving tests may be used to verify the consistency of the product from lot to lot.

An adhesive or sealant specification, like all material specifications, is a document that specifies values for all the important properties, together with limits of variability and methods for determining these values. There are many adhesives and sealant specifications, of which the most prominent are the industrial and government specifications. These describe and establish the technical and physical characteristics or performance requirements of adhesive materials. The most common sources of standards and specifications for the adhesives and sealants industry are the following:

- American Society of Testing and Materials (ASTM)
- International Organization for Standardization (ISO)
- U.S. Department of Defense (military and federal specifications and standards)
- National Aeronautics and Space Administration (NASA)
- Society of Automotive Engineers (SAE)
- Technical Association of the Pulp and Paper Industry (TAPPI)

Relevant ASTM specifications and standards are presented in App. G. A variety of federal and military specifications describing adhesives and test methods have been prepared. Selected government specifications are also described in App. G. Table 19.4 identifies several other sources of specifications for adhesives, sealants, and related equipment.

Certain specifications and standards provide excellent tutorials on adhesives and sealants. For example, MIL-HDBK-691 offers a complete handbook on adhesive bonding, and MIL-HDBK-725 provides a guide to the properties and uses of adhesives. ASTM C 962 provides an excellent source of information regarding sealant joint design and the types of sealants that are appropriate for various substrates. Although this specification is primarily for construction sealants, much of the information that it contains is generally useful for other sealant applications.

TABLE 19.4 Sources of Common Specifications and Standards for Adhesives and Sealants

Military Naval Publication and Forms Center 5801 Tabor Ave. Philadelphia, PA 19120	Society of Automotive Engineers Inc. 400 Commonwealth Drive Warren, PA 14096
Federal Standards Specification Sales 3FRSBS Bldg. 197 Washington Navy Yard General Services Administration Washington, DC 20407	American Society for Testing and Materials (ASTM) 1916 Race Street Philadelphia, PA 19103 American Society for Nondestructive Testing 1916 Arlingate Plaza Columbus, OH 43228

REFERENCES

1. Hagemaier, D. J. "End-Product Nondestructive Evaluation of Adhesive Bonded Metal Joint," in *Adhesives and Sealants*, vol. 3, *Engineered Materials Handbook*, ASM International, Materials Park, OH, 1990.
2. DeFrayne, G. O., "Adhesive Specification and Quality Control," in *Adhesives in Manufacturing*, G. L. Schneberger, ed., Marcel Dekker, New York, 1983.
3. DeFrayne, "Adhesive Specifications and Quality Control."

CHAPTER 20
TESTING

20.1 INTRODUCTION

Adhesive tests are of interest to both the formulator and the end user. They are generally used to do the following:

1. Choose among materials or processes, such as adhesive, adherend, or joint design.
2. Monitor the quality of production materials to make certain that they have not changed since the last time they were verified for use in the bonding process.
3. Confirm the effectiveness of a bonding process, such as surface cleaning or curing.
4. Investigate parameters or process variables that may lead to measured differences in the performance of the bond.

There are two categories of common tests for adhesives: fundamental property tests and end-use tests. End-use tests, such as peel and shear, are those that try to simulate the type of loading and service conditions to which a joint will be subjected. These tests are relatively straightforward, but experience is required to establish the correct sample type and testing procedures, judge the reliability of the resulting data, and interpret the results and apply them to a practical application.

The determination of a fundamental or bulk property, such as viscosity, hardness, or settling rate, is usually simpler and more reproducible than end-use tests. However, a correlation between fundamental properties and the results of end-use testing is very difficult. Fundamental property tests are usually employed to assess the consistency of the incoming raw materials, adhesive formulation, or substrates. Often fundamental property testing is done after there is a failure or unexplained occurrence to determine if a change in the incoming material may be the possible culprit.

A number of standard tests for adhesives and sealants have been specified by the American Society for Testing and Materials (ASTM) and other professional organizations such as the U.S. Department of Defense and the Society of Automotive Engineers. By far, ASTM standards are the most commonly referenced test methods. Selected ASTM standards are presented in App. G. The properties usually reported by suppliers of adhesives and displayed in the technical literature generally reference ASTM standards.

Adhesive testing can also be classified as to the condition of the epoxy adhesive at the time of test. Tests are conducted on the

- Raw materials, uncured epoxy mix, and cured epoxy (bulk property tests)
- Adhesive in various states of cure
- Cured adhesive and final test specimen or production joint

20.2 TESTS ON THE EPOXY RESIN

The most commonly used tests for properties of the materials used in epoxy adhesive formulation and uncured mixed adhesives are viscosity, shelf life, percent of solids, and moisture content. These are generally applied to the resinous components.

ASTM has developed a *Standard Guide for Testing Epoxy Resins*, which covers several procedures that can be conducted on the same starting sample (e.g., viscosity, color, and density). Other contents of the standard include test methods for nonvolatile content, acid value, epoxy content, hydrolyzable chlorine, and total chlorine. The most important of these test methods are described below.

20.2.1 Viscosity

Viscosity is defined as the resistance of a liquid material to flow. Viscosity can be used as an indirect determination of the epoxy resin's molecular weight and chemical structure. It is usually measured in fundamental units of poise or centipoise. The unit of centipoise (cP) is sometimes confusing unless one is familiar with these particular units. Practical comparisons of the viscosities of common fluids may be of assistance in understanding the viscosity of epoxy resins measured in centipoise.

Common fluid	Viscosity, cP
Water	1.0
#10 Motor oil	400
Castor oil	1,000
Karo syrup	3,500
#40 Motor oil	4,500
Hershey chocolate syrup	25,000

There are also a number of specialty viscosity tests that employ their own relative units. These tests have been developed for specific industries such as paints and coatings. Generally, these viscosity units can be directly converted to poise.

Adhesive viscosity is an indication of how easily the product can be pumped or spread onto a surface. It reveals information, together with the liquid's setting rate and surface tension, that is pertinent to the wetting characteristics of the adhesive. The viscosity also reveals information regarding the age and compounding of the adhesive. Through the relatively easy measurement of viscosity, changes in density, stability, solvent content, and molecular weight can be noticed. Viscosity measurements for free-flowing adhesives or sealants are usually based on one of the methods described in ASTM D 1084.

The most popular viscosity test for products ranging in viscosity from 50 to 200,000 cP uses a rotating spindle instrument such as the Brookfield viscometer. The equipment used for this measurement is shown in Fig. 20.1. The instrument measures the resistance of the fluid to a spindle of certain size that is rotating at a predetermined rate. The method is relatively simple and quick. It can be adapted to either the laboratory or production floor.

Another test for determining the viscosity of liquid adhesives measures the time it takes the test liquid to flow by gravity completely out of a cup with a certain size hole in the bottom. These consistency cups are designed to expel 50 mL of sample in 30 to 100 s under controlled temperature and relative humidity conditions. The number of seconds for complete flow-out of the sample is determined. There are different cup volumes and hole sizes that can be used, and conversions exist for relating the viscosity measured in one cup to another. This test is commonly used in the paint industry for adjusting the solvent content in paint systems.

The viscosity of thixotropic materials that exhibit a shear rate dependency is usually determined by the procedure described in ASTM D 2556. The viscosity is determined at different shear rates, and from this plot, apparent viscosity associated with a particular rotational speed and spindle shape can be obtained. Materials with thixotropic characteristics include Vaseline jelly and toothpaste. They are materials that tend to have very high viscosity characteristics and exhibit no flow at low shear rates. However, when pressure is applied (higher shear rates), the material flows easily, exhibiting a characteristic of lower viscosity. Such materials are very common in the adhesive and sealant industries. Thixotropic materials can be pumped through a nozzle, mixed, or applied to a surface with little resistance. However, when applied to a vertical surface, they will not flow under their own weight.

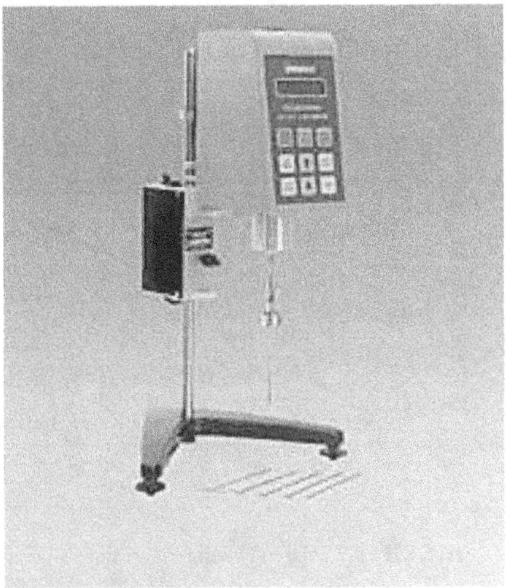

FIGURE 20.1 Digital Brookfield viscometer allows the measurement of viscosity and torque. (*Photo courtesy of Brookfield Laboratories.*)

The viscosity of nonflowable products is determined by an extrusion test. A Semco 440 nozzle or its equivalent is attached to a standard adhesive/sealant cartridge filled with the material to be tested. The cartridge is then placed in an air-operated sealant gun set for a pressure of 90 to 95 psi. The weight of material that is extruded in 10 s is measured, and the extrusion rate is reported in grams per minute (g/min). Care must be taken to clear trapped air from the cartridge and nozzle.

20.2.2 Softening Point

Resins melting above room temperature are often characterized by softening point rather than by solution viscosity or elevated-temperature (melt) viscosity. Uncured, solid epoxy resins may be qualitatively graded by determining the temperature at which they begin to soften. Typical softening point values for DGEBA epoxy resins as a function of molecular weight are given in Fig. 20.2.

There are several methods commonly employed to measure the softening point, such as the ring-and-ball method and Durran's mercury methods. These measure the temperature at which the resin reaches an arbitrary softness.

Indentation tests, such as the ring and ball method, are the most common methods used for determining the softening point. In Durran's mercury method, a known weight of the resin is melted in a standard test tube and cooled, and a known quantity of clean mercury is placed on top of the solidified sample. The tube and contents are again heated, and the temperature at which the molten resin rises to the top of the mercury is recorded as the melting point. Another approach for measuring the softening point is to employ a dilatometer or thermal analysis test to measure second-order transition temperatures (i.e., sharp break points in the thermal expansion characteristics as a function of temperatures).

Since epoxy resins are not single crystalline structures and have a broad molecular weight distribution, sharp melting points are generally not noticeable. Any value for the

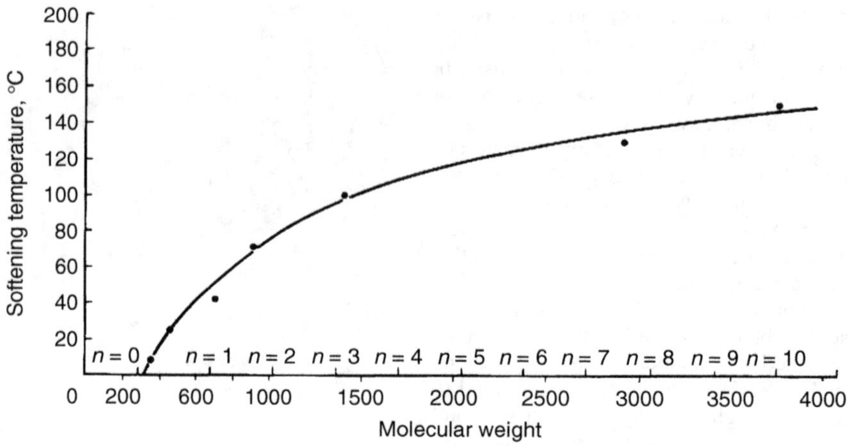

FIGURE 20.2 Typical softening point values for uncured DGEBA epoxy versus molecular weight.[1]

melting point is empirical and determined partly by the type of test and conditions under which it is performed.

20.2.3 Epoxy and Hydroxyl Content

Next to viscosity, epoxy content and hydroxyl content are the prime properties of importance in formulating epoxy-based adhesive systems. Epoxy content allows one to determine the correct amount of curing agent to use. With certain curing agents it may also be important to know the hydroxyl content. These values determine the reactivity of the system.

Epoxy content is generally determined by wet analytical techniques. ASTM D 1652 is widely used. The most common method is based on the addition of hydrogen halide (e.g., hydrogen chloride, hydrogen bromide, or hydrogen iodide) to the epoxy group. The difference between the amount of acid added and the amount unconsumed, determined by titration with standard base, is a measure of the epoxy content. There are numerous epoxy assay methods based on this technique. The specific method used depends on the resin type being analyzed as well as on the extent to which side reactions can occur.

Epoxy content can also be determined by infrared spectroscopy. In going from lower- to higher-molecular-weight species of the same epoxy resin, bands in the infrared spectrum change in intensity as a percentage composition of the various groups changes. The specific epoxy resin determines which bands should be used for the calculation of epoxy equivalent weight.

The hydroxyl equivalent weight of epoxy resins can be determined by several methods. The most common is esterification with acids, reaction with acetyl chloride, and reaction with lithium aluminum hydride. Infrared spectroscopy may also be used.

The hydroxyl content is of importance in estimating average molecular weight. To an extent, it also determines system functionality since the hydroxyl groups could be reactive sites. Certain curing agents may preferentially react with the hydroxyl groups.

20.2.4 Shelf Life

When a polymeric resin is stored for a considerable time, physical and chemical changes may occur within the material that will affect its performance as an adhesive or sealant.

The shelf life is generally defined as the time that an adhesive or sealant can be stored, preferably under controlled conditions, and remain unchanged, at least from a fundamental property point of view. ASTM D 1337 provides a method for determining the shelf life of an adhesive or sealant. The change in consistency (viscosity) or bond strength is measured after various storage periods at a specified temperature, as shown in Fig. 17.4. The shelf life of the product is generally determined to be the point at which it is no longer workable or its viscosity has increased by a factor of 2.

All the materials to be tested for shelf life are stored in unopened containers. Storage in containers that have been opened exposes the resin to oxygen and humidity that, depending on the type of resin, could drastically reduce the shelf life and affect final properties. Most polymeric adhesives and sealants and their components have a shelf life greater than 6 months at room temperature. However, some one-component adhesives need to be stored at refrigerated conditions to have a practical shelf life.

Certain adhesives or sealants may also have limited life when stored at cold temperatures or when exposed to repeated freeze-thaw cycles. Generally, these are water-based products, but freeze-thaw stability problems are not restricted to them. The resistance to freeze-thaw cycles in low-temperature storage is measured using procedures similar to the shelf life measurement discussed above. Even though an adhesive or sealant may have good low-temperature storage properties, it should be brought to room temperature before mixing or application.

20.2.5 Solids Content

The solids content of an adhesive or sealant should be checked to ensure that formulation or dilution errors have not been made. *Solids* can refer to the nonvolatile component of the adhesive or the inorganic component of the adhesive.

The nonvolatile solids content or percent of solids is usually determined by weighing a small amount of material in a clean container, heating or curing until a constant weight is obtained, and then weighing the container again. The percent of solids may be determined as the ratio of the sample weights before and after curing times 100. The percent of solids is an indirect measurement of the amount of volatile material in the sample that was driven off during the heating cycle. The volatile material may be solvent, water, or other additives.

ASTM D 1489 offers a test method for determining the nonvolatile content of aqueous adhesives. The solids content (or conversely the volatile content) is a measurement of the actual nonvolatile materials that are contained within the formulation. Addition of solvent can be used to "correct" the viscosity of adhesives or sealants to account for variability in the manufacturing process or inconsistent base materials in the resin formulation.

The ash test is a measure of the total inorganic content in a sample. In this test, a weighed sample is placed in a muffle furnace at approximately 540°C for 30 min or at a temperature and time long enough to completely pyrolize any organic matter. The remaining material is inorganic fillers, reinforcement, etc. ASTM D 5040 provides a method of measuring the ash content of adhesives. Ash content will determine the amount of inorganic fillers in an adhesive or sealant sample. The manufacturer can adjust the filler content, like the solids content, to correct for errors or inconsistency in formulation. Higher concentrations of solvent or filler are often used to lower the cost of the adhesives or sealant system.

20.2.6 Specific Gravity

The *specific gravity* of epoxy resins and other raw materials is the weight of the materials compared with the weight of an equal volume of water. Specific gravity can be determined either on the uncured, liquid epoxy resin or on the cured, solid epoxy. The specific gravity of the cured specimen is generally higher than that of the uncured specimen due to cure shrinkage. Changes in specific gravity are most noticeable when the inorganic filler content changes.

A variety of gravimetric methods may be used for the determination of specific gravity. Common instruments used with the liquid resins are the hydrometer and the psychometer. ASTM D 901 provides a method for measuring the specific gravity of low-viscosity liquids.

20.2.7 Color

Color of the uncured epoxy resins is an indication of contamination, incomplete polymerization, oxidation, and other factors. Aromatic resins with lower molecular weight generally have better (lower) color. Color is usually reported in terms of the Gardner liquid color standards or the Hellige color comparator or by means of the APHA number.

Refractive index is defined as the ratio of the speed of light of a given wavelength in vacuum to the speed of light at the same wavelength in the substance measured. The refractive index of epoxy resins is seldom reported. However, the property is important for cured epoxy adhesives that are used in the assembly of optical devices.

20.2.8 Chlorine Content

Epoxy resins will contain some hydrolyzable or active chlorine depending on the completion of the epichlorohydrin reaction. It may also contain less active chlorine via chlorinated groups along the polymeric chain. Hydrolyzable chlorine may be determined by ASTM D 1726 and total chlorine by ASTM D 1847.

The active chlorine is capable, under certain conditions, of blocking reactions with Lewis base catalysts. The main effect of high chlorine content is to reduce system reactivity and crosslink density. Chlorine may also cause corrosion of delicate substrates. Chlorine content is a common test for epoxy resins in the electrical and electronic industries.

20.3 PROPERTIES OF ADHERENDS

Fundamental property tests are also commonly used to characterize the substrate and provide a "signature" for lot-to-lot comparisons. In all too many cases, the substrate is considered to be a constant throughout the life of a production part. In practice, there are a number of common reasons for substrates to change.

Metallic substrates change primarily due to the nature of their surfaces. They can form different oxide layers that are dependent on the chemistry of the surface and the way it has been treated during fabrication of the part. The degree of surface cleaning or surface roughness may also change throughout a production run due to differences in processing chemicals or procedures or due to contamination of the materials that are used.

For organic substrates, the possible types of change are numerous and potentially more significant. Polymeric substrates may change in modulus due to aging, loss of plasticizer, or continued chemical reaction. They may change because of slight formulation variations that take place from lot to lot. These changes may not be sufficient to cause a change in the bulk physical properties. Therefore, the change may go unnoticed by the quality control personnel because they are looking only at bulk properties and not at subtle formulation alterations that may affect the adhesion properties. However, these substrate modifications could result in drastic changes in the surface properties of the material.

A case in point is an elastomeric substrate, such as compounded nitrile rubber. Generally, the user will specify the rubber part by compressive strength, extension, hardness, or other such standard bulk property tests. However, the supplier can formulate virtually thousands of compositions that will meet these specified properties. In this process she or he may use an incompatible plasticizer, or a low-molecular-weight extender, or a new filler that is more hydroscopic. These small changes could have a very large effect on the resulting joint strength. Therefore, any change in substrate formulation must be reverified with regard to its adhesion characteristics.

Test methods used to determine the uniformity of substrates are numerous and vary with the type of material. They are generally the same tests used to characterize the material or to determine its fundamental physical properties. Tests that are commonly employed are hardness, tensile strength, modulus, and surface characteristics such as roughness or contact angle with a standard liquid. Often a test similar to the nonvolatile test mentioned above is used to determine if there are any compounds in the substrate that are capable of outgassing on exposure to elevated temperatures. Moisture content of certain hydroscopic polymers, such as nylon and polycarbonate, is also known to affect adhesion.

20.4 TESTS ON THE CURING ADHESIVE

Many different test methods are used to measure the reactivity or cure rate of the epoxy adhesive. Some of these, such as working life or pot life, are very practical and are used to plan the production process. Others, such as exotherm, are used to determine reaction kinetics. Still others are used to characterize the epoxy network as it cures for the purposes of determining the degree of crosslinking and the rheological properties of the curing adhesive.

20.4.1 Working Life

The working life of an adhesive spans the time from when the adhesive is ready for use (i.e., mixed and ready to apply to a substrate) to the time when the system is no longer usable because the setting mechanism has progressed to such an extent that the adhesive is no longer workable. This characteristic is also known as the *pot life* of the adhesive. The term *pot life* usually defines a specific mass of adhesive and a specific temperature. Whereas *working life* does not denote such specifics. ASTM D 1338 establishes two procedures for determining the working life. One method uses viscosity change, and the other uses shear strength tests as the criteria for determining when the effective working life has expired.

Working life is usually determined on a volume of adhesive that is practical and normally used in production. The mass of the tested material must be defined in the test report because many adhesives have a working life that is dependent on sample mass.

20.4.2 Cure Rate

Structural adhesives usually require curing by the application of heat, the addition of a catalyst, the addition of pressure, or a combination of the three. The strength developed in the adhesive joint at various times during the curing process may be determined by lap shear tensile specimens. This test is commonly used to determine when an adhesive or sealant is fully cured or when the system reaches a "handling" strength so that the assembled product can be moved with moderate care.

There are also several methods of determining cure rate on the bulk adhesive material. These are generally analytical procedures that are common in most polymeric materials laboratories. With these methods, such fundamental properties as dielectric loss factor, mechanical damping characteristics, and exotherm are measured as a function of time and temperature.

Cure time is very important for sealants as well as adhesives. Often the sealant will be required to function as a barrier or resist the movement of substrates very soon after it is applied. With construction sealants, for example, it may not be possible to delay the environmental conditions until after the adhesive cures. Thus, curing time becomes a critical parameter in selection of the sealant. ASTM C 679 covers a method for determining the time that a mechanic can work the sealant into the joint before the sealant starts to skin or solidify.

Simple Solvent Wipe Test. The simplest test method to determine the extent of cure is to rub a cotton swab that has been soaked in a suitable solvent (e.g., methyl ethyl ketone) against the surface of the cured adhesive. If the adhesive softens, it is very far from a fully cured condition, and the degree of softness is a very gross indication of the degree of cure.

The cotton swab will remove any uncured material. The contents of the swab can be analyzed for traces of unreacted material. Solvent extraction can be used to chemically remove unreacted components from the swab for analytical measurement.

Hardness. Hardness of the adhesive or sealant may be used as an indication of cure. It may also be used as a quality control check on certain substrates. Hardness can be determined in several ways: (1) resistance to indentations, (2) rebound efficiency, and (3) resistance to scratching or abrasion. The first method is the most commonly used technique.

There are several ways of measuring indentation, but they differ only in the type of equipment used. Basically, they all measure the size of indentation produced by a hardened steel or diamond tool under a defined pressure. A durometer is an instrument for measuring hardness by pressing a needlelike instrument into the specimen. Durometers are available in several scales for measuring relatively hard, brittle materials to soft elastomers. The two types appropriate for most cured adhesives and sealants are the Shore Type A and Shore Type D. ASTM C 661 offers a method for measuring indentation hardness of elastomeric-type sealants.

Lower hardness readings than expected may be an indication of undercure or of a formulation change in the adhesive. They may also be an indication of entrapped air in the adhesive or sealant or an unwanted chemical reaction with the environment. Higher hardness readings than expected may be an indication of overcure.

A simple, but not very quantitative, hardness test has been used for hundreds of years—the fingernail indentation test. The indentation that a fingernail makes in the edge of an adhesive bond or in the body of a sealant can often be used as an approximate indication of hardness of the material.

Cure Rate as Measured by Strength Development of Prototype Joints. Cure rate is an important factor when the expense of jigs and fixturing equipment is high or fast production rates are critical. It is also used as a quality control test to determine if the curing mechanism within the adhesive has changed from lot to lot or if it may have been spoiled by storage, moisture contamination, etc.

ASTM D 1144 provides a recommended practice for determining the rate of bond strength development for either tensile or lap shear specimens. However, peel and canteliever tests can also be used effectively. Measured bond strength values of partially cured test specimens are compared with those of a reference (i.e., fully cured adhesive joint) to assess the extent of cure. This method may suit some applications, but it is limited in accuracy because it does not directly measure the degree of cure in the adhesive, and the effect on the joint design and substrates may override the effect of cure development.

Direct Methods of Assessing Degree of Cure. The test methods described above are indirect methods for measuring the degree of cure. They are generally used for simple quality control tests. For more sophisticated analysis, direct methods are used to measure the degree of cure of epoxy adhesives. These direct methods include spectroscopy, dielectric measurements, glass transition temperature monitoring, rheological measurements, and optical effects.

Nuclear magnetic resonance (NMR), infrared, and Raman spectroscopy are the most common methods of direct monitoring of the cure of an epoxy adhesive. Each of these techniques produces a unique profile or spectrum of the materials being investigated. Features in the spectrum of the adhesive (normally peaks or shoulders) are monitored for specific changes that relate to the formulation of new chemical bonds or the disappearance of functional groups. These features are seen to grow, shrink, or move position on the spectrum. The distance over which the change occurs is measured to provide a quantitative measure of the state of cure.

Dielectric test methods are used to measure the cure of epoxy adhesives between two conducting electrodes. This method is especially appropriate for metal-to-metal joints because the substrates themselves can be used as the electrode. The adhesive is treated as a capacitor during the test. Its response (dielectric constant, dissipation factor, etc.) over a range of electrical frequencies is measured as a function of curing time.

As the adhesive cures, its measured dielectric constant and dissipation factor change. Generally the dielectric constant gradually decreases until the adhesive finally cures, at which point the dielectric constant remains constant with cure time. The dissipation factor generally goes through an early peak during cure, which represents the process of the liquid adhesive becoming lower in viscosity (due to cure temperature or exotherm effects), and then increasing in viscosity until gelation. The dissipation factor then gradually decreases until full cure is achieved as ionic and polar groups are becoming restricted through crosslinking.

Dielectric monitoring has been used to determine the optimum time at which pressure can be applied on adhesives and composites that are very pressure-dependent. This is especially important for films and prepregs that have a high degree of retained solvent. Optimum pressure should not be applied until the solvent has had a chance to escape the adhesive when it is in the liquid state. Pressure is then applied when the solvent has escaped and the viscosity of the adhesive is high enough to prevent squeeze-out from the joint.

Rheological measurements can also be made to determine the curing rate of an epoxy adhesive. The viscosity will change over time. The change in viscosity is measured as a function of the resistance to rotation of a sample placed between two plates. Cone and plate rheometry using disposable plates are employed to monitor the mechanical loss and storage modulus of adhesive as a function of curing time. This method is more suitable for measuring the early stages of cure.

The rotational viscosity method described above to measure working life or pot life is a form of rheological measurement of cure. However, cone and plate rheometry is preferred for accurate measurements because the specimen size and geometry are similar to those that occur in an adhesive joint.

20.5 TESTS ON THE BONDED PRODUCT (STANDARD TEST SPECIMENS AND PROTOTYPE JOINTS)

A number of tests on finished bonds are useful in checking the assembly processes and in determining the adequacy of a specific adhesive for an application. These tests are generally in the form of standard test specimens or on prototype joints of the actual product.

A word of caution is due here. End-use data are sufficient to compare strengths of various bonding systems. The data can be used to compare the relative effectiveness of different adhesives, surface treatments, curing schedules, and so forth. They can also be used to separate and quantitatively define the many variables that ultimately determine the performance of a joint. However, end-use test results cannot be readily translated into specific strength values for an actual production joint. Actual joints generally have a complex geometry that is significantly different from the test specimen geometry. Under the specific environments and the kinds, frequency, and severity of the stress that the joint actually sees in service, an adhesive or sealant may perform spectacularly better or worse than what is represented by ASTM tests on a supplier's data sheet.

The most noticeable differences between standard test results and the strength from an actual joint in service are due to several factors.

1. Joint design is seldom the same.
2. The mode and application of stress loading in service are usually different and more complex than they are in practice.
3. Environmental aging is usually less severe in service (laboratory tests tend to "accelerate" aging so that the testing can be completed in a reasonable time). However, the effects of the actual service environment are generally more complex. For example, there may be simultaneous exposure to cyclic stress, cyclic temperature, and humid environments.
4. Laboratory test specimens are usually made in "controlled" environments. This control pertains to the equipment, the cleanliness (less weak boundary layer opportunity in the lab), and the personnel (training and awareness).
5. Sample population is limited with actual production parts because of expense. Even with laboratory specimens, a full design-of-experiment statistical process is difficult to achieve because of the many production variables that can affect the joint strength.

Thus, the most reliable test is to measure the strength of an actual assembly under actual operating conditions. Unfortunately, such tests are often expensive or impractical. The next best method is to measure the strength of an actual assembly under simulated operating conditions that do not stress the joint significantly differently than it would normally be stressed in service.

The usefulness and limitations of standard testing methods should be understood clearly before a testing program is established. The end user must choose the test geometry, procedure, and methodology that best serve the application. To do this, one needs to understand the differences in the various test methods and the outside parameters that will affect the data. Once the advantages and limitations of the various standard tests are understood, the end user may find it necessary to devise his or her own methods to test specific combinations of loads and environments that are anticipated.

It should be apparent by now that numerous parameters can affect the performance of a joint, and many combinations of those parameters are possible. Therefore, a prime rule in any adhesive or sealant testing program is to standardize and document test variables as thoroughly as possible. The adhesive formulator, supplier, and end user should all utilize the exact same procedures and specimen construction. One should make every effort to ensure that similar tests that are performed at different locations in the company and by different personnel are identical duplicates of one another.

20.5.1 Standard Test Methods

The physical testing of standard adhesive joints provides a standard of comparison for materials and processes that are being evaluated for a specific bonding application. Standard tests also provide a measure to control the adequacy of the bonding process, once it is established, and of assessing its conformance to specification.

Standard test methods are useful only if they can be reproduced. It is important that the same results be measured by both the adhesive developer and the end user. It is also important that the results be reproducible with time and with different testing personnel. The accuracy and reproducibility of test results depend on the conditions under which the bonding process is performed. The following variables must be strictly controlled.

1. Procedures for cleaning, etching, and drying the surfaces of the substrates prior to application of adhesive.
2. The time between surface preparation and application of adhesive and the environmental conditions present during this period. This includes the temperature and percent of relative humidity. Usually standard atmospheric conditions are specified (25°C and 50 ± 4 percent relative humidity).
3. Complete procedures for mixing the adhesive components, if more than one.
4. Conditions and methods for application of the adhesive to the substrate surface.
5. Curing conditions, including the pressure, temperature, and time of the curing cycle. It should be specified whether the temperature is measured within the glue line, or on some other point on the substrate, or at some location within the curing oven. The temperatures could vary significantly at these different locations depending on the weight and size of the assembly. When an adhesive producer specifies a temperature and time for cure, it refers to the conditions of the actual adhesive within the bond line.
6. Conditioning procedures for specimens after curing and prior to testing.
7. The rate at which the sample is loaded during test. Peel and impact tests especially are very dependent on the speed at which the sample is tested.

A standard test report usually documents the resulting measurements, such as tensile shear strength and peel strength. It should also indicate all the pertinent conditions that are required to ensure reproducibility in subsequent testing. It is often very useful to describe the failure mode of the tested specimens. An analysis of the type (or mode) of failure is an extremely valuable tool to determine the cause of adhesive failure. The failed joint should be visually examined to determine where and to what extent failure occurred. The percent of the failure that is in the adhesion mode and that in the cohesion mode should be provided. A description of the failure mode itself (location, percent coverage, uniformity, etc.) is often quite useful. The purpose of this exercise is to establish the weak link in the joint to better understand the mechanism of failure.

Numerous standard test methods have been developed by various government, industrial, and university investigators. Many of these have been prepared or adopted under the auspices of the ASTM Committee D 14 on Adhesives or other professional societies. Reference to the appropriate standards will adequately equip one with the background necessary to conduct the test or a version of it. Several of the more common standard tests are described in this section. Numerous variations exist for specific applications or materials. In these descriptions, the emphasis is on understanding of the reasons for the test, its relationship to a specific adhesive property, advantages and limitations of the test, and possible variations or extrapolations of the test method. The detailed description of the test mechanics is kept to a minimum, since they are adequately covered in the existing standards and specifications.

Tensile Tests. The tensile strength of an adhesive joint is seldom reported in the adhesive supplier's literature because pure tensile stress is not often encountered in actual production. An exception to this is the tensile test of the bonds between the skins and core of a honeycomb or composite sandwich. However, not only is the tensile test useful as a quality control test for metal and sandwich adhesives, but also it can be employed to yield fundamental and uncomplicated tensile strain, modulus, and strength data for the adhesive.

The ASTM D 897 tensile button test is widely used to measure the tensile strength of a butt joint made with cylindrical specimens (Fig. 20.3). The tensile strength of this bond is defined as the maximum tensile load per unit area required to break the bond (measured in pounds per squre inch). The cross-sectional bond area is usually specified to be equal to 1 in.^{2}. The specimen is loaded by means of two grips that are designed to keep the loads axially in line. The tensile test specimen requires considerable machining to ensure parallel surfaces.

A similar specimen design uses a sandwich construction with a dissimilar material bonded between the two cylindrical halves of the button specimen. This design is commonly used to measure the tensile strength of adhesives between dissimilar materials or if the adherend does not have the strength or characteristics to be machined into the shape of the button specimen. With some modifications in the dimensions, the button tensile test has also been adapted for testing adherence of honeycomb-cover sheets to the core (ASTM C 297).

Although developed and used primarily for testing wood joints, the cross-type tension test, shown in Fig. 20.4, can be used to test other substrates or honeycomb specimens (Fig. 20.4*c*). This simple cross-lap specimen is described in ASTM D 1344. This test method is attractive in that it does not involve significant machining or high specimen cost, as the button tensile test specimens do. It is very important that the specimens be thick and rigid enough to resist bending. With only moderate bending, the loads will quickly go into peel or cleavage stress. Because of a high degree of variation arising out of possible bending modes, a sample population of at least 10 is recommended for this test method.

Generally, the high-modulus adhesives possess the highest tensile strengths. There appears to be an inverse relationship between thickness and tensile bond strength for particular adhesive-adherend combinations. Thin bonds, without adhesive starvation, are best. Thicker bonds give lower strengths because cleavage forces more readily occur due to nonaxial loading, bending of the specimens, or internal stresses in the adhesive.

Tensile Shear Tests. The lap shear or tensile shear test measures the strength of the adhesive in shear. It is the most common adhesive test because the specimens are inexpensive,

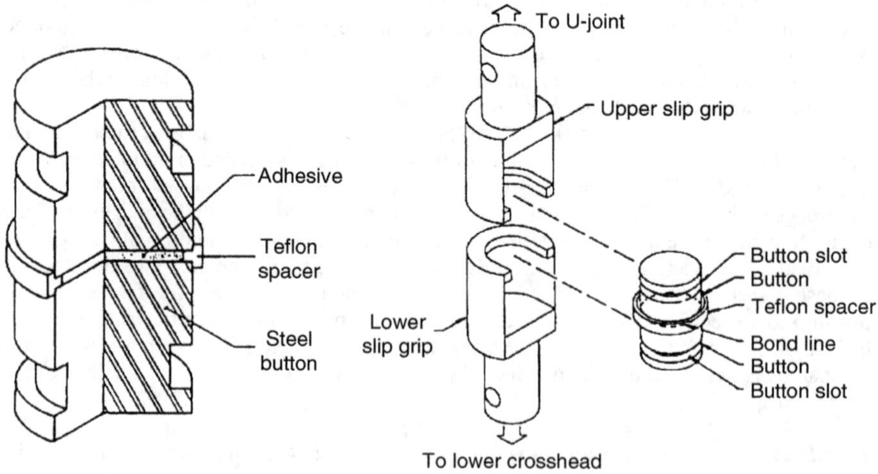

FIGURE 20.3 Standard button tensile test specimen and tensile test grips.[2]

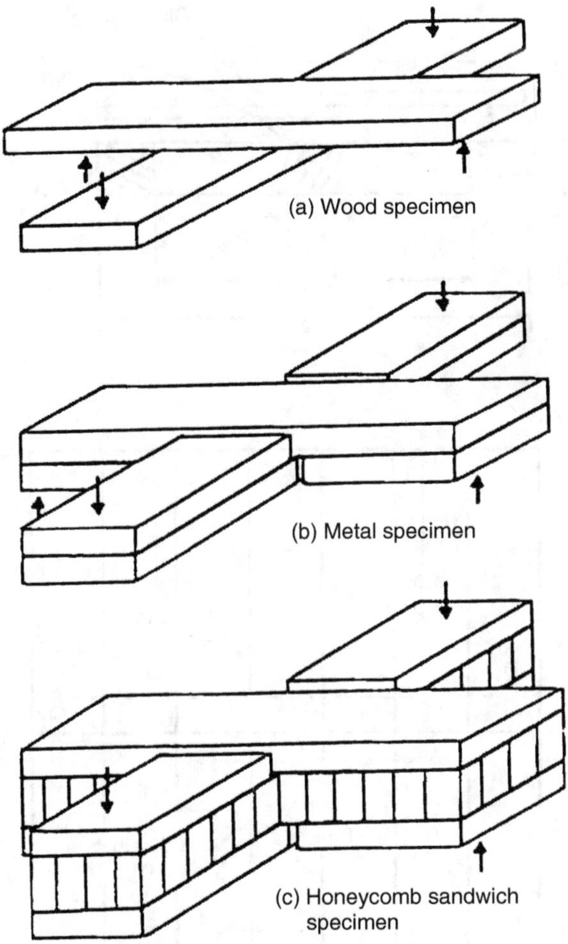

FIGURE 20.4 Cross-lap tension test specimen.[3]

easy to fabricate, and simple to test. However, at times it is difficult to minimize or eliminate bending stresses in common shear joint specimens. Because the standard lap shear tests introduce some degree of cleavage or peel into the adhesive joint, values obtained for the lap shear strength of epoxy adhesives may average 4000 to 5000 psi, whereas values for bulk tensile strengths have been reported up to 12,000 psi.[4]

The common lap shear test method is described in ASTM D 1002, and the standard test specimen in shown in Fig. 20.5. This is the most commonly used shear test for structural adhesives. However, due to the stress distribution in the adhesive arising from the joint configuration, the failure strength values are of little use for engineering design purposes. Testing is carried out by pulling the two ends of the overlap in tension, causing the adhesive to be stressed in shear. Hence, this test is frequently called the tensile shear test. Since the test calls for a sample population of 5, specimens can be made and cut from larger test panels, illustrated in Fig. 20.5*b*.

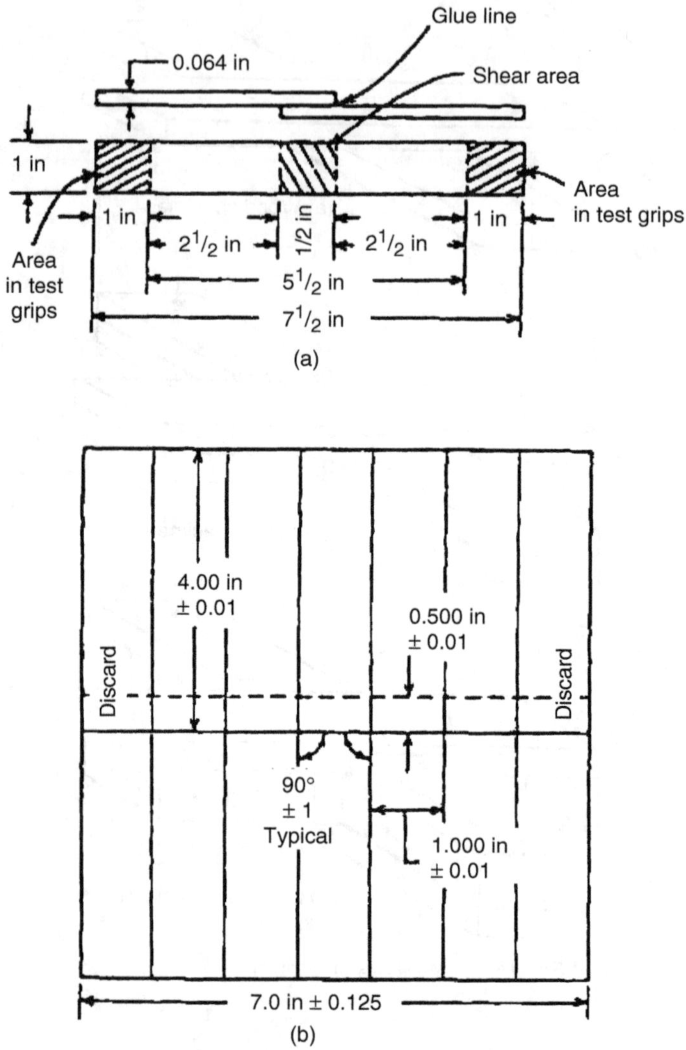

FIGURE 20.5 Standard lap shear test specimen design: (*a*) form and dimension of lap shear specimen; (*b*) standard test panel of five lap shear specimens. (*From ASTM D 1002.*)

The width of the lap shear specimen is generally 1 in. The recommended length of overlap, for metal substrates of 0.064-in thickness, is 0.5 ± 0.05 in; however, it is recommended that the overlap length be chosen so that the yield point of the substrate is not exceeded. In lap shear specimens, an optimum adhesive thickness exists. For maximum bond strengths, the optimum thickness varies with adhesives of different moduli (from about 2 mils for high-modulus adhesives to about 6 mils for low-modulus adhesives).[5]

The lap shear specimen can be used for determining the shear strength of dissimilar materials in a manner similar to that described for the laminated button tension specimen.

Thin or relatively weak materials such as plastics, rubber, or fabrics are sandwiched between the adherends and tested.

Two other variations are used to avoid the bending forces that occur with simple ASTM D 1002 specimens: the laminated lap shear specimen (ASTM D 3165) shown in Fig. 20.6*a* and the double-lap specimen (ASTM D 3528) shown in Fig. 20.6*b*. These specimens minimize the joint eccentricity and provide higher strength values than does the single-overlap specimen. For the specimen in Fig. 20.6*a*, the overlap joint can be made from saw cuts in the top and bottom substrates of a bonded laminate. This process negates the effects of extruded adhesive at the edges of the lap and the sheared edge of the standard type of lap shear specimen. As a result, the chances of deformation and uneven surface preparation are lessened.

Compression shear tests are also commonly used. ASTM D 2182 describes a simple compression specimen geometry and the compression shear test apparatus. The compression shear design also reduces bending and, therefore, peeling at the edges of the laps. Higher and more realistic strength values are obtained with the compression shear specimen over the standard lap shear specimen.

Peel Test. A well-designed joint minimizes peel stress, but not all peel forces can be eliminated. Because epoxy adhesives are notoriously weak in peel, tests to measure peel resistance are very important. Peel tests involve stripping away a flexible adherend from another adherend that may be flexible or rigid. The specimen is usually peeled at an angle of 90° or 180°. The most common types of peel test are the t-peel, the floating roller peel, the 180° peel, and the climbing drum methods. Representative test specimens are shown in Fig. 20.7. The values resulting from each test method can be substantially different; hence it is important to specify the test method employed.

Peel values are recorded in pounds per inch of width of the bonded specimen. They tend to fluctuate more than any other adhesive test result because of the extremely small area at which the stress is localized during loading. Even during the test, the peel strength values tend to fluctuate depending on the type of adhesive, adherend, and condition of the test. In preparing the samples, care must be taken to produce void-free laminated bond lines.

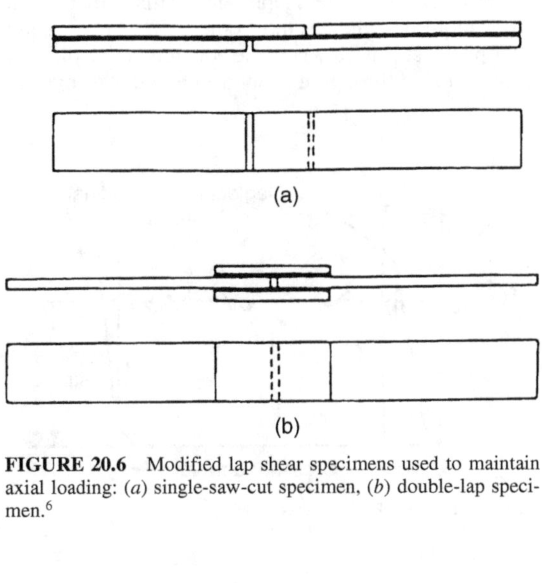

(a)

(b)

FIGURE 20.6 Modified lap shear specimens used to maintain axial loading: (*a*) single-saw-cut specimen, (*b*) double-lap specimen.[6]

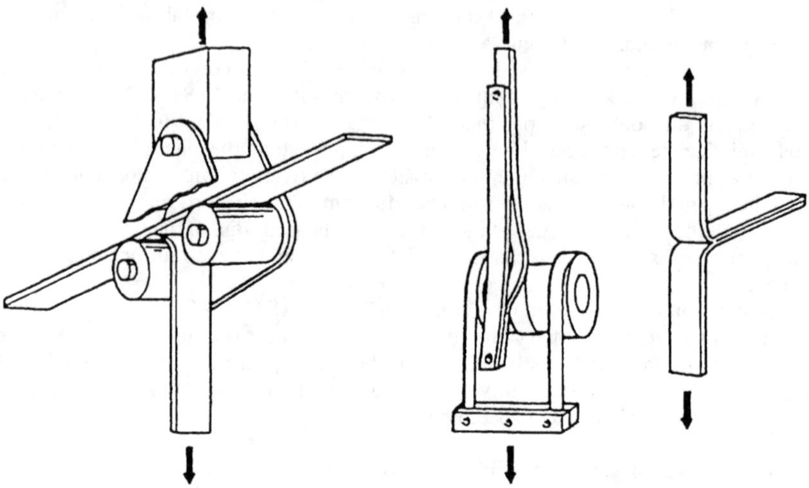

FIGURE 20.7 Common types of adhesive peel test: (left) floating roller peel, (center) climbing drum peel, (right) t-peel.[7]

A typical load curve of the t-peel test is shown in Fig. 20.8. Peel strength is taken as the average value of the center portion of the curve, usually over at least a 5-in length.

The rate of peel loading is more important than in lap shear loading, and it should be known and controlled as closely as possible. The rate at which the load is applied is usually specified in the ASTM test procedure. Adhesive thickness also has a significant effect on peel strength values, as does the angle of peeling.

The t-peel test is described in ASTM D 1876 and is the most popular of all peel tests. The t-peel specimen is shown in Fig. 20.7. Generally, this test method is used when both adherends are flexible. Because the angle of peel is controlled by the properties of the adherends and the adhesive, the test is less reproducible than other peel tests. A sample population of at least 10 is required, whereas most other ASTM tests require a minimum of 5.

The floating roller peel test is used when one adherend is flexible and the other is rigid. The flexible member is peeled through a spool arrangement to maintain a constant angle

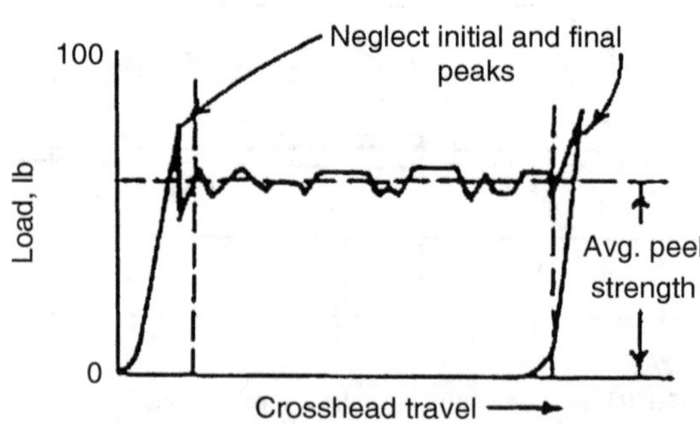

FIGURE 20.8 Peel test record.[8]

of peel. Thus, the values obtained are generally more reproducible. The floating roller peel resistance test is designated ASTM D 3167.

The climbing drum peel specimen is described in ASTM D 1781. This test method is intended primarily for determining peel strength of thin metal facings on honeycomb cores, although it can be used for joints where at least one member is flexible. The fixtures of the floating roller peel and drum peel tests help stabilize the angle of peel so that they generally provide higher and more reproducible peel values for a given adhesive than the t-peel method.[9,10]

A variation of the t-peel test is a 180° stripping test, described in ASTM D 903. This method is commonly used when one adherend is flexible enough to permit 180° turn near the point of loading. This test offers more reproducible results than the t-peel test because the angle of peel is kept constant, although it is dependent on the nature of the adherend.

Cleavage Test. Cleavage tests are conducted by prying apart one end of a rigid bonded joint and measuring the load necessary to cause rupture. Cleavage tests are usually used in place of peel tests when both adherends are rigid. The test is a subjective measure of the cleavage strength of an adhesive and a qualitative measure of the fracture toughness of the adhesive. Data obtained are adaptable to engineering design when the service stress applied to the functional joint is in the cleavage mode.

The cleavage test utilizes a specimen where the load is intentionally placed on one edge of the bonded area. The test method is described in ASTM D 1062. The specimen is usually loaded at a rate of 0.05 in/min until failure occurs. The failing load is reported as breaking load per unit area in pounds per square inch. A standard test specimen is illustrated in Fig. 20.9. Because cleavage test specimens involve considerable machining, peel tests are usually preferred where possible.

Fatigue Test. Fatigue testing places a given load repeatedly on a bonded joint. Lap shear or other specimens are tested on a machine capable of inducing cyclic loading

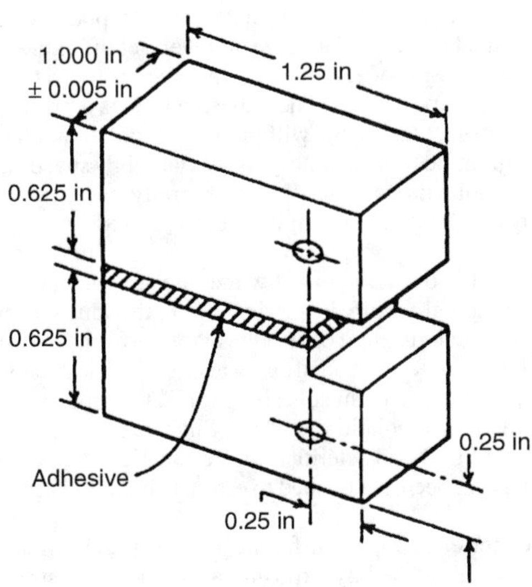

FIGURE 20.9 Cleavage test specimen. (*From ASTM D 1062.*)

(usually in tension but also in bending) on the joint. ASTM D 3166 provides procedures for testing and measurement of the fatigue strength of lap specimens.

The fatigue strength of an adhesive is reported as the number of cycles of a known load necessary to cause failure. Cycles to failure and the corresponding loads are plotted on coordinates of stress versus the logarithm of the number of cycles. The point at which the smooth curve connecting the points of minimum stress crosses the 10 million cycle line is usually reported as the fatigue strength. Fatigue strength is dependent on the adhesive, curing conditions, joint geometry, mode of stressing, magnitude of stress, and frequency and amplitude of load cycling.

Lap shear fatigue data are limited in engineering design because of the complex stress distribution patterns of the lap shear specimen configuration relative to most practical joint designs. However, fatigue testing of the actual part or of the functional joint design itself provides useful engineering design values.

Impact Test. Impact testing is of importance because adhesives, like most polymeric materials, are sensitive to high rates of application of force. The resistance of an adhesive to impact can be determined by ASTM D 950. This test is analogous to the Izod impact test method used for impact studies on materials.

The specimen is mounted in a grip, shown in Fig. 20.10, and placed in a standard impact machine. One adherend is struck with a pendulum hammer traveling at 11 ft/s, and the energy of impact is reported in pounds per square inch of bonded area. It is often difficult to achieve reproducible results with impact testing, and as a result, the test is not widely used in production situations. Impact data indicate that as the thickness of the adhesive film increases, its apparent strength also increases.[11] This suggests that a portion of the energy required to rupture the bond is absorbed by the adhesive layer and is independent of adhesion.

The impact measurements of a viscoelastic adhesive are affected by the rate at which the impact occurs. Often it is very difficult to achieve high rates of impact with laboratory testing. One example of this is an adhesive system used to bond shock mounting pads to electrical equipment aboard submarines. Impact testing of specimens done in a laboratory using pendulum or drop-weight impact fixtures showed that the adhesive would not fail under the loads expected. However, the proof test was to place the electrical equipment, with shock mounting attached, aboard a barge and then set off explosive charges at various depths under the barge. The proof test showed that the viscoelastic adhesive failed when high rates of impact were experienced. The adhesive acts as a brittle polymer at high rates of loading because the molecular chains within the material do not have time to slip by one another and absorb a great deal of the energy. Special, high-speed impact tests have been developed for certain applications. These tests generally use chemical explosive force or electromagnetic energy to establish the high speeds required.

Environmental Tests. It is desirable to know the rate at which an adhesive bond will lose strength due to environmental factors in service. Strength values determined by short-term tests do not give an adequate indication of an adhesive's performance during continuous environmental exposure. Laboratory-controlled aging tests seldom last longer than a few thousand hours. To predict the permanence of an adhesive over a 20-year product life requires accelerated test procedures and extrapolation of data. Such extrapolations are extremely risky because the causes of adhesive bond deterioration are complex (see Sec. 15.2.2). Unfortunately no universal method has yet been established to estimate bond life accurately from short-term aging data.

Adhesives may experience many different and exotic environments. Environmental aging of adhesives is accomplished by exposing a stressed or unstressed joint to simulated operating conditions. Exposure is typically to elevated temperature, water, salt spray, or various chemical solutions that simulate the service conditions. A number of standard chemicals that are used to soak bonded specimens for 7 days at room temperature are

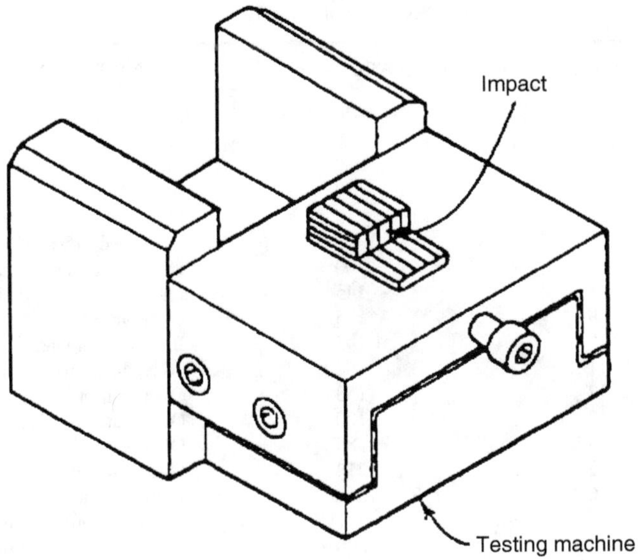

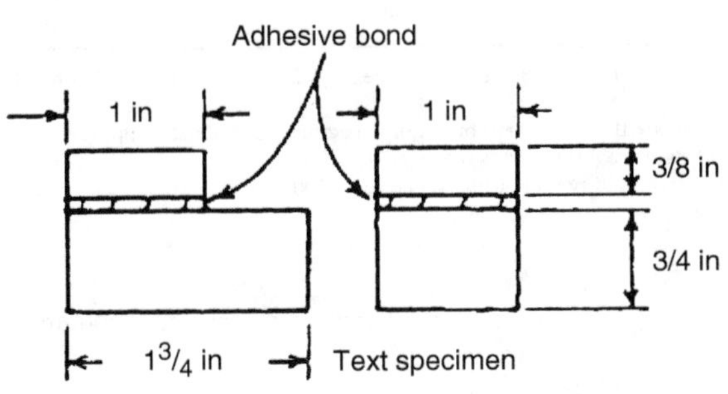

FIGURE 20.10 Impact test specimen. (*From ASTM D 950.*)

described in ASTM D 896 (see Table 20.1). The method merely outlines the environment in which the test specimens are conditioned before testing. After suitable exposure, bonds may be tested in whatever manner seems appropriate.

The type of stress and environment should be selected to be a close approximation of real-life conditions. Two simultaneous environments, such as heat and moisture, may cause the adhesive to degrade much faster than when it is exposed in any single environment because one condition could accelerate the effect of the other.

Stress aging tests are important because they more accurately simulate service conditions than simple tests where the specimens are merely hung or placed in a test environment. The Alcoa Stress Test Fixture (Fig. 20.11) is a device for applying stresses of various magnitudes to lap shear specimens and then aging these stressed specimens in different environments. A plot of stress to time of failure can give information about the loss of bond strength under stressed conditions. This aging phenomenon and a typical stress-to-failure plot are shown in Chap. 15.

TABLE 20.1 Standard Test Exposure of Adhesives[12]

Test exposure number	Test temperature,* °F	Moisture conditions
1	–70	As conditioned
2	–30	As conditioned
3	–30	Presoaked†
4	32	As conditioned
5	73.4	50 percent relative humidity
6	73.4	Immersed in water
7	100	88 percent relative humidity
8	145	Oven, uncontrolled humidity
9	145	Over water‡
10	145	Immersed in water
11	158	Oven, uncontrolled humidity
12	158	Over water‡
13	180	Oven, uncontrolled humidity
14	180	Over water‡
15	212	Oven, uncontrolled humidity
16	212	Immersed in water
17	221	Oven, uncontrolled humidity
18	300	Oven, uncontrolled humidity
19	400	Oven, uncontrolled humidity
20	500	Oven, uncontrolled humidity
21	600	Oven, uncontrolled humidity

*The tolerance for test temperature shall be ±1.8°F up to 180°F and ±1 percent for temperatures over 180°F.

†Presoaking shall consist of submerging specimen in water and applying vacuum at 20 in of mercury until weight equilibrium is reached.

‡The relative humidity will ordinarily be 95 to 100 percent.

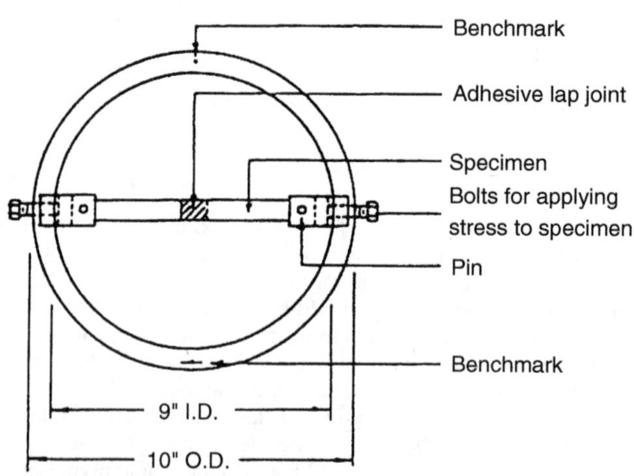

FIGURE 20.11 Alcoa stressing fixture for measuring simultaneous effect of stress and environmental conditions.[13]

There are several analytical tools that provide methods of extrapolating test data. One of these tools is the Williams, Landel, Ferry (WLF) transformation.[14] This method uses the principle that the work expended in deforming a flexible adhesive is a major component of the overall practical work of adhesion. The materials used as flexible adhesives are usually viscoelastic polymers. As such, the force of separation is highly dependent on their viscoelastic nature and is, therefore, rate- and temperature-dependent. Test data, taken as a function of rate and temperature, can be expressed in the form of master curves obtained by WLF transformation. This offers the possibility of studying adhesive behavior over a sufficient range of temperatures and rates for most practical applications. High rates of strain may be simulated by testing at lower rates of strain and lower temperatures.

20.5.2 Testing of Prototype Parts

There are several ways that actual joints can be tested. For quality control purposes, a proof test is commonly used, as described above. This test imposes a stress on the specimen but limits it to a point well below where any destruction of the joint can occur. This type of test only looks for serious flaws in the bonding processes such as interface contamination, air entrapment in the joint, or undercured adhesive. The prototype joints can also be tested to destruction by using similar test methods described above for standardized testing. This, however, is generally not done at great frequency because of the cost. Generally, most actual joint tests are performed to ascertain the cause of failure (i.e., forensic analysis of the failed joint) or for nondestructive determination of the adequacy of the bond.

Examination of Failed Joints for Cause of Failure. Close examination of failed joints can sometimes lead to an explanation of why the specimen failed. With visual or microscopic examination it is sometimes evident that adhesive failure occurred at the original interface due to improper wetting or at some new interface, leaving behind a thin layer of adhesive. In comparing surface features after bond failure with the original adherend surface, the maximum resolution of about 1×10^{-8} m (100 Å) for scanning electron microscopes may not always be sufficient to detect a thin film of adhesive closely reproducing the original surface profile. Optical and staining methods described by Brett[15] to determine the presence of such films are mainly applicable to fairly thick films, since optical techniques use interference phenomenon. Films a few angstroms thick are still largely undetected.

In recent years the use of highly specialized surface characterization tools has greatly improved the opportunity for deducing the surface composition. These tools have been developed for the purpose of analyzing both the adherend and the adhesive. For adherends, analytical examination generally centers on either (1) the surface chemistry for prepared surfaces by elemental analysis, chemical species, or analysis of contaminants and boundary layers or (2) analysis of failed surface for evidence of interfacial failure, failure within the adherend (e.g., metal oxide, composite matrix or fiber), or failure within a primer or other boundary layer. Analysis of the adhesive generally consists of characterizing the cured film, curing agents, and failed specimen surface chemistry.

The sciences of microphotography and holography have been used for analysis of adhesive bonds. Magnification and photography of the failed substrate often lead to useful clues to the cause of failure. Holography is a method of producing photographic images of flaws and voids using coherent light such as that produced by a laser. The major advantage of holography is that it photographs successive "slices" through the scene volume. A true three-dimensional image of a defect or void can then be reconstructed.

Over the last 20 years, analytical tools have become available that allow for the characterization of the elemental and chemical composition of solid surfaces. The application of these analytical tools has increased our understanding of surface properties and successfully characterized surface layers.

The most popular of these are secondary ion mass spectroscopy (SIMS), ion scattering spectroscopy (ISS), and Auger electron spectroscopy (AES) which have been developed by investigators such as Baun, McDevitt, and Solomon.[16–20] These tools have proved practical even when the surface films are only on the order of atomic dimensions or when the failure occurred near the original interface and included parts of both the adhesive and the adherend.

By itself, SIMS has been shown to be a powerful tool for elemental surface characterization by Benninghoven[21] and Schubert and Tracy;[22] however, uncertain or rapidly changing secondary ion yield due to changes in chemical bonding makes quantitative analysis virtually impossible using SIMS alone.[23,24] SIMS is most helpful when combined with other techniques, such as ISS and AES, which use a beam of ions of correct energy for combined use with SIMS.

The greatest strength of electron spectroscopy for chemical analysis (ESCA) lies in its ability to provide information on the surface chemistry of polymers or organics. The excitation of the surface is relatively nondestructive. AES is a useful high-spatial-resolution technique for the analysis of metals, alloys, and inorganic materials. Polymer and organic surfaces pose problems because of beam damage and sample charging. ISS is an elemental analysis technique with single atomic layer resolution. ISS is often used in conjunction with other surface techniques.

Nondestructive Testing. Nondestructive testing (NDT) is far more economical than destructive test methods, and every assembly can be tested if desired. Several nondestructive test methods are used to check the appearance and quality of structures made with adhesives or sealants. The main methods are simple ones such as visual inspection, tap, proof, and more advanced physical monitoring such as ultrasonic or radiographic inspection. The most difficult defects to find are those related to improper curing and surface treatments. Therefore, great care and control must be exercised in surface preparation procedures and shop cleanliness.

A universal NDT method for evaluating bonded structures is not currently available. Generally, the selection of a test method is based on

- Part configuration and materials of construction
- Types and sizes of flaws to be detected
- Accessibility to the inspection area
- Availability and qualifications of equipment and personnel
- Throughput rate required of the NDT process
- Required documentation of the process and test results

Ultrasonic Inspection. The success of the tap test, although limited, led to the use of ultrasonics to determine bond quality. Ultrasonic methods are at present the most popular NDT technique for use on adhesive joints. Ultrasonic testing measures the response of the bonded joint to loading by low-power ultrasonic energy. Short pulses of ultrasonic energy can be introduced on one side of the structure and detected on the other side. This is called *through-transmission testing*. A void or high damped adhesive (undercured or filled with air) also prevents the ultrasonic energy from passing efficiently through the structure.

A number of different types of ultrasonic inspection techniques using pulsed ultrasound waves from 2.25 to 10 MHz can be applied to bonded structures.[25] The most common methods are as follows:

- Contact pulse echo—the ultrasonic signal is transmitted and received by a single unit.
- Contact through transmission—the transmitting search unit is on one side of the bonded structure, and the receiving unit is on the other.
- Immersion method—the assembly is immersed in a tank or water, and the water acts as a coupling mechanism for the ultrasonic signal.

Pulse echo techniques are perhaps the easiest to use in production. The sound is transmitted through the part, and reflections are obtained from voids at the bond interface. The result is generally considered only qualitative because a poorly bonded joint will show as a good joint as long as it is acoustically coupled. A thin layer of oil or water at the interface may also act as a coupling and disguise an unbonded area. Shear waves can be introduced into the structure with a wedge-shaped transducer.[26] This technique is effective in analyzing sandwich structures.

Bonded structures that are ultrasonically tested by the immersion methods often use a C-scan recorder to record the test. This recorder is an electric device that accepts signals from the pulser-receiver and prints out a plan view of the part. The ultrasonic search unit is automatically scanned over the part. The ultrasonic signals for bond or unbond are detected from built-in reference standards. C-scan NDT techniques are used extensively by aircraft manufacturers to inspect bonded parts.

One of the oldest and best known ultrasonic testing systems for NDT is the Fokker Bond Tester. This method uses a sweep frequency resonance method of ultrasonic inspection. Some degree of quantitative analysis is claimed with the Fokker Bond Tester in the aircraft industry.

Other NDT Methods. *Radiography (x-ray) inspection* can be used to detect voids or discontinuities in the adhesive bond. This method is more expensive and requires more skilled experience than ultrasonic methods. The adhesive must contain some metal powder or other suitable filler to create enough contrast to make defects visible. This method is applicable to honeycomb sandwich structures as well as metal and nonmetal joints.

Thermal transmission methods are relatively new techniques for adhesive inspection. Heat flow is determined by monitoring the surface temperature of a test piece a short time immediately after external heating or cooling has been applied. Subsurface anomalies alter the heat flow pattern and, thereby, affect the surface temperature. The surface temperature difference can be detected by thermometers, thermocouples, or heat-sensitive coatings. Liquid crystals applied to the joint can make voids visible if the substrate is heated.

Thermal transmission testing is an excellent way of detecting various types of anomalies such as surface corrosion under paint before the corrosion becomes visually evident. Thin, single-layer structures, such as aircraft skin panels, can be inspected for surface and subsurface discontinuities. This test is simple and inexpensive, although materials with poor heat-transfer properties are difficult to test, and the joint must be accessible from both sides. For nonmetallic materials, the defect diameter must be on the order of 4 times its depth below the surface to obtain a reliable thermal indication. For metals, the defect diameter must be approximately 8 times its depth. Some bright surfaces such as bare copper and aluminum do not emit sufficient infrared radiation and may require a darkening coating on their surface.

With *thermal surface impedance testing*, heat is injected into the test objects' surface from a hot gas pulse. The resulting surface temperature transient is analyzed to determine the bond quality in nearly real time. The surface temperature transients are sensed using an innovative noncontacting, emissivity-independent infrared sensor.[27] This method is not adversely affected by surface blemishes or roughness.

REFERENCES

1. Lee, H., and Neville, K., *Handbook of Epoxy Resins*, McGraw-Hill, New York, 1968, p. 4.26.
2. Anderson, G. P., et al., "Effect of Removing Eccentricity from Button Tensile Adhesive Tests," *Adhesively Bonded Joints: Testing , Analysis, and Design*, ASTM STP 981, W. S. Johnson, ed., American Society of Testing and Materials, Conshohocken, PA, 1988.
3. Elliot, S. Y., "Techniques for Evaluation of Adhesives," Chapter 31 in *Handbook of Adhesive Bonding*, C. V. Cagle, ed., McGraw-Hill, New York, 1973.

4. Mathews, D. H., and Silver, I., *Methods for Testing the Strength Properties of Adhesives and Test Data*, NEVOID Report 3923, 1962.
5. Elliot, "Techniques for Evaluation of Adhesives."
6. Elliot, "Techniques for Evaluation of Adhesives."
7. DeLollis, N. J., *Adhesive for Metals Theory and Technology*, Industrial Press, New York, 1970.
8. Elliot, "Techniques for Evaluation of Adhesives."
9. DeLollis, N. J., *Adhesives, Adherends, Adhesion*, Robert E. Krieger Publishing Co., Huntington, NY, 1980.
10. Gent, A. N., "More on the Peel Test Riddle," *Adhesives Age*, February 1997, p. 59.
11. Silver, I., "Shear Impact and Shear Tensile Properties of Adhesives," *Modern Plastics*, vol. 26, no. 9, 1949, p. 95.
12. "How to Test Adhesive Properties," *Materials Engineering*, March 1972, pp. 60–64.
13. Miniford, J. D., "Permanence of Adhesive-Bonded Aluminum Joints," Chapter 23 in *Adhesives in Manufacturing*, G. L. Schneberger, ed., Marcel Dekker, New York, 1983.
14. Miniford, "Permanence of Adhesive-Bonded Aluminum Joints."
15. Brett, C. L., "The Detection of Thin Films of Epoxy Resin on Metal Surfaces," *Journal of Applied Polymer Science*, vol. 18, 1974, p. 315.
16. Baun, W. L., et al., "Chemistry of Metal and Alloy Adherends by Secondary Ion Mass Spectroscopy, Ion Scattering Spectroscopy, and Auger Electron Spectroscopy," *ASTM STP 596*, American Society of Testing and Materials, Conshohocken, PA, March 1975.
17. McDevitt, N. T., and Baun, W. L., "Some Observations of the Relation Between Chemical Surface Treatments and the Growth of Anodic Barrier Layer Films," Air Force Materials Laboratory Technical Repot 76-74, June 1976.
18. McDevitt, N. T., et al., "Surface Studies of Anodic Aluminum Oxide Layers Formed in Phosphoric Acid Solutions," Air Force Materials Laboratory Technical Report 77-55, May 1977.
19. McDevitt, N. T., et al., "Accelerated Corrosion of Adhesively Bonded 7075 Aluminum Using Wedge Crack Specimens," Air Force Materials Laboratory Technical Report 77-184, October 1977.
20. Baun, W. L., et al. "Pitting Corrosion and Surface Chemical Properties of a Thin Oxide Layer on Anodized Aluminum," in Air Force Materials Laboratory Technical Report 78-128, September 1978.
21. Benninghoven, A., "Statistical Method of Secondary Ion Spectroscopy," *Surface Science*, vol. 28, 1971, p. 541.
22. Schubert, R., and Tracy, J. C., "Simple, Inexpensive SIMS Apparatus," *Review of Scientific Instruments*, 44:487 (1973).
23. Schubert, R., "The Analysis of 301 Stainless Steel by SIMS," *Journal of Vacuum Science Technology*, vol. 12, no. 1, 1975, p. 505.
24. Werner, R., *Developments in Applied Spectroscopy*, vol. 7A, 1969, p. 297.
25. Hagemaier, D. J., "End Product Nondestructive Evaluation of Adhesive Bonded Metal Joints," *Adhesives and Sealants*, vol. 3, *Engineered Materials Handbook*, ASM International, Materials Park, OH, 1990.
26. Botsco, R. T., and Anderson, R. T., "Nondestructive Testing: Assuring Reliability in Critically Bonded Structures," *Adhesives Age*, May 1984, pp. 19–21.
27. Thomas, R. L., and Favro, L. D., "Thermal Wave Inspection of Adhesive Disbonding," *41st Annual SAMPE Symposium*, March 1996.

APPENDIX A

TRADE NAMES AND MANUFACTURERS: EPOXY ADHESIVES, EPOXY RESINS, CURING AGENTS AND CATALYSTS, ADDITIVES AND MODIFIERS

TABLE A.1 Trade Names and Manufacturers of Selected Epoxy Adhesive Formulators

Trade name	Manufacturer	Web site
Able-	Ablestik Laboratories	www.ablestik.com
Allaco Allibond	Bacon Industries, Inc.	www.baconindustries.com
Ablebond Amicon Eccobond	Emerson and Cuming, Inc.; National Starch and Chemicals	www.emersoncuming.com
Conapoxy	Conap, Inc.; Division of Cytek	www.conap.com
Elastolock	B.F. Goodrich, Engineered Polymer Products	www.epp.goodrich.com
Epo-Tek	Epoxy Technology, Inc.	
Epocap Epocure Eposet Epoweld Monoepoxy Hardman	Elementis Specialties	www.elementis-specialties.com
Epoxi-Patch EA- Hysol Megabond	Loctite-Henkel	www.loctite.com
Scotch Weld	3M Company	www.3M.com
Smooth-On	Smooth-On, Inc.	www.smooth-on.com
TraCon	Tra-Con, Inc.	www.tra-con.com
Fusor Versilok	Lord Corp.	www.lord.com
Resiweld	H.B. Fuller Company Devcon; Illinois Tool Works	www.hbfuller.com www.devcon.com
Magnobond	Magnolia	www.magnapoxy.com
Permabond	Permabond International	www.permabond.com
Araldite	Huntsman	www.araldite.com
	Systems Three	www.systemsthree.com
Tru-Bond	Devcon	www.devcon.com
	Cytec	www.cytec.com
	Master Bond	www.masterbond.com
	Nbond	www.nbond.com
Betamate	Dow	www.dow.com
	Sovereign Specialty Chemicals	www.sovereignsc.com
	Epic Resins	www.epicresins.com

TABLE A.2 Trade Names and Manufacturers of Epoxy Resins

Trade names	Manufacturer	Web site
DEN DER DOW	Dow Chemical Company	www.dow.com
EPON EPI-REZ HELOXY	Resolution Performance Products, LLC	www.resins.com
ERL	Bakelite AG	www.bakelite.de
EPALLOY	CVC Specialty Chemicals Inc.	www.cvcchem.com
Araldite	Huntsman	www.araldite.com
Epotuf	Reichold	www.reichold.com
Hetron	Ashland Specialty Chemical Company	www.ashchem.com
	Magnolia Plastics	www.magnaepoxy.com
Chem Res	Cognis	www.cognis.com
	Nan Ya Plastics Corporation	www.npc.com.tw
Epotec	Thai Epoxy & Allied Products Co.	www.thaiepoxy.com
	UPPC AG	www.uppc.de
Uvacure	UCB Chemicals	www.chemicals.ucb-group.com
Incorez	Industrial Copolymers Ltd.	www.incorez.com
	Ciba Specialty Chemicals Corp.	www.cibasc.com

TABLE A.3 Trade Names and Manufacturers of Curing Agents

Trade name	Manufacturer	Web site
Ancamine Ancamide	Air Products and Chemicals	www.airproducts.com
DEH	Dow	www.dow.com
Epi-Cure	Resolution Performance Products, LLCC	www.resins.com
Ajicure	Ajinomoto, Inc.	www.ajinomoto-usa.com
	Akzo Nobel Inc.	www.akzonobel-polymerchemicals.com
Ethacure	Albemarle Corp	www.albemarle.com
	Ciba Specialty Chemicals Corp.	www.cibasc.com
Epamine	Epic Resins	www.epicresin.com
Jeffamine Aradur	Huntsman Corp.	www.huntsman.com
Leecure	Leepoxy Plastics, Inc.	www.leepoxy.com
	Lonza, Inc.	www.lonza-us.com
	Milliken Chemical Div., Milliken & Company	www.milliken.com
	Rohm and Hass (Morton International, Inc.)	www.rohmhass.com
Epotuf	Reichold Chemicals, Reactive Polymers Division	www.reichold.com
Uni-Rez	Arizona Chemical (Union Camp Corp.)	www.arizonachemical.com
Tonox	Crompton Uniroyal Uniqema	www.crompton.com www.uniqema.com
Capcure	Cognis	www.cognis.com
Versamid	Elementis	www.elementis-specialties.com
Versamine	UCB Chemicals Corp.	www.chemicals.ucb-group.com
	SKW Corporation	www.skwchem.com
Dyhard (dicyandiamide)	Degussa	www.degussa.com
	Pacific Epoxy	www.pacificepoxy.com
	CVC Specialty Industries	www.cvcchem.com
	SIQ-Kunstharze GmbH	www.siq-kunstharze.com
Dytek	Invista, Inc.	www.dytek.invista.com
	UPPC AG	www.uppc.de
	Thai Epoxy & Allied Products Co.	www.thaiepoxy.com
	BASF	www.basf.com

TABLE A.4 Trade Names and Manufacturers of Epoxy Additives

Trade name	Manufacturer	Web site
Adhesion promoters	Degussa Corporation	www.degussa.com
	Gelest	www.gelest.com
	Crompton OSi Specialties	www.cromptoncorp.com
	Dow Corning Crop.	www.dowcorning.com
	NuSil Technology	www.nusil.com
Defoamers	Air Products and Chemicals, Inc.	www.airproducts.com
	Ashland Specialty Chemical Co.; Drew Industrial	www.drewindustrial.com
	Cognis Corp.	www.cognis.com
Dispersing agents	Ultra Additives Inc.	www.ultraadditives.com
Extenders and fillers	Albemarle Corporation	www.albermarle.com
	Degussa Corporation	www.degussa.com
	Engelhard	www.engelhard.com
	K-T Clay Company	www.k-tclay.com
	R.T. Vanderbilt Company	www.rtvanderbilt.com
	Wacker Chemical Corporation	www.wackersilicones.com
	Franklin Industrial Minerals	www.frankmin.com
	Imerys	www.imerys.com
	Specialty Minerals Inc.	www.mineralstech.com
	Cabot Corporation	www.cabot-corp.com
	Alcoa	www.alcoa.com
	Reynolds Metals Company	www.rmc.com
	Rheox Div. of Elementis	www.elementis-specialties.com
	Thompson, Weinman, & Co.	
	DuPont	www.dupont.com
	C.K. Williams Company	
	Excalibur Minerals Inc.	www.excalibur.com
	Malvern Minerals	
	J.J. Huber Corp., Engineered Materials Division	
Flame and smoke retardants	Albemarle Corporation	www.albemarle.com
	Akrochem Corporation	www.akrochem.com
	Clariant Corporation	www.clariant-northamerica.com
	Akzo Nobel Chemicals	www.akzo-nobel.com
Dyes and pigments	Elementis Specialties	www.elementis-specialties.com
Plasticizers	Nova Specialty Chemicals Division	www.novachem.com
	Bayer Corporation	www.bayer.com
	Eastman Chemicals	www.eastman.com
Solvents	Ashland Distribution Company	www.ashdist.com
	Union Carbide Corporation	www.unioncarbide.com
	Solutia Inc.	www.solutia.com
	Dow Chemicals Company	www.dow.com
	Calumet Lubricants Co.	www.calumetlub.com
	Hoechst Celanese Corporation	www.hoechst.com
Elastomeric modifiers	Noveon (formerly B.F. Goodrich)	www.noveon.com
	Toray	www.toray.com

APPENDIX B

PROPERTIES OF SELECTED COMMERCIAL EPOXY ADHESIVE FORMULATIONS

TABLE B.1 Two-Component Epoxy Adhesives

Adhesive trade name	Adhesive supplier	Mix ratio, by weight	Curing conditions	Properties
Master Bond EP 35	Master Bond	100/70	24–48 h at RT 1–2 h at 95°C	High strength at 450–500°F
Master Bond 21 LV	Master Bond	100/100	24–48 h at RT 1–2 h at 95°C	General-purpose adhesive; good chemical resistance
Master Bond EP45HT	Master Bond	100/30	1 h at 65°C plus 2–3 h at 125–150°C	Excellent chemical and temperature resistance; sterilization-resistant
Araldite AY103/Hardener HY 991	Huntsman	100/40	24 h at RT 15 min at 100°C	Low-viscosity, transparent liquid
Araldite AV 144-2/Hardener HV 997	Huntsman	100/60	10 h at RT 6 min at 100°C	Paste; tough high strength, good chemical resistance
Araldite XB 5308/Hardener XB 5309-1	Huntsman	100/100	10 h at RT 7 min at 100°C	Thixotropic, nonsagging adhesive that forms resilient bonds to FRP
Stycast E 1000 A/B	Emerson and Cuming; National Starch	100/120	3 h at 80°C	Low-viscosity, flexible adhesive; –40 to +65°C service temperature
Stycast W 19/Catalyst 9	Emerson and Cuming; National Starch	100/12	16–24 h at RT 1–2 h at 95°C	Very low viscosity (250 cP)
Bondmaster 4E90 A/B	Bondmaster	100/35	24 h at RT 4 h at 65°C	Unfilled, general-purpose adhesive; transparent
Scotch-Weld 2216	3M Company	2/3	24–48 h at RT	General-purpose, flexible epoxy; 8–12 h handling strength
Scotch-Weld 3501	3M Company	1/1	2 h at RT	Fast cure; 20–30 min handling strength; rigid
Scotch-Weld 1751	3M Company	3/2	24 h at RT	Long working life; 8–12 h handling strength; rigid
Lord 310 A/B	Lord		24 h at RT 1 h at 83°C	Good bond strength to composites; heat resistance to 200°C
PolyBond 33	Nbond Adhesives International	1/1	24 h at RT	Good bond strength to many plastics; t-peel of 20 lb/in of width
Titan Bond Plus	Bostik Findley	1/1	24 h at RT	Good adhesion to ABS and PVC; bonds to wet or dry surfaces

7522	Bostik Findley	1/1	48–72 h at RT	General performance, clear adhesive
7575	Bostik Findley	1/1	1 h at RT	Fast-set (5-min) epoxy
Hysol EA 956	Loctite-Henkel	100/58	24 h at RT 1 h at 71°C	Low-viscosity adhesive with high-temperature (149°C) performance
Hysol EA 9300	Loctite-Henkel	100/33	5–7 days at RT 1 h at 82°C	High peel strength and excellent environmental durability
Hysol EA 9309.3NA	Loctite-Henkel	100/22	5–7 days at RT 1 hr at 82°C	Toughened adhesive with excellent peel strength; contains glass beads for bond line control
FE 7004	H.B. Fuller Company	100/100 (flexible) 200/100 (rigid)	24–48 h at RT	General-purpose adhesive; 60-min gel time
FE 7017	H.B. Fuller Company	100/100	24–48 h at RT	Sprayable adhesive; bonds to oily metals; gels in 35 min
A1177B1/B2	Sovereign Engineered Adhesives LLC	100/100	24–48 h at RT	Provides a hard, durable bond conforming to several federal specifications
EL2995A/B	Sovereign Engineered Adhesives LLC	100/100	24–48 h at RT	Flexible, impact-resistant; long pot life
Tra Bond 2101	Tra-Con		24 h at RT	Low hardness (90D); service temperature –76 to +248°F

TABLE B.2 One-Component Epoxy Adhesives

Adhesive trade name	Adhesive supplier	Curing conditions	Properties
Supreme 10HT	Master Bond	60 min at 125°C 35–45 min at 150°C	High-strength adhesive with good chemical and temperature resistance (to 200°C)
Supreme 10 HT/S	Master Bond	60 min at 125°C 35–45 min at 150°C	Silver-filled version of Supreme 10HT for electrical conductivity
EP 36	Master Bond	Melts at 82°C; cures in 90–120 min at 125–150°C	B-stage epoxy; good temperature resistance (to 250°C) with flexibility and elongation. Good thermal cycling properties
Araldite AV 118	Huntsman	45 min at 125°C	Good peel strength; good heat and chemical resistance
Araldite AV 8553	Huntsman	30 min at 125°C	Thixotropic epoxy with good impact strength
Araldite AV 4600	Huntsman	15 min at 150°C	Quick-cure, one-component epoxy
Hysol 9346.5	Loctite-Henkel	1 h at 121°C	Moderate viscosity, high peel and shear strengths; excellent hot/wet properties
Hysol 9434NA	Loctite-Henkel	30 min at 125°C	High-temperature, nonsag epoxy adhesive
Scotch-Weld 2214	3M Company	60 min at 121°C	Aluminum-filled, general-purpose adhesive
Scotch-Weld 1386	3M Company	60 min at 177°C	For metal-to-metal bonding; high strength and impact resistance; conforms to federal specifications
Scotch-Weld 2290	3M Company	30 min at 177°C	A 21% solids epoxy solution; B-stageable; excellent for laminated steel cores
Magnobond 6297	Magnolia Plastics	40 min at 125°C	High-temperature service; good adhesion to FRP
ESP308	Permabond International	45 min at 150°C	Service temperature from –65 to +350°F; 3800–6000 psi tensile shear strength

TABLE B.3 Epoxy Film Adhesives

Adhesive trade name	Adhesive supplier	Curing conditions	Properties
Hysol EA 9602.3	Loctite-Henkel	1 h at 121°C	Modified epoxy film, tacky, available supported and unsupported; exceeds MMM-A-132 and MIL-A-25463
Hysol EA 9696	Loctite-Henkel	1 h at 121°C	Moisture-resistant, toughened adhesive; service temperature to 121°C
Hysol EA 9673	Loctite-Henkel	1 h at 177°C	BMI modified epoxy; superior strength to 288°C; moisture-resistant
Hysol EA 9628H	Loctite-Henkel	1 h at 121°C	Modified epoxy film, high peel strength; good environmental resistance and temperature strength to 121°C
AF-42	3M Company	1 h at 177°C	Very high peel strength (70 lb/in of width) and tensile shear strength (4800 psi); transparent
AF-191	3M Company	1 h at 177°C	High peel and shear strengths. High strength at 177°C; MIL-A-132; supported film
AF-111	3M Company	1 h at 125°C	Bonds metal to metal and metal to honeycomb; low-temperature cure
FM 73	Cytec Engineered Materials	60–90 min at 121°C	Toughened epoxy film; preferred system on the PABST program for primary bonding of aluminum
FM 94	Cytec Engineered Materials	60–90 min at 121°C	Modified epoxy; provides metal-to-metal and metal-to-honeycomb bonds
FM 1000	Cytec Engineered Materials	1 h at 177°C	High peel and elongation; tensile shear strength of 2900–6000 psi; unsupported film

APPENDIX C
SELECTED EPOXY RESINS

Resin	Epoxy equivalent weight	Viscosity, cP, at 25°C unless indicated otherwise	Comment	Supplier
Liquid bisphenol A type				
DER 317	192–203	16,000–25,000	Fast cure with polyamine-type curing agents	Dow
DER 330	176–185	7,000–1,000	Standard undiluted DGEBA	Dow
DER 331	182–192	11,000–14,000	For general purposes up to 204°C; undiluted	Dow
DER 332	172–176	4,000–6,000	High-purity DGEBA	Dow
DER 364	190–210	4,000–7,000	Noncrystallizing	Dow
DER 383	176–183	9,000–10,500	Extended pot life	Dow
EPON 825	175–180	5,000–6,500	High-purity resin with strong crystallization tendency	Resolution
EPON 826	178–186	6,500–9,500	Provides good chemical resistance; subject to crystallization when cold	Resolution
EPON 828	185–192	11,000–15,000	Standard undiluted DGEBA	Resolution
EPON 8280	185–195	11,000–15,000	Antisettling and antifoaming resin	Resolution
EPON 830	170–225	17,000–22,500	Minimal crystallization tendency	Resolution
EPR 164	188	10,000		Bakelite AG
EPR 154	180	8,500		Bakelite AG
Epalloy 7200			Modified DGEBA; high reactivity for low-temperature cures	CVC
Liquid bisphenol F type				
EPR 161	172	3,850		Bakelite AG
EPR 166	184	8,000	Blend with bisphenol A	Bakelite AG
EPR 144	173	4,200	Blend with bisphenol A	Bakelite AG
DER 354	158–175	3,000–5,500		Dow
DER 354 LV	160–170	2,000–3,000	Low-viscosity bisphenol F	Dow
EPON 862	166–177	2,500–4,500	Low-viscosity bisphenol F	Resolution

Multifunctional resins				
EPON 1031	195–230	Solid	Tetraglycidyl ether of tetraphenolethane; functionality of 3.5	Resolution
EPON HPT 1050	176–181	31,000–40,000	Epikote 154; epoxy novolac; functionality of 3.6	Resolution
EPON SU-3	187–207	Semisolid	Epoxy novolac; functionality of 3	Resolution
EPON SU-8	195–230	Semisolid	Epikote 157; epoxy novolac; functionality of 8	Resolution
EPON 161	169–178	18,000–28,000	Epoxy novolac; functionality of 2.5	
DEN 431	172–179	1,000–1,700 at 52°C	Epoxy novolac	Dow
DEN 438	176–181	31,000–40,000 at 52°C	Epoxy novolac	Dow
Epalloy 8370			Low–viscosity epoxy novolac; functionality 3.9	CVC
Solid bisphenol A type				
DER 661	500–560	400–800 at 150°C	High chemical resistance	Dow
DER 662	575–685	900–2,000 at 150°C	Used as an additive to reduce melt viscosity	Dow
DER 664	875–955	4,000–8,000 at 150°C	Thermally cured; used in primer formulation	Dow
DER 669	3,500–5,500	Semisolid	Excellent flexibility	Dow
EPON 834	230–280	Semisolid	Used for adhesive property modification	Resolution
EPON 1001F	525–550	Solid	Used to provide improved toughness, flexibility	Resolution
EPON 1004F	800–950	Solid	Medium MW; containing hydroxyl groups	
Diluted epoxy resins				
DER 324	195–204	600–800	Aliphatic glycidyl ether diluent in DGEBA	Dow
DER 353	190–200	800–1,000	Aliphatic glycidyl ether diluent in bisphenol F	Dow
EPR 355	167	500	Bisphenol A/bisphenol F blend	Bakelite AG
EPR 362	178	500	Bisphenol A	Bakelite AG
EPR 275	253	550	Bisphenol A	Bakelite AG

(Continued)

(*Continued*)

Resin	Epoxy equivalent weight	Viscosity, cP, at 25°C unless indicated otherwise	Comment	Supplier
		Waterborne epoxy dispersions		
XU 19000.02	240–260	—	Modified DGEBA	Dow
EPI-REZ WD-510	190–205	8,000–12,000	Water-dispersible bisphenol A resin	Resolution
EPI-REZ 3510-W-60	185–215	500–5,000	Waterborne dispersion of bisphenol A resin similar to EPON 828	Resolution
EPI-REZ 3519-W-50	560–660	5,000–15,000	Waterborne dispersion of a CTBN modified epoxy resin	Resolution
EPI-REZ 3522-W-60	615–715	8,000–18,000	Waterborne dispersion of a solid bisphenol A	Resolution
EPI-REZ 5003-W-55	195–215	5,000–15,000	Waterborne dispersion of an epoxy novolac resin with functionality of 3 (EPON SU-3 type)	Resolution
EPI-REZ 5520-W-60	480–560	5,000–15,000	Waterborne dispersion of urethane modified bisphenol A resins	Resolution
		Flexible epoxy resins		
DER 732	310–220	55–75	Polyglycol-epichlorohydrin; viscosity reducer for standard resins	Dow
DER 736	175–205	30–60	Polyglycol-epichlorohydrin	Dow
DER 755	260–300	1,000–2,000	50/50 Blend of DGEBA and polyglycol	
Heloxy Modifier 505	300–500	550–650	Polyglycidyl ether of castor oil; generally used as an additive	Resolution
Heloxy Modifier 71	390–470	400–900	Dimer acid diglycidyl ester; provides flexilbility, improved impact strength and toughness when used as an additive	Resolution
		Brominated resins		
DER 542	305–355	Semisolid	Generally blended with other DGEBA resins	Dow
EPON 1163	380–410	Semisolid	Low-melting; containing 48% bromine	Resolution
EPON 1124-A-80	425–443	Acetone solution	Conventional brominated epoxy for FR-4 applications	Resolution

Epoxy acrylate resins				
EPA 113	175	1200		Bakelite AG
EPON 8021	150	85–115	Very low viscosity	Resolution
EPON 8101	192	400–800		Resolution
EPON 8111	140	800–1,100	Rapid reaction rate	Resolution
EPON 8121	130	2,700–3,700	Very high reactivity when combined with aliphatic amines	Resolution
EPON 8161	177	1,800–2,400	Performance similar to EPON 828	Resolution
Toughened resins				
EPR 02306/S05	230	22,000		Bakelite AG
EPR 03161	206	16,000		Bakelite AG
EPR 03356	230	23,000		Bakelite AG
EPR 04593	335	6,600 at 60°C		Bakelite AG
EPON 58003	285–330	1,500–3,000 at 52°C	CTBN elastomer modified bisphenol F resin	Resolution
EPON Resin 58006	330–360	1,500–3,000 at 52°C	CTBN elastomer modified bisphenol A resin	Resolution
EPON 58901	195–210	1,000–5,000 at 52°C	CTBN elastomer epoxy functional adduct with EPON 828; provides tack and early green strength to adhesives	Resolution
EPON 58135	208–220	250–500 at 52°C	Elastomer modified bisphenol A specifically designed for adhesion to vinyl	Resolution

APPENDIX D
EPOXY CURING AGENTS

Curing agent	Type	Active hydrogen equivalent weight	Viscosity, cP, at 25°C	Mix ratio, pph (DGEBA EEW: 185)	Curing conditions; comments	Supplier
			Polyamines			
DEH 20	DETA	20.6	6	10.9	RT	Dow
DEH 24	TETA	24.4	20	12.9	RT	Dow
DEH 26	TEPA	27.1	50	14.3	RT	Dow
DEH 29	Polyethylene polyamine	28.8	75–160 at 40°C	15.2	RT	Dow
DEH 39	Aminoethylpiperazine	43	10	22.8	RT; improved impact and shorter pot life	Dow
DEH 52	Amine adduct (DETA modified with liquid epoxy)	42–47	5,000–7,000	28	RT	Dow
DEH 58	Amine blend; DETA modified bisphenol A	30	85–130	15.9		Dow
DEH 85	Phenolic curing agent	250–280	290–470 at 150°C			Dow
EPI CURE 3223	DETA	20.7	1500–4000	10.9	RT; gel time 25 min	Resolution
EPI CURE 3234	TETA	24.5	25	12.9	RT; gel time 30 min	Resolution
EPI CURE 3200	*N*-aminoethylpiperazine	43	20	22.6	RT; gel time 19 min	Resolution
EPI CURE 3300	Isophoronediamine (IPDA)	42.6	13–18	22.7	RT and ET; gel time 128 min at RT	Resolution
EPI CURE 3282	Amine adduct	38	2900–4900	20	RT; gel time 15 min	Resolution
			Aromatic amines			
MPDA	Metaphenelyene diamine	27	Melting point 60°C	14.3	ET	
DADPS or DDS	Diamino diphenyl sulfone	57	Melting point 175°C	30.2	ET; good B-stage shelf life; can be accelerated with BF_3-MEA or aliphatic amines	

	Diethyltoluene diamine	44.6		23.6	Low-viscosity liquid aromatic diamine; low exotherm and longer pot life than other aromatic amines	
Ancamine Z	Aromatic amine eutectic		2000			Resolution
Ancamine Y	Aromatic amine				Increased pot life over Ancamine Z	Resolution
			Polyetheramines			
JEFFAMINE D-230	Polyoxypropylene			32	ET	Huntsman
			Anhydrides			
NMA	Nadic methyl anhydride			60–90	Liquid anhydride having long pot life at room temperature	
HHPA	Hexahydrophthalic anhydride			60–75	Low-melting-point (35°C) solid; soluble in DGEBA at RT	
TMA	Trimellitic anhydride			60–90	Good high-temperature properties; reacts rapidly at high temperatures	
DDSA	Dodecenyl succinic anhydride			95–130	Liquid anhydride; imparts flexibility to cured composition	
PA	Phthalic anhydride			40–65	Solid anhydride with melting point of 128°C; low exotherm and long pot life	
BTDA	Benzophenonetetracarboxylic dianhydride			40–50	Good high-temperature properties; solid at RT	

(Continued)

(*Continued*)

Curing agent	Type	Active hydrogen equivalent weight	Viscosity, cP, at 25°C	Mix ratio, pph (DGEBA EEW: 185)	Curing conditions; comments	Supplier
			Catalysts			
EPI CURE 3253	DMP-30, tri (*s*-dimethylaminoethyl) phenol	575–625	180–380	10		Resolution
	DMP 10, dimethy-laminomethyl phenol					Resolution
BDMA	Benzyldimethylamine					
Ancamine K61B	Tertiary amine salt					Resolution
CAPCURE EH-30	Tertiary amine					Cognis
Versamine EH-30	Tertiary amine					UCB chemical
			Polyamides			
Vesamide 100			Semisolid	70–110	RT; several days to full cure	Elementis
Ancamide 100						Air Products and Chemicals
Versamid 115			High-viscosity liquid	60–100	RT; several days to full cure; can be cured at ET in 1–2 h at 100°C	Elementis
Ancamide 115						Air Productions and Chemicals
Versamid 125			Intermediate-viscosity liquid	50–100	RT; several days to full cure; can be cured at ET in 1–2 h at 100°C	Elementis

Ancamid 260A						Air Productions and Chemicals
Versamid 140			Low-viscosity liquid	30–70	RT; several days to full cure; can be cured at ET in 1–2 h at 100°C	Elementis
Ancamid 350A						Air Productions and Chemicals
EPI CURE 3115		230–246	50,000–70,000 at 40°C	82	Gel time 5.5 h	Resolution
EPI CURE 3140		360–390	3,000–4,000 at 40°C	50	Gel time 2 h	Resolution
EPI CURE 3160		90–96	1,000–3,500			Resolution
			Amidoamines			
EPI CURE 3010		95	400–700	50	Gel time 76 min	Resolution
EPI CURE 3025		95	200–400	50	Gel time 100 min	Resolution
EPI CURE 3055		90	150–300	48	Gel time 240 min	Resolution
EPI CURE 3072		65	500–900	35	Gel time 40 min	Resolution
EPI CURE 3046		90	120–280	47	Gel time 255 min	Resolution
			Mercaptans			
CAPCURE 3830-81	Catalyzed mercaptan				Gel time 3–5 min	Cognis
CAPCURE 3-800	Uncatalyzed mercaptan					Cognis
			Imidazoles			
EMI-24	2-Ethyl-4-methyl imidazole					Air Products and Chemicals

APPENDIX E
INDEX TO FORMULATIONS

The following is an index to complete epoxy adhesive formulations that are presented in this book. Generic or less complete formulations can be also found in many of the tables and figures.

Table	Figure	Formulation
7.4		Epoxy-phenolic adhesive compositions of commercially available types
7.5		Epoxy-nylon adhesive composition
7.7		The effect of LP-3 polymer on the physical properties of liquid epoxy resin
7.8		Typical formulations and properties of polysulfide-epoxy adhesives
7.10		Polybutadiene toughened epoxy system
8.2		Typical flexible epoxy formulations and their effect on properties
8.4		Nonfiller silyated polyether-epoxy blend
8.5		Properties of a CTBN modified epoxy adhesive
8.6		Formulas for two-part CTBN modified epoxy adhesive system
8.7		Two-part ATBN modified epoxy adhesive
8.8		Typical two-part CTBN and ATBN toughened epoxy adhesives
9.6		Cellulose fibers in an epoxy adhesive
11.5		General-purpose epoxy adhesive with amidoamine curing agents
11.6		Effect of fillers on tensile shear strength of polyamide cured epoxy adhesives
11.7		Typical formulation for an epoxy adhesive cured with an amidoamine
11.8		Typical epoxy adhesive cured with triethylenetetramine (TETA)
11.9		Room temperature cure, general-purpose epoxy adhesive cured with triethylenetetramine (TETA)
11.11		Quick-curing epoxy adhesive cured with polymercaptan
11.12		Starting formulations for rapid-setting, room temperature curing epoxy adhesives
11.13		Starting formulation for a flexible epoxy adhesive containing polyamide curing agent and reactive diluent
11.14		Starting formulation for a high-peel-strength adhesive
11.15		Epoxy-polysulfide adhesive formulation
11.16		Formulation for a flexible epoxy-polysulfide adhesive
11.17		Epoxy-polysulfide adhesive formulations
11.18		The effect of LP-3 polymer on the physical properties of liquid epoxy resins
11.19		Typical formulations and properties of epoxy-terminated and nonterminated polysulfide-epoxy adhesive
11.20		Formulation for a two-part CTBN modified epoxy adhesive system
11.21		Starting formulation for a fast-setting, toughened epoxy adhesive

(Continued)

(*Continued*)

Table	Figure	Formulation
11.22		High-temperature epoxy adhesive that cures at room temperature
11.23		H_2SO_4 resistant epoxy adhesive formulations
	11.3	Tensile strength development at 25°C of various adhesives (steel-to-concrete)
	11.4	Tensile strength development at −18°C of various adhesives (steel-to-concrete)
	11.5	Effect of a flexibilizer on tensile shear and peel strength of epoxy adhesive formulation
	11.6	Tensile shear strength retention of weathered aluminum and copper test specimens
12.2		General-purpose, two-component epoxy adhesive cured with DEAPA
12.3		Elevated-temperature curing epoxy adhesive formulations
12.4		Starting formulation for epoxy adhesive cured with MPDA
12.6		Formulation for an anhydride cured epoxy adhesive
12.7		One-component epoxy adhesive cured with dicyandiamide
12.9		Formulations for a dicyandiamide and modified aliphatic amine cured epoxy adhesive
12.10		Starting formulation for an epoxy adhesive cured with BF_3 amine catalyst
12.11		High-temperature epoxy adhesive utilizing epoxy novolac resin
12.12		BTDA curing agent in a high-temperature epoxy adhesive
12.14		CTBN toughened adhesive formulation
	12.1	High-temperature epoxy adhesives
13.4		Starting formulations for epoxy tape and film adhesives
13.5		Starting formulations for epoxy powder and stick adhesives
13.6		Unsupported film adhesive cast from solvent solution
14.3		Typical two-component epoxy-polyamide emulsion starting point formulation
14.4		Epoxy emulsion adhesive formulations
14.5		Waterborne epoxy-dicyandiamide adhesive tensile shear data
15.6		First commercial epoxy-phenolic adhesive systems
15.11		Effect of heat aging on the tensile shear strength of BTDA epoxy adhesives
16.5		Starting formulations for epoxy adhesive for bonding aluminum
16.7		Starting formulations for several epoxy adhesives recommended for bonding plastic substrates
16.13		Starting formulations for an epoxy adhesive for polyolefin, polyester, and thermoplastic substrates
16.16		Starting formulations for epoxy adhesives to bond concrete

APPENDIX F

SURFACE PREPARATION METHODS FOR COMMON SUBSTRATE MATERIALS

The following sources were used in compiling this appendix:

Cagle, C. V., *Adhesives Bonding Techniques and Applications*, McGraw-Hill, New York, 1968.

DeLollis, N. J., *Adhesives for Metals Theory and Technology*, Industrial Press, New York, 1970.

Schields, J., *Adhesives Handbook*, CRC Press, Boca Raton, FL, 1970.

Guttman, W. H., *Concise Guide to Structural Adhesives*, Reinhold, New York, 1961.

Preparing the Surface for Adhesive Bonding, Hysol Division, Dexter Corp., Bulletin Gl-600.

ASTM D 2093, *Preparation of Surfaces of Plastics Prior to Adhesive Bonding*, American Society for Testing and Materials, Conshohocken, PA.

ASTM D 2651, *Preparation of Metal Surfaces for Adhesive Bonding*, American Society for Testing and Materials, Conshohocken, PA.

Adhesives and Sealants, vol. 3, *Engineered Materials Handbook*, ASM International, Materials Park, OH, 1990.

Handbook of Plastics Joining, Plastics Design Library, Norwich, NY, 1997.

Note: All formulations in the following tables of this appendix are presented on a parts-by-weight basis unless otherwise indicated.

TABLE F1 Surface Preparation Methods for Metal Substrates

Substrates	Cleaning methods	Substrate treatments	Comments
Aluminum and aluminum alloys	Trichloroethylene	1. Sandblast or 100-grit emery cloth followed by solvent degreasing.	Medium- to high-strength bonds, suitable for noncritical applications
		2. Immerse for 10 min at 70–82°C in a commercial alkaline cleaner or Sodium metasilicate 3.0 Sodium hydroxide 1.5 Sodium dodecylbenzene sulfonate, such as Nacconol 90G (Stephan Co., Nothfield, IL) 1.5 Water (distilled) 128.0 Wash in water below 65°C and etch for 12–15 min at 66–71°C in Sodium dichromate 1.0 Sulfuric acid (96%) 10.0 Water (distilled) 30.0 Rinse in distilled water after washing in tap water and dry in air.	Optimum bond strength per FPL etch process. Specified in ASTM D 2651 and MIL-A-9067. Solvent degrease may replace alkaline cleaning.
		3 Alkaline clean as described in item 2 above. Then immerse for 10–12 min at 60–65°C in the following solution: Sulfuric acid (6.5 to 9.5 *N*) 27–36 % by wt. Ferric sulfate 135 to 165 g/L Rinse as described in item 2 above.	P-2 Etch specified in ASTM D 2651.

		4. Degrease with solvent. Abrade lightly with mildly abrasive cleaner. Rinse in deionized water; wipe; or air-dry. Etch 20 min at RT in Sodium dichromate 2 Sulfuric acid (96%) 7 Rinse thoroughly in deionized water; dry at 70°C for 30 min.	Room temperature etch
		5. Form a paste using sulfuric acid–sodium dichromate solution and finely divided silica or fuller's earth. Apply; do not permit paste to dry. Time depends on degree of contamination (usually greater than 10 min at RT). Wash very thoroughly with deionized water, and air-dry.	Paste form of acid etch, useful when part cannot be immersed.
		6. Phosphoric acid anodizing can be preformed in accordance with ASTM D 3933.	
		7. Sulfuric acid anodizing can be performed in accordance with MIL-A-8625.	Found suitable for bare aluminum (nonclad), machined, or chemically milled parts which must be corrosion-protected
Brass and bronze (see also copper and copper alloys)	Trichloroethylene	Etch for 5 min at 20°C in Zinc oxide 20 Sulfuric acid (96%) 460 Nitric acid (69%) 360 Rinse in water below 65°C, and reetch in the acid solution for 5 min at 49°C. Rinse in distilled water after washing, and dry in air.	Temperature must not exceed 65°C when washing and drying.

(Continued)

TABLE F1 Surface Preparation Methods for Metal Substrates (*Continued*)

Substrates	Cleaning methods	Substrate treatments	Comments
Chromium	Trichloroethylene	1. Abrasion. Grit or vapor blast, or 100-grit emery cloth, followed by solvent degreasing.	Suitable for general-purpose bonding
		2. Etch for 1–5 min at 90–95°C in Hydrochloric acid (37%) 17 Water 20 Rinse in distilled water after cold/hot water washing, and dry in hot air.	For maximum bond strength
Copper and copper alloys	Trichloroethylene	1. Abrasion. Sanding, wire brushing, or 100-grit emery cloth, followed by vapor or solvent degreasing.	Solution for general-purpose bonding. Use 320-grit emery cloth for foil
		2. Etch for 10 min at 66°C in Ferric sulfate 1 Sulfuric acid (96%) 75 Water 8 Wash in water at 20°C, and etch in cold solution of Sodium dichromate 1 Sulfuric acid (96%) 2 Water 17 Etch until a bright clean surface has been obtained. Rinse in water, dip in ammonium hydroxide (s.g. 0.88) and wash in tap water. Rinse in distilled water and dry in warm air.	For maximum bond strength. Suitable for brass and bronze. ASTM D 2651 sulfuric acid–dichromate–ferric sulfate etch

3. Etch for 1–2 min at 20°C in Ferric chloride (42% solution in water) 15 Nitric acid (69%) 30 Water 197 Rinse in distilled water after cold water wash and dry in air at 20°C.	Room temperature etch. ASTM D 2651 nitric acid–ferric chloride etch
4. Etch for 30 s at 20°C in Ammonium persulfate 1 Water 4 Rinse in distilled water after cold water wash and dry in air at 20°C.	Alternative etching solution to above where fast processing is required
5. Solvent-degrease. Immerse 30 s at RT in Nitric acid (69%) 30 Deionized water 90 Rinse in running water and transfer immediately to next solution; immerse for 1–2 min at 98°C in Ebonol C (Enthone, Inc., New Haven, CT) 24 oz and equivalent water to make 1 gal of solution. Rinse in deionized water and air-dry.	For copper alloys containing over 95% copper. Stable surface for hot bonding
6. Solvent-degrease in nitric acid solution as in item 5 above. Immerse immediately for 2–3 min at 93–102°C in the following solution in 1 gal of water: Sodium chlorite (technical) 4.01 oz Trisodium phosphate 1.34 oz Sodium hydroxide 0.67 oz Rinse thoroughly in running water until a neutral test is produced when touched with indicator paper. Dry. Bond as soon as possible, but within the same working day.	ASTM D 2651 nitric acid–sodium chlorite etch. Suitable for copper alloys containing over 95% copper. Not suitable for adhesives containing chlorides or for hot-bonding polyethylene

(*Continued*)

TABLE F1 Surface Preparation Methods for Metal Substrates (*Continued*)

Substrates	Cleaning methods	Substrate treatments	Comments
Gold	Trichloroethylene	Solvent or vapor degrease after light abrasion with a fine emery cloth	
Iron		See steel (mild)	
Lead and lead-based solders	Trichloroethylene	Abrasion. Grit or vapor blast, or 100-grit emery cloth followed by solvent degreasing	
Magnesium and magnesium alloys	Trichloroethylene	1. Abrasion with 100-grit emery cloth followed by solvent degreasing	Apply the adhesive immediately after abrasion
		2. Vapor-degrease. Immerse for 10 min at 60–70°C in Deionized water 95 Sodium metasilicate 2.5 Trisodium pyrophosphate 1.1 Sodium hydroxide 1.1 Sodium dodecylbenzene sulfonate, such as Nacconol 90G 0.3 Rinse in water and dry below 60°C.	Medium to high bond strength. ASTM D 2651 alkaline detergent treatment
		3. Vapor-degrease. Immerse for 10 min at 71–88°C in Water 4 Chromic acid 1 Rinse in water and dry below 60°C.	High bond strength. ASTM D 2651 chromic acid etch

		4. Vapor-degrease. Immerse for 5–10 min at 63–80°C in Water 12 Sodium hydroxide 1 Rinse in water. Immerse for 5–15 min at RT in Water 123 Chromic acid 24 Calcium nitrate 1.8 Rinse in water and dry below 60°C.	ASTM D 2651 sodium hydroxide–chromic acid etch
		5. Light anodic treatment and various corrosion preventive treatments have been developed by magnesium producers.	Details may be obtained from the magnesium alloy producers or from *ASM Handbook*, vol. 5, and Military Specification MIL-M-4502, Type I, Classes 1, 2, and 3.
		6. Some dichromate conversion coatings and wash primers designed for corrosion prevention are suitable for adhesive bonding.	Preliminary tests should be conducted to determine suitability of these processes before acceptance. Details of the processes are found in the above-referenced ASME *Handbook* and also in MIL-M-3171
Nickel	Trichloroethylene	1. Abrasion with 100-grit emery cloth followed by solvent degreasing.	For general-purpose bonding
		2. Etch for 5 s at 20°C in nitric acid (69%). Wash in cold and hot water followed by a distilled water rinse and air-dry at 40°C.	For general-purpose bonding
Silver	Trichloroethylene	Abrasion with 320-grit emery cloth followed by solvent degreasing.	

(*Continued*)

TABLE F1 Surface Preparation Methods for Metal Substrates (*Continued*)

Substrates	Cleaning methods	Substrate treatments	Comments
Steel (stainless)	Trichloroethylene	1. Abrasion with 100-grit emery cloth, grit or vapor blast, followed by solvent degreasing.	Dry grit or sandblasting tends to warp thin-sheet materials; these methods are suited only for thick section parts.
		2. Solvent-degrease and abrade with grit paper. Degrease again. Immerse for 10 min at 65–71°C in the following solution: Water 90 Sulfuric acid (s.g. 1.84) 37 Sodium dodecylbenzene Sulfonate, such as Nacconol 90G 0.2 (Stephan Co., Northfield, IL) Rinse thoroughly and remove smut with a stiff brush if necessary. Immerse for 10 min at RT in the following bright dip solution: Water 88 Nitric acid (s.g. 1.42) 15 Hydrofluoric acid (35.35%, s.g. 1.15) 2	ASTM D 2651 sulfuric, nitric, hydrofluoric etch
		3. Immerse for 2 min at approximately 93°C in the following solution heated by a boiling water bath: Hydrochloric acid (s.g. 1.2) 200 Orthophosphoric acid (s.g. 1.8) 30 Hydrofluoric acid (s.g. 1.15) 10	For maximum resistance to heat and environment. ASTM D 2651 hydrochloric, orthophosphoric, hydrofluoric acid etch

4. Etch for 15 min at 63°C in Saturated sodium dichromate solution 30 Sulfuric acid 100 Remove carbon residue with nylon brush while rinsing. Rinse in distilled water and dry in warm air at 93°C.	ASTM D 2651 sulfuric, sodium dichromate etch
5. Vapor-degrease for 10 min and pickle for 10 min at 20°C in Nitric acid (69%) 10 Hydrofluoric acid (48%) 2 Water 88 Dry in air under 70°C.	Room temperature etch. Treatment may be followed by passivation for 20 min in 5–10% w/v chromic acid solution
6. Immerse for 15 min at 63 ± 3°C in the following solution: Water 47.2 Sodium metasilicate 1.0 Anionic surfactant, such as Triton X2000 (Rohm and Haas, Philadelphia, PA 1.8	ASTM D 2651 sodium metasilicate treatment
7. Immerse for 10 min at 60–65°C in the following solution: Water 45 Hydrochloric acid (s.g. 1.2) 50 Formalin solution (40%) 10 Hydrogen peroxide (30–35%) 2 Rinse thoroughly, and then immerse for 5 min at 50–65°C in sulfuric acid dichromate solution used for aluminum. Rinse thoroughly, and then dry at not over 93°C.	ASTM D 2651 hydrochloric sulfuric–dichromate etch.

(Continued)

TABLE F1 Surface Preparation Methods for Metal Substrates (*Continued*)

Substrates	Cleaning methods	Substrate treatments	Comments
Steel (mild, iron, and ferrous metals other than stainless)	Trichloroethylene	1. Abrasion. Grit or vapor blast followed by solvent degreasing with water-free solvents.	Xylene or toluene is preferred to acetone and ketone, which may be moist enough to cause rusting
		2. Etch in the following solution for 5 min at 23°C: Deionized water 64.99% by volume Surfactant 0.01% by volume Phosphoric acid (85%) 30% by volume Nitric acid (40 Baume) 5% by volume	ASTM D 2651 nitric-phosphoric acid etch. Bonding should follow immediately after etching treatments since ferrous metals are prone to rusting. Abrasion is more suitable for procedure where bonding is delayed
		3. Etch for 5–10 min at 20°C in Hydrochloric acid (37%) 1 Water 1 Rinse in distilled water after cold water wash and dry in warm air for 10 min at 93°C.	
		4. Etch for 10 min at 60°C in Orthophosphoric acid (85%) 1 Ethyl alcohol (denatured) 2 Brush off carbon residue with nylon brush while washing in running water. Rinse with deionized water and heat for 1 h at 120°C.	For maximum strength
Tin	Trichloroethylene	Solvent or vapor-degrease after light abrasion with a fine emery cloth (320-grit).	

Titanium and titanium alloys	Trichloroethylene	1. Abrasion. Grit or vapor blast or 100-grit emery cloth; followed by solvent degrease; or scour with a nonchlorinated cleaner, rinse, and dry.	For general-purpose bonding
		2. Etch for 5–10 min at 20°C in Sodium fluoride 2 Chromium trioxide 1 Sulfuric acid (96%) 10 Water 50 Rinse in water and distilled water. Dry in air at 93°C.	
		3. Etch for 2 min at RT in Hydrofluoric acid (60%) 63 mL Hydrochloric acid (37%) 841 mL Orthophosphoric acid (85%) 89 mL Rinse in water and distilled water. Dry in air at 93°C.	Suitable for alloys to be bonded with high-temperature adhesives (e.g., polybenzimidazole). Bond within 10 min of treatment. ASTM D 2651
		4. Etch for 10–15 min at 38–52°C in Nitric acid (69%) 6 Hydrofluoric acid (60%) 1 Water 20 Rinse with water and distilled water. Dry in oven at 71–82°C for 15 min.	Alternative etch for alloys to be bonded with polyimide adhesives is nitric : hydrofluoric : water in a ratio of 5 : 1 : 27 by wetght. Etch 30 s at 20°C.
		5. Commercial etching liquids and pastes (Plasa-Jell 107C, Semco).	

(*Continued*)

TABLE F1 Surface Preparation Methods for Metal Substrates (*Continued*)

Substrates	Cleaning methods	Substrate treatments	Comments
		6. Immerse for 15 min at 76°C in the following solution to make 1 gal (3.6 L): Caustic cleaner, such as Vitro-Klene (Turco Purex Industries, Carson, CA) 6–8 oz Water Remainder Rinse in cold tap water, and immerse for 5 min at RT in the following solution: Nitric acid (s.g. 1.5) 48 Ammonium bifluoride (technical) 3 Water 49 Rinse in cold tap water and air-dry at room temperature. Immerse for 2 min at RT in the following solution with sufficient water to make 1 gal (3.6 L): Trisodium phosphate (technical) 50.0 g Sodium fluoride (technical) 8.9 g Hydrofluoric acid (48%) 26.0 mL Air-dry at room temperature.	ASTM D 2651 nitric hydrofluoric etch
		8. Stainless steel surface processes have been generally found satisfactory for titanium.	Sulfuric, sodium dichromate etch; sodium metasilicate; and hydrochloric sulfuric–dichromate etch processes for stainless steel

Zinc and zinc alloys	Trichloroethylene	1. Abrasion. Grit or vapor blast or 100-grit emery cloth followed by solvent degreasing.	For general-purpose bonding
		2. Etch for 2–4 min at 20°C in Hydrochloric acid (37%) 10–20 Water 90–80 Rinse with warm water and distilled water. Dry in air at 66–71°C for 30 min.	Glacial acetic acid is an alternative to hydrochloric acid
		3. Etch for 3–6 min at 38°C in Sulfuric acid (96%) 2 Sodium dichromate 1 Water 8	Suitable for freshly galvanized metal

TABLE F2 Surface Preparation Methods for Plastic Substrates

Substrates	Cleaning method	Substrate treatments	Comments
Acetal (copolymer)	Acetone	1. Abrasion. Grit or vapor blast, or medium-grit emery cloth followed by solvent degreasing.	For general-purpose bonding
		2. Etch in the following acid solution for 10 s at 25°C: Potassium dichromate 75 Distilled water 120 Sulfuric acid (96%) 1500 Rinse in distilled water, dry in air at RT.	For maximum bond strength. ASTM D 2093
Acetal (homopolymer)	Acetone	1. Abrasion. Sand with 280-grit emery cloth followed by solvent degreasing.	For general-purpose bonding
		2. Satinizing technique. Immerse the part in the following for 5–30 s at 80–120°C: Perchloroethylene 96.85 1,4 Dioxane 3.0 *p*-Toluenesulfonic acid 0.05 Cab-o-Sil (Cabot Corp.) 0.10 Transfer the part immediately to an oven at 120°C for 1 min. Wash in hot water. Dry in air at 120°C.	For maximum bond strength. Recommended by DuPont
Acrylonitrile butadiene styrene	Acetone	1. Abrasion. Grit or vapor blast, or 220-grit emery cloth, followed by solvent degreasing.	
		2. Etch in chromic acid solution for 20 min at 60°C.	Recipe 2 for methyl pentene

Cellulosics: cellulose, cellulose acetate, cellulose acetate butyrate, cellulose nitrate, cellulose propionate, ethyl cellulose	Methanol, isopropanol	1. Abrasion. Grit or vapor blast or 220-grit emery cloth followed by solvent degreasing. 2. After procedure 1, dry the part at 100°C for 1 h and apply adhesive before the plastic cools to room temperature.	For general-purpose bonding
Diallyl phthalate, diallyl isophthalate	Acetone, methyl ethyl ketone	Abrasion. Grit or vapor blast or 100-grit emery cloth followed by solvent degreasing.	Steel wool may be used for abrasion
Epoxy	Acetone, methyl ethyl ketone	Abrasion. Grit or vapor blast or 100-grit emery cloth followed by solvent degreasing.	Sand or steel shot is suitable abrasive
Ethylene vinyl acetate	Methanol	Prime with epoxy adhesive and fuse into the surface by heating for 30 min at 100°C.	
Furane, ionomer, melamine resins, SAN, polysulfone, and rigid vinyl	Acetone, methyl ethyl ketone	Abrasion. Grit or vapor blast or 100-grit emery cloth followed by solvent degreasing.	
Methyl pentene	Acetone	1. Abrasion. Grit or vapor blast or 100-grit emery cloth followed by solvent degreasing. 2. Immerse for 1 h at 60°C in Potassium chromate 7.5 Water 27.5 Sulfuric acid (96%) 65.0 Rinse in water and distilled water. Dry in warm air.	For general-purpose bonding

(Continued)

TABLE F2 Surface Preparation Methods for Plastic Substrates (*Continued*)

Substrates	Cleaning methods	Substrate treatments	Comments
		3. Immerse for 5–10 min at 90°C in potassium permanganate (saturated solution), acidified with sulfuric acid (96%). Rinse in water and distilled water. Dry in warm air. 4. Prime surface with a lacquer based on urea formaldehyde resin diluted with carbon tetrachloride.	Coatings (dried) offer excellent bonding surfaces without further treatment
Phenolic and phenolic melamine resins	Acetone, methyl ethyl ketone, detergent	1. Abrasion. Grit or vapor blast or abrade with 100-grit emery cloth followed by solvent degreasing. 2. Removal of surface layer of one ply of fabric previously placed on surface before curing. Expose fresh bonding surface by tearing off the ply prior to bonding.	Steel wool may be used for abrasion. Sand or steel shot is suitable abrasive. Glass fabric decorative laminate may be degreased with detergent solution
Polyamide (nylon)	Acetone, methyl ethyl ketone	1. Abrasion. Grit or vapor blast or abrade with 100-grit emery cloth followed by solvent degreasing. 2. Prime with a spreading dough based on the type of rubber to be bonded in an admixture with isocyanate. 3. Prime with resorcinol formaldehyde adhesives.	Sand or steel shot is suitable abrasive Suitable for bonding polyamide textiles to natural and synthetic rubbers Good adhesion to primer coat with epoxy adhesives in metal-to-plastic joints
Polycarbonate	Methanol, isopropanol, detergent	Abrasion. Grit or vapor blast or 100-grit emery cloth followed by solvent degreasing.	Sand or steel shot is suitable abrasive

Fluorocarbons: polychlorotrifluoroethylene, polytetrafluoroethylene, polyvinyl fluoride, polymonochlorotrifluoroethylene	Trichloroethylene	1. Wipe with solvent and treat with the following for 15 min at RT: Naphthalene (128 g) dissolved in tetrahydrofuran (1 L) to which is added sodium (23 g) during a stirring period of 2 h Wash in acetone to remove excess organic materials, and subsequently wash with distilled or deionized water. Before bonding, dry the treated adherends in an air circulating oven at 37 ± 3°C for about 1 h.	Sodium-treated surface must not be abraded before use. Hazardous etching solutions require skillful handling. Proprietary etching solutions are commercially available (see procedure 2). PTFE film available in etched condition from various suppliers
		2. Wipe with solvent and treat as recommended in one of the following commercial etchings: Bond Aid (W. S. Shamban and Co.) Fluoroetch (Action Technologies) Tetraetch (W. L. Gore Associates)	
		3. Prime with epoxy adhesives and fuse into the surface by heating for 10 min at 370°C followed by 5 min at 400°C.	Bond within 15 min of treatment
		4. Expose to one of the following gases activated by corona discharge: Air (dry) for 5 min Air (wet) for 5 min Nitrous oxide for 10 min Nitrogen for 5 min	
		5. Expose to electric discharge from a tesla coil (50 kV ac) for 4 min.	Bond within 15 min of treatment

(*Continued*)

TABLE F2 Surface Preparation Methods for Plastic Substrates (*Continued*)

Substrates	Cleaning methods	Substrate treatments	Comments
Polyesters, polyethylene terphthalate (Mylar)	Detergent, acetone, methyl ethyl ketone	1. Abrasion. Grit or vapor blast or 100-grit emery cloth followed by solvent degreasing. 2. Immerse for 10 min at 70–95°C in Sodium hydroxide 2 Water 8 Rinse in hot water and dry in hot air.	For general-purpose bonding For maximum bond strength. Suitable for Mylar films
Chlorinated polyether	Acetone, methyl ethyl ketone	Etch for 5 min at 71 ± 3°C in Potassium dichromate 75 Water 120 Sulfuric acid (96%) 1500 Rinse in water and distilled water. Dry in air.	Suitable for film materials such as Pentane. ASTM D 2093
Polyethylene, polypropylene, polyformaldehyde	Acetone, methyl ethyl ketone	1. Solvent degreasing. 2. Expose surface to gas burner flame (or oxyacetylene oxidizing flame) until substrate is glossy. 3. Etch in the following: Potassium dichromate 75 Water 120 Sulfuric acid (96%) 1500 Polyethylene and polypropylene: 60 min at 25°C or 1 min at 71°C Polyformaldehyde: 10 s at 25°C.	Low-bond-strength applications, generally only with a pressure-sensitive or contact adhesives For maximum bond strength. ASTM D 2093

Polymethylmethacrylate (acrylic)	Acetone, methyl ethyl ketone, detergent, methanol, trichloroethylene, isopropanol	Abrasion. Grit or vapor blast or 100-grit emery cloth followed by solvent degreasing.	For maximum strength, relieve stresses by annealing plastic for 5 h at 100°C
Polyphenylene	Trichloroethylene	Abrasion. Grit or vapor blast or 100-grit emery cloth followed by solvent degreasing.	
Polyphenylene oxide	Methanol	Solvent degrease.	Plastic is soluble in xylene and may be primed with adhesive in xylene solvent
Polystyrene	Methanol, isopropanol, detergent	Abrasion. Grit or vapor blast or 100-grit emery cloth followed by solvent degreasing.	Suitable for rigid plastic
Polysulfone	Methanol	Vapor degrease.	
Polyurethane	Acetone, methyl ethyl ketone	Abrade with 100-grit emery cloth and solvent degrease.	
Polyvinylchloride, polyvinylidene chloride, polyvinyl fluoride	Trichloroethylene, methyl ethyl ketone	1. Abrasion grit or vapor blast or 100-grit emery cloth followed by solvent degreasing. 2. Solvent wipe with ketone.	Suitable for rigid plastic. For maximum strength, prime with nitrile phenolic adhesive Suitable for plasticized material
Styrene acrylonitrile	Trichloroethylene	Solvent degrease.	
Urea formaldehyde	Acetone, methyl ethyl ketone	Abrasion. Grit or vapor blast or 100-grit emery cloth followed by solvent degreasing.	

TABLE F3 Surface Preparation Methods for Elastomeric Substrates

Substrates	Cleaning methods	Substrate treatments	Comments
Natural rubber	Methanol, isopropanol	1. Abrasion followed by brushing. Grit or vapor blast or 280-grit emery cloth, followed by solvent wipe.	For general-purpose bonding
		2. Treat the surface for 2–10 min with sulfuric acid (96%) at RT. Rinse thoroughly with cold water/hot water. Dry after rinsing in distilled water. Residual acid may be neutralized by soaking for 10 min in 10% ammonium hydroxide after hot water washing.	Adequate pretreatment is indicated by the appearance of hairline surface cracks on flexing the rubber. Suitable for many synthetic rubbers when given 10–15 min etch at RT. Unsuitable for use on butyl, polysulfide, silicone, chlorinated polyethylene, and polyurethane rubbers
		3. Treat surface for 2–10 min with paste made from sulfuric acid and barium sulfate. Apply paste with stainless steel spatula and follow procedure 2 above.	
		4. Treat surface for 2–10 min in Sodium hypochlorite 6 Hydrochloric acid (37%) 1 Water 200 Rinse with cold water and dry.	Suitable for those rubbers amenable to treatments 2 and 3
Butadiene styrene	Toluene	1. Abrasion followed by brushing. Grit or vapor blast or 280-grit emery cloth followed by solvent wipe.	Excess toluene results in swollen rubber. A 20-min drying time will restore the part to its original dimensions
		2. Prime with butadiene styrene adhesive in an aliphatic solvent.	
		3. Etch surface for 1–5 min at RT following method 2 for natural rubber.	
Butadiene nitrile	Methanol	1. Abrasion followed by brushing. Grit or vapor blast or 280-grit emery cloth followed by solvent wipe.	
		2. Etch surface for 10–45 s at RT, following procedure 2 for natural rubber.	

Butyl and chlorobutyl rubber	Toluene	1. Solvent wipe. 2. Immerse in the following solution at 21–32°C for 90–150 s: Hydrochloric acid (37%) 0.3 Sodium hypochlorite (5.25%) 3.0 Distilled waters 97.0 Rinse in tap water followed by distilled water. Dry at 65°C maximum. 3. Prime with butyl rubber adhesive in an aliphatic solvent.	Solution life is 4 h maximum
Chlorosulfonated polyethylene	Acetone, methyl ethyl ketone	Abrasion followed by brushing. Grit or vapor blast or 280-grit emery cloth followed by solvent wipe.	General-purpose bonding
Fluorosilicones	Methanol	Application of fluorosilicone primer to metal where intention is to bond unvulcanized rubber.	Primer available from Dow Corning
Polyacrylic	Methanol	Abrasion followed by brushing. Grit or vapor blast or 100-grit emery cloth followed by solvent wipe.	For general-purpose bonding
Polybutadiene	Methanol	Solvent wipe.	General-purpose bonding
Polychloroprene (neoprene)	Toluene, methanol, isopropanol	1. Abrasion followed by brushing. Grit or vapor blast or 100-grit emery cloth followed by solvent wipe. 2. Etch surface for 5–10 min at RT following procedure 2 for natural rubber.	Adhesion improved by abrasion with 280-grit emery cloth followed by solvent wipe
Polysulfide	Methanol	Immerse overnight in strong chlorine water, wash, and dry.	

(*Continued*)

TABLE F3 Surface Preparation Methods for Elastomeric Substrates (*Continued*)

Substrates	Cleaning methods	Substrate treatments	Comments
Polyurethane	Methanol	1. Abrasion followed by brushing. Grit or vapor blast or 280-grit emery cloth followed by solvent wipe. 2. Incorporation of a chlorosilane into the adhesive elastomer system; 1% by weight is usually sufficient.	Chlorosilane is available commercially. Addition to adhesive eliminates need for priming and improves adhesion to glass and metals. Silane may be used as a surface primer
Silicone	Acetone, methanol	1. Application of primer (e.g., Chemlok 607, Lord Chemical Co.). 2. Exposure to oxygen gas activated by corona discharge for 10 min.	

TABLE F4 Surface Preparation Methods for Miscellaneous Substrates

Substrates	Cleaning methods	Substrate treatments	Comments
Brick and fired nonglazed building materials	Methyl ethyl ketone	Abrade surface with a wire brush; remove all dust and contaminants.	
Carbon graphite	Acetone	Abrasion. Abrade with 220-grit emery cloth and solvent-degrease after dust removal.	For general-purpose bonding
Glass and quartz (nonoptical)	Acetone, detergent	1. Abrasion grit blast with carborundum and water slurry, and solvent-degrease. Dry for 30 min at 100°C. Apply the adhesive before the glass cools to RT. 2. Immerse for 10–15 min at 20°C in Sodium dichromate 7 Water 7 Sulfuric acid (96%) 400 Rinse in water and distilled water. Dry thoroughly.	For general-purpose bonding. Drying process improves bond strength For maximum strength
Glass (optical)	Acetone, detergent	Clean in an ultrasonically agitated detergent bath. Rinse: dry below 38°C.	
Ceramics and porcelain	Acetone	1. Abrasion grit blast with carborundum and water slurry and solvent-degrease. 2. Solvent-degrease or wash in warm aqueous detergent, rinse, and dry. 3. Immerse for 15 min at 20°C in Sodium dichromate 7 Water 7 Sulfuric acid (96%) 400 Rinse in water and distilled water. Oven-dry at 66°C.	

(*Continued*)

TABLE F4 Surface Preparation Methods for Miscellaneous Substrates (*Continued*)

Substrates	Cleaning methods	Substrate treatments	Comments
Concrete, granite, stone	Perchloroethylene	1. Abrasion. Abrade with a wire brush, degrease with detergent, and rinse with hot water before drying.	For general-purpose bonding
		2. Etch with 15% hydrochloric acid until effervescence ceases. Wash with water until surface is litmus-neutral. Rinse with 1% ammonia and water. Dry thoroughly before bonding.	Applied by stiff bristle brush. Acid should be prepared in a plastic pail. The 10–12% hydrochloric or sulfuric acids are alternative etchants; 10% w/w sodium bicarbonate solution may be used instead of ammonia for acid neutralization
Wood, plywood		Abrasion. Dry wood is smoothed with a suitable emery paper. Sand plywood along the direction of the grain.	For general-purpose bonding
Painted surface	Detergent	1. Clean with detergent solution, abrade with a medium emery cloth, final wash with detergent.	Bond generally as strong as the paint
		2. Remove paint by solvent or abrasion, and pretreat exposed base.	For maximum adhesion

APPENDIX G
SPECIFICATIONS AND STANDARDS

TABLE G1 American Society for Testing and Materials (ASTM)

ASTM standard	
	Adhesives and adhesion
C 297	Flatwise Tensile Strength of Metal to Honeycomb Core Bonds
C 557	Specification for Adhesives for Fastening Gypsum Wallboard to Wood Framing
D 229	Test Method for Shear Strength and Shear Modulus of Structural Adhesives (Napkin Ring Style Test Piece)
D 411	Methods for Testing Shellac Used for Electrical Insulation
D 570	Test Methods Relating to Moisture Absorption
D 696	Test Methods Relating to Coefficient of Thermal Expansion
D 816	Test Methods for Testing Rubber Cements
D 896	Test Method for Resistance of Adhesive Bonds to Chemical Reagents
D 898	Test Method for Applied Weight per Unit Area of Liquid Adhesive
D 899	Test Method for Applied Weight per Unit Area of Liquid Adhesive
D 903	Test Method for Peel or Stripping Strength of Adhesive Bonds
D 904	Practice for Exposure of Adhesive Specimens to Artificial (Carbon Arc Type) and Natural Light
D 905	Test Method for Strength Properties of Adhesive Bonds in Shear by Compression Loading
D 906	Test Method for Strength Properties of Adhesives in Plywood Type Construction in Shear by Tension Loading
D 907	Terminology of Adhesives
D 950	Test Method for Impact Strength of Adhesive Bonds
D 997	Test Method for Tensile Properties of Adhesive Bonds
D 1000	Test Methods for Pressure Sensitive Adhesive Coated Tapes Used for Electrical and Electronic Applications
D 1002	Test Method for Strength Properties of Adhesives in Shear by Tension Loading (Metal-to-Metal)
D 1062	Test Method for Cleavage Strength of Metal-to-Metal Adhesive Bonds
D 1084	Test Methods for Viscosity of Adhesives
D 1101	Test Methods for Integrity of Glue Joints in Structural Laminated Wood Products for Exterior Use
D 1144	Practice for Determining Strength Development of Adhesive Bonds

(*Continued*)

TABLE G1 American Society for Testing and Materials (ASTM) (*Continued*)

ASTM standard	
	Adhesives and adhesion (*Continued*)
D 1146	Test Method for Blocking Point of Potential Adhesive Layers
D 1151	Test Method for Effect of Moisture and Temperature on Adhesive Bonds
D 1183	Test Methods for Resistance of Adhesive to Cyclic Laboratory Aging Conditions
D 1184	Test Method for Flexural Strength of Adhesive Bonded Laminated Assemblies
D 1304	Methods of Testing Adhesives Relative to Their Use as Electrical Insulation
D 1337	Test Method for Storage Life of Adhesive by Consistency and Bond Strength
D 1338	Test Method for Working Life of Liquid or Paste Adhesives by Consistency and Bond Strength
D 1344	Method of Testing Cross-Lap Specimens for Tensile Properties of Adhesives (Discontinued in 1987)
D 1382	Test Method for Susceptibility of Dry Adhesive Films to Attack by Roaches
D 1383	Test Method for Susceptibility of Dry Adhesive Films to Attack by Laboratory Rats
D 1488	Test Method for Amylaceous Matter in Adhesives
D 1489	Test Method for Non-volatile Content Aqueous Adhesives
D 1490	Test Method for Non-volatile Content of Urea Formaldehyde of Resin Solutions
D 1579	Test Method for Filler Content of Phenol, Resorcinol, and Melamine Adhesives
D 1580	Specification for Liquid Adhesives for Automatic Machine Labeling of Glass Bottles
D 1581	Test Method for Bonding Permanency of Water- or Solvent-Soluble Liquid Adhesives for Labeling Glass Bottles (Intent to Withdraw)
D 1582	Test Method for Non-volatile Content of Phenol, Resorcinol, and Melamine Adhesives
D 1583	Test Method for Hydrogen Ion Concentration of Dry Adhesive Films
D 1584	Test Method for Water Absorptiveness of Paper Labels (Intent to Withdraw)
D 1713	Test Method for Bonding Permanency of Water- or Solvent-Soluble Liquid Adhesives for Automatic Machine Sealing Top Flaps of Fiberboard Specimens (Intent to Withdraw)
D 1714	Test Method for Water Absorptiveness of Fiberboard Specimens of Adhesives (Intent to Withdraw)
D 1779	Specification for Adhesive for Acoustical Materials
D 1780	Practice for Conducting Creep Tests of Metal-to-Metal Adhesives
D 1781	Method for Climbing Drum Peel Test for Adhesives
D 1828	Practice for Atmospheric Exposure of Adhesive Bonded Joints and Structures
D 1874	Specification for Water- or Solvent-Soluble Liquid Adhesives for Automatic Machine Scaling of Top Flaps of Fiberboard Shipping Cases
D 1875	Test Method for Density of Adhesives in Fluid Form
D 1876	Test Method for Peel Resistance of Adhesives (T-Peel Test)
D 1879	Practice for Exposure of Adhesive Specimens to High Energy Radiation
D 1916	Test Method for Penetration of Adhesives
D 1994	Test Method for Determination of Acid Numbers of Hot Melt Adhesives
D 1995	Test Method for Multi-Modal Strength of Autoadhesives (Contact Adhesives)
D 2093	Practice for Preparation of Surfaces of Plastics Prior to Adhesive Bonding
D 2094	Practice of Preparation of Bar and Rod Specimens for Adhesion Tests
D 2095	Test Method for Tensile Strength of Adhesives by Means of Bar and Rod Specimens

TABLE G1 American Society for Testing and Materials (ASTM) (*Continued*)

ASTM standard	
	Adhesives and adhesion (*Continued*)
D 2183	Test Method for Flow Properties of Adhesives in Shear by Compression Loading (Metal-to-Metal)
D 2293	Test Method for Creep Properties of Adhesives in Shear by Compression Loading (Metal-to-Metal)
D 2294	Test Method for Creep Properties of Adhesives in Shear by Tension Loading (Metal-to-Metal)
D 2295	Test Method for Strength Properties of Adhesives in Shear by Tension Loading at Elevated Temperatures (Metal-to-Metal)
D 2301	Specification for Vinyl Chloride Plastic Pressure Sensitive Electrical Insulating Tape
D 2339	Test Method for Strength Properties of Adhesives in Two-Ply Wood Construction in Shear by Tension Loading
D 2484	Specification for Polyester Film Pressure Sensitive Electrical Insulating Tape
D 2556	Test Method for Apparent Viscosity of Adhesives Having Shear Rate Dependent Flow Properties
D 2557	Test Method for Strength Properties of Adhesive in Shear by Tension Loading in the Temperature Range from –267.8 to –55°C (–450 to –67°F)
D 2558	Test Method for Evaluation Peel Strength of Shoe Sole Attaching Adhesives (Intent to Withdraw)
D 2559	Specification for Adhesive for Structural Laminated Wood Products for Use Under Exterior (Wet Use) Exposure Conditions
D 2651	Practice for Preparation of Metal Surfaces for Adhesive Bonding
D 2674	Methods of Analysis of Sulfochromate Etch Solution Used in Surface Preparation of Aluminum
D 2686	Specification for Polytetrafluoroethylene Backed Pressure Sensitive Electrical Insulating Tape
D 2739	Test Method for Volume Resistivity of Conductive Adhesives
D 2754	Specification for High Temperature Glass Cloth Pressure Sensitive Electrical Insulating Tape
D 2851	Specification for Liquid Optical Adhesive
D 2918	Practice for Determining Durability of Adhesive Joints Stressed in Peel
D 2919	Test Method for Determining Durability of Adhesive Joints Stressed in Shear by Tension Loading
D 2979	Test Method for Pressure Sensitive Tack of Adhesive Using an Inverted Probe Machine
D 3005	Specification of Low Temperature Resistant Vinyl Chloride Plastic Pressure Sensitive Electrical Insulating Tape
D 3006	Specification of Polyethylene Plastic Pressure Sensitive Electrical Insulating Tape
D 3024	Specification for Protein Base Adhesive for Structural Laminated Wood Products for Use Under Interior (Dry Use) Exposure Conditions (Discontinued 1991)
D 3110	Specification for Adhesive Used in Nonstructural Glued Lumber Products
D 3111	Test Method for Flexibility Determination of Hot Melt Adhesives by Mandrel Bend Test Method
D 3121	Test Method for Tack of Pressure Sensitive Adhesives by Rolling Ball
D 3163	Test Method for Determining the Strength of Adhesively Bonded Rigid Plastic Lap Shear Joints in Shear by Tension Loading

(*Continued*)

TABLE G1 American Society for Testing and Materials (ASTM) (*Continued*)

ASTM standard	
	Adhesives and adhesion (*Continued*)
D 3164	Test Method for Determining the Strength of Adhesively Bonded Plastic Lap Shear Sandwich Joints in Shear by Tension Loading
D 3165	Test Method for Strength Properties of Adhesives in Shear by Tension Loading of Laminated Assemblies
D 3166	Test Method for Fatigue Properties of Adhesives in Shear by Tension Loading (Metal-to-Metal)
D 3167	Test Method for Floating Roller Peel Resistance of Adhesives
D 3310	Test Method for Determining Corrosivity of Adhesive Materials
D 3418	Test Methods Relating to Glass Transition Temperature
D 3433	Practice for Fracture Strength in Cleavage of Adhesives in Bonded Joints
D 3434	Practice for Multiple Cycle Accelerated Aging Test (Automatic Boil Test) for Exterior Wet Use Wood Adhesives
D 3482	Test Method for Determining Electrolytic Corrosion of Copper by Adhesives
D 3498	Specification for Adhesives for Field Gluing Plywood to Lumber Framing for Wood Systems
D 3528	Test Method for Strength Properties of Double Lap Shear Adhesive Joints by Tension Loading
D 3535	Test Method for Resistance to Deformation Under Static Loading for Structural Wood Laminating Adhesives Used Under Exterior (Wet Use) Exposure Conditions
D 3632	Practice for Accelerated Aging of Adhesive Joints by the Oxygen Pressure Method
D 3658	Test Method for Determining the Torque Strength of Ultraviolet (UV) Light Cured Glass-to-Metal Adhesive Joints
D 3706	Test Method for Hot Tack of Wax Polymer Blends by the Flat Spring Test
D 3747	Specification for Emulsified Asphalt Adhesive for Adhering Roof Insulation
D 3762	Test Method for Adhesive Bonded Surface Durability of Aluminum (Wedge Tests)
D 3807	Test Method for Strength Properties of Adhesive in Cleavage Peel by Tension Loading (Engineering Plastics-to-Engineering Plastics)
D 3808	Practice for Qualitative Determination of Adhesion of Adhesives to Substrate by Spot Adhesion Test Method
D 3929	Practice for Evaluating the Stress Cracking of Plastics by Adhesives Using the Bent-Beam Method
D 3930	Specification for Adhesives for Wood Based Materials for Construction of Manufactured Homes
D 3931	Test Method for Determining Strength of Gap Filling Adhesive Bonds in Shear by Compression Loading
D 3932	Practice for the Control of the Application of Structural Fasteners When Attached by Hot Melt Adhesives
D 3933	Practice for Preparation of Aluminum Surface for Structural Adhesive Bonding (Phosphoric Acid Anodizing)
D 3983	Test Method for Measuring the Strength and Shear Modulus of Non-rigid Adhesives by the Thick Adherend Tensile Lap Specimen
D 4027	Test Method for Measuring Shear Properties of Structural Adhesive by the Modified Rail Test
D 4299	Test Methods for Effect of Bacterial Contamination on Permanence of Adhesive Preparations and Adhesive Films (Discontinued 1990; replaced by D 4300 and D 4783)

TABLE G1 American Society for Testing and Materials (ASTM) (*Continued*)

ASTM standard	
	Adhesives and adhesion (*Continued*)
D 4300	Test Methods for the Ability of Adhesive Films to Support or Resist the Growth of Fungi
D 4317	Specification for Polyvinyl Acetate Based Emulsion Adhesives
D 4338	Test Method for Flexibility Determination of Supported Adhesive Films by Mandrel Bend Test Method
D 4339	Test Method for the Determination of Odor of Adhesives
D 4426	Test Method for Determination of Percent Non-volatile Content of Liquid Phenolic Resins Used for Wood Laminating
D 4497	Test Method for Determining the Open Time of Hot Melt Adhesives (Manual Method)
D 4498	Test Method for Heat-Fail Temperature in Shear of Hot Melt Adhesives
D 4499	Test Method for Heat Stability of Hot Melt Adhesives
D 4500	Test Method for Determining Grit, Lumps, or Undissolved Matter in Water Borne Adhesives
D 4501	Test Method for Shear Strength of Adhesive Bonds Between Rigid Substrates by the Block-Shear Method
D 4502	Test Method for Heat and Moisture Resistance of Wood Adhesives Joints
D 4562	Test Method for Shear Strength of Adhesives Using Pin-and-Collar Specimen
D 4680	Test Method for Creep and Time to Failure of Adhesives in Static Shear by Compression Loading (Wood-to-Wood)
D 4688	Test Methods for Evaluating Structural Adhesives for Fingerjointing Lumber
D 4689	Specification for Adhesives, Casein Type
D 4690	Specification for Urea-Formaldehyde Resin Adhesives
D 4783	Test Methods for Resistance of Adhesives Preparation in Container to Attack by Bacteria, Yeast and Fungi
D 4800	Guide for Classifying and Specifying Adhesives
D 4896	Guide for the Use of Adhesive Bonded Single Lap-Joint Specimen Test Results
D 5040	Test Methods for Ash Content of Adhesives
D 5041	Test Method for Fracture Strength in Cleavage of Adhesives in Bonded Joints
D 5113	Test Method for Determining Adhesive Attack on Rigid Cellular Polystyrene Foam
D 5215	Test Method for Instrumental Evaluation of Staining of Vinyl Flooring by Adhesives
D 5266	Practice for Estimating the Percentage of Wood Failure in Adhesive Bonded Joints
D 5267	Test Method for Determination of Extrudability of Cartridge Adhesives
D 5330	Specification for Tape, Pressure Sensitive, Packaging, Filament Reinforced
D 5375	Test Methods for Liner Removal at High Speeds from Pressure Sensitive Label Stock
D 5570	Test Method for Water Resistance of Tape and Adhesives Used as a Box Closure
D 5574	Test Methods for Establishing Allowable Mechanical Properties of Wood Bonding Adhesives for Design of Structural Joints
D 5677	Specification for Fiberglass (Glass Fiber Reinforced Thermosetting Resin) Pipe and Pipe Fittings, Adhesive Bonded Joint Type, for Aviation Jet Turbine Fuel Lines

(*Continued*)

TABLE G1 American Society for Testing and Materials (ASTM) (*Continued*)

ASTM standard	
	Adhesives and adhesion (*Continued*)
D 5686	Standard Specification for Fiberglass (Glass Fiber Reinforced Thermosetting Resin) Pipe and Pipe Fittings, Adhesive Bonded Joint Type Epoxy Resin, for Condensate Return Lines
D 5749	Standard Specification for Reinforced and Plain Gummed Tape for Sealing and Securing
D 5751	Standard Specification for Adhesives Used for Laminate Joints in Nonstructural Lumber Products
D 5793	Standard Test Method for Binding Sites per Unit Length or Width of Pile Yarn Floor Coverings
D 5824	Determining Resistance to Delamination of Adhesive Bonds in Overlay Wood Core Laminates Exposed to Heat and Water
D 5999	Test Method for Noninterference of Adhesives in Repulping
D 6004	Test Method for Determining Adhesive Shear Strength of Carpet Adhesives
D 6005	Test Method for Determining Slump Resistance of Carpet Adhesives
D 6105	Practice for Application of Electrical Discharge Surface Treatment (Activation) of Plastics for Adhesive Bonding
E 229	Test Method for Shear Strength and Shear Modulus of Structural Adhesives
E 864	Practice for Surface Preparation of Aluminum Alloys to be Adhesively Bonded in Honeycomb Shelter Panels
E 866	Specification for Corrosion Inhibiting Adhesive Primer for Aluminum Alloys to be Adhesively Bonded in Honeycomb Shelter Panels
E 874	Practice for Adhesive Bonding of Aluminum Facings to Nonmetallic Honeycomb Core for Shelter Panels
E 900	Specification for Core Splice Adhesive for Honeycomb Sandwich Structural Panels
E 1307	Surface Preparation and Structural Adhesive Bonding of Precured, Nonmetallic Composite Facings to Structural Core for Flat Shelter Panels
E 1512	Test Methods for Testing Bond Performance of Adhesive Bonded Anchors
E 1555	Specification for Structural Paste Adhesive for Sandwich Panel Repair
E 1793	Standard Practice for Preparation of Aluminum Alloy for Bonding in Foam and Beam Type Transportable Shelters
E 1794	Standard Specification for Adhesive for Bonding Foam Cored Sandwich Panels (200°F Elevated Humidity Service), Type II Panels
E 1800	Specification for Adhesive for Bonding Foam Cored Sandwich Panels (160°F Elevated Humidity Service), Type I Panels
E 1801	Practice for Adhesive Bonding of Aluminum Facings in Foam and Beam Type Shelters
E 1826	Specification for Low Volatile Organic Compound (VOC) Corrosion Inhibiting Adhesive Primer for Aluminum Alloys to be Adhesively Bonded
	Sealants and sealing
C 510	Test Method for Staining and Color Change of Single or Mulicomponent Joint Sealants
C 603	Test Method for Extrusion Rate and Application Life of Elastomeric Sealants
C 711	Test Method for Low Temperature Flexibility and Tenacity of One Part Elastomeric Solvent Release Type Sealants
C 712	Test Method for Bubbling of One Part Elastomeric Solvent Release Type Sealants

TABLE G1 American Society for Testing and Materials (ASTM) (*Continued*)

ASTM standard	
	Sealants and sealing (*Continued*)
C 719	Test Method for Adhesion and Cohesion of Elastomeric Joint Sealants Under Cyclic Movement (Hockman Cycle)
C 794	Test Method for Adhesion in Peel of Elastomeric Joint Sealants
C 811	Recommended Practice for Surface Preparation of Concrete for Application of Chemical Resistance Resin Monolithic Surfaces
C 879	Method of Testing Release Papers Used with Preformed Tape Sealants
C 906	Test Method for T-Peel Strength of Hot Applied Sealants
C 907	Test Method for Tensile Adhesive Strength of Preformed Tape Sealants by Disk Method
C 920	Specification for Elastomeric Joint Sealants
C 972	Test Method of Compression Recovery of Tape Sealant
C 1016	Standard Test Method for Determination of Water Absorption of Sealant Backing (Joint Filler Material)
C 1087	Standard Test Method for Determining Compatibility of Liquid Applied Sealants with Accessories Used in Structural Glazing Systems
C 1135	Determining Tensile Adhesion Properties of Structural Sealants
C 1184	Specification for Structural Silicone Sealants
C 1247	Test Method for Durability of Sealants Exposed to Continuous Immersion in Liquids
C 1248	Test Method for Staining of Porous Substances by Joint Sealants
C 1249	Guide for Secondary Seal for Sealed Insulating Glass Units for Structural Sealant Glazing Applications
C 1253	Test Method for Determining the Outgassing Potential of Sealant Backing
C 1257	Test Method for Accelerated Weathering of Solvent Release Type Sealants
C 1265	Standard Test Method for Determining the Tensile Properties of an Insulating Glass Edge Seal for Structural Glazing Applications
C 1294	Standard Test Method for Compatibility of Insulating Glass Edge Sealants with Liquid Applied Glazing Materials
C 1311	Specification of Solvent Release Sealants
C 1330	Specification for Cylindrical Sealant Backing for Use with Cold Liquid Applied Sealants
C 1369	Specification for Secondary Edge Sealants for Structurally Glazed Insulating Glass Units
C 1375	Guide for Substrates Used In Testing Building Seals and Sealants
C 1392	Guide for Evaluating Failure of Structural Sealant Glazing
C 1394	Guide of In-Situ Structural Silicone Glazing Evaluation
D 1985	Practice for Preparing Concrete Blocks for Testing Sealant for Joints and Cracks
D 2202	Test Method for Slump of Sealants
D 2203	Test Method for Staining from Sealants
D 2377	Test Method for Tack Free Time of Caulking Compounds and Sealants
D 2828	Specification for Nonbituminous Inserts for Contraction Joints in Portland Cement Concrete Airfield Pavements, Sawable Type
D 3406	Specification for Joint Sealant, Hot Poured, Elastomeric Type, for Portland Cement Concrete Pavements
D 3538	Standard Test Method for Strength Properties of Double Lap Shear Adhesives Joints by Tension Loading

(*Continued*)

TABLE G1 American Society for Testing and Materials (ASTM) (*Continued*)

ASTM standard	
	Sealants and sealing (*Continued*)
D 3569	Specification for Joint Sealant, Hot Applied, Elastomeric, Jet Fuel Resistant Type for Portland Cement Concrete Pavements
D 3581	Specification for Joint Sealant, Hot Applied, Jet Fuel Resistant Type, for Portland Cement Concrete & Tar-Concrete Pavements
D 3910	Practice for Design, Testing, and Construction of Slurry Seal
D 4070	Specification for Adhesive Lubricant for Installation of Preformed Elastomeric Bridge Compression Seals in Concrete Structures
D 4259	Practice for Abrading Concrete
D 4260	Practice for Acid Etching Concrete
D 5167	Practice for Melting of Hot Applied Joint and Crack Sealant and Filler for Evaluation
D 5249	Specification for Backer Materials for Use with Cold and Hot Applied Joint Sealants in Portland Cement Concrete and Asphalt Joints
D 5749	Standard Specification for Reinforced and Plain Gummed Tape for Sealing and Securing
D 5893	Specification for Cold Applied, Single Component, Chemically Curing Silicone Joint Sealant for Portland Cement Concrete Pavements

TABLE G2 U.S. Federal Specifications and Standards

Specification	
A-A-272	Caulking Compounds
A-A-373	Glazing Compound
DOD-C-24176	Cement Epoxy, Metal Repair
HH-C-536	Caulking Compound for Gasket Connections
MIL-R-46082	Retaining Compounds, Single Component
MIL-A-101	Adhesive, Water Resistant
MIL-A-1154	Adhesive-Bonds Vulcanized Synthetic Rubber Parts
MIL-A- 13883	Adhesive, Synthetic Rubber
MIL-A-14042	Adhesive Epoxy
MIL-A-21016	Linoleum and Tile Adhesive
MIL-A-22010	Adhesive, Solvent Type Polyvinylchloride
MIL-A-22397	Phenol and Resorcinol Resin Base
MIL-A- 22895	Adhesives, Metal Identification Plate
MIL-A-24179	Adhesive, Flexible
MIL-A-25463	Adhesive Film Form, Metallic Structure Sandwich Construction
MIL-A-3167	Adhesives for Plastic Inhibitors
MIL-A-3316	Fire Resistant Adhesive
MIL-A-374	Adhesive, Paste for Demolition Charges
MIL-A-3920	Adhesive, Optical, Thermosetting
MIL-A-3941	Water Resistant Label Adhesives
MIL-A-43316	Adhesive, Patching
MIL-A-43365	Adhesive, Repair of Radome, Air
MIL-A-46050	Adhesive, Special Rapid Room Temperature
MIL-A-46051	Adhesives, Room Temperature and Intermediate Temperature
MIL-A-46091	Adhesive, Brake Lining to Metal
MIL-A-46106	Adhesive, Sealant Silicone, General Purpose
MIL-A-46146	Adhesive, Sealants, Silicone, RTV, Noncorrosive
MIL-A-4833	Adhesive, Cellulose Nitrate Base
MIL-A-5090	Adhesive, Heat Resistant Metal to Metal
MIL-A-5092	Light Stress Bonding Only
MIL-A-52194	Adhesive, Epoxy
MIL-A-5540	Polychloroprene Adhesive
MIL-A-60091	Adhesive for Bonding Demolition Charges
MIL-A-81236	Adhesive, Epoxy Resins
MIL-A-82484	Adhesives and Sealing Compounds
MIL-A-83377	Adhesive Bonding (Structural) for Aerospace and Other Systems, Requirements for
MIL-A-8576	Adhesive, Acrylic Monomer Base
MIL-A-9117	Synthetic Elastomeric Sealant
MIL-A-12850	Natural Liquid Rubber Cement
MIL-C-14064	Grinding Disk Cement
MIL-C-15705	Caulking Compound
MIL-C-18255	Caulking Compound with Synthetic Rubber Base
MIL-C-18969	Caulking Compound—Watertight Exterior Hull Seams of Vessels
MIL-C-23092	Cement, Natural Rubber
MIL-C-27315	Coating Systems, Elastomeric
MIL-C-27725	Coating, Corrosion, Preventative, Air Fuel Tanks
MIL-C-5539	Natural Rubber Cement
MIL-C-7438	Core Material, Aluminum, for Sandwich Construction

(Continued)

TABLE G2 U.S. Federal Specifications and Standards (*Continued*)

Specification	
MIL-C-8073	Core Material, Plastic Honeycomb, Laminated Glass Fabric Base, for Aircraft Structural Applications
MIL-C-81986	Core, Material, Plastic Honeycomb, Nylon Paper Base: for Aircraft Structural Applications
MIL-C-83019	Protective Coating, Integral Fuel Tank
MIL-C-83231	Coating, Polyurethane, Rain, Erosion Resistant
MIL-C-8514	Resin-Acid Metal Pretreatment Compound for Aircraft
MIL-C-897	Cement Rubber
MIL-D-17951	Sealing Compound for Use with Deck Covering
MIL-G-413	Marine Glue
MIL-P-20628	Putty, Sealing
MIL-P-23236	Paint Coating, Ship Fuel, Salt Tanks
MIL-P-46276	Primer, Bonding
MIL-R-17882	Epoxy Resin for Pipe Repair
MIL-S-11030	Non-curing, Polysulfide Base Sealing Compound
MIL-S-11031	Adhesive, Curing Compound for Bonding Metal to Metal
MIL-S-12935	Synthetic Resin Lumber Knot Sealing Compound
MIL-S-15204	High Temperature Sealing Compound
MIL-S-17377	Compound Composed of Plastics or Resinous Binders
MIL-S-20541	Liquid Rubber Cement
MIL-S-22473	Anaerobic Single Component Sealing Compound
MIL-S-23498	Sealing Compound, Bearing Preservation
MIL-S-24340	Sealing, Ship Deck Polyurethane
MIL-S-3927	Sealing Compound, Thread
MIL-S-4383	Sealing Compound Synthetic Rubber Base
MIL-S-45180	Thread and Gasket Sealing Compounds
MIL-S-46163	Sealing, Lubricating and Wicking Compounds Thread Locking
MIL-S-7502	High Adhesion Sealant
MIL-S-7916	Thread and Gasket Sealing Compound
MIL-S-81732	Sealing Compound, Electrical
MIL-S-81733	Sealing, Coating, Corrosion Inhibitive
MIL-S-83315	Sealing, Aluminum Structure
MIL-S-83318	Sealant, Quick Repair Integral Fuel Tank
MIL-S-83430	Sealing Compound, Integral Fuel Tanks and Fuel Cell Cavities
MIL-S-8516	Sealing Compound—Protects Electrical Components
MIL-S-8784	Sealing Compound for Aircraft Fuel Tank
MIL-S-8802	High Temperature Sealant
MIL-T-5542	Thread Compound, Antiseize and Sealing
MIL-T-83483	Thread Compound, Antiseize, Molybdenum Disulfide Petrolatum
MMM-A-001058	Adhesive, Rubber Base Pressurized Dispensers
MMM-A-001993	Adhesive, Epoxy, Flexible, Filled
MMM-A-100	Animal Glue
MMM-A-115	Asphalt Tile Cement
MMM-A-121	Adhesive, Binding
MMM-A-122	High Strength Adhesive
MMM-A-125	Casein Glue in Powder Form
MMM-A-132	Adhesive, Heat Resistant
MMM-A-134	Adhesive, Epoxy Resin, Metal to Metal Structural Bonding
MMM-A-137	Strong Linoleum Cement
MMM-A-138	Adhesive, Metal to Wood

TABLE G2 U.S. Federal Specifications and Standards (*Continued*)

Specification	
MMM-A-139	Adhesive, Natural or Synthetic
MMM-A-150	Paste for Fibrous Acoustical Materials
MMM-A-1617	Adhesive, Rubber Base, General Purpose
MMM-A-1754	Adhesive and Sealing Compound, Epoxy, Metal Filled
MMM-A-177	Adhesive Paste
MMM-A-178	Adhesive, Paper Label
MMM-A-179	Adhesive, Paper Label
MMM-A-180	Thermoplastic Synthetic Resin
MMM-A-181	Phenol, Melamine or Resorcinol Resin Base
MMM-A-182	Adhesive, Innertube Repair
MMM-A-187	Epoxy Resin Paste
MMM-A-188	Thermosetting Urea Resin
MMM-A-189	Synthetic Rubber for Hot and Cold Bonding
MMM-A-193	Adhesive, Vinyl Acetate Resin Emulsion
MMM-A-1931	Adhesive, Epoxy, Silver Filled, Conductive
MMM-A-250	Water Resistant Rubber Base Liquid Adhesive
MMM-A-260	Water Resistant Liquid Adhesive
MMM-A-132	Adhesives, Heat Resistant, Airframe Structural, Metal to Metal
MMM-B-00350	Binder Adhesive, Epoxy Resin—Flexible
SS-S-1996	Sealer, Water and Weather Resistant
TT-C-00598	Caulking Compound
TT-C-1796	Caulking Compounds, Metal Seam and Wood Seam
TT-F-320	Filler for Cracks in Wood, Metal, Concrete and Cement
TT-F-322	Metal Surface Dent Filler
TT-F-336	Filler for Wood
TT-F-340	Plastic Wood Filler
TT-P-1536	Plumbing Fixture Setting Compound
TT-P-781	Putty and Elastic Compound
TT-P-791	Wood Sash Glazing Putty
TT-S-00230	Single Component Synthetic Rubber Base Joint Sealant
TT-S-0227	Sealing Compound
TT-S-1732	Sealing Compound, Pipe Joint and Thread
VV-S-190	Solid Form Sealing Compound for Overseas Shipments
MIL-STD-401	Sandwich Construction and Core Materials: General Test Methods
Federal Test Standard 175	Adhesives, Method for Testing

TABLE G3 Other Industry Specification and Standards

Society of Automotive Engineers	
ARP 1524	Surface Preparation and Priming of Aluminum Alloy Parts for High Durability Structural Adhesive Bonding, Phosphoric Acid Anodizing
APR 1575	Surface Preparation and Priming of Aluminum Alloy Parts for High Durability Structural Adhesive Bonding, Hand Applied Phosphoric Acid Anodizing
ARP 1843	Surface Preparation for Structural Adhesive Bonding Titanium Alloy Parts
ARP 4069	Aerospace Recommended Practice for Sealing Integral Fuel Tanks
AMS 1320	Decal Adhesive Remover
AMS 3374	Sealing Compound, One Part Silicone, Aircraft Firewall
AMS 3375	Adhesive/Sealant, Fluorosilicone Aromatic Fuel Resistant, One Part Room Temperature Vulcanizing
AMS 3376	Sealing Compound, Noncuring, Groove Injection, Temperature and Fuel Resistant
AMS 3491	Surface Treatment of Polytetrafluoroethylene
AMS 3106	Primer, Adhesive, Corrosion Inhibiting, –67 to 200°F
AMS 3107	Primer, Adhesive, Corrosion-Inhibiting, for High Durability Structural Adhesive Bonding
AMS 3681	Adhesive, Electrically Conductive, Silver Organic Base
AMS 3685	Adhesive, Synthetic Rubber, Buna N Type
AMS 3686	Adhesive Polyimide Resin, Film and Paste, High Temperature Resistant, 315°C or 600°F
AMS 3687	Adhesive Film, Humidity Resistant, for Sandwich Panels
AMS 3688	Adhesive, Foaming, Honeycomb Core Splice, Structural, –55 to 82°C
AMS 3689	Adhesive, Foaming, Honeycomb Core Splice, Structural, –55 to 177°C
AMS 3690	Adhesive Compound, Epoxy, Room Temperature Curing Adhesive Compound, Epoxy, Room Temperature Curing
AMS 3691	Adhesive Compound Epoxy, Medium Temperature Application
AMS 3692	Adhesive Compound, Epoxy, High Temperature Application
AMS 3693	Adhesive, Modified Epoxy, Moderate Heat Resistant, 120°C Curing, Film Type
AMS 3695	Adhesive Film, Epoxy Base for High Durability Structural Adhesive Bonding
AMS 3696	Aerodynamic Fairing Compound, –55 to 85°C
AMS 3697	Aerodynamic Fairing Compound, –55 to 150°C
AMS 3698	Adhesive Film, Hot Melt, Addition Type Polyimide, for Foam Sandwich Structure, –67 to 450°F
AMS 3704	Adhesive, Contact Chloroprene, Resin Modified
J 1523	Recommended Practice for Metal to Metal Overlap Shear Strength Test for Automotive Type Adhesives
J 1525	Recommended Practice for Lap Shear for Automotive Type Adhesives for Fiber Reinforced (FRP) Bonding
American Architectural Manufacturers Association	
AAMA 1407.1	Voluntary Specification for a Single Component Sealant for Residential Sheet Products
AAMA 850	Penetration Sealants Guide Manual
AAMA CW-13	Curtain Wall Manual No. 13—Structural Sealant Glazing Systems
American Concrete Institute	
ACI 504R	Guide to Joint Sealants for Concrete Structures

TABLE G4 Other Standards Organizations

Acronym	Publisher	Example
	International standards organizations	
AS	Australian Standards—Australia	AS 2990
BS	British Standard Institute—United Kingdom	BS 3924
CGSB	Canadian General Standards Board—Canada	CGSB 25.14
DEF	UK Ministry of Defense Standards—United Kingdom	DEF 103
DIN	Deutsche Institut fur Normung—Germany	DIN 52451
EN	European Committee for Standardization—Europe	EN 60454
JIS	Japanese Industrial Standards—Japan	JIS Z9900
ISO	International Standard Organization—Switzerland	ISO 9653
NEN	Netherlands Normalisatie Instituut	NEN 10244-7
NF	Norme Française (AFNOR)—France	NF T76-103
NS	Norges Standardiseringsforbund—Norway	NS 4828
SEN, SIS, SMS	Standardiserigns-Kommissionen I Sverige—Sweden	SEN 01 03 45
	Other industrial and professional organizations	
AA	Aluminum Association	AA53
ACI	American Concrete Institute	ACI 211.1
AISC	American Institute of Steel Construction	AISC M011
AMS	Aerospace Materials Specifications, Society of Automotive Engineers	AMS 1374
ANSI	American National Standards Institute	ANSI X12.27
ASCE	American Society of Civil Engineers	ASCE 10-90
ASME	American Society of Mechanical Engineers	ASME B31.3
ASQ	American Society for Quality	ASQ Q90
NACE	National Association of Corrosion Engineers	NACE MR 01 75
SPI	Society of Plastics Institute	SPI B 151.21
TAPPI	Technical Association of the Pulp and Paper Industry	TAPPI 207
UL	Underwriters' Laboratories	UL 94

APPENDIX H
CONVERSION FACTORS

To convert	To	Multiply by
	Area	
in^2	mm^2	$6.451600 \times 10^{+2}$
in^2	cm^2	6.451600
ft^2	m^2	9.290304×10^{-2}
	Force	
dyn	N	1.000000×10^{-5}
lbf	N	4.44822
	Force per unit area	
lb/in^2 (psi)	MPa	6.894757×10^{-3}
MN/m^2	MPa	1.000000
kg/mm^2	MPa	9.806650
dyn/cm^2	Pa	1.000000
	Force per unit length	
lb/in (piw)	N/m	$1.751268 \times 10^{+2}$
dyne/cm	mJ/m^2	1.000000
lb/ft	kg/m	1.488
	Impact energy	
ft · lbf	J	1.355818
N · m	J	1.000000
	Length	
Å (Angstroms)	m	1.000000×10^{-10}
in	mm	25.4
mil	in	1.000000×10^{-3}
	Mass	
lb	kg	4.5359237×10^{-1}
oz	gm	28.35
	Pressure	
atm	Pa	$1.013250 \times 10^{+5}$
atm	torr	760
psi	Pa	$6.894757 \times 10^{+3}$
torr (mmHg @0°C)	Pa	$1.333220 \times 10^{+2}$
	Temperature	
°F	°C	$5/9 \times (°F - 32)$
°C	°F	$(°C \times 9)/5 + 32$
	Viscosity	
P	Pa · s (pascal-second)	1.000000×10^{-1}
P	g/cm · s	1.000000
m^2/s	St	$1.000000 \times 10^{+4}$

INDEX

ABOUT THE AUTHOR

Edward M. Petrie has been active in the adhesives industry for over 37 years—both in the formulation and the end-use sides of the industry. He has BS (Chemical Engineering) and MS (Polymer Science) degrees from Carnegie Mellon University and an MBA from Duquesne University.

Throughout his industrial career, Mr. Petrie was employed by two major global corporations (ABB and Westinghouse) as an internal consultant on adhesives and polymeric materials. In both corporations he was part of the Central Research Laboratories and assisted all of the operating divisions within the organization.

His expertise includes material and process selection, testing, formulating, substrate preparation, and quality control with emphasis on the end-user application and performance requirements. Major projects consist of problem solving, cost reduction, formulation optimization, selection of joint design and production processes, and failure analysis. Applications include structural as well as non-structural joining of all materials. Mr. Petrie is also an expert in joining polymeric materials by methods other than adhesives including mechanical fastening, self-fastening, and heat and solvent welding.

Mr. Petrie has authored over 100 papers on adhesives as well as the popular *Handbook of Adhesives and Sealants* (McGraw-Hill, 2000) and *Plastic Materials and Processes—A Concise Encyclopedia* (John Wiley & Sons, 2003). Mr. Petrie has also given seminars on adhesives and adhesion to corporations, universities, government organizations, and professional societies.

Currently, Mr. Petrie is an independent consultant and sole proprietor of EMP Solutions. He is also technical advisor and consultant to SpecialChem4Adhesives.com, an online service platform dedicated entirely to the adhesives and sealants industry.

www.ingramcontent.com/pod-product-compliance
Lightning Source LLC
LaVergne TN
LVHW010620300926
849162LV00001B/18
* 9 7 8 0 0 7 1 4 5 5 4 4 2 *